BIOZONE

Student Workbook

IB
BIOLOGY
SECOND EDITION

IB BIOLOGY
Student Workbook

Second edition 2014

Thirteenth printing

ISBN 978-1-927173-93-0

Copyright © **2014** Richard Allan
Published by **BIOZONE International Ltd**

**Printed by REPLIKA PRESS PVT LTD using
paper from renewable and waste materials**
www.replikapress.com

Purchases of this workbook may be made
direct from the publisher:

www.the**BIOZONE**.com

BIOZONE International Ltd.
Telephone: +64 7-856 8104
Fax: +64 7-856 9243
Toll FREE phone: 1-855-246-4555 (USA-Canada only)
Toll FREE fax: 1-855-935-3555 (USA-Canada only)
Email: sales@biozone.co.nz
Website: www.the**BIOZONE**.com

Cover Photograph

The snowy owl (*Bubo scandiacus*) is a migratory bird
inhabiting the Arctic tundra in warmer months, and migrating
south to North America, Europe, and Asia in winter. It is
diurnal, hunting its primary food source (lemmings) day and
night. Snowy owls breed on the Arctic tundra and are highly
territorial, defending the nest vigorously against much larger
animals (including wolves). Its magnificent plumage ranges
from snowy white (in older males), to white with dark bars
and spots (in females and juveniles).
PHOTO: iStock © Adam Bennie

Note to the Student

This second edition of IB Biology has been specifically structured and written to meet the content and skills requirements of Biology for the IB Diploma. Content for SL (core) and HL is covered by a wealth of activities which provide both consolidation and extension of prior knowledge. Learning objectives for each chapter provide you with a concise guide to required understandings, applications, and skills and *Theory of Knowledge* and *International-mindedness* are supported throughout. We have provided a wide range of activities that will enable you to build on what you know already, explore new topics, work with your classmates, and practise your skills in data handling and interpretation, and in answering free response questions. We hope that you find the workbook valuable and that you make full use of its features.

To use this workbook most effectively, take note of the features outlined in this introduction. Understanding the activity coding system, and using the external resources we provide will help to scaffold your learning so that you understand the principles and processes involved in each topic of study.

Look out for these features and know how to use them:

▶ The chapter introduction provides you with a summary of the understandings, applications, and skills for the topic, phrased as a set of learning aims. Use the check boxes to identify and then mark off the points as you complete them. The chapter introduction also provides you with a list of key terms for the chapter, from which you can construct your own glossary as you work through the activities. See *Using the Workbook* on page 1 for more information.

▶ The activities form the bulk of this workbook. They have codes assigned to them according to the skills they emphasize. Each has a short introduction to the topic and provides some relevant background. Most of the information is associated with pictures and diagrams, and your understanding of the content is reviewed through the questions. Some of the activities involve modelling and group work.

▶ The free response questions allow you to use the information on the page to answer questions about the content of the activity. They may require you to apply your understanding to a new situation where the same principles operate.

▶ A coding system on the page tab identifies the type of activity. For example, it might focus on understanding a concept (*KNOW*), an application of that understanding (*APP*), or demonstration of a skill (*SKILL*). A full list of codes is given on page 1, but the codes themselves are relatively self explanatory.

▶ *TEST* activities enable you to test your understanding of the key terms used in the chapter and their use.

▶ *LINK* tabs at the bottom of the activity page identify connections across the curriculum (utilizations) and pages covering similar principles that may be applied to different situations.

▶ *WEB* tabs at the bottom of the activity page alert the reader to the **Weblinks** resource, which provides external, online support material for the activity in the form of an animation, video clip, or quiz. Bookmark the Weblinks page (see details on page 2) and visit it frequently as you progress through the workbook.

SPELLING: This workbook uses Oxford English as per the IB Diploma Programme syllabus document.

Meet the Writing Team

Tracey Greenwood
I have been writing resources for students since 1993. I have a Ph.D in biology, specializing in lake ecology and I have taught both graduate and undergraduate biology.

Tracey
Senior Author

Lissa Bainbridge-Smith
I worked in industry in a research and development capacity for eight years before joining BIOZONE in 2006. I have an M.Sc from Waikato University.

Lissa
Author

Kent Pryor
I have a BSc from Massey University majoring in zoology and ecology and taught secondary school biology and chemistry for 9 years before joining BIOZONE as an author in 2009.

Kent
Author

Richard Allan
I have had 11 years experience teaching senior secondary school biology. I have a Masters degree in biology and founded BIOZONE in the 1980s after developing resources for my own students.

Richard
Founder & CEO

Thanks also to:

The staff at BIOZONE, including Gemma Conn and Julie Fairless for design and graphics support, Paolo Curray for IT support, Debbie Antoniadis and Tim Lind for office handling and logistics, and the BIOZONE sales team.

Contents

Activity is marked: to be done;   ✔ when completed

Contents

Activity is marked: ☐ to be done; ☑ when completed

Contents

Activity is marked: 🔲 to be done; ☑ when completed

Using This Workbook

▶ The outline of the chapter structure below will help you to navigate through the material in each chapter.

Introduction
- A check list of understandings, applications, and skills for the chapter.
- A list of key terms.

Activities
- The KEY IDEA provides your focus for the activity.
- Annotated diagrams help you understand the content.
- Questions explore the content of the page.

Review
- Create your own summary for review.
- Hints help you to focus on what is important.
- Your summary will consolidate your understanding of the content in the chapter.

Literacy
- Activities are based on, but not restricted to, the introductory key terms list.
- Several types of activities test your understanding of the concepts and biological terms in the chapter.

Link ideas in separate activities within and between chapters

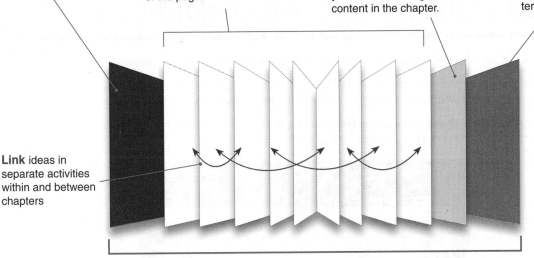

Structure of a chapter

▶ The activities make up most of this workbook. Each one has a similar structure and they are organized through the chapter in a way that unpacks the material in a series of steps.

Activity number
Activities are numbered to make it easy to navigate through the workbook and to find related activities.

Key idea
The key idea provides the message for the activity. It helps you to understand what is important on the page.

Page tabs
The **page tabs** at the bottom of an activity page indicate the type of activity, related activities and external weblinks.

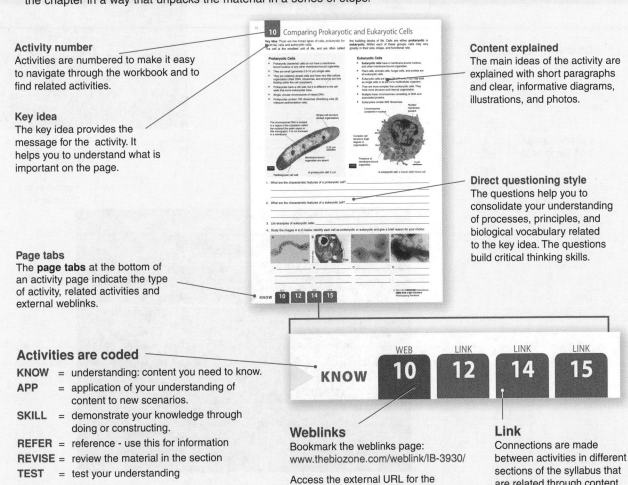

Content explained
The main ideas of the activity are explained with short paragraphs and clear, informative diagrams, illustrations, and photos.

Direct questioning style
The questions help you to consolidate your understanding of processes, principles, and biological vocabulary related to the key idea. The questions build critical thinking skills.

Activities are coded

KNOW	=	understanding: content you need to know.
APP	=	application of your understanding of content to new scenarios.
SKILL	=	demonstrate your knowledge through doing or constructing.
REFER	=	reference - use this for information
REVISE	=	review the material in the section
TEST	=	test your understanding

Weblinks
Bookmark the weblinks page:
www.thebiozone.com/weblink/IB-3930/

Access the external URL for the activity by clicking on the link next to its number.

Link
Connections are made between activities in different sections of the syllabus that are related through content or because they build on prior knowledge.

BIOZONE's Online Resources

WEBLINKS is an online resource compiled by BIOZONE to enhance or extend the content provided in the activities, largely though explanatory animations and short videos. All external websites have been selected for their suitability and accuracy and regularly checked. From this page, you can also access a wide range of annotated 3D models provided by BIOZONE and check for any errata or clarifications to the book or model answers since printing.

www.thebiozone.com/weblink/IB-3930/

▶ This WEBLINKS page provides links to external websites and **3D models** supporting the activities.

▶ The external websites are, for the most part, narrowly focussed animations and video clips directly relevant to some aspect of the activity on which they are cited. They provide great support to help your understanding.

▶ The comprehensive collection of annotated 3D models provides a different way to visualize and understand theoretical content. Choose those models relevant to your programme or interests.

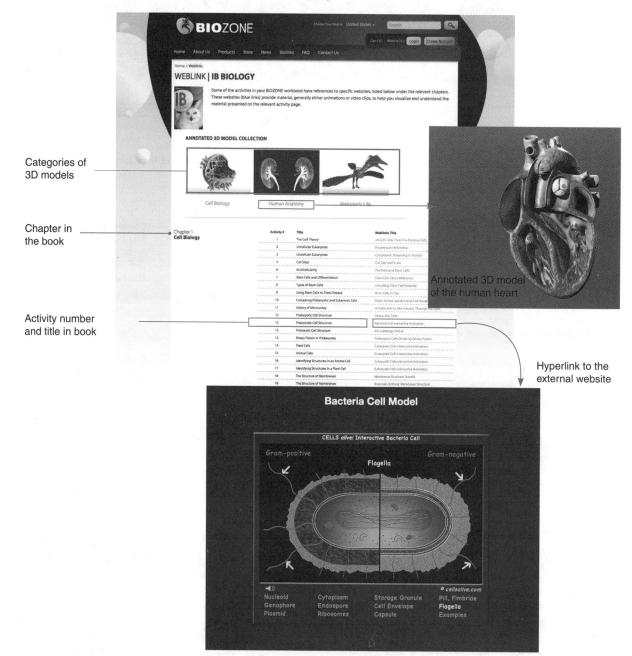

Categories of 3D models

Chapter in the book

Activity number and title in book

Annotated 3D model of the human heart

Hyperlink to the external website

Bookmark weblinks by typing in the address: it is not accessible directly from BIOZONE's website
Corrections and clarifications to current editions are always posted on the weblinks page

Topic 1

Cell Biology

Key terms

active transport
amphipathic
binary fission
cell cycle
cell differentiation
cell theory
concentration gradient
cyclin
diffusion
electron microscope
endocytosis
endosymbiotic theory
eukaryotic cell
exocytosis
facilitated diffusion
fluid mosaic model
interphase
ion pump
light microscope
metastasis
mitosis
mitotic index
multicellular
mutagen
oncogene
organelle
osmosis
passive transport
phospholipid
plasma membrane
prokaryotic cell
specialized cell
stem cell
tumour

1.1 Introduction to cells

Understandings, applications, skills

Activity number

☐ 1 Outline the cell theory and the evidence supporting it. Use examples to show that the cell theory is a generalization that applies to most but not all organisms.

1

> **TOK** *How do we distinguish living from non-living environments?*

☐ 2 Describe the criteria for life as demonstrated by unicellular organisms. Investigate life functions using *Paramecium* and *Scenedesmus*.

1 2

☐ 3 Explain the significance of surface area to volume ratio to cell size.

3

☐ 4 Calculate the magnification of drawings and the size of cell structures in light and electron micrographs and in drawings.

4 5

☐ 5 Explain how multicellularity results in the emergence of new properties. Explain how specialized tissues develop by cell differentiation during development.

6 7

☐ 6 Describe the properties of stem cells and explain their role in embryonic development. Explain how stem cells can be used to treat disease. Discuss the ethics of producing and using stem cells for therapeutic use.

8 9

1.2 Ultrastructure of cells

Understandings, applications, skills

Activity number

☐ 1 Describe the structure and function of a prokaryotic cell, e.g. *E. coli*. Draw the ultrastructure of a prokaryotic cell based on electron micrographs.

10 12

☐ 2 Describe the process and purpose of binary fission in prokaryotes.

13

☐ 3 Describe the structure and function of a eukaryotic cell, e.g. liver cell. Compare and contrast the structure of typical plant and animal cells.

10 14 15

☐ 4 Explain the higher resolution of electron microscopes relative to light microscopes and relate this to the greater cellular detail that can be seen. Draw the ultrastructure of a eukaryotic cell based on electron micrographs. Use electron micrographs to identify cellular structures and deduce the function of specialized cells.

11 16 17

> **TOK** *Are knowledge claims based on observations made using technology as valid as those made without technological assistance?*

1.3 Membrane structure

Understandings, applications, skills

Activity number

☐ 1 Describe the fluid mosaic model of the plasma membrane, explaining why the phospholipids form a bilayer. Draw a diagram to illustrate the fluid mosaic model, including cholesterol and embedded proteins.

18

☐ 2 Describe the diversity and roles of proteins in the plasma membrane.

18

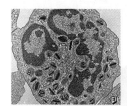

□ 3 Describe how cholesterol regulates membrane fluidity and permeability   **18**

□ 4 Analyse evidence from electron microscopy supporting the current fluid mosaic model of membrane structure (and falsification of previous models).   **19**

> **TOK** *The models for plasma membrane structure have changed as a result of new evidence and ways of analysis. Why learn about discredited models?*

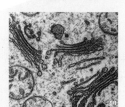

1.4   Membrane transport
Understandings, applications, skills   Activity number

□ 1 Describe and explain how particles move across membranes by diffusion, facilitated diffusion, osmosis, and active transport.   **20 21 23**

□ 2 Explain why tissues used in medical procedures must be bathed in solutions with the same osmolarity as the cytoplasm.   **21**

□ 3 Demonstrate the effect of osmosis using hypertonic and hypotonic solutions.   **22**

□ 4 Describe active transport using the sodium-potassium pump and facilitated diffusion using potassium channels in axons.   **20 24 26**

□ 5 Describe how endocytosis and exocytosis are possible because of the fluid nature of the plasma membrane. Describe how vesicles move material around within the cell.   **25 26**

1.5   Origins of cells
Understandings, applications, skills   Activity number

□ 1 Understand that cells can only form by division of pre-existing cells. Explain how Pasteur's experiments dispelled the idea of spontaneous generation   **1 27**

□ 2 Explain how the first cells might have originated and describe any supporting evidence.   **28**

□ 3 Explain the endosymbiotic theory for the origin of eukaryotic cells and the evidence for it. Know that the almost universal nature of the genetic code indicates a common origin of life.   **29 30**

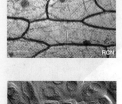

1.6   Cell division
Understandings, applications, skills   Activity number

□ 1 Describe the outcome of mitotic division and explain its role in eukaryotes.   **31**

□ 2 Describe mitosis as a continuous process, with distinct stages. Recognize and describe the events in the following stages in mitosis: prophase, metaphase, anaphase, telophase.   **32**

□ 3 Recognize stages in the eukaryotic cell cycle: interphase, mitosis, cytokinesis. Describe the events occurring during interphase stages: G1, S, and G2.   **32**

□ 4 Identify phases of mitosis from micrographs. Determine the miotic index of a cell from micrographs.   **33**

□ 5 Explain the regulation of the cell cycle by cyclins.   **34**

> **TOK** *Cyclins were discovered by 'accident' when researchers were studying development in marine invertebrates. To what extent are new discoveries the result of intuition rather than luck?*

□ 6 Explain how mutagens and oncogenes are involved in the development of primary tumours. Explain the role of metastasis in the spread of cancer and the development of secondary tumours. Discuss the correlation between smoking and the incidence of cancer.   **35**

1 Cell Theory

Key Idea: All living organisms are composed of cells. The cell is the basic unit of life.

The cell theory is a fundamental idea of biology. This idea, that all living things are composed of cells, developed over many years and is strongly linked to the invention and refinement of the microscope in the 1600s. The term cell was coined by Robert Hooke after observing a thin piece of cork under a microscope in which he saw walled compartments that reminded him of the cells a monk might live in.

The Cell Theory

The idea that cells are fundamental units of life is part of the cell theory. The basic principles of the theory (as developed by early biologists) are:

▶ All living things are composed of cells and cell products.

▶ New cells are formed only by the division of pre-existing cells.

▶ The cell contains inherited information (genes) that are used as instructions for growth, functioning, and development.

▶ The cell is the functioning unit of life; all chemical reactions of life take place within cells.

Homeostasis: Cells maintain a stable internal environment by carrying out a continuousl series of chemical reactions.

Metabolism: Life is a continual series of chemical reactions. Cells sustain these reactions by using the energy in food molecules (e.g. glucose).

Response: All cells respond to their environment. Receptors in the plasma membrane detect molecules in the environment and send signals to the internal machinery of the cell.

Life Functions

All cells show the functions of life. They use food (e.g. glucose) to maintain a stable internal environment, grow, reproduce, and produce wastes.

Nutrition: All cells require food to provide energy to power chemical reactions and nutrients to build cell components.

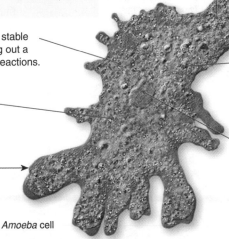
Amoeba cell

Growth: Cells grow bigger over time. When they get big enough and acquire enough materials they may divide.

Reproduction: Cells divide to produce new cells. Unicellular organisms divide to produce a genetically identical daughter cell. In multicellular diploid organisms, germline cells produce haploid gametes.

Exceptions to the Cell Theory

Mucor sp. hyphae

The alga *Caulerpa* consists of one multi-nucleated cell, yet it grows to the size of a large plant. Its shape is maintained by the cell wall and microtubules, but there are no separate cells.

Muscle fibres form from the fusion of many myoblasts (individual muscle stem cells), producing a large multi-nucleated fibre. These fibres can be 20 cm or more long.

Some **fungi** produce hyphae that lack cross walls dividing the hyphae into cells. They are known as aseptate hyphae (as opposed to septate hyphae that do contain cross walls).

1. Cells are the fundamental unit of life. Explain what this means: _____

2. To what extent is an organism such as *Caulerpa* an exception to the cell theory?_____

WEB
1 KNOW

2 Unicellular Eukaryotes

Key Idea: Unicellular organisms are able to perform all life functions, although there is a large amount of diversity in the way they do so.

Unicellular (single-celled) eukaryotes comprise the majority of the diverse kingdom, Protista. They are found almost anywhere there is water, including within larger organisms (as parasites or symbionts). The protists are a very diverse group, exhibiting some features typical of generalized eukaryotic cells, as well as specialized features. *Paramecium* is heterotrophic, ingesting food particles. *Scenedesmus* is autotrophic (synthesizes its own food). Most protistans reproduce asexually by binary fission (distinct from prokaryotic binary fission). Most can also reproduce sexually, most commonly by fusion of gametes to produce a zygote.

Paramecium

Paramecium is a common protozoan in freshwater and marine environments. It feeds on bacteria, algae, and yeasts, sweeping them into the oral groove with its cilia. There are numerous species of *Paramecium*, ranging in size from 50 μm to 300 μm long.

Size: 240 x 80 μm
Habitat: Freshwater, sea water

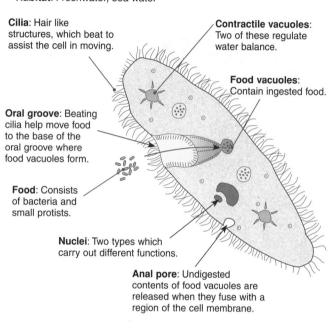

Cilia: Hair like structures, which beat to assist the cell in moving.

Contractile vacuoles: Two of these regulate water balance.

Food vacuoles: Contain ingested food.

Oral groove: Beating cilia help move food to the base of the oral groove where food vacuoles form.

Food: Consists of bacteria and small protists.

Nuclei: Two types which carry out different functions.

Anal pore: Undigested contents of food vacuoles are released when they fuse with a region of the cell membrane.

Scenedesmus

Scenedesmus is a freshwater green alga (autotrophic protist) that forms colonies of 4, 8, or sometimes 16 cells. Its colonial existence and the outer spines give it protection from grazers (e.g. *Daphnia*). Spines normally only grow from the outer most cells in the colony.

Size: 12.5 x 5 μm
Habitat: Freshwater

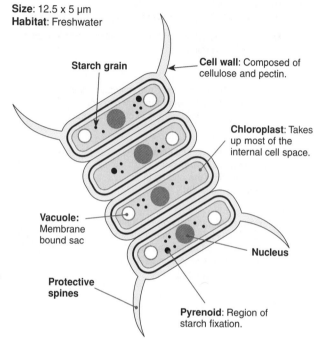

Starch grain

Cell wall: Composed of cellulose and pectin.

Chloroplast: Takes up most of the internal cell space.

Vacuole: Membrane bound sac

Nucleus

Protective spines

Pyrenoid: Region of starch fixation.

1. Identify two ways in which *Scenedesmus* defends itself against grazers: _____

   _____

2. Suggest why *Scenedesmus* colonies commonly consist of 4, 8 or 16 cells: _____

   _____

   _____

3. Explain how the life function of nutrition is carried out by:

   (a) *Paramecium*: _____

   _____

   _____

   _____

   (b) *Scenedesmus*: _____

   _____

   _____

4. Suggest why *Paramecium* needs to be particularly mobile: _____

   _____

© 2012-2014 **BIOZONE** International
ISBN: 978-1-927173-93-0
Photocopying Prohibited

3 Surface Area and Volume

Key Idea: Diffusion is less efficient in cells with a small surface area relative to their volume than in cells with a large surface area relative to their volume.

When an object (e.g. a cell) is small it has a large surface area in comparison to its volume. Diffusion is an effective way to transport materials (e.g. gases) into the cell. As an object becomes larger, its surface area compared to its volume is smaller. Diffusion is no longer an effective way to transport materials to the inside. This places a physical limit on the size a cell can grow, with the effectiveness of diffusion being the controlling factor. Larger organisms overcome this constraint by becoming multicellular.

Diffusion in Organisms of Different Sizes

Single-Celled Organisms

Single-celled organisms (e.g. *Amoeba*), are small and have a large surface area relative to the cell's volume. The cell's requirements can be met by the diffusion or active transport of materials into and out of the cell (below).

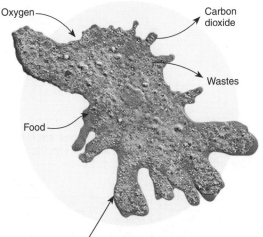

Oxygen

Carbon dioxide

Wastes

Food

The **plasma membrane**, which surrounds every cell, regulates movements of substances into and out of the cell. For each square micrometre of membrane, only so much of a particular substance can cross per second.

Multicellular Organisms

Multicellular organisms (e.g. plants and animals) are often quite large and large organisms have a small surface area compared to their volume. They require specialized systems to transport the materials they need to and from the cells and tissues in their body.

In a multicellular organism, such as an elephant, the body's need for respiratory gases cannot be met by diffusion through the skin.

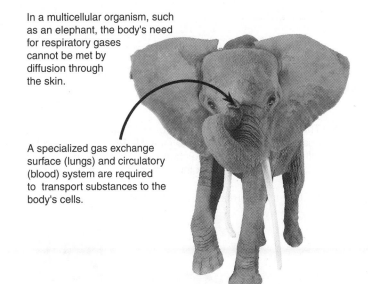

A specialized gas exchange surface (lungs) and circulatory (blood) system are required to transport substances to the body's cells.

The diagram below shows four hypothetical cells of different sizes. They range from a small 2 cm cube to a 5 cm cube. This exercise investigates the effect of cell size on the efficiency of diffusion.

2 cm cube

3 cm cube

4 cm cube

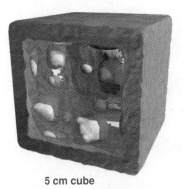

5 cm cube

1. Calculate the volume, surface area and the ratio of surface area to volume for each of the four cubes above (the first has been done for you). When completing the table below, show your calculations.

Cube size	Surface area	Volume	Surface area to volume ratio
2 cm cube	$2 \times 2 \times 6 = 24\,cm^2$ (2 cm x 2 cm x 6 sides)	$2 \times 2 \times 2 = 8\,cm^3$ (height x width x depth)	24 to 8 = 3:1
3 cm cube			
4 cm cube			
5 cm cube			

© 2012-2014 **BIOZONE** International
ISBN: 978-1-927173-93-0
Photocopying Prohibited

LINK 20   LINK 6   **KNOW**

2. Create a graph, plotting the surface area against the volume of each cube, on the grid on the right. Draw a line connecting the points and label axes and units.

3. Which increases the fastest with increasing size: the **volume** or the **surface area**?

_____

4. Explain what happens to the ratio of surface area to volume with increasing size.

_____

_____

_____

_____

_____

_____

_____

5. The diffusion of molecules into a cell can be modelled by using agar cubes infused with phenolphthalein indicator and soaked in sodium hydroxide (NaOH). Phenolphthalein turns a pink colour when in the presence of a base. As the NaOH diffuses into the agar, the phenolphthalein changes to pink and thus indicates how far the NaOH has diffused into the agar. By cutting an agar block into cubes of various sizes, it is possible to show the effect of cell size on diffusion.

(a) Use the information below to fill in the table on the right:

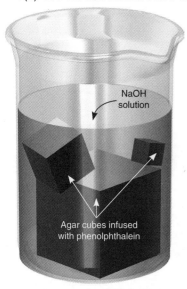

Agar cubes infused with phenolphthalein

NaOH solution

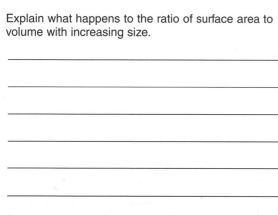

Cube 1

2 cm

Cube 2

1 cm

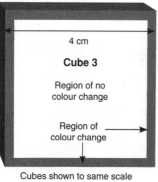

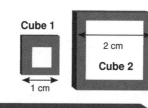

4 cm

Cube 3

Region of no colour change

Region of colour change

Cubes shown to same scale

Cube	1	2	3
1. Total volume (cm^3)			
2. Volume not pink (cm^3)			
3. Diffused volume (cm^3) (subtract value 2 from value 1)			
4. Percentage diffusion			

(b) Diffusion of substances into and out of a cell occurs across the plasma membrane. For a cuboid cell, explain how increasing cell size affects the ability of diffusion to provide the materials required by the cell:

_____

_____

_____

_____

6. Explain why a single large cell of 2 cm x 2 cm x 2 cm is less efficient in terms of passively acquiring nutrients than eight cells of 1 cm x 1 cm x 1 cm:

_____

_____

_____

4 Cell Sizes

Key Idea: Cells vary in size (2-100 µm), with prokaryotic cells being approximately 10 times smaller than eukaryotic cells. Cells can only be seen properly when viewed through the magnifying lenses of a microscope. The images below show a variety of cell types, including a multicellular microscopic animal and a virus (non-cellular) for comparison. For each of these images, note the scale and relate this to the type of microscopy used.

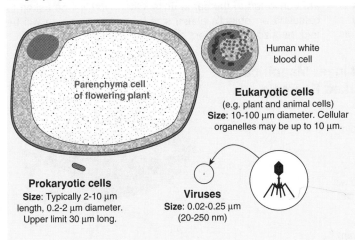

Parenchyma cell of flowering plant

Human white blood cell

Eukaryotic cells
(e.g. plant and animal cells)
Size: 10-100 µm diameter. Cellular organelles may be up to 10 µm.

Prokaryotic cells
Size: Typically 2-10 µm length, 0.2-2 µm diameter. Upper limit 30 µm long.

Viruses
Size: 0.02-0.25 µm (20-250 nm)

Unit of Length (International System)		
Unit	**Metres**	**Equivalent**
1 metre (m)	1 m	= 1000 millimetres
1 millimetre (mm)	10^{-3} m	= 1000 micrometres
1 micrometre (µm)	10^{-6} m	= 1000 nanometres
1 nanometre (nm)	10^{-9} m	= 1000 picometres

Micrometres are sometime referred to as microns. Smaller structures are usually measured in nanometres (nm) e.g. molecules (1 nm) and plasma membrane thickness (10 nm).

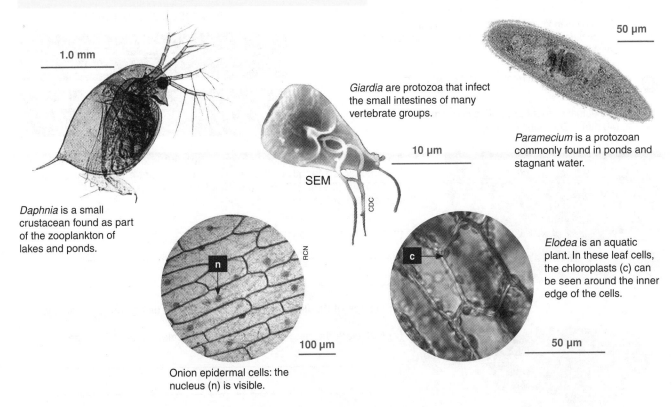

1.0 mm

Daphnia is a small crustacean found as part of the zooplankton of lakes and ponds.

Giardia are protozoa that infect the small intestines of many vertebrate groups.

SEM

10 µm

50 µm

Paramecium is a protozoan commonly found in ponds and stagnant water.

n

Onion epidermal cells: the nucleus (n) is visible.

100 µm

c

Elodea is an aquatic plant. In these leaf cells, the chloroplasts (c) can be seen around the inner edge of the cells.

50 µm

1. Using the measurement scales provided on each of the photographs above, determine the longest dimension (length or diameter) of the cell/animal/organelle indicated in µm and mm. Attach your working:

(a) *Daphnia*: _____ µm _____ mm   (d) *Elodea* leaf cell: _____ µm _____ mm

(b) *Giardia*: _____ µm _____ mm   (e) Chloroplast: _____ µm _____ mm

(c) Nucleus _____ µm _____ mm   (f) *Paramecium*: _____ µm _____ mm

2. (a) List a-f in question 1 in order of size, from the smallest to the largest:

_____

(b) Study your ruler. Which one of the above could you see with your unaided eye?_____

3. Calculate the equivalent length in millimetres (mm) of the following measurements:

(a) 0.25 µm: _____   (b) 450 µm: _____   (c) 200 nm: _____

LINK
10

WEB
4

KNOW

5 Calculating Linear Magnification

Key Idea: Magnification is how much larger an object appears compared to its actual size. It can be calculated from the ratio of image height to object height.

Microscopes produce an enlarged (magnified) image of an object allowing it to be observed in greater detail than is possible with the naked eye. **Magnification** refers to the number of times larger an object appears compared to its actual size. Linear magnification is calculated by taking a ratio of the image height to the object's actual height. If this ratio is greater than one, the image is enlarged. If it is less than one, it is reduced. To calculate magnification, all measurements are converted to the same units. Often, you will be asked to calculate an object's actual size, in which case you will be told the size of the object and the magnification.

Calculating Linear Magnification: A Worked Example

1.0 mm

1. Measure the body length of the bed bug image (right). Your measurement should be 40 mm (*not* including the body hairs and antennae).

2. Measure the length of the scale line marked 1.0 mm. You will find it is 10 mm long. The magnification of the scale line can be calculated using equation 1 (below right).

   The magnification of the scale line is **10** (10 mm / 1 mm)

   NB: The magnification of the bed bug image will also be 10x because the scale line and image are magnified to the same degree.

3. Calculate the actual (real) size of the bed bug using equation 2 (right):

   The actual size of the bed bug is **4 mm** (40 mm / 10 x magnification)

Microscopy Equations

1. $\text{Magnification} = \dfrac{\text{measured size of the object}}{\text{actual size of the object}}$

2. $\text{Actual object size} = \dfrac{\text{size of the image}}{\text{magnification}}$

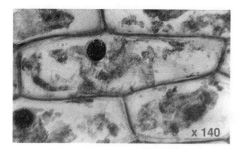

x 140

1. The bright field microscopy image on the left is of onion epidermal cells. The measured length of the onion cell in the centre of the photograph is 52,000 µm (52 mm). The image has been magnified 140 x. Calculate the actual size of the cell:

   _____

   _____

   _____

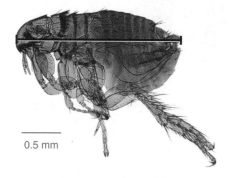

0.5 mm

2. The image of the flea (left) has been captured using light microscopy.

   (a) Calculate the magnification using the scale line on the image:

   _____

   _____

   _____

   (b) The body length of the flea is indicated by a line. Measure along the line and calculate the actual length of the flea:

   _____

   _____

   _____

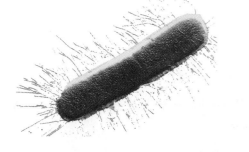

3. The image size of the *E.coli* cell (left) is 43 mm, and its actual size is 2 µm. Using this information, calculate the magnification of the image:

   _____

   _____

   _____

LINK

4

© 2012-2014 **BIOZONE** International
ISBN: 978-1-927173-93-0
Photocopying Prohibited

6 Multicellularity

Key Idea: Specialized cells and tissues arise through cell differentiation, which is regulated through differential gene expression. The complex interactions of cells in multicellular organisms results in the emergence of new properties.

The cell is the functioning unit structure from which living organisms are made. In multicellular organisms, cell differentiation produces specialized cells with specific functions. Cells with related functions associate to form tissues, and tissues are organized into organs. With each step in this hierarchy of biological order, new properties emerge that were not present at simpler levels of organization. Life is an emergent property of billions of chemical reactions that are driven by the input of energy that produces work and results in decreased entropy (disorder) within the system.

How can it be that all of an organism's cells have the same genetic material, but the cells have a wide variety of shapes and functions? The answer is through **cell differentiation**.

During development, the unspecialized stem cells of the zygote develop along germline, endoderm, mesoderm, or ectoderm lines to form specialized cells. Although each cell has the same genetic material (genes), differences in gene expression determine which type of cell forms. Once the developmental pathway of a cell is determined, it cannot then change into another cell type.

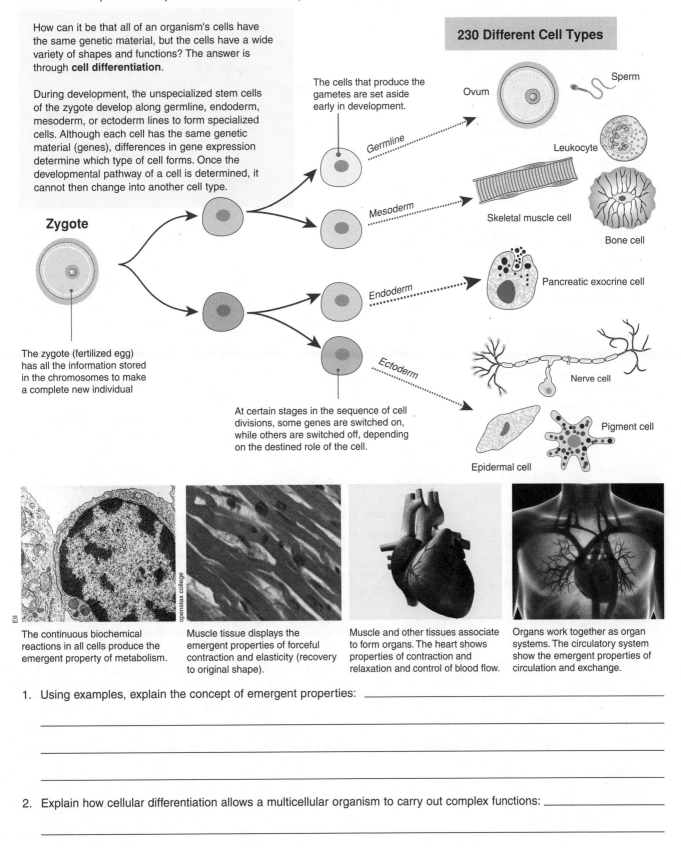

230 Different Cell Types

The cells that produce the gametes are set aside early in development.

Germline — Ovum, Sperm, Leukocyte

Mesoderm — Skeletal muscle cell, Bone cell

Endoderm — Pancreatic exocrine cell

Ectoderm — Nerve cell, Pigment cell, Epidermal cell

Zygote

The zygote (fertilized egg) has all the information stored in the chromosomes to make a complete new individual

At certain stages in the sequence of cell divisions, some genes are switched on, while others are switched off, depending on the destined role of the cell.

The continuous biochemical reactions in all cells produce the emergent property of metabolism.

Muscle tissue displays the emergent properties of forceful contraction and elasticity (recovery to original shape).

Muscle and other tissues associate to form organs. The heart shows properties of contraction and relaxation and control of blood flow.

Organs work together as organ systems. The circulatory system show the emergent properties of circulation and exchange.

1. Using examples, explain the concept of emergent properties: _____

_____

_____

_____

2. Explain how cellular differentiation allows a multicellular organism to carry out complex functions: _____

_____

_____

LINK 7   WEB 6   **KNOW**

7 Stem Cells and Differentiation

Key Idea: Stem cells are undifferentiated cells found in multicellular organisms. They are characterized by the properties of self renewal and potency.

A zygote can differentiate into many different types of cells because early on it divides into stem cells. Stem cells are unspecialized and can give rise to the many cell types that make up the tissues and organs of a multicellular organism. The differentiation of multipotent stem cells in bone marrow gives rise to all the cell types that make up blood, a fluid connective tissue. Multipotent (or adult) stem cells are found in most body organs, where they replace old or damaged cells and replenish the body's cells throughout life.

Stem Cells and Blood Cell Production

New blood cells are produced in the red bone marrow, which becomes the main site of blood production after birth, taking over from the fetal liver. All types of blood cells develop from a single cell type: called a **multipotent stem cell**. These cells are capable of mitosis and of differentiation into 'committed' precursors of each of the main types of blood cell. Each of the different cell lines is controlled by a specific **growth factor**. When a stem cell divides, one of its daughters remains a stem cell, while the other becomes a precursor cell, either a **lymphoid cell** or **myeloid cell**. These cells continue to mature into the various specialized cell types.

Properties of Stem Cells

Self renewal: The ability to divide many times while maintaining an unspecialized state.

Potency: The ability to differentiate into specialized cells.

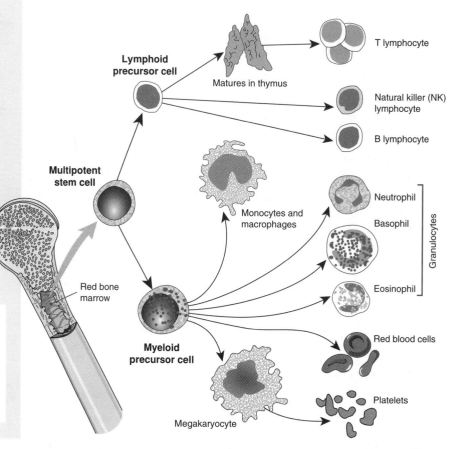

Categories of Stem Cells

Totipotent stem cells

These stem cells can differentiate into all the cells in an organism. Example: In humans, the zygote and its first few divisions. The meristematic tissue of plants is also totipotent.

Pluripotent stem cells

These stem cells can give rise to any cells of the body, except extra-embryonic cells (e.g. placenta and chorion). Example: Embryonic stem cells.

Multipotent stem cells

These adult stem cells can give rise a limited number of cell types, related to their tissue of origin. Example: Bone marrow stem cells, epithelial stem cells, bone stem cells (osteoblasts).

1. Describe the two defining features of stem cells:

    (a) _____

    (b) _____

2. Explain the role of stem cells in the development of specialized tissues in multicellular organisms: _____

    _____

    _____

    _____

    _____

© 2012-2014 **BIOZONE** International
ISBN: 978-1-927173-93-0

8 Types of Stem Cells

Key Idea: The potency of stem cells depends on their origin. Both embryonic and adult stem cells can be used to replace diseased and damaged tissue.

The properties of self renewal and potency make stem cells suitable for a wide range of applications. Stem cells from early stage embryos (embryonic stem cells) are pluripotent and can potentially be cultured to provide a renewable source of cells for studies of human development and gene regulation, for tests of new drugs and vaccines, for monoclonal antibody production, and for treating any type of diseased or damaged tissue. Adult stem cells from bone marrow or umbilical cord blood can give rise to a more limited number of cell types. Although their potential use is more restricted, there are fewer ethical issues associated with their use.

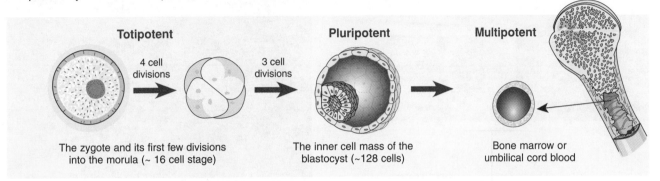

Totipotent — 4 cell divisions → 3 cell divisions → **Pluripotent** → **Multipotent**

The zygote and its first few divisions into the morula (~ 16 cell stage)

The inner cell mass of the blastocyst (~128 cells)

Bone marrow or umbilical cord blood

Embryonic Stem Cells

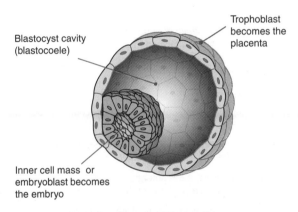

Trophoblast becomes the placenta

Blastocyst cavity (blastocoele)

Inner cell mass or embryoblast becomes the embryo

Embryonic stem cells (**ESC**) are derived from the inner cell mass of blastocysts (above). Blastocysts are embryos that are about five days old and consist of a hollow ball of 50-150 cells. Cells derived from the inner cell mass are **pluripotent**. They can become any cells of the body, with the exception of placental cells. When cultured without any stimulation to differentiate, ESC retain their potency through multiple cell divisions. This means they have great potential for therapeutic use in regenerative medicine and tissue replacement. However, the use of ESC involves the deliberate creation and destruction of embryos and is therefore is ethically unacceptable to many people.

Adult Stem Cells

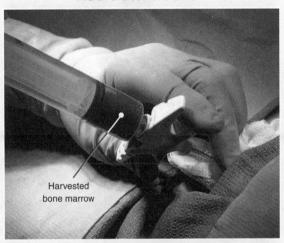

Harvested bone marrow

Adult stem cells (ASC) are undifferentiated cells found in several types of tissues (e.g. brain, bone marrow, fat, and liver) in adults, children, and umbilical cord blood. Unlike ESCs, they are **multipotent** and can only differentiate into a limited number of cell types, usually related to the tissue of origin. There are fewer ethical issues associated with using ASC for therapeutic purposes, because no embryos are destroyed. For this reason, ASC are already widely used to treat a number of diseases including leukemia and other blood disorders.

1. (a) Distinguish between embryonic stem cells and adult stem cells with respect to their **potency**:

_____

_____

_____

(b) What is the significance of this difference to their use in the treatment of disease: _____

_____

_____

_____

LINK **9**   WEB **8**   **APP**

Embryonic Stem Cell (ESC) Cloning

ESC can come from embryos that have been fertilized *in vitro* and then donated for research. These cell lines will not be patient-matched because each new embryo is unique. However, ESC can also come from cloned embryos created using somatic cell nuclear transfer using a donor nucleus from the patient, as shown below. These ESC lines will patient-matched.

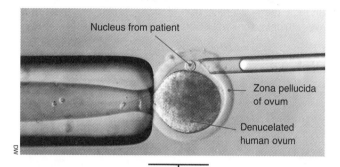

Nucleus from patient

Zona pellucida of ovum

Denucelated human ovum

A mild electric shock induces the development of a pre-embryo cell containing the patient's DNA.

After 5 days development, the inner cell mass of about 30 stem cells is removed.

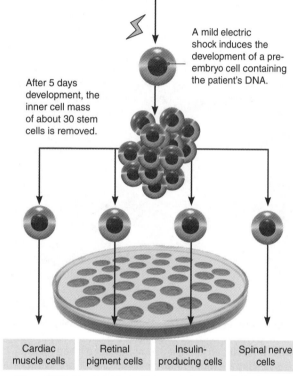

| Cardiac muscle cells | Retinal pigment cells | Insulin-producing cells | Spinal nerve cells |

When ESCs are provided with appropriate growth factors and conditions, they will differentiate into specific specialized cell types.

Issues in ESC Cloning

For all tissue transplants, e.g. blood transfusions and bone marrow transplants, tissues must be matched for histocompatibility between different individuals. If donor material is poorly matched to the recipient, the recipient's immune system rejects the donor cells. Stem cell cloning (**therapeutic cloning**) provides a way around this problem. Stem cell cloning produces genetically matched stem cells that can be turned into any cell type in the human body.

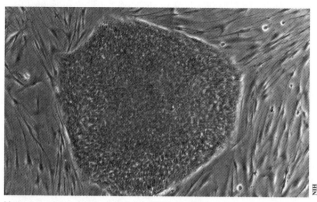

Human embryonic stem cells (hESC) growing on mouse embryonic fibroblasts. The mouse fibroblasts act as feeder cells for the culture, releasing nutrients and providing a surface for the ESCs to grow on.

ESC therapy has enormous potential to make life changing improvements to the health of people with diseased or damaged organs. Organs or tissues derived from a patient's ESC could be transplanted back into that patient without fear of tissue rejection or the need for ongoing immunosuppressive drug therapies. Despite this, many groups oppose the use of therapeutic cloning for many reasons including:

▸ The technology used to create the embryo could be used for reproductive cloning, i.e. creating a clone of the original human.

▸ The creation of stem cell line requires the destruction of a human embryo and thus human life.

▸ Human embryos have the potential to develop into an individual and thus have the same rights of the individual.

▸ Saving or enhancing the quality of life of an individual does not justify the destruction of the life of another (i.e. the embryo).

▸ ESC research has not produced any viable long term treatment, while other techniques (e.g. adult stem cells) have.

▸ There are other stem cell techniques that do not require the creation of an embryo but achieve similar results (e.g. cell lines grown from adult stem cells or umbilical cord blood).

2. (a) In your opinion, what is the one most important ethical issue associated with the use of ESC in medicine?

_____

_____

(b) What advantage does therapeutic cloning offer over conventional therapeutic use of embryonic stem cells?

_____

_____

_____

3. Umbilical cord blood is promoted as a rich source of multipotent stem cells for autologous (self) transplants. Can you see a problem with the use of a baby's cord blood to treat a disease in that child at a later date?

_____

_____

_____

© 2012-2014 **BIOZONE** International
ISBN: 978-1-927173-93-0

9. Using Stem Cells to Treat Disease

Key Idea: Embryonic stem cells have been used to treat Stargardt's disease with apparent success. New techniques make it possible to produce pluripotent cells for widespread therapeutic use in an ethically acceptable way.

The therapeutic use of embryonic stem cells (ESC) to replace lost retinal cells in patients with Stargardt's disease (an eye disease) has shown that such therapies are able to restore function to diseased organs. Future treatments using induced pluripotent cells derived from adult tissues will create patient-matched cell lines and bypass the need for embryos.

Stem Cells for Stargardt's Disease

Stargardt's disease is an inherited form of juvenile macular degeneration (a loss of the central visual field of the eye). The disease is associated with a number of mutations and results in dysfunction of the retinal pigment epithelium (RPE) cells, which nourish the retinal photoreceptor cells and protect the retina from excess light. Dysfunction of the RPE causes deterioration of the photoreceptor cells in the central portion of the retina and progressive loss of central vision. This often begins between ages 6 and 12 and continues until a person is legally blind. Trials using stem cells have obtained promising results in treating the disease.

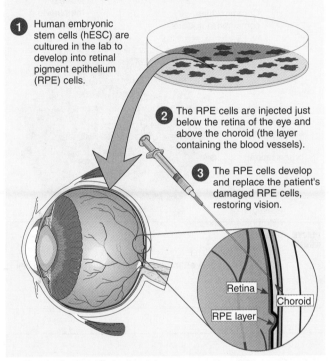

1. Human embryonic stem cells (hESC) are cultured in the lab to develop into retinal pigment epithelium (RPE) cells.

2. The RPE cells are injected just below the retina of the eye and above the choroid (the layer containing the blood vessels).

3. The RPE cells develop and replace the patient's damaged RPE cells, restoring vision.

Retina
Choroid
RPE layer

Stem Cells for Type 1 Diabetes?

Type 1 diabetes results from the body's own immune system attacking and destroying the insulin producing beta cells of the pancreas. In theory, new beta cells could be produced using stem cells. Research is focused on how to obtain the stem cells and deliver them effectively to the patient. Many different techniques are currently being investigated. Most techniques use stem cells from non-diabetics, requiring recipients to use immunosuppressant drugs so the cells are not rejected.

A study published in 2014 described a method for treating type 1 diabetes in mice using fibroblast cells taken from the skin of mice.

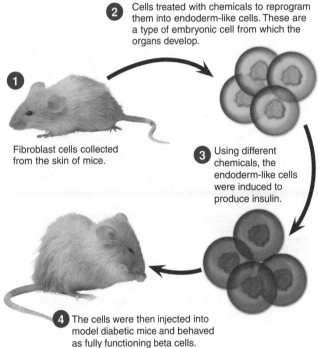

2. Cells treated with chemicals to reprogram them into endoderm-like cells. These are a type of embryonic cell from which the organs develop.

1. Fibroblast cells collected from the skin of mice.

3. Using different chemicals, the endoderm-like cells were induced to produce insulin.

4. The cells were then injected into model diabetic mice and behaved as fully functioning beta cells.

1. Describe one potential advantage of using embryonic stem cells for tissue engineering technology:

_____

_____

2. (a) Explain the basis for correcting Stargardt's disease using stem cell technology: _____

_____

_____

(b) Suggest why researchers derived the RPE cells from embryos rather than by reprogramming a patient's own cells:

_____

_____

(c) What advantage is there in reprogramming a patient's own cells and when would this be a preferable option?

_____

_____

_____

LINK 201   LINK 110   WEB 9   **APP**

10 Comparing Prokaryotic and Eukaryotic Cells

Key Idea: There are two broad types of cells, prokaryotic (or bacterial) cells and eukaryotic cells.
The cell is the smallest unit of life, and are often called the building blocks of life. Cells are either **prokaryotic** or **eukaryotic**. Within each of these groups, cells may vary greatly in their size, shape, and functional role.

Prokaryotic Cells

▶ Prokaryotic (bacterial) cells do not have a membrane-bound nucleus or any other membrane-bound organelles.

▶ They are small (generally 0.5-10 µm) single cells.

▶ They are relatively simple cells and have very little cellular organization (their DNA, ribosomes, and enzymes are free floating within the cell cytoplasm).

▶ Prokaryotes have a cell wall, but it is different to the cell walls that some eukaryotes have.

▶ Single, circular chromosome of naked DNA.

▶ Prokaryotes contain 70S ribosomes (Svedberg units (S) measure sedimentation rate).

Eukaryotic Cells

▶ **Eukaryotic cells** have a membrane-bound nucleus, and other membrane-bound organelles.

▶ Plant cells, animals cells, fungal cells, and protists are all eukaryotic cells.

▶ Eukaryotic cells are large (30-150 µm). They may exist as single cells or as part of a multicellular organism.

▶ They are more complex than prokaryotic cells. They have more structure and internal organization.

▶ Multiple linear chromosomes consisting of DNA and associated proteins.

▶ Eukaryotes contain 80S ribosomes.

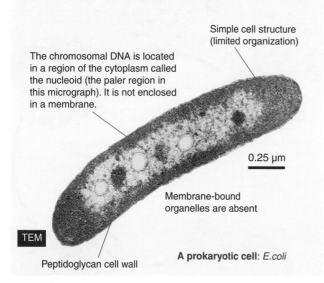

The chromosomal DNA is located in a region of the cytoplasm called the nucleoid (the paler region in this micrograph). It is not enclosed in a membrane.

Simple cell structure (limited organization)

0.25 µm

Membrane-bound organelles are absent

TEM

Peptidoglycan cell wall

A prokaryotic cell: *E.coli*

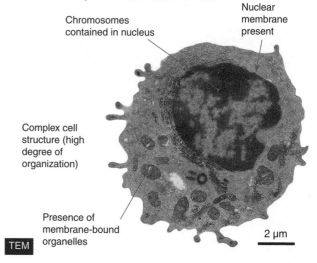

Chromosomes contained in nucleus

Nuclear membrane present

Complex cell structure (high degree of organization)

Presence of membrane-bound organelles

2 µm

TEM

A eukaryotic cell: a human white blood cell

1. What are the characteristic features of a prokaryotic cell? _____

_____

_____

2. What are the characteristic features of a eukaryotic cell? _____

_____

_____

3. List examples of eukaryotic cells: _____

4. Study the images A to D below. Identify each cell as prokaryotic or eukaryotic and give a brief reason for your choice:

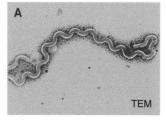 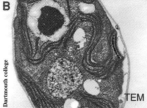 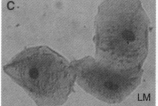 

A _____ B _____ C _____ D _____

© 2012-2014 **BIOZONE** International
ISBN: 978-1-927173-93-0
Photocopying Prohibited

11 History of Microscopy

Key Idea: Microscopes are used to view objects that cannot be viewed in detail with the naked eye. Microscopes have become increasingly sophisticated over time with improvements in both magnification and resolution.

Lenses of various descriptions have been used for around 4000 years to view objects, but it is only in the last few hundred years that techniques have developed to build sophisticated devices for viewing microscopic objects. Early microscopes suffered from image distortion such as chromatic aberration (the production of images with the light split into the different colours). The development of more sophisticated techniques in lens and microscope production reduced this problem. The development of electron microscopes has made it possible to image objects to the atomic level.

Milestones in Microscopy

1500	Convex lenses with a magnification greater than x5 became available.
1595	**Zacharias Janssen** of Holland has been credited with the first compound microscope (more than one lens).
1662	**Robert Hooke** of England used the term 'cell' in describing the microscopic structure of cork.
1675	**Antoni van Leeuwenhoek** of Holland produced over 500 single lens microscopes that had a magnification of 270 times.
1800s	The discovery that lenses combining two types of glass reduced chromatic aberration (the production of images with the light split into the different colours) allows clear images to be viewed.
1830	**Joseph Jackson Lister** demonstrated that spherical aberration (the focussing of light rays at different points due to the curve of the lens) could be reduced by using different lenses at precise distances from each other.
1878	**Ernst Abbe** produced a formula for correlating resolution to the wavelength of light, and so describes the maximum resolution of a light microscope.
1903	**Richard Zsigmondy** developed the ultramicroscope allowing objects smaller than the wavelength of light to be viewed.
1932	**Frits Zernike** invented the phase-contrast microscope making transparent or colourless objects easier to view.
1938	**Ernst Ruska** developed the transmission electron microscope (TEM). Electrons pass through an object and are focused by magnets. The short wavelength of electrons allows study of incredibly small objects. **Manfred von Ardenne** developed the scanning electron microscope (SEM) around the same time allowing the surface of objects to be imaged.
1981	**Gerd Binning** and **Heinrich Rohrer** invented the scanning tunneling electron microscope (STM), producing three dimensional images at the atomic level.

1595
The first compound microscope (the Janssen microscope, left) consisted of three draw tubes with lenses inserted into the tubes. The microscope was focussed by sliding the draw tube in or out.

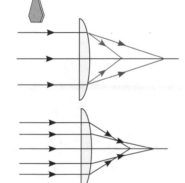

Single lens.

Pointed spike which is the specimen holder.

Screw threads adjusted the position of the specimen (up, down, and focus)

1675
A Leeuwenhoek microscope c. 1673 (right) was only a glorified magnifying glass by today's standards.

1800s
Chromatic aberration. Blue light refracts more than red light producing different focal points.

1830
Spherical aberration. Light entering at the edge of the lens focuses closer to the lens than light entering near the centre of the lens.

1932

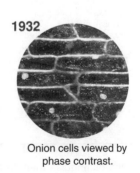

Onion cells viewed by phase contrast.

1981

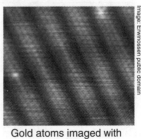

Image: Erwinossen public domain

Gold atoms imaged with an STM

1938   **TEM**

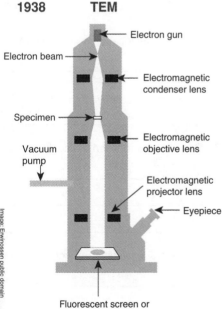

Electron gun

Electron beam

Electromagnetic condenser lens

Specimen

Vacuum pump

Electromagnetic objective lens

Electromagnetic projector lens

Eyepiece

Fluorescent screen or photographic plate

12 Prokaryotic Cell Structure

Key Idea: Prokaryotic cells have a simpler cell structure than eukaryotic cells.

Prokaryotic (bacterial) cells are much smaller than eukaryotic cells and lack many eukaryotic features, such as a distinct nucleus and membrane-bound cellular organelles. The cell wall is an important feature. It is a complex, multi-layered structure and has a role in the organism's ability to cause disease. A generalized prokaryote, *E. coli*, is shown below.

E. coli Structure

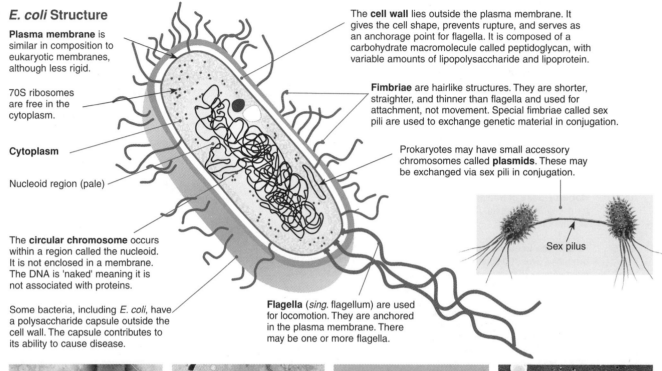

Plasma membrane is similar in composition to eukaryotic membranes, although less rigid.

70S ribosomes are free in the cytoplasm.

Cytoplasm

Nucleoid region (pale)

The **circular chromosome** occurs within a region called the nucleoid. It is not enclosed in a membrane. The DNA is 'naked' meaning it is not associated with proteins.

Some bacteria, including *E. coli*, have a polysaccharide capsule outside the cell wall. The capsule contributes to its ability to cause disease.

The **cell wall** lies outside the plasma membrane. It gives the cell shape, prevents rupture, and serves as an anchorage point for flagella. It is composed of a carbohydrate macromolecule called peptidoglycan, with variable amounts of lipopolysaccharide and lipoprotein.

Fimbriae are hairlike structures. They are shorter, straighter, and thinner than flagella and used for attachment, not movement. Special fimbriae called sex pili are used to exchange genetic material in conjugation.

Prokaryotes may have small accessory chromosomes called **plasmids**. These may be exchanged via sex pili in conjugation.

Sex pilus

Flagella (*sing*. flagellum) are used for locomotion. They are anchored in the plasma membrane. There may be one or more flagella.

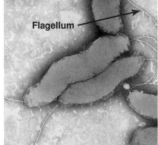

Flagellum

A spiral shape is one of four bacterial shapes (the others being rods, commas, and spheres). These *Campylobacter* cells also have flagella.

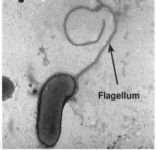

Flagellum

Helicobacter pylori, is a comma-shaped vibrio bacterium that causes stomach ulcers in humans. It moves by means of polar flagella.

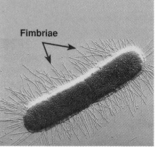

Fimbriae

Flagellum

Escherichia coli is a rod-shaped bacterium, common in the human gut. The fimbriae surrounding the cell are used to adhere to the intestinal wall.

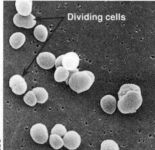

Dividing cells

Bacteria usually divide by binary fission. During this process, DNA is copied and the cell splits into two cells, as in these round (cocci) cells.

1. Describe the location, role, and general composition of the bacterial cell wall: _____

   _____

   _____

   _____

2. Describe the function of flagella in bacteria and distinguish them from fimbriae: _____

   _____

   _____

   _____

3. How are sex pili different from fimbriae and what is their significance? _____

   _____

   _____

4. On a separate sheet of paper, draw and annotate a generalized prokaryotic cell to include plasma membrane, cell wall, capsule, plasmid DNA, chromosome, flagella, and fimbriae. Staple it into your workbook.

© 2012-2014 **BIOZONE** International
ISBN: 978-1-927173-93-0
Photocopying Prohibited

13 Binary Fission in Prokaryotes

Key Idea: Binary fission involves division of the parent body into two, fairly equal, parts to produce two identical cells. Binary fission is a form of asexual reproduction carried out by most prokaryotes, some eukaryotic organelles, such as chloroplasts, and some unicellular eukaryotes (although the process is somewhat different in eukaryotic cells). The time required for a bacterial cell to divide, or for a population of bacterial cells to double, is called the generation time. Generation times may be quite short (20 minutes) in some species and as long as several days in others.

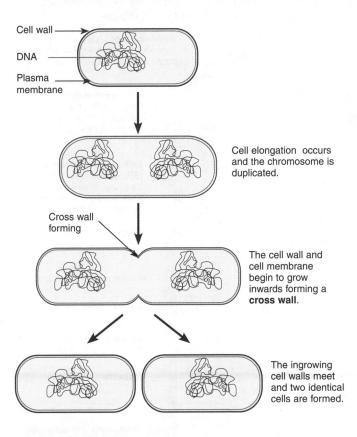

Cell wall
DNA
Plasma membrane

Cell elongation occurs and the chromosome is duplicated.

Cross wall forming

The cell wall and cell membrane begin to grow inwards forming a **cross wall**.

The ingrowing cell walls meet and two identical cells are formed.

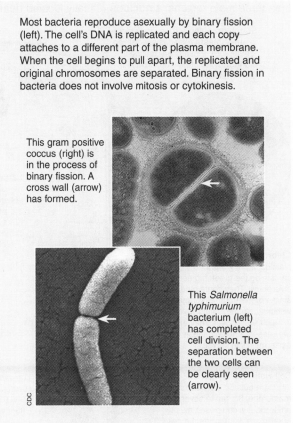

Most bacteria reproduce asexually by binary fission (left). The cell's DNA is replicated and each copy attaches to a different part of the plasma membrane. When the cell begins to pull apart, the replicated and original chromosomes are separated. Binary fission in bacteria does not involve mitosis or cytokinesis.

This gram positive coccus (right) is in the process of binary fission. A cross wall (arrow) has formed.

This *Salmonella typhimurium* bacterium (left) has completed cell division. The separation between the two cells can be clearly seen (arrow).

CDC

Generation time (minutes)	Population size
0	1
20	2
40	4
60	8
80	
100	
120	
140	
160	
180	
200	
220	
240	
260	
280	
300	
320	
340	
360	

1. What is **binary fission**? _____

_____

_____

_____

2. Explain why the formation of the **cross wall** is important in binary fission:

_____

_____

_____

3. Explain the term **generation time**: _____

_____

_____

4. A species of bacteria reproduces every 20 minutes. Complete the table (left) by calculating the number of bacteria present at 20 minute intervals.

5. State how many bacteria were present after:

(a) 1 hour: _____

(b) 3 hours: _____

(c) 6 hours: _____

14 Plant Cells

Key Idea: Plant cells are eukaryotic cells. They have many features in common with animal cells, but they also have several unique features.

Eukaryotic cells have a similar basic structure, although they may vary tremendously in size, shape, and function. Certain features are common to almost all eukaryotic cells, including their three main regions: a **nucleus** (usually located near the centre of the cell), surrounded by a watery **cytoplasm**, which is itself enclosed by the **plasma membrane**. Plant cells are enclosed in a cellulose cell wall, which gives them a regular and uniform appearance. The cell wall protects the cell, maintains its shape, and prevents excessive water uptake. It provides rigidity to plant structures but permits the free passage of materials into and out of the cell.

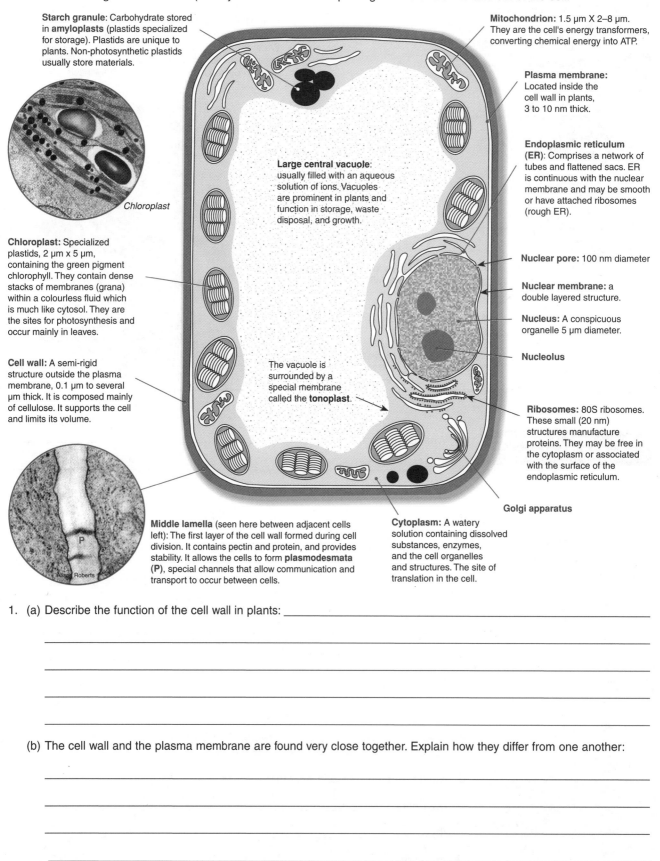

Starch granule: Carbohydrate stored in **amyloplasts** (plastids specialized for storage). Plastids are unique to plants. Non-photosynthetic plastids usually store materials.

Chloroplast

Chloroplast: Specialized plastids, 2 μm x 5 μm, containing the green pigment chlorophyll. They contain dense stacks of membranes (grana) within a colourless fluid which is much like cytosol. They are the sites for photosynthesis and occur mainly in leaves.

Cell wall: A semi-rigid structure outside the plasma membrane, 0.1 μm to several μm thick. It is composed mainly of cellulose. It supports the cell and limits its volume.

Mitochondrion: 1.5 μm X 2–8 μm. They are the cell's energy transformers, converting chemical energy into ATP.

Plasma membrane: Located inside the cell wall in plants, 3 to 10 nm thick.

Endoplasmic reticulum (ER): Comprises a network of tubes and flattened sacs. ER is continuous with the nuclear membrane and may be smooth or have attached ribosomes (rough ER).

Nuclear pore: 100 nm diameter

Nuclear membrane: a double layered structure.

Nucleus: A conspicuous organelle 5 μm diameter.

Nucleolus

Large central vacuole: usually filled with an aqueous solution of ions. Vacuoles are prominent in plants and function in storage, waste disposal, and growth.

The vacuole is surrounded by a special membrane called the **tonoplast**.

Ribosomes: 80S ribosomes. These small (20 nm) structures manufacture proteins. They may be free in the cytoplasm or associated with the surface of the endoplasmic reticulum.

Golgi apparatus

Middle lamella (seen here between adjacent cells left): The first layer of the cell wall formed during cell division. It contains pectin and protein, and provides stability. It allows the cells to form **plasmodesmata (P)**, special channels that allow communication and transport to occur between cells.

Cytoplasm: A watery solution containing dissolved substances, enzymes, and the cell organelles and structures. The site of translation in the cell.

Alison Roberts

1. (a) Describe the function of the cell wall in plants: _____

_____

_____

_____

_____

(b) The cell wall and the plasma membrane are found very close together. Explain how they differ from one another:

_____

_____

_____

_____

© 2012-2014 **BIOZONE** International
ISBN: 978-1-927173-93-0
Photocopying Prohibited

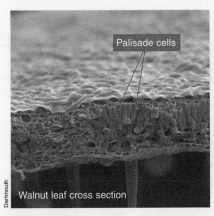
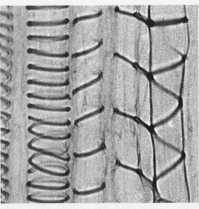
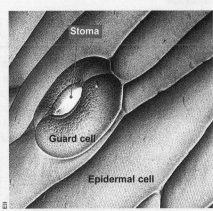

Plants require many different cell types to carry out specific functions. The palisade cells (shown above) contain chloroplasts for photosynthesis.

Xylem tissue makes up part of the vascular tissue in a plant. Vessels (above) and tracheids are specialized cells for carrying water from the roots to the leaves.

Guard cells regulate the opening and closing of the stoma, thus regulating gas exchange and water loss.

2. Explain how organelles increase the efficiency of the cell: _____

_____

_____

_____

3. Identify the organelle in the plant cell diagram on the previous page that is most commonly found in stems and leaves:

_____

4. Explain why palisade cells are found in the upper region of the leaf: _____

_____

_____

_____

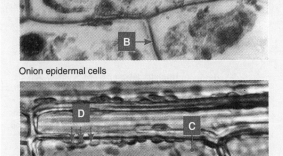

Onion epidermal cells

Elodea cells

5. The two photographs (left) show plant cells as seen by a light microscope. Identify the basic features labelled **A-D**:

A: _____

B: _____

C: _____

D: _____

6. Cytoplasmic streaming is a feature of eukaryotic cells, often clearly visible with a light microscope in plant (and algal) cells.

(a) Explain what is meant by cytoplasmic streaming:

_____

_____

(b) For the *Elodea* cell (lower, left), draw arrows to indicate cytoplasmic streaming movements.

7. Describe three structures/organelles present in generalized plant cells but absent from animal cells:

(a) _____

(b) _____

(c) _____

15 Animal Cells

Key Idea: Animal cells are eukaryotic cells. They have many features in common with plant cells, but also have a number of unique features.

Although plant and animal cells have many features in common, animal cells do not have a regular shape and some (such as phagocytic white blood cells) are quite mobile. The diagram below shows the ultrastructure of a **liver** cell (hepatocyte). It contains organelles common to most relatively unspecialized human cells. Hepatocytes make up 70-80% of the liver's mass. They are metabolically active, with a large central nucleus, many mitochondria, and large amounts of rough endoplasmic reticulum. Thin, cellular extensions called microvilli increase surface area of the cell, increasing its capacity for absorption.

Structures and Organelles in a Liver Cell

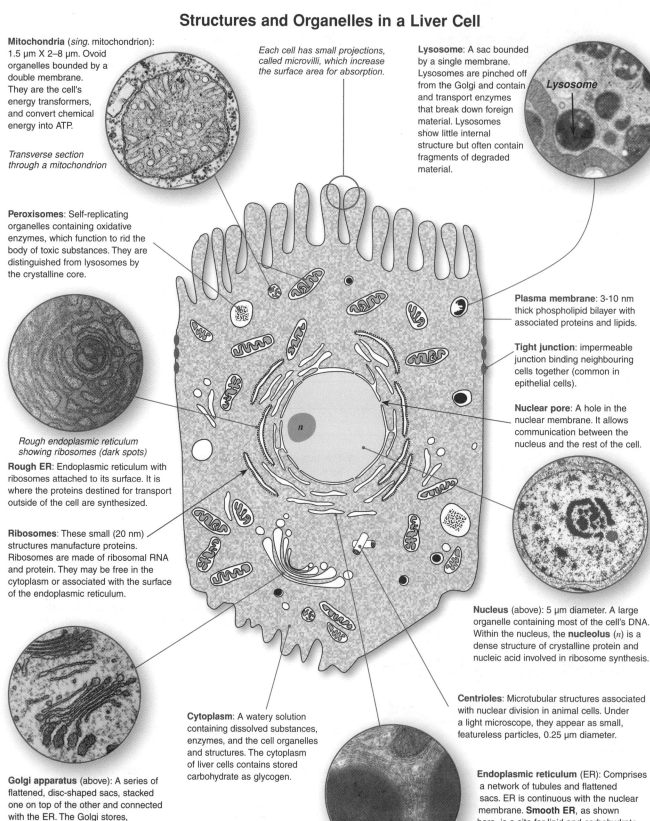

Mitochondria (*sing.* mitochondrion): 1.5 µm X 2–8 µm. Ovoid organelles bounded by a double membrane. They are the cell's energy transformers, and convert chemical energy into ATP.

Transverse section through a mitochondrion

Each cell has small projections, called microvilli, which increase the surface area for absorption.

Lysosome: A sac bounded by a single membrane. Lysosomes are pinched off from the Golgi and contain and transport enzymes that break down foreign material. Lysosomes show little internal structure but often contain fragments of degraded material.

Lysosome

Peroxisomes: Self-replicating organelles containing oxidative enzymes, which function to rid the body of toxic substances. They are distinguished from lysosomes by the crystalline core.

Rough endoplasmic reticulum showing ribosomes (dark spots)

Rough ER: Endoplasmic reticulum with ribosomes attached to its surface. It is where the proteins destined for transport outside of the cell are synthesized.

Ribosomes: These small (20 nm) structures manufacture proteins. Ribosomes are made of ribosomal RNA and protein. They may be free in the cytoplasm or associated with the surface of the endoplasmic reticulum.

Plasma membrane: 3-10 nm thick phospholipid bilayer with associated proteins and lipids.

Tight junction: impermeable junction binding neighbouring cells together (common in epithelial cells).

Nuclear pore: A hole in the nuclear membrane. It allows communication between the nucleus and the rest of the cell.

Nucleus (above): 5 µm diameter. A large organelle containing most of the cell's DNA. Within the nucleus, the **nucleolus** (*n*) is a dense structure of crystalline protein and nucleic acid involved in ribosome synthesis.

Centrioles: Microtubular structures associated with nuclear division in animal cells. Under a light microscope, they appear as small, featureless particles, 0.25 µm diameter.

Golgi apparatus (above): A series of flattened, disc-shaped sacs, stacked one on top of the other and connected with the ER. The Golgi stores, modifies, and packages proteins. It 'tags' proteins so that they go to their correct destination.

Cytoplasm: A watery solution containing dissolved substances, enzymes, and the cell organelles and structures. The cytoplasm of liver cells contains stored carbohydrate as glycogen.

Endoplasmic reticulum (ER): Comprises a network of tubules and flattened sacs. ER is continuous with the nuclear membrane. **Smooth ER**, as shown here, is a site for lipid and carbohydrate metabolism, including hormone synthesis.

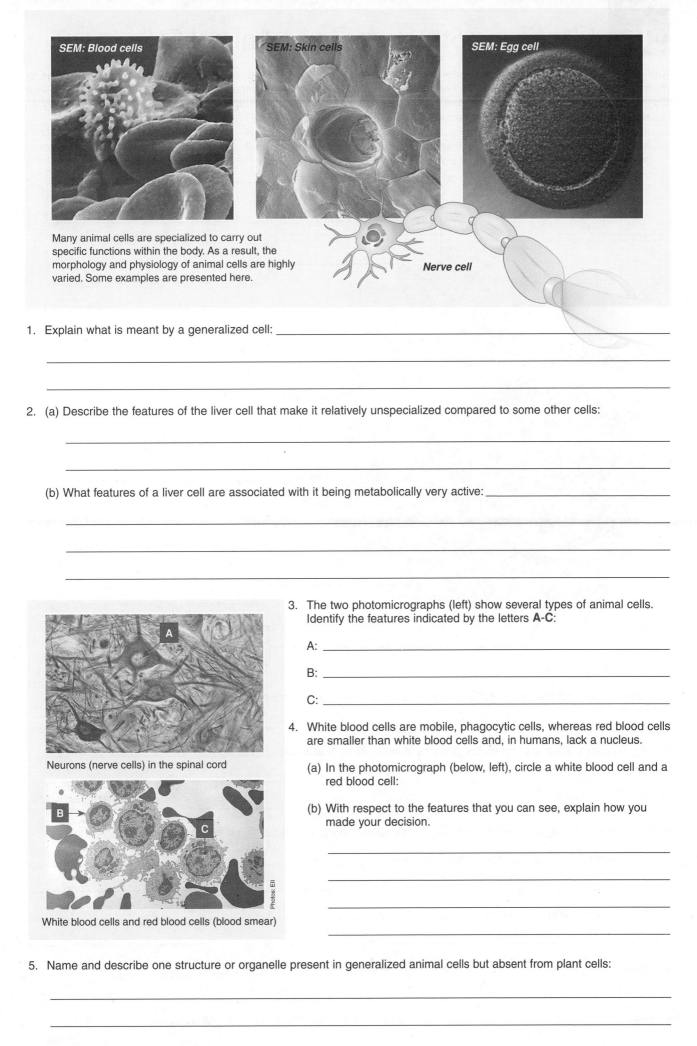

SEM: Blood cells

SEM: Skin cells

SEM: Egg cell

Many animal cells are specialized to carry out specific functions within the body. As a result, the morphology and physiology of animal cells are highly varied. Some examples are presented here.

Nerve cell

1. Explain what is meant by a generalized cell: _____

_____

_____

2. (a) Describe the features of the liver cell that make it relatively unspecialized compared to some other cells:

_____

_____

(b) What features of a liver cell are associated with it being metabolically very active: _____

_____

_____

_____

3. The two photomicrographs (left) show several types of animal cells. Identify the features indicated by the letters **A-C**:

A: _____

B: _____

C: _____

Neurons (nerve cells) in the spinal cord

4. White blood cells are mobile, phagocytic cells, whereas red blood cells are smaller than white blood cells and, in humans, lack a nucleus.

(a) In the photomicrograph (below, left), circle a white blood cell and a red blood cell:

(b) With respect to the features that you can see, explain how you made your decision.

_____

_____

_____

_____

White blood cells and red blood cells (blood smear)

Photos: Eli

5. Name and describe one structure or organelle present in generalized animal cells but absent from plant cells:

_____

_____

© 2012-2014 **BIOZONE** International
ISBN: 978-1-927173-93-0

16 Identifying Structures in an Animal Cell

Key Idea: The position of the organelles in an electron micrograph can result in variations in their appearance.
Our current knowledge of cell ultrastructure has been made possible by the advent of electron microscopy. Transmission electron microscopy is the most frequently used technique for viewing cellular organelles. When viewing TEMs, the cellular organelles may appear to be quite different depending on whether they are in transverse or longitudinal section.

1. Identify and label the structures in the animal cell below using the following list of terms: *cytoplasm, plasma membrane, rough endoplasmic reticulum, mitochondrion, nucleus, centriole, Golgi apparatus, lysosome*

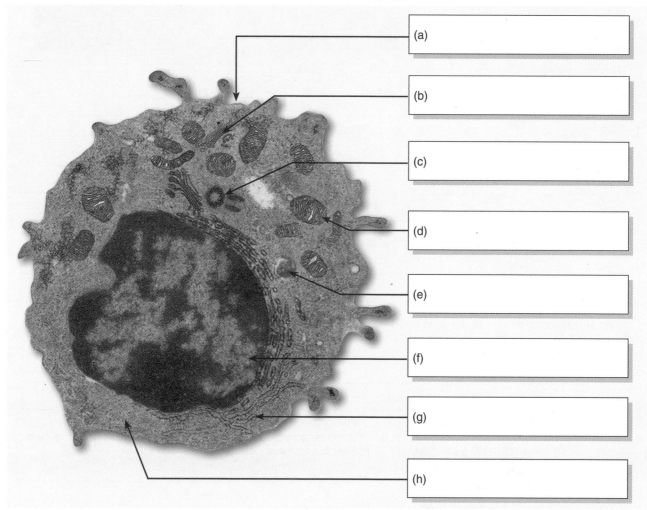

(a)

(b)

(c)

(d)

(e)

(f)

(g)

(h)

2. Which of the organelles in the EM above are clearly obvious in both transverse and longitudinal section?

_____

3. Why do plants lack any of the mobile phagocytic cells typical of animals? _____

_____

_____

4. The animal cell pictured above is a lymphocyte. Describe the features that suggest to you that:

(a) It has a role in producing and secreting proteins: _____

_____

(b) It is metabolically very active: _____

_____

5. What features of the lymphocyte cell above identify it as eukaryotic? _____

_____

6. Draw a generalized animal cell to include the features noted above. Staple it into your workbook.

© 2012-2014 **BIOZONE** International
ISBN: 978-1-927173-93-0
Photocopying Prohibited

17 Identifying Structures in a Plant Cell

1. Study the diagrams on the other pages in this chapter to familiarize yourself with the structures found in plant cells. Identify and label the ten structures in the cell below using the following list of terms: *nuclear membrane, cytoplasm, endoplasmic reticulum, mitochondrion, starch granules, chromosome, vacuole, plasma membrane, cell wall, chloroplast*

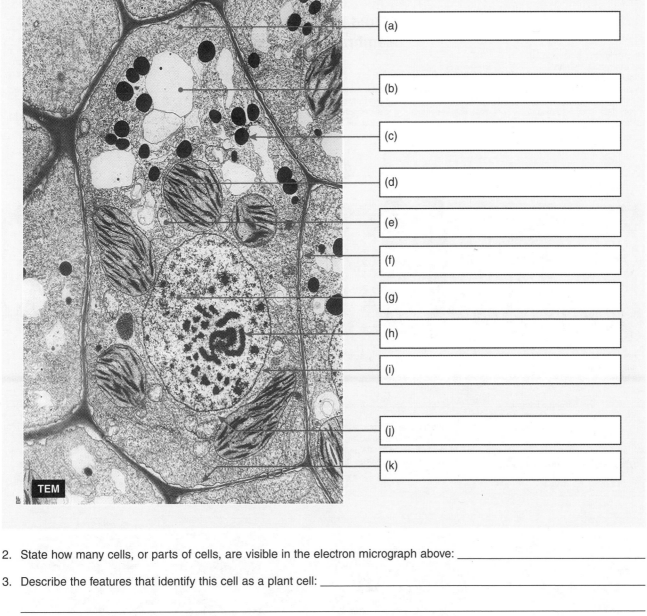

(a)

(b)

(c)

(d)

(e)

(f)

(g)

(h)

(i)

(j)

(k)

TEM

2. State how many cells, or parts of cells, are visible in the electron micrograph above: _____

3. Describe the features that identify this cell as a plant cell: _____

_____

4. (a) Explain where cytoplasm is found in the cell : _____

_____

(b) Describe what cytoplasm is made up of: _____

_____

5. Describe two structures, pictured in the cell above, that are associated with storage:

(a) _____

_____

(b) _____

_____

6. Draw a generalized plant cell to include the features noted above. Staple it into your workbook.

© 2012-2014 **BIOZONE** International
ISBN: 978-1-927173-93-0
Photocopying Prohibited

LINK
44

WEB
17

SKILL

18 The Structure of Membranes

Key Idea: The plasma membrane is composed of a lipid bilayer with proteins moving freely within it.

All cells have a plasma membrane that forms the outer limit of the cell. Bacteria, fungi, and plant cells have a cell wall that is quite distinct and outside the plasma membrane. Membranes are also found inside eukaryotic cells as part of membranous organelles. Current knowledge of membrane structure has been built up from observations and experiments. The now-accepted model of membrane structure is the **fluid mosaic model** described below.

The Fluid Mosaic Model of Membrane Structure

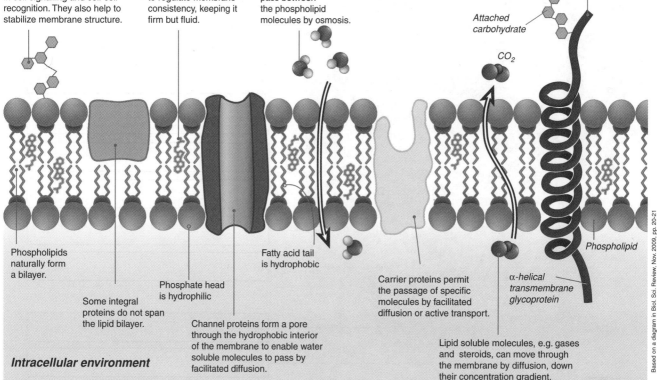

Glycolipids in membranes are phospholipids with attached carbohydrate. Like glycoproteins, they are involved in cell signalling and cell-cell recognition. They also help to stabilize membrane structure.

Cholesterol is a packing molecule and interacts with the phospholipids to regulate membrane consistency, keeping it firm but fluid.

Water molecules pass between the phospholipid molecules by osmosis.

Glycoproteins are proteins with attached carbohydrate. They are important in membrane stability, in cell-cell recognition, and in cell signalling, acting as receptors for hormones and neurotransmitters.

Attached carbohydrate

CO_2

Phospholipid

Phospholipids naturally form a bilayer.

Fatty acid tail is hydrophobic

α-helical transmembrane glycoprotein

Phosphate head is hydrophilic

Some integral proteins do not span the lipid bilayer.

Channel proteins form a pore through the hydrophobic interior of the membrane to enable water soluble molecules to pass by facilitated diffusion.

Carrier proteins permit the passage of specific molecules by facilitated diffusion or active transport.

Lipid soluble molecules, e.g. gases and steroids, can move through the membrane by diffusion, down their concentration gradient.

Intracellular environment

Based on a diagram in Biol. Sci. Review, Nov. 2009, pp. 20-21

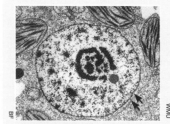

The **nuclear membrane** that surrounds the nucleus helps to control the passage of genetic information to the cytoplasm. It may also serve to protect the DNA.

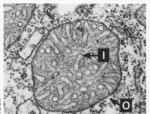

Mitochondria have an outer membrane (**O**) which controls the entry and exit of materials involved in aerobic respiration. Inner membranes (**I**) provide attachment sites for enzyme activity.

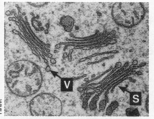

The **Golgi apparatus** comprises stacks of membrane-bound sacs (**S**). It is involved in packaging materials for transport or export from the cell as secretory vesicles (**V**).

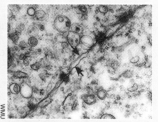

The **plasma membrane** (arrowed above) surrounds the cell and controls the movement of most substances into and out of the cell.

1. Identify the component(s) of the plasma membrane involved in:

   (a) Facilitated diffusion: _____

   (b) Active transport: _____

   (c) Cell signalling: _____

   (d) Regulating membrane fluidity: _____

2. How do the properties of phospholipids contribute to their role in forming the structural framework of membranes?

   _____

   _____

3. (a) Describe the modern fluid mosaic model of membrane structure: _____

_____

_____

(b) Explain how the fluid mosaic model accounts for the observed properties of cellular membranes: _____

_____

_____

_____

_____

4. Discuss the various functional roles of membranes in cells: _____

_____

_____

_____

5. (a) Name a cellular organelle that possesses a membrane: _____

(b) Describe the membrane's purpose in this organelle: _____

_____

_____

_____

6. Describe the purpose of cholesterol in plasma membranes: _____

_____

_____

7. List three substances that need to be transported **into** all kinds of animal cells, in order for them to survive:

(a) _____ (b) _____ (c) _____

8. List two substances that need to be transported **out** of all kinds of animal cells, in order for them to survive:

(a) _____ (b) _____

9. Use the symbol for a phospholipid molecule (below) to draw a **simple labelled diagram** to show the structure of a plasma membrane (include features such as lipid bilayer and various kinds of proteins):

Symbol for phospholipid

19 How Do We Know? Membrane Structure

Key Idea: The freeze-fracture technique for preparing and viewing cellular membranes has provided evidence to support the fluid mosaic model of the plasma membrane.

Cellular membranes play many extremely important roles in cells and understanding their structure is central to understanding cellular function. Moreover, understanding the structure and function of membrane proteins is essential to understanding cellular transport processes, and cell recognition and signalling. Cellular membranes are far too small to be seen clearly using light microscopy, and certainly any detail is impossible to resolve. Since early last century, scientists have known that membranes were composed of a lipid bilayer with associated proteins. The original model of membrane structure, proposed by Davson and Danielli, was the unit membrane (a lipid bilayer coated with protein). This model was later modified by Singer and Nicolson after the discovery that the protein molecules were embedded *within* the bilayer rather than coating the outside. But how did they find out just how these molecules were organized?

The answers were provided with electron microscopy, and one technique in particular – **freeze fracture**. As the name implies, freeze fracture, at its very simplest level, is the freezing of a cell and then fracturing it so the inner surface of the membrane can be seen using electron microscopy. Membranes are composed of two layers of phospholipids held together by weak intermolecular bonds. These split apart during fracture.

The procedure involves several steps:

▶ Cells are immersed in chemicals that alter the strength of the internal and external regions of the plasma membrane and immobilize any mobile macromolecules.

▶ The cells are passed through a series of glycerol solutions of increasing concentration. This protects the cells from bursting when they are frozen.

▶ The cells are mounted on gold supports and frozen using liquid propane.

▶ The cells are fractured in a helium-vented vacuum at -150° C. A razor blade cooled to -170° C acts as both a cold trap for water and the fracturing instrument.

▶ The surface of the fractured cells may be evaporated a little to produce some relief on the surface (known as etching) so that a three-dimensional effect occurs.

▶ For viewing under an electron microscope (EM), a replica of the cells is made by coating them with gold or platinum to ~3 nm thick. A layer of carbon around 30 nm thick is used to provide contrast and stability for the replica.

▶ The samples are then raised to room temperature and placed into distilled water or digestive enzymes, which separates the replica from the sample. The replica is then rinsed in distilled water before it is ready for viewing.

The freeze fracture technique provided the necessary supporting evidence for the current fluid mosaic model of membrane structure. When cleaved, proteins in the membrane left impressions that showed they were embedded into the membrane and not a continuous layer on the outside as earlier models proposed.

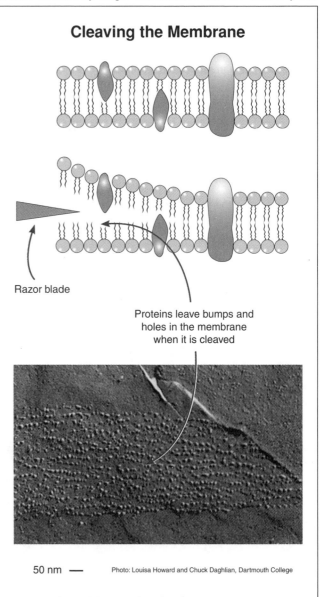

Cleaving the Membrane

Razor blade

Proteins leave bumps and holes in the membrane when it is cleaved

50 nm —   Photo: Louisa Howard and Chuck Daghlian, Dartmouth College

1. Explain how freeze-fracture studies provided evidence for our current model of membrane structure:

_____

_____

_____

2. The Davson and Danielli model of membrane structure was the unit membrane; a phospholipid bilayer with a protein coat. Explain how the freeze-fracture studies showed this model to be flawed:

_____

_____

_____

© 2012-2014 **BIOZONE** International
ISBN: 978-1-927173-93-0
Photocopying Prohibited

SKILL

20 Diffusion

Key Idea: Diffusion is the movement of molecules from higher concentration to a lower concentration (i.e. down a concentration gradient).

The molecules that make up substances are constantly moving about in a random way. This random motion causes molecules to disperse from areas of high to low concentration. This movement is called **diffusion**. Each type of molecule moves down its own concentration gradient. Diffusion is important in allowing exchanges with the environment and in the regulation of cell water content.

What is Diffusion?

Diffusion is the movement of particles from regions of high concentration to regions of low concentration. Diffusion is a passive process, meaning it needs no input of energy to occur. During diffusion, molecules move randomly about, becoming evenly dispersed.

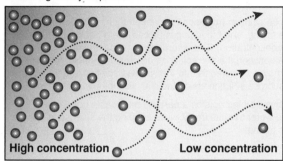

Concentration gradient →

If molecules can move freely, they move from high to low concentration (down a concentration gradient) until evenly dispersed.

Factors Affecting the Rate of Diffusion

Concentration gradient	The rate of diffusion is higher when there is a greater difference between the concentrations of two regions.
The distance moved	Diffusion over shorter distance occurs at a greater rate than over a larger distance.
The surface area involved	The larger the area across which diffusion occurs, the greater the rate of diffusion.
Barriers to diffusion	Thick barriers have a slower rate of diffusion than thin barriers.
Temperature	Particles at a high temperature diffuse at a greater rate than at a low temperature.

Types of Diffusion

Simple Diffusion
Molecules move directly through the membrane without assistance. Example: O_2 diffuses into the blood and CO_2 diffuses out.

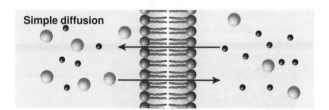

Simple diffusion

Facilitated Diffusion
Carrier-Mediated Facilitated Diffusion
Carrier proteins allow large lipid-insoluble molecules that cannot cross the membrane by simple diffusion to be transported into the cell. Example: the transport of glucose into red blood cells.

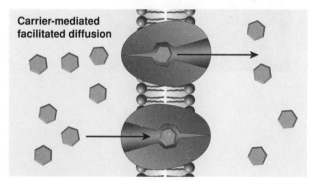

Carrier-mediated facilitated diffusion

Channel-Mediated Facilitated Diffusion
Channels (hydrophilic pores) in the membrane allow inorganic ions to pass through the membrane. Example: K^+ ions exiting nerve cells to restore resting potential.

1. What is diffusion? _____

_____

_____

_____

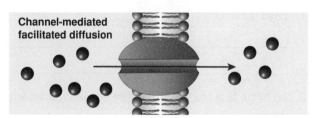

Channel-mediated facilitated diffusion

2. What do the three types of diffusion described above all have in common?

_____

_____

3. What would happen in each of the following diffusion situations? In each situation a 1 cm³ volume of salt is placed into 50 mL of tap water at 20°C.

   (a) The water is heated to 40°C: _____

   (b) The water has two teaspoons of salt already dissolved in it before the salt is added: _____

   _____

21 Osmosis

Key Idea: Osmosis is the diffusion of water molecules from a lower solute concentration to a higher solute concentration across a partially permeable membrane.

The partially permeable membrane allows some molecules, but not others, to pass through. Water molecules will diffuse across a partially permeable membrane until an equilibrium is reached and net movement is zero. The plasma membrane of a cell is an example of a partially permeable membrane. Osmosis is a passive process and does not require any energy input.

Osmotic Potential

The presence of solutes (dissolved substances) in a solution increases the tendency of water to move into that solution. This tendency is called the osmotic potential or osmotic pressure. The greater a solution's concentration (i.e. the more total dissolved solutes it contains) the greater the osmotic potential.

Osmosis is important when handling body tissues for medical transport or preparation. The tissue must be bathed a solution with an osmolarity (solute concentration) equal to the tissue's to avoid a loss or gain of fluid in the tissue.

The red blood cells below were placed into a solution with lower osmolarity than the internal environment of the cells (a **hypertonic** solution). As a result the cells have lost water and begun to shrink, losing their usual discoid shape.

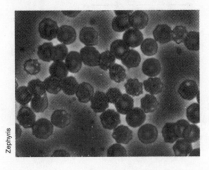

Zephyris

Demonstrating Osmosis

Osmosis can be demonstrated using dialysis tubing in a simple experiment (described below). Dialysis tubing, like all cellular membranes, is a partially permeable membrane. ·

A sucrose solution (high solute concentration) is placed into dialysis tubing, and the tubing is placed into a beaker of water (low solute concentration). The difference in concentration of sucrose (solute) between the two solutions creates an osmotic gradient. Water moves by osmosis into the sucrose solution and the volume of the sucrose solution inside the dialysis tubing increases.

The dialysis tubing acts as a partially permeable membrane, allowing water to pass freely, while keeping the sucrose inside the dialysis tubing.

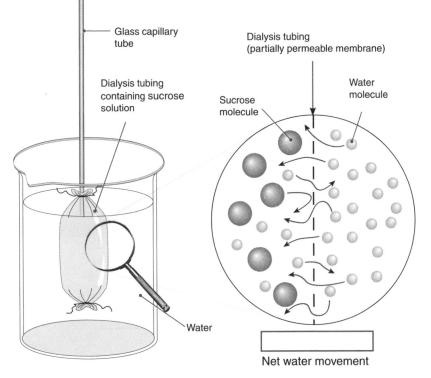

Glass capillary tube

Dialysis tubing containing sucrose solution

Dialysis tubing (partially permeable membrane)

Water molecule

Sucrose molecule

Water

Net water movement

1. What is osmosis? _____

_____

_____

2. (a) In the blue box on the diagram above, draw an arrow to show the direction of net water movement.

(b) Why did water move in this direction? _____

_____

_____

3. What would happen to the height of the water in the capillary tube if the sucrose concentration was increased?

_____

© 2012-2014 **BIOZONE** International
ISBN: 978-1-927173-93-0
Photocopying Prohibited

22 Estimating Osmolarity

Key Idea: A cell placed in a hypotonic solution will gain water while a cell placed in a hypertonic solution will lose water. The osmolarity (a measure of solute concentration) of a cell or tissue can be estimated by placing part of the cell or tissue into a series of solutions of known concentration and observing if the tissue loses (hypertonic solution) or gains (hypotonic solution) water. The solution in which the tissue remains unchanged indicates the osmolarity of the tissue.

The Aim

To investigate the osmolarity of potatoes by placing cubes of potato in varying solutions of sucrose, $C_{12}H_{22}O_{11}$ (table sugar).

The Method

Fifteen identical 1.5 cm³ cubes of potato where cut and weighed in grams to two decimal places. Five solutions of sucrose were prepared in the following range (in mol L⁻¹): 0.00, 0.25, 0.50, 0.75, 1.00. Three potato cubes were placed in each solution for two hours, stirring every 15 minutes. The cubes were then retrieved, patted dry on blotting paper and weighed again.

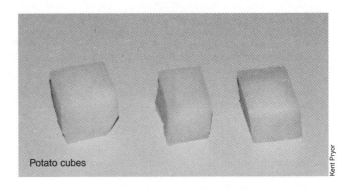
Potato cubes

The Results

	Potato sample	Initial mass (I) (g)	Final mass (F) (g)
[Sucrose] 0.00 molL⁻¹	1	5.11	6.00
	2	5.15	6.07
	3	5.20	5.15
Total			
Change (C) (F-I) (g)			
% Change (C/I x 100)			
[Sucrose] 0.25 molL⁻¹	1	6.01	4.98
	2	6.07	5.95
	3	7.10	7.00
Total			
Change (C) (F-I) (g)			
% Change (C/I x 100)			
[Sucrose] 0.50 molL⁻¹	1	6.12	5.10
	2	7.03	6.01
	3	5.11	5.03
Total			
Change (C) (F-I) (g)			
% Change (C/I x 100)			
[Sucrose] 0.75 molL⁻¹	1	5.03	3.96
	2	7.10	4.90
	3	7.03	5.13
Total			
Change (C) (F-I) (g)			
% Change (C/I x 100)			
[Sucrose] 1.00 molL⁻¹	1	5.00	4.03
	2	5.04	3.95
	3	6.10	5.02
Total			
Change (C) (F-I) (g)			
% Change (C/I x 100)			

1. Complete the table (left) by calculating the total mass of the potato cubes, the total change in mass, and the total % change in mass for all the sucrose concentrations:

2. Use the grid below to draw a line graph of the sucrose concentration vs total % change in mass:

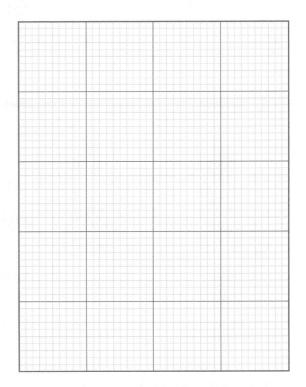

3. Use the graph to estimate the osmolarity of the potato (the point where there is no change in mass):

_____

_____

4. Identify which of the solutions are hypotonic and which are hypertonic.

_____

_____

_____

_____

© 2012-2014 **BIOZONE** International
ISBN: 978-1-927173-93-0
Photocopying Prohibited

LINK
21

SKILL

23 Active Transport

Key Idea: Active transport uses energy to transport molecules against their concentration gradient across a partially permeable membrane.

Active transport is the movement of molecules (or ions) from regions of low concentration to regions of high concentration across a cellular membrane by a transport protein. Active transport needs energy to proceed because molecules are being moved against their concentration gradient.

▶ The energy for active transport comes from **ATP** (adenosine triphosphate). Energy is released when ATP is hydrolysed (water is added) forming ADP (adenosine diphosphate) and inorganic phosphate (Pi).

▶ Transport (carrier) proteins in the membrane are used to actively transport molecules from one side of the membrane to the other (below).

▶ Active transport can be used to move molecules into and out of a cell.

▶ Active transport can be either primary or secondary. Primary active transport directly uses ATP for the energy to transport molecules. In secondary active transport, energy is stored in a concentration gradient. The transport of one molecule is coupled to the movement of another down its concentration gradient, ATP is not directly involved in the transport process.

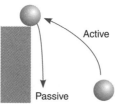

A ball falling is a passive process (it requires no energy input). Replacing the ball requires active energy input.

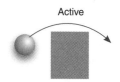

It requires energy to actively move an object across a physical barrier.

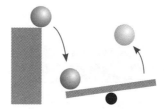

Sometimes the energy of a passively moving object can be used to actively move another. For example, a falling ball can be used to catapult another (left).

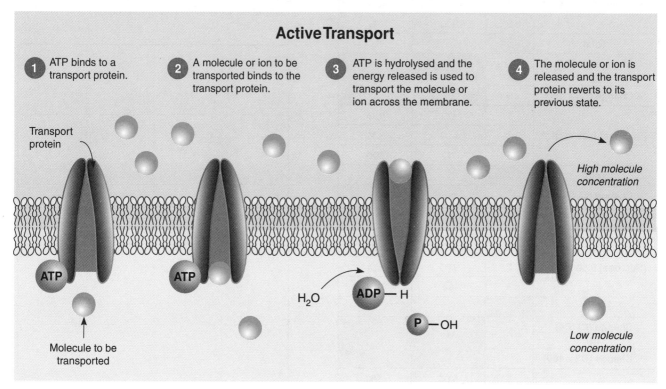

Active Transport

1 ATP binds to a transport protein.

2 A molecule or ion to be transported binds to the transport protein.

3 ATP is hydrolysed and the energy released is used to transport the molecule or ion across the membrane.

4 The molecule or ion is released and the transport protein reverts to its previous state.

Transport protein

High molecule concentration

ATP

ATP

H_2O

ADP — H

P — OH

Low molecule concentration

Molecule to be transported

1. What is **active transport**? _____

_____

_____

_____

2. Where does the energy for active transport come from? _____

3. What is the difference between primary active transport and secondary active transport? _____

_____

_____

_____

© 2012-2014 **BIOZONE** International
ISBN: 978-1-927173-93-0
Photocopying Prohibited

24 Ion Pumps

Key Idea: Ion pumps are transmembrane proteins that use energy to move ions and molecules across a membrane against their concentration gradient.

Sometimes molecules or ions are needed in concentrations that diffusion alone cannot supply to the cell, or they cannot diffuse through the plasma membrane. In this case ion pumps move ions (and some molecules) across the plasma membrane. The sodium-potassium pump (below, right) is found in almost all animal cells and is common in plant cells also. The concentration gradient created by ion pumps is often coupled to the transport of other molecules such as glucose across the membrane.

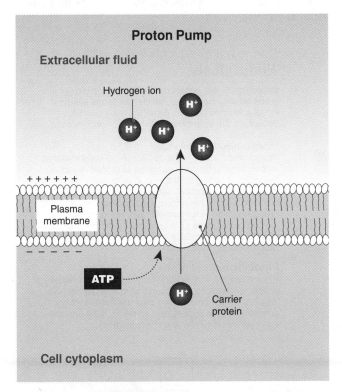

Proton Pump

Extracellular fluid

Hydrogen ion

H^+

Plasma membrane

ATP

Carrier protein

Cell cytoplasm

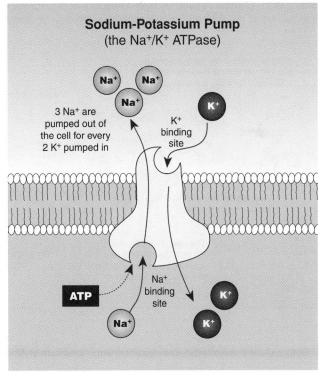

Sodium-Potassium Pump
(the Na^+/K^+ ATPase)

3 Na^+ are pumped out of the cell for every 2 K^+ pumped in

K^+ binding site

Na^+ binding site

ATP

Proton pumps create a potential difference across the membrane by using energy (ATP or electrons) to move H^+ from the inside of a cell to the outside. This difference can be coupled to the transport of other molecules. In cellular respiration and the light reactions of photosynthesis, the energy for moving the H^+ comes from electrons, and the return of H^+ drives the synthesis of ATP by ATP synthase. A proton pump also drives sucrose transport in the phloem.

The sodium-potassium pump is a specific protein in the membrane that uses energy in the form of ATP to exchange sodium ions (Na^+) for potassium ions (K^+) across the membrane. The unequal balance of Na^+ and K^+ across the membrane creates large concentration gradients that can be used to drive transport of other substances (e.g. cotransport of glucose). The Na^+/K^+ pump also helps to maintain ion balance and so helps regulate the cell's water balance.

1. Why is ATP required for membrane pump systems to operate? _____

_____

_____

2. Explain why diffusion is not always able to meet the ion or molecule needs of the cell: _____

_____

_____

3. Describe two consequences of the extracellular accumulation of sodium ions: _____

_____

_____

_____

4. Explain how a potential difference of H^+ ions can be used to do work: _____

_____

_____

_____

LINK 233 LINK 166 LINK 23 WEB 24 APP

25 Exocytosis and Endocytosis

Key Idea: Endocytosis and exocytosis are active transport processes. Endocytosis involves the cell engulfing material. Exocytosis involves the cell expelling material.

Most cells carry out **cytosis**, a type of active transport in which the plasma membrane folds around a substance to transport it across the plasma membrane. The ability of cells to do this is a function of the flexibility of the plasma membrane. Cytosis results in bulk transport of substances into or out of the cell and is achieved through the localized activity of the cell cytoskeleton. **Endocytosis** involves material being engulfed and taken into the cell. It typically occurs in protozoans and some white blood cells of the mammalian defence system (phagocytes). **Exocytosis** is the reverse of endocytosis and involves expelling material from the cell in vesicles or vacuoles that have fused with the plasma membrane. Exocytosis is common in cells that export material (secretory cells).

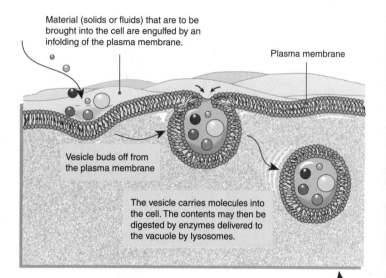

Material (solids or fluids) that are to be brought into the cell are engulfed by an infolding of the plasma membrane.

Plasma membrane

Vesicle buds off from the plasma membrane

The vesicle carries molecules into the cell. The contents may then be digested by enzymes delivered to the vacuole by lysosomes.

Both endocytosis and exocytosis require energy in the form of ATP.

Endocytosis

Endocytosis (left) occurs by invagination (infolding) of the plasma membrane, which then forms vesicles or vacuoles that become detached and enter the cytoplasm. There are two main types of endocytosis:

Phagocytosis: 'cell-eating'
Phagocytosis involves the cell engulfing **solid material** to form large vesicles or vacuoles (e.g. food vacuoles). Examples: Feeding in *Amoeba*, phagocytosis of foreign material and cell debris by neutrophils and macrophages. Some endocytosis is **receptor mediated** and is triggered when receptor proteins on the extracellular surface of the plasma membrane bind to specific substances. Examples include the uptake of lipoproteins by mammalian cells.

Pinocytosis: 'cell-drinking'
Pinocytosis involves the non-specific uptake of **liquids** or fine suspensions into the cell to form small pinocytic vesicles. Pinocytosis is used primarily for absorbing extracellular fluid. Examples: Uptake in many protozoa, some cells of the liver, and some plant cells.

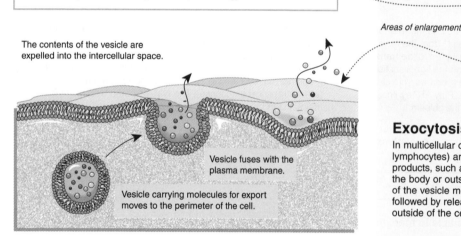

The contents of the vesicle are expelled into the intercellular space.

Vesicle fuses with the plasma membrane.

Vesicle carrying molecules for export moves to the perimeter of the cell.

Areas of enlargement

Exocytosis

In multicellular organisms, several types of cells (e.g. lymphocytes) are specialized to manufacture and export products, such as proteins, from the cell to elsewhere in the body or outside it. Exocytosis (left) occurs by fusion of the vesicle membrane and the plasma membrane, followed by release of the vesicle's contents to the outside of the cell.

1. Distinguish between **phagocytosis** and **pinocytosis**: _____

   _____

2. Describe an example of phagocytosis and identify the cell type involved: _____

   _____

3. Describe an example of exocytosis and identify the cell type involved: _____

   _____

4. How does each of the following substances enter a living macrophage:

   (a) Oxygen: _____ (c) Water: _____

   (b) Cellular debris: _____ (d) Glucose: _____

© 2012-2014 **BIOZONE** International
ISBN: 978-1-927173-93-0
Photocopying Prohibited

26 Active and Passive Transport Summary

Key Idea: Cells move materials into and out of the cell by either passive transport, which does not use energy, or by active transport which requires energy, usually as ATP.
Cells need to move materials into and out of the cell. Molecules needed for metabolism must be accumulated from outside the cell, where they may be scarce, and waste products and molecules for use elsewhere must be exported

from the cell. Some materials (e.g. gases and water) move into and out of the cell by passive transport processes, down their concentration gradients, without energy expenditure. The movement of other molecules against their concentration gradients involves active transport. Active transport processes involve the expenditure of energy in the form of ATP, and therefore use oxygen.

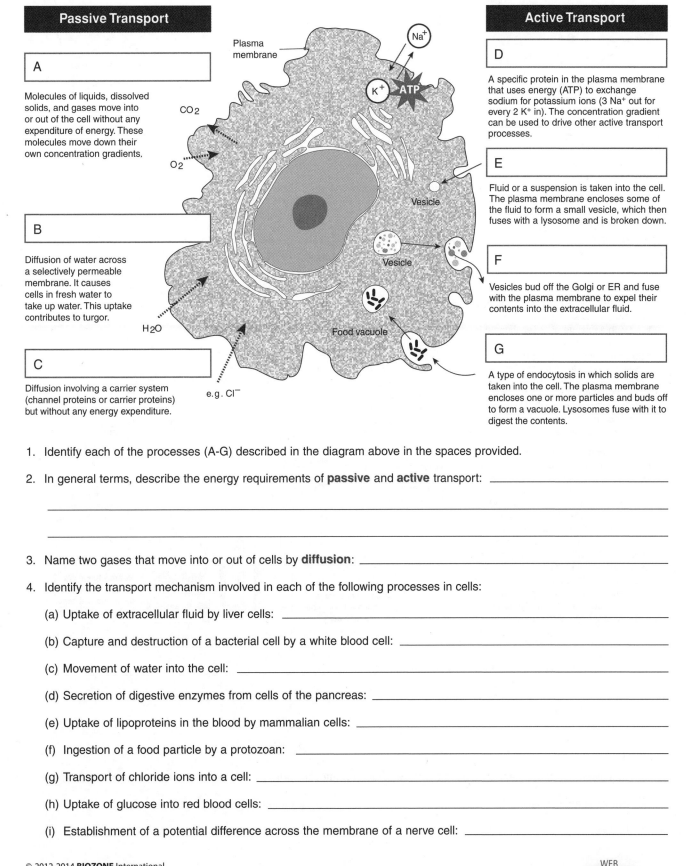

Passive Transport

A

Molecules of liquids, dissolved solids, and gases move into or out of the cell without any expenditure of energy. These molecules move down their own concentration gradients.

B

Diffusion of water across a selectively permeable membrane. It causes cells in fresh water to take up water. This uptake contributes to turgor.

C

Diffusion involving a carrier system (channel proteins or carrier proteins) but without any energy expenditure.

Plasma membrane

CO_2

O_2

H_2O

e.g. Cl^-

Na^+

K^+ ATP

Vesicle

Vesicle

Food vacuole

Active Transport

D

A specific protein in the plasma membrane that uses energy (ATP) to exchange sodium for potassium ions (3 Na^+ out for every 2 K^+ in). The concentration gradient can be used to drive other active transport processes.

E

Fluid or a suspension is taken into the cell. The plasma membrane encloses some of the fluid to form a small vesicle, which then fuses with a lysosome and is broken down.

F

Vesicles bud off the Golgi or ER and fuse with the plasma membrane to expel their contents into the extracellular fluid.

G

A type of endocytosis in which solids are taken into the cell. The plasma membrane encloses one or more particles and buds off to form a vacuole. Lysosomes fuse with it to digest the contents.

1. Identify each of the processes (A-G) described in the diagram above in the spaces provided.

2. In general terms, describe the energy requirements of **passive** and **active** transport: _____

_____

_____

3. Name two gases that move into or out of cells by **diffusion**: _____

4. Identify the transport mechanism involved in each of the following processes in cells:

    (a) Uptake of extracellular fluid by liver cells: _____

    (b) Capture and destruction of a bacterial cell by a white blood cell: _____

    (c) Movement of water into the cell: _____

    (d) Secretion of digestive enzymes from cells of the pancreas: _____

    (e) Uptake of lipoproteins in the blood by mammalian cells: _____

    (f) Ingestion of a food particle by a protozoan: _____

    (g) Transport of chloride ions into a cell: _____

    (h) Uptake of glucose into red blood cells: _____

    (i) Establishment of a potential difference across the membrane of a nerve cell: _____

27 Investigating the Origin of Life

Key Idea: The origin of life is not yet known, but experiments have shown it cannot just appear from nowhere.

For a large part of human history, people believed that life could be generated spontaneously from elements in the environment. In 1862, Louis Pasteur showed experimentally that this was not the case (below). In the 1950s, Stanley Miller and Harold Urey attempted to recreate the conditions of primitive Earth and produce the biological molecules that preceded the development of the first cells. Their experiments have helped us to understand the conditions under which life first arose.

Pasteur's Experiment

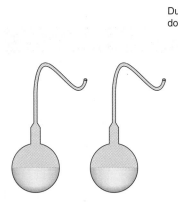
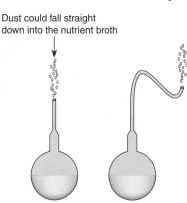
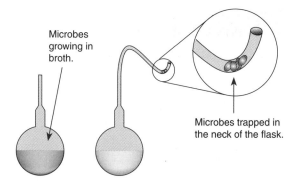

Dust could fall straight down into the nutrient broth

Microbes growing in broth.

Microbes trapped in the neck of the flask.

Louis Pasteur disproved the idea of spontaneous generation with a simple experiment. He filled two swan necked flasks with a nutrient broth and then boiled them to kill any microbes present.

He then broke off the neck of one of the flasks allowing air and dust (on which microbes are carried) to fall straight down onto the broth. The neck of the other flask prevented dust falling onto the broth

The broth in the broken flask eventually turned dark, indicating microbial growth. The broth in the unbroken flask remained unchanged. Pasteur concluded that spontaneous generation could not have occurred or both flasks would have turned dark.

The Miller-Urey Experiment

Miller and Urey set up a reaction vessel filled with a mixture of gases thought to be present in Earth's early atmosphere. The gases were heated and exposed to simulated lightning (right). The experiment was run for a week, after which samples were taken from the collection trap for analysis. Up to 4% of the carbon (from methane) had been converted to amino acids.

From this and subsequent experiments it has been possible to form all 20 amino acids commonly found in organisms, along with nucleic acids, several sugars, lipids, adenine, and even ATP (if phosphate is added to the flask). Researchers now believe that the early atmosphere may be similar to the vapours given off by modern volcanoes: carbon monoxide (CO), carbon dioxide (CO_2), and nitrogen (N_2), however even if the reaction mixture is adjusted for these gases the outcome is largely the same. Note the absence of free oxygen.

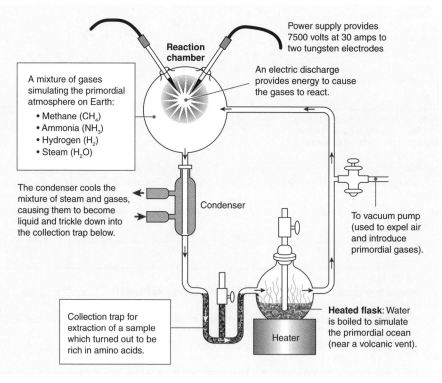

Power supply provides 7500 volts at 30 amps to two tungsten electrodes

Reaction chamber

An electric discharge provides energy to cause the gases to react.

A mixture of gases simulating the primordial atmosphere on Earth:
• Methane (CH_4)
• Ammonia (NH_3)
• Hydrogen (H_2)
• Steam (H_2O)

The condenser cools the mixture of steam and gases, causing them to become liquid and trickle down into the collection trap below.

Condenser

To vacuum pump (used to expel air and introduce primordial gases).

Collection trap for extraction of a sample which turned out to be rich in amino acids.

Heater

Heated flask: Water is boiled to simulate the primordial ocean (near a volcanic vent).

1. Explain the reasoning behind Pasteur's conclusion: _____

_____

_____

_____

2. In the Miller-Urey experiment simulating the conditions on primeval Earth, identify parts of the apparatus equivalent to:

(a) Primeval atmosphere: _____   (c) Lightning: _____

(b) Primeval ocean: _____   (d) Volcanic heat: _____

28 The First Cells

Key Idea: They first cells evolved in a number of small steps, probably preceded by self replicating RNA molecules.

A key problem in understanding how life began is how biological information was first stored, copied, and replicated. Modern life requires many complex molecules for replication that did not exist in life's early history. The discovery of ribozymes in 1982 (more than fifteen years after they were first hypothesized) helped to solve this problem, at least in part. Ribozymes are enzymes formed from RNA, which itself can store biological information. The ribozymes can catalyse the replication of the original RNA molecule. This mechanism for self replication has led to the theory of an "**RNA world**".

An RNA World

RNA is able to act as a vehicle for both information storage and catalysis. It therefore provides a way around the problem that genes require enzymes to form and enzymes require genes to form. The first stage of evolution may have proceeded by RNA molecules performing the catalytic activities necessary to assemble themselves from a nucleotide soup. RNA molecules could then begin to synthesize proteins. However, there is a problem with RNA as a prebiotic molecule because the ribose is unstable. This has led to the idea of a pre-RNA world in which molecules similar to, but simpler and more stable than RNA (such as PNA, right), preceded RNA as the first catalysts and template molecules.

Dynamics of an RNA World

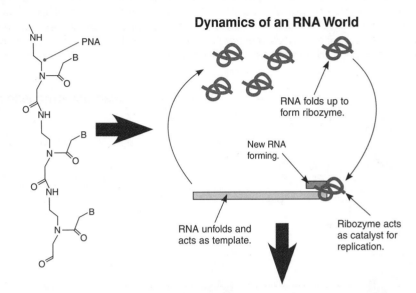

RNA folds up to form ribozyme.

New RNA forming.

RNA unfolds and acts as template.

Ribozyme acts as catalyst for replication.

The Formation of Proto-cells

Certain types of organic molecule (e.g. fatty acids) spontaneously form micelles when placed in an aqueous solution. Micelles are a loosely bound aggregation of molecules.

Several micelles can interact to form vesicles large enough to contain other molecules. Mutually cooperating RNAs and proteins trapped inside a vesicle would be able to replicate without moving away from each other.

Vesicle growth by the attraction of micelles would eventually cause it to be unstable and split in two. Each vesicle would take a random number of RNAs with it.

Mutation and Competition

After the establishment of self replicating RNA molecules there would have been competition of a sort. Incorrect copies of the original RNA produced new varieties of RNA.

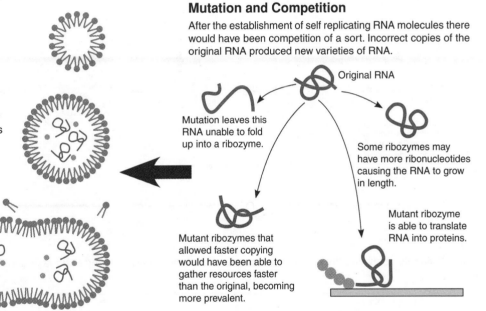

Original RNA

Mutation leaves this RNA unable to fold up into a ribozyme.

Some ribozymes may have more ribonucleotides causing the RNA to grow in length.

Mutant ribozymes that allowed faster copying would have been able to gather resources faster than the original, becoming more prevalent.

Mutant ribozyme is able to translate RNA into proteins.

1. Why did the discovery of ribozymes add weight to the RNA world hypothesis? _____

_____

_____

_____

2. Explain how mutations in RNA templates led to the first form of evolution: _____

_____

_____

_____

29 The Origin of Eukaryotes

Key Idea: Eukaryotes probably formed when a small prokaryote-like cell was engulfed by a larger one and formed an endosymbiotic relationship.

The first eukaryotes were unicellular and occur only rarely in microfossils. The first fossil evidence dates to 2.1 bya, but molecular evidence suggests that the eukaryotic lineage is much more ancient and closer to the origin of life. The original endosymbiotic theory (Margulis, 1970) proposed that eukaryotes arose as a result of an endosymbiosis between two prokaryotes, one of which was aerobic and gave rise to the mitochondrion. The hypothesis has since been modified

to recognize that eukaryotes probably originated with the appearance of the nucleus and flagella, with later acquisition of mitochondria and chloroplasts by endosymbiosis. Primitive eukaryotes probably acquired mitochondria by engulfing purple bacteria. Similarly, chloroplasts may have been acquired by engulfing primitive cyanobacteria. In both instances, the organelles produced became dependent on the nucleus of the host cell to direct some of their metabolic processes. Unlike mitochondria, chloroplasts were probably acquired independently by more than one organism, so their origin is polyphyletic.

The origin of eukaryotic organelles

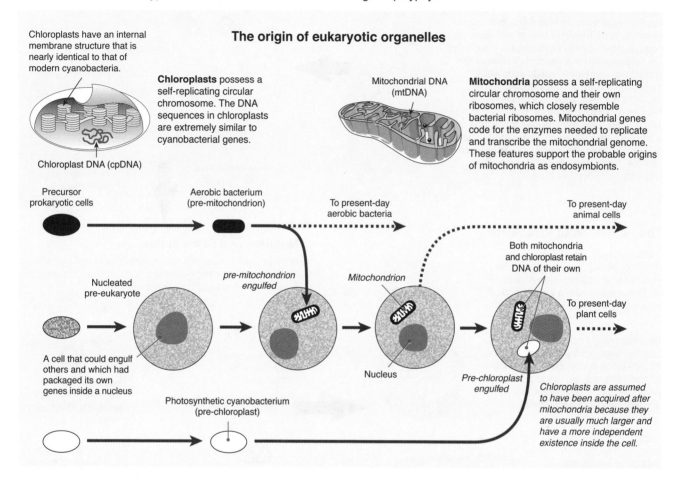

Chloroplasts have an internal membrane structure that is nearly identical to that of modern cyanobacteria.

Chloroplasts possess a self-replicating circular chromosome. The DNA sequences in chloroplasts are extremely similar to cyanobacterial genes.

Chloroplast DNA (cpDNA)

Mitochondrial DNA (mtDNA)

Mitochondria possess a self-replicating circular chromosome and their own ribosomes, which closely resemble bacterial ribosomes. Mitochondrial genes code for the enzymes needed to replicate and transcribe the mitochondrial genome. These features support the probable origins of mitochondria as endosymbionts.

Precursor prokaryotic cells

Aerobic bacterium (pre-mitochondrion)

To present-day aerobic bacteria

To present-day animal cells

Nucleated pre-eukaryote

pre-mitochondrion engulfed

Mitochondrion

Both mitochondria and chloroplast retain DNA of their own

To present-day plant cells

A cell that could engulf others and which had packaged its own genes inside a nucleus

Photosynthetic cyanobacterium (pre-chloroplast)

Nucleus

Pre-chloroplast engulfed

Chloroplasts are assumed to have been acquired after mitochondria because they are usually much larger and have a more independent existence inside the cell.

1. Distinguish between the two possible sequences of evolutionary change suggested in the endosymbiosis theory:

_____

_____

_____

2. How does the endosymbiotic theory account for the origins of the following eukaryotic organelles?

(a) Mitochondria: _____

_____

(b) Chloroplasts: _____

_____

3. What evidence from modern mitochondria and chloroplasts supports the endosymbiotic theory? _____

_____

_____

© 2012-2014 **BIOZONE** International
ISBN: 978-1-927173-93-0
Photocopying Prohibited

30 The Common Ancestry of Life

Key Idea: The ancestry of all organisms on Earth today can be traced back to a common ancestor.

Traditional schemes for classifying the living world were based primarily on morphological (structural) comparisons. These have been considerably revised with the increased use of molecular techniques, which compare the DNA, RNA, and proteins of organisms to establish evolutionary relationships. On the basis of molecular evidence, scientists have been able to clarify the very earliest origins of eukaryotes and to recognize two prokaryote domains (rather than one prokaryote superkingdom). Powerful evidence for the common ancestry of all life comes from the commonality in the genetic code, and from the similarities in the molecular machinery of all cells.

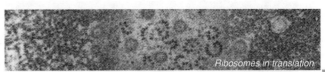

Ribosomes in translation

Most organisms share the same genetic code, i.e. the same combination of three DNA bases code for the same amino acid, although there are some minor variations (e.g. in mitochondria). Evidence suggests the code was subject to selection pressure which acted to minimize the effect of point mutations or errors in translation.

In all living systems, the genetic machinery consists of self-replicating DNA molecules. Some DNA is transcribed into RNA, some of which is translated into proteins. The machinery for translation (above) involves proteins and RNA. Ribosomal RNA analyses support a universal common ancestor.

The Tree of Life

Genetic evidence shows there are three main groups or domains of life. Only the Eukarya have representatives that are multicellular. There is evidence that there has been multiple gene transfers between the various domains of life.

DOMAIN BACTERIA

Other bacteria · Cyanobacteria · Proteobacteria (many pathogens) · Hyperthermophillic bacteria

DOMAIN ARCHAEA

Sulfobacteria · Methanogens, extreme thermophiles, halobacteria · Korarchaeota

Bacteria that gave rise to chloroplasts
Bacteria that gave rise to mitochondria

DOMAIN EUKARYA

Animals · Fungi · Plants · Algae · Ciliates · Other single-celled eukaryotes

AN ARCHAEAN ORIGIN FOR EUKARYOTES
Archaeal RNA polymerase, which transcribes DNA, more closely resembles the equivalent molecule in eukaryotes than in bacteria. The protein components of archaeal ribosomes are also more like those in eukaryotes. These similarities in molecular machinery indicate that the eukaryotes diverged from the archaeans, not the bacteria.

Last Universal Common Ancestor (LUCA)

EUKARYOTES CARRY BACTERIAL GENES!
New evidence indicates the presence of bacterial genes in eukaryotes that are unrelated to photosynthesis or cellular respiration. These could be explained by gene transfers during evolution. A revised tree might include a complex network of connections to indicate single and multiple gene transfers across domains.

Adapted from: Uprooting the tree of life

1. Explain the role of molecular phylogenetics in revising the traditional classification schemes (pre-1980):

2. Describe the evidence for the archaean origin of eukaryotic cells:

3. What evidence is there for a Last Universal Common Ancestor?

LINK 151 · LINK 29 · WEB 30 · **KNOW**

31 Why Cells Need To Divide

Key Idea: Mitosis has three primary functions: growth of the organism, replacement of damaged or old cells, and asexual reproduction (in some organisms).

Mitotic cell division produces daughter cells that are genetically identical to the parent cell. It has three purposes: growth, repair, and reproduction. Multicellular organisms grow from a single fertilized cell into a mature organism that may consist of several thousand to several trillion cells. Repair occurs by replacing damaged and old cells with new cells. Some unicellular eukaryotes (such as yeasts) and some multicellular organisms (e.g. *Hydra*) reproduce asexually by mitotic division.

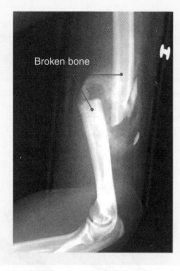

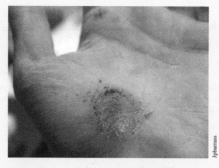

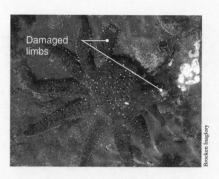

Broken bone

Damaged limbs

Repair

Mitosis is vital in the repair and replacement of damaged cells. When you break a bone, or graze your skin, new cells are generated to repair the damage. Some organisms, like this sea star (above right) are able to generate new limbs if they are broken off.

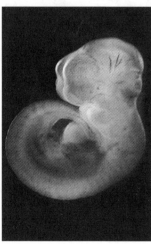

Growth

Multicellular organisms develop from a single cell. Organisms, such as this 12 day old mouse embryo (left), grow by increasing their cell number. Cell growth is highly regulated and once the mouse reaches its adult size (above), physical growth stops.

Asexual reproduction

Some simple eukaryotic organisms reproduce asexually by mitosis. Yeasts (such as baker's yeast, used in baking) can reproduce by budding. The parent cell buds to form a daughter cell (right). The daughter cell continues to grow, and eventually separates from the parent cell.

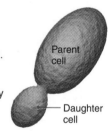

Parent cell

Daughter cell

1. Use examples to explain the role of mitosis in:

   (a) Growth of an organism: _____

   _____

   _____

   (b) Replacement of damaged cells: _____

   _____

   _____

   (c) Asexual reproduction: _____

   _____

   _____

2. If a cell with 24 chromosomes undergoes mitosis, how many chromosomes will be in each of the daughter cells?

   _____

© 2012-2014 **BIOZONE** International
ISBN: 978-1-927173-93-0
Photocopying Prohibited

32 Mitosis and the Cell Cycle

Key Idea: Mitosis is an important part of the cell cycle in which the replicated chromosomes are separated and the cell divides, producing two new identical cells.

Mitosis (or M-phase) is part of the **cell cycle** in which an existing cell (the parent cell) divides into two (the daughter cells). Unlike meiosis, mitosis does not result in a change of chromosome numbers and the daughter cells are identical to the parent cell. Although mitosis is part of a continuous cell cycle, it is often divided into stages to help differentiate the processes occurring. Mitosis is one of the shortest stages of the cell cycle. When a cell is not undergoing mitosis, it is said to be in interphase. Interphase accounts for 90% of the cell cycle. Cytokinesis (the division of the newly formed cells) is distinct from nuclear division.

The Cell Cycle

Interphase

Cells spend most of their time in interphase. Interphase is divided into three stages (right):

▶ The first gap phase.
▶ The S-phase.
▶ The second gap phase.

During interphase the cell grows, carries out its normal activities, and replicates its DNA in preparation for cell division.
Interphase is not a stage in mitosis.

Mitosis and cytokinesis (M-phase)

Mitosis and cytokinesis occur during M-phase. During mitosis, the cell nucleus (containing the replicated DNA) divides in two equal parts. Cytokinesis occurs at the end of M-phase. During cytokinesis the cell cytoplasm divides, and two new daughter cells are produced.

S phase: Chromosome replication (DNA synthesis).

Second gap phase: Rapid cell growth and protein synthesis. Cell prepares for mitosis.

Mitosis: Nuclear division

First gap phase: Cell increases in size and makes the mRNA and proteins needed for DNA synthesis.

Cytokinesis: The cytoplasm divides and the two cells separate. Cytokinesis is part of M phase but distinct from nuclear division.

An Overview of Mitosis

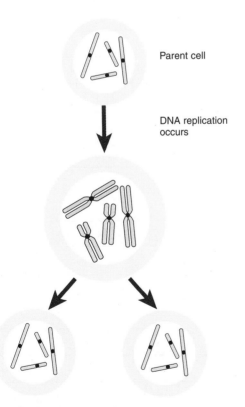

Parent cell

DNA replication occurs

The cell divides forming two identical daughter cells. The chromosome number remains the same as the parent cell.

Cytokinesis

In plant cells (below top), cytokinesis (division of the cytoplasm) involves construction of a cell plate (a precursor the new cell wall) in the middle of the cell. The cell wall materials are delivered by vesicles derived from the Golgi. The vesicles coalesce to become the plasma membranes of the new cell surfaces. Animal cell cytokinesis (below bottom) begins shortly after the sister chromatids have separated in anaphase of mitosis. A contractile ring of microtubular elements assembles in the middle of the cell, next to the plasma membrane, constricting it to form a cleavage furrow. In an energy-using process, the cleavage furrow moves inwards, forming a region of abscission (separation) where the two cells will separate.

Plant (onion) cells

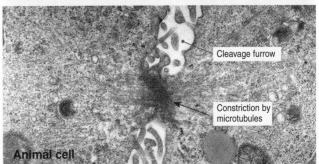

Cleavage furrow

Constriction by microtubules

Animal cell

dsworth Center- New York State Department of Health

LINK **34**   LINK **33**   LINK **31**   WEB **32**   **KNOW**

The Cell Cycle and Stages of Mitosis

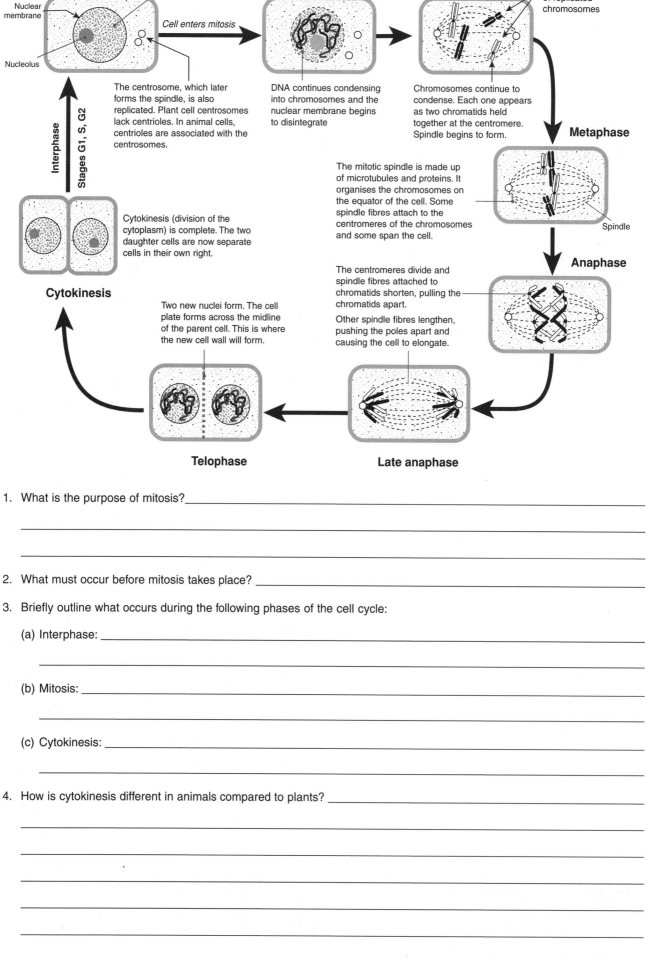

1. What is the purpose of mitosis?_____

_____

_____

2. What must occur before mitosis takes place? _____

3. Briefly outline what occurs during the following phases of the cell cycle:

   (a) Interphase: _____

   _____

   (b) Mitosis: _____

   _____

   (c) Cytokinesis: _____

   _____

4. How is cytokinesis different in animals compared to plants? _____

_____

_____

_____

_____

_____

© 2012-2014 **BIOZONE** International
ISBN: 978-1-927173-93-0

33 Recognizing Stages in Mitosis

Key Idea: The stages of mitosis can be recognized by the organization of the cell and chromosomes.

Although mitosis is a continuous process it is divided into four stages (prophase, anaphase, metaphase, and telophase) to more easily describe the processes occurring during its progression.

The Mitotic Index

The mitotic index measures the ratio of cells in mitosis to the number of cells counted. It is a measure of cell proliferation and can be used to diagnose cancer. In areas of high cell growth the mitotic index is high such as in plant apical meristems or the growing tips of plant roots. The mitotic index can be calculated using the formula:

$$\text{Mitotic index} = \frac{\text{Number of cells in mitosis}}{\text{Total number of cells}}$$

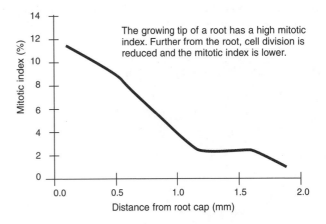

The growing tip of a root has a high mitotic index. Further from the root, cell division is reduced and the mitotic index is lower.

1. Use the information on the previous page to identify which stage of mitosis is shown in each of the photographs below:

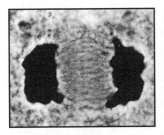   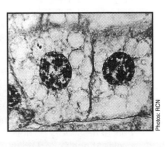

(a) _____ (b) _____ (c) _____ (d) _____

2. (a) The light micrograph (right) shows a section of cells in an onion root tip. These cells have a cell cycle of approximately 24 hours. The cells can be seen to be in various stages of the cell cycle. By counting the number of cells in the various stages it is possible to calculate how long the cell spends in each stage of the cycle. Count and record the number of cells in the image that are in mitosis and those that are in interphase. Cells in cytokinesis can be recorded as in interphase. Estimate the amount of time a cell spends in each phase.

Stage	No. of cells	% of total cells	Estimated time in stage
Interphase			
Mitosis			
Total		100	

(b) Use your counts from 2(a) to calculate the mitotic index for this section of cells.

_____

_____

3. What would you expect to happen to the mitotic index of a population of cells that loses the ability to divide as they mature?

_____

_____

Onion Root Tip Cells

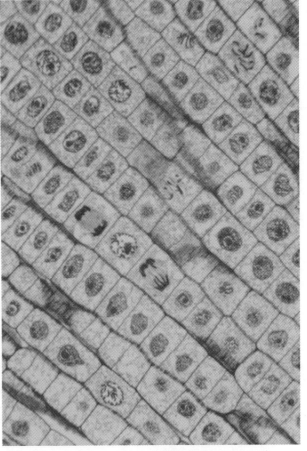

LINK 32 WEB 33 **SKILL**

34 Regulation of the Cell Cycle

Key Idea: The cell cycle is regulated to ensure cells only divide as and when required.

Mitosis is virtually the same for all eukaryotes but aspects of the cell cycle can vary enormously between species and even between cells of the same organism. For example, the length of the cell cycle varies between cells such as intestinal and liver cells. Intestinal cells divide around twice a day, while cells in the liver divide once a year. However, if these tissues are damaged, cell division increases rapidly until the damage is repaired. Variation in the length of the cell cycle are controlled by regulatory mechanisms that slow down or speed up the cell cycle in response to changing conditions.

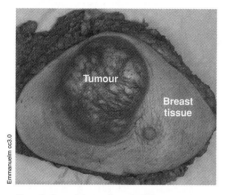

Regulation of the cell cycle is important in detecting and repairing genetic damage, and preventing uncontrolled cell division. Tumours and cancers, such as this breast cancer (above) are the result of uncontrolled cell division.

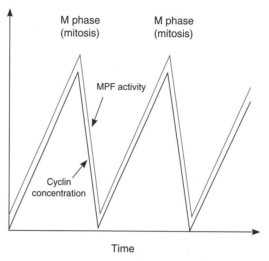

A substance called an M-phase promoting factor (MPF) controls cell regulation. MPF is made up of two regulatory molecules, **cyclins** and **cyclin-dependent kinases** (CDKs).

Cyclins are proteins that control the progression of cells through the cell cycle by activating CDKs (which are enzymes).

CDKs phosphorylate other proteins to signal a cell is ready to proceed to the next stage in the cell cycle. Without cyclin, CDK has little kinase activity; only the cyclin-CDK complex is active. CDK is constantly present in the cell, cyclin is not.

Checkpoints During the Cell Cycle

There are three checkpoints during the cell cycle. A **checkpoint** is a critical regulatory point in the cell cycle. At each checkpoint, a set of conditions determines whether or not the cell will continue into the next phase. For example, cell size is important in regulating whether or not the cell can pass through the G₁ checkpoint.

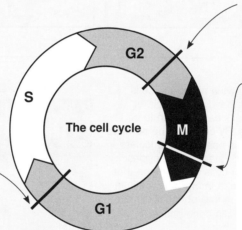

G₂ Checkpoint:
Pass this checkpoint if:
▶ Cell size is large enough.
▶ Replication of chromosomes has been successfully completed.

Metaphase checkpoint
Pass this checkpoint if:
▶ All chromosomes are attached to the mitotic spindle.

G₁ checkpoint
Pass this checkpoint if:
▶ Cell size is large enough.
▶ Sufficient nutrients are available.
▶ Signals from other cells have been received.

The discovery of cyclins was accidental. While studying embryonic development in sea urchins the early 1980s, Joan Ruderman and Tim Hunt discovered the cyclins involved in regulating the cell cycle.

1. What would happen if the cell cycle was not regulated? _____

_____

_____

2. (a) Suggest why the cell cycle is shorter in epithelial cells (such as intestinal cells) than in liver cells:

_____

_____

_____

   (b) Describe another situation in which the cell cycle shortens to allow for a temporary rapid rate of cell division:

_____

_____

_____

© 2012-2014 **BIOZONE** International
ISBN: 978-1-927173-93-0
Photocopying Prohibited

35 Cancer: Cells Out of Control

Key Idea: Cancerous cells have lost their normal cellular control mechanisms. Cancer may be caused by carcinogens. Cells that become damaged beyond repair will normally undergo a programmed cell death (**apoptosis**), which is part of the cell's normal control system. Cancer cells evade this control and become immortal, continuing to divide without any regulation. **Carcinogens** are agents capable of causing cancer. Roughly 90% of carcinogens are also **mutagens**, i.e. they damage DNA. Long-term exposure to carcinogens accelerates the rate at which dividing cells make errors. Any one of a number of cancer-causing factors (including defective genes) may interact to induce cancer.

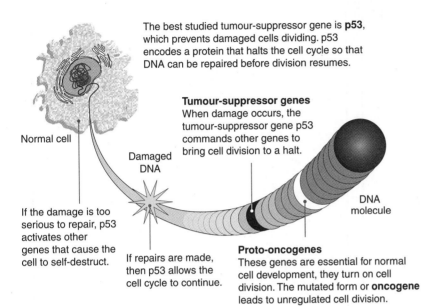

The best studied tumour-suppressor gene is **p53**, which prevents damaged cells dividing. p53 encodes a protein that halts the cell cycle so that DNA can be repaired before division resumes.

Normal cell

Damaged DNA

DNA molecule

Tumour-suppressor genes
When damage occurs, the tumour-suppressor gene p53 commands other genes to bring cell division to a halt.

If the damage is too serious to repair, p53 activates other genes that cause the cell to self-destruct.

If repairs are made, then p53 allows the cell cycle to continue.

Proto-oncogenes
These genes are essential for normal cell development, they turn on cell division. The mutated form or **oncogene** leads to unregulated cell division.

Cancer: Cells out of Control

Two types of gene are involved in controlling the cell cycle: **proto-oncogenes**, which start the cell division process and **tumour-suppressor genes**, which switch off cell division. In their normal form, both work together to perform vital tasks such as repairing defective cells and replacing dead ones.

Mutations (a change in the DNA sequence) in these genes can stop them operating normally. Proto-oncogenes, through mutation, can give rise to **oncogenes**; genes that lead to uncontrollable cell division.

Cancerous cells result from changes in the genes controlling normal cell growth and division. The resulting cells become immortal and no longer carry out their functional role. Mutations to tumour-suppressor genes initiate most human cancers.

How Tumours Spread

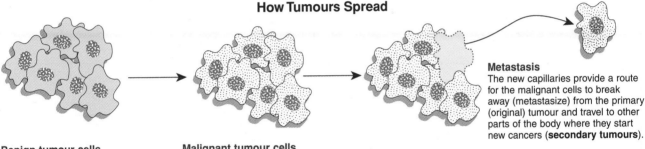

Benign tumour cells
Mutations cause the formation of a benign (harmless) tumour. The formation of new cells is matched by cell death. These cells do not spread.

Malignant tumour cells
More mutations may cause the cells to become malignant (harmful) forming a **primary tumour**. Changes to the cell chemistry encourage capillary formation. New capillaries grow into the tumour, providing it with nutrients so it can grow rapidly.

Metastasis
The new capillaries provide a route for the malignant cells to break away (metastasize) from the primary (original) tumour and travel to other parts of the body where they start new cancers (**secondary tumours**).

1. (a) How do proto-oncogenes and tumour suppresor genes normally regulate the cell cycle? _____

_____

_____

(b) How do oncogenes disrupt the normal cell cycle regulatory mechanisms? _____

_____

_____

2. A study was carried out to determine if there is a correlation between smoking and lung cancer. Analyse the graph (right) and state if you think there is a correlation. Give a reason to support your answer:

_____

_____

_____

_____

Cigarettes smoked per person per year

Lung cancer deaths per 100,000

Cigarette consumption (men)

Lung cancer (men)

4000
3000
2000
1000

150
100
50

1900   1920   1940   1960   1980
Year

LINK
WEB

34   35   **KNOW**

36 Chapter Review

Summarize what you know about this topic under the headings provided. You can draw diagrams or mind maps, or write short notes to organize your thoughts. Use the images and hints included to help you:

Ultrastructure of cells

HINT: Compare prokaryotic and eukaryotic cells, and plant and animal cells.

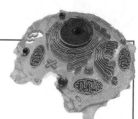

Introduction to cells

HINT: Life functions, surface area: volume ratio, magnification, and cellular differentiation.

Membrane structure

HINT: Define the fluid mosaic model and draw a cell membrane.

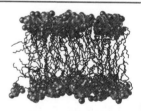

© 2012-2014 **BIOZONE** International
ISBN: 978-1-927173-93-0
Photocopying Prohibited

REVISE

Membrane transport

HINT: Define diffusion, osmosis, and active transport.

The origin of cells

HINT: Review the endosymbiotic theory and the common ancestry of life.

Cell division

HINT: Draw and label the process of mitosis.

37 KEY TERMS: Did You Get It?

1. Match each term to its definition, as identified by its preceding letter code.

active transport ....

diffusion ....

endocytosis ....

ion pump ....

mitosis ....

organelle ....

osmosis ....

passive transport ....

plasma membrane ....

A A partially-permeable phospholipid bilayer forming the boundary of all cells.

B The movement of substances across a biological membrane without energy expenditure.

C The passive movement of molecules from high to low concentration.

D A transmembrane protein that moves ions across a plasma membrane against their concentration gradient.

E The phase of a cell cycle resulting in nuclear division.

F The energy-requiring movement of substances across a biological membrane against a concentration gradient.

G Active transport in which molecules are engulfed by the plasma membrane, forming a phagosome or food vacuole within the cell.

H Passive movement of water molecules across a partially permeable membrane down a concentration gradient.

I A structural and functional part of the cell, usually bound within its own membrane. Examples include the mitochondria and chloroplasts.

2. (a) Identify organelle 1: _____

   (b) The organelle in (a) is found in a plant cell / animal cell / both plant and animal cells (circle the correct answer).

   (c) Identify organelle 2: _____

   (d) The organelle in (c) is found in a plant / animal cell / plant and animal cell (circle the correct answer).

3. Match the statements in the table below to form a complete paragraph. The left hand column is in the correct order, the right hand column is not.

   (a) Cells are the basic...

   A cell is enclosed by a plasma membrane...

   A phospholipid is made up of a...

   Proteins are embedded...

   Eukaryotic cells contain many different types of organelle...

   Each organelle carries out a specific function in the cell...

   ...such as photosynthesis or respiration.

   ...hydrophilic head and a hydrophobic tail.

   ...units of life.

   ...in the plasma membrane.

   ...made of a phospholipid bilayer.

   ...some of which are composed of membranes.

   (b) Transport of molecules though the plasma membrane...

   Active transport requires the input of energy...

   Passive transport involves the movement of molecules from...

   Simple diffusion can occur...

   Facilitated diffusion involves proteins in the plasma membrane...

   Active transport involves membrane proteins, which couple the energy provided by ATP...

   ....to the movement of molecules or ions against their concentration gradient.

   ...high concentration to low concentration (down a concentration gradient).

   ...can be active or passive.

   ...directly across the membrane.

   ...which help molecules or ions to move through.

   ...whereas passive transport does not.

TEST

Molecular Biology

Key terms

absorption spectrum
action spectrum
amino acid
anabolism
anticodon
ATP
catabolism
catalyst
cell respiration
chlorophyll
condensation
denaturation
dipole
disaccharide
DNA
DNA polymerase
DNA replication
enzyme
fermentation
helicase
hydrolysis
lipid
metabolism
molecular biology
monosaccharide
nucleic acid
nucleotide
photosynthesis
polypeptide
polysaccharide
protein
ribosome
RNA
RNA polymerase
semi-conservative
sugar
transcription
transfer RNA
translation

2.1 Molecules to metabolism

Understandings, applications, skills

Activity number

☐ 1 Define molecular biology. Explain why carbon is able to form such a wide diversity of stable compounds. Describe the organic compounds around which life is based, including carbohydrates, lipids, protein, and nucleic acids. — 38

☐ 2 Draw molecular diagrams of glucose (ring form), ribose (ring form), a saturated fatty acid, and a generalized amino acid. — 38 42 47

☐ 3 Identify biochemicals as sugars, lipids, or amino acids from molecular diagrams. — 38 42 46

☐ 4 Explain how the artificial synthesis of urea, which is produced by living organisms, helped to falsify vitalism (Nature of Science: falsification of theories). — 38

☐ 5 Explain the role of enzymes in metabolism. Distinguish between anabolism and catabolism, including the role of condensation and hydrolysis in these. — 42 45-47

2.2 Water

Understandings, applications, skills

Activity number

☐ 1 Describe the structure of water. Show in a diagram how hydrogen bonds form between water molecules. Compare the thermal properties of water and methane. — 30

☐ 2 Explain the cohesive, adhesive, thermal, and solvent properties of water and their importance to living organisms. Explain why sweat cools. Explain how the solubility of substances in water affects the way they are transported in blood. — 40

> **TOK** *Claims about the memory of water have been categorized as pseudo-scientific. How do we distinguish scientific from pseudoscientific claims?*

2.3 Carbohydrates and lipids

Understandings, applications, skills

Activity number

☐ 1 Explain how monosaccharide monomers are linked together in condensation reactions to form disaccharides and polysaccharides. — 42

☐ 2 Describe the structure and function of storage polysaccharides: cellulose and starch in plants and glycogen in humans and other mammals. Compare cellulose, starch, and glycogen using molecular visualization software. — 43 44

☐ 3 Explain the role of lipids in energy storage. Explain how triglycerides are formed by condensation from fatty acids and glycerol. Describe the structure of fatty acids, distinguishing saturated, monounsaturated, and polyunsaturated fatty acids. Distinguish between *cis*- and *trans*- fatty acids. — 45

☐ 4 Outline the evidence for the health risks of *trans* fats and saturated fatty acids. Evaluate the evidence for health claims about lipids. — 46

> **TOK** *How do we decide between conflicting viewpoints about dietary fat?*

2.4 Proteins

Understandings, applications, skills

Activity number

☐ 1 Explain how the 20 different amino acids found in polypeptides are joined by condensation reactions. Draw molecular diagrams to show formation of the peptide bond. Explain how variation in the sequence of amino acids gives rise to variation in the polypeptides synthesized on ribosomes. Explain how the amino acid sequence of polypeptides is encoded by genes. — 47

☐ 2 Use examples to show that the amino acid sequence determines the 3-D conformation of a protein and that a functional protein may consist of one or more polypeptides. Explain protein denaturation by heat and extremes of pH. — 48

38 Organic Molecules

Key Idea: Organic molecules make up most of the chemicals found in living organisms.

Molecular biology is a branch of science that studies the molecular basis of biological activity. All life is based around carbon, which is able to combine with many other elements to form a large number of carbon-based (or organic) molecules. Specific groups of atoms, called functional groups, attach to a C-H core and determine the specific chemical properties of the molecule. The organic macromolecules that make up living things can be grouped into four classes: carbohydrates, lipids, proteins, and nucleic acids. These larger molecules are built up from smaller components in anabolic reactions and broken down by catabolic reactions. The sum total of anabolic and catabolic reactions in the cell or organism is metabolism.

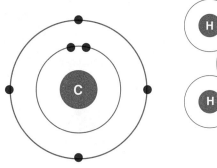

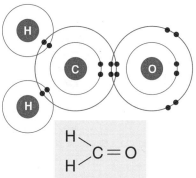

Organic macromolecule	Structural unit	Elements
Carbohydrates	Sugar monomer	C, H, O
Proteins	Amino acid	C, H, O, N, S
Lipids	Not applicable	C, H, O
Nucleic acids	Nucleotide	C, H, O, N, P

A carbon atom (above) has four electrons that are available to form up to four **covalent bonds** with other atoms. A covalent bond forms when two atoms share a pair of electrons. The number of covalent bonds formed between atoms in a molecule determines the shape and chemical properties of the molecule.

Methanal (molecular formula CH_2O) is a simple organic molecule. A carbon (C) atom bonds with two hydrogen (H) atoms and an oxygen (O) atom. In the structural formula (blue box), the bonds between atoms are represented by lines. Covalent bonds are very strong, so the molecules formed are very stable.

The most common elements found in organic molecules are carbon, hydrogen, and oxygen, but organic molecules may also contain other elements, such as nitrogen, phosphorus, and sulfur. Most organic macromolecules are built up of one type of repeating unit or 'building block', except lipids, which are quite diverse in structure.

What is Metabolism?

Metabolism describes the sum total of the biochemical reactions that sustain life in a cell or organism. These reactions are brought about by catalytic proteins called enzymes and occur in pathways often involving many steps with various intermediates. Each intermediate is the substrate for the next step in the pathway. Metabolism is usually divided into catabolic and anabolic pathways.

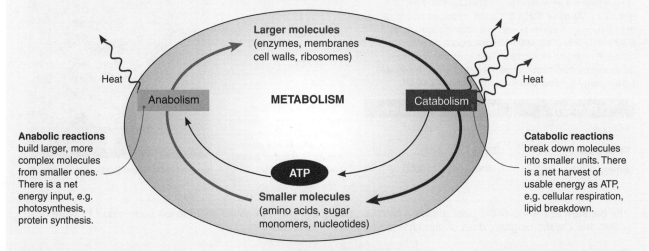

Anabolic reactions build larger, more complex molecules from smaller ones. There is a net energy input, e.g. photosynthesis, protein synthesis.

Catabolic reactions break down molecules into smaller units. There is a net harvest of usable energy as ATP, e.g. cellular respiration, lipid breakdown.

Organic Molecules Can Be Synthesized

In 1828, in an attempt to prepare ammonium cyanate, Friedrich Wöhler treated silver cyanate with ammonium chloride and obtained **urea** (molecular formula $(NH_2)_2CO$), an organic molecule which had been known as a component of urine since the 1700s. Wöhler's production of urea, by lucky accident, was the first artificial synthesis of an organic compound from inorganic reactants. It helped to discredit vitalism – a mainstream theory at the time proposing that organic molecules could only be made by living organisms and could not be synthesized artificially.

Right: Friedrich Wöhler, a German chemist and the reaction that produced urea.

$AgNCO + NH_4Cl \rightarrow (NH_2)_2CO + AgCl$

1. On the diagram of the carbon atom top left, mark with arrows the electrons that are available to form covalent bonds with other atoms.

2. Investigate Wöhler's synthesis of urea and discuss how it differs from the metabolic pathway that produces urea in living organisms. In what way was Wöhler's synthesis of urea serendipitous? Why did it take some time before vitalism was discredited? What further evidence accumulated to refute it? Make a summary of your findings and attach it to this page.

ISBN: 978-1-927173-93-0

KNOW

39 Water

Key Idea: Water forms bonds between other water molecules and also with ions allowing water to act as a medium for transporting molecules.

Water (H_2O) is the main component of living things, and typically makes up about 70% of any organism. Water is important in cell chemistry as it takes part in, and is a common product of, many reactions. Water can form bonds with other water molecules, and also with other ions (charged molecules). Because of this chemical ability, water is regarded as the universal solvent.

Water Forms Hydrogen Bonds

A water molecule is polar, meaning it has a positively and a negatively charged region. In water, oxygen has a slight negative charge and each of the hydrogens have a slight positive charge. Water molecules have a weak attraction for each other, forming large numbers of weak hydrogen bonds with other water molecules (far right).

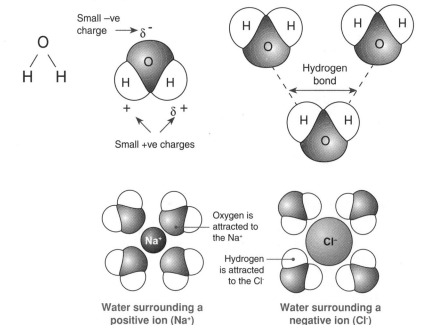

Intermolecular bonds between water and other polar molecules or ions are important for biological systems. Inorganic ions may have a positive or negative charge (e.g sodium ion is positive, chloride ion is negative). The charged water molecule is attracted to the charged ion and surrounds it (right). This formation of intermolecular bonds between water and the ions is what keeps the ions dissolved in water. Polar molecules such as amino acids and carbohydrates also dissolve readily in water.

Oxygen is attracted to the Na⁺

Hydrogen is attracted to the Cl⁻

Water surrounding a positive ion (Na⁺)

Water surrounding a negative ion (Cl⁻)

Comparing Water and Methane

Water and methane are both small molecules, but have very different chemical properties because of their chemistry. Methane (CH_4) is a small hydrocarbon consisting of four hydrogen atoms bound to a carbon atom (right). Methane is non-polar and hydrogen bonds do not form between methane molecules. The molecules are much more weakly held together compared to water molecules. Little energy is needed to force the molecules apart.

Property	Methane	Water
Formula	CH_4	H_2O
Melting point	−182°C	0°C
Boiling point	−160°C	100°C

The bonds water forms with other water molecules requires a lot of energy to break. This is why water has a much higher boiling point than methane.

Kochendes_wasser02

1. The diagram at the top of the page shows a positive sodium ion and a negative chloride ion surrounded by water molecules. On the diagram, draw on the charge of the water molecules.

2. Explain the formation of hydrogen bonds between water and other polar molecules:

_____

_____

_____

_____

_____

3. Why does methane have a much lower melting point and boiling point than water?_____

_____

_____

© 2012-2014 **BIOZONE** International
ISBN: 978-1-927173-93-0
Photocopying Prohibited

40 The Properties of Water

Key Idea: Water's chemical properties influence its physical properties and its ability to transport molecules in solution. Water's cohesive, adhesive, thermal, and solvent properties come about because of its polarity and ability to form hydrogen bonds with other polar molecules. These physical properties allow water, and water based substances (such as blood), to transport polar molecules in solution. The ability of substances to dissolve in water varies. **Hydrophilic** (water-loving) substances dissolve readily in water (e.g. salts, sugars). **Hydrophobic** (water-hating) substances (e.g. oil) do not dissolve in water. Blood must transport many different substances, including hydrophobic ones.

Cohesive Properties

Water molecules are cohesive, they stick together because hydrogen bonds form between water molecules. Cohesion allows water to form drops and allows the development of surface tension. Example: The cohesive and adhesive properties of water allow it to be transported as an unbroken column through the xylem of plants.

Adhesive Properties

Water is attracted to other molecules because it forms hydrogen bonds with other polar molecules. Example: The adhesion of water molecules to the sides of a capillary tube is responsible for a meniscus (the curved upper surface of a liquid in a tube).

Solvent Properties

Other substances dissolve in water because water's dipolar nature allows it to surround other charged molecules and prevent them from clumping together. Example: Mineral transport through a plant.

Thermal Properties

▶ Water has the highest heat capacity of all liquids, so it takes a lot of energy before it will change temperature. As a result, water heats up and cools down slowly, so large bodies of water maintain a relatively stable temperature.

▶ Water is liquid at room temperature and has a high boiling point because a lot of energy is needed to break the hydrogen bonds. The liquid environment supports life and metabolic processes.

▶ Water has a high latent heat of vaporization, meaning it takes a lot of energy to transform it from the liquid to the gas phase. In sweating, the energy is provided by the body, so sweat has a cooling effect.

Transporting Substances in Blood

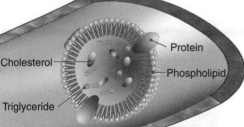

Cholesterol — Protein — Phospholipid — Triglyceride

Sodium chloride	Glucose	Amino acids	Oxygen	Fats and Cholesterol
Sodium chloride (NaCl) is highly soluble. NaCl dissolves in blood plasma into the ions Na⁺ and Cl⁻.	Glucose is a polar molecule so readily dissolves into blood plasma for transport around the body.	All amino acids have both a positive or negative charge, and are highly soluble in blood. However, their variable R-chain can alter the solubility of amino acids slightly.	Oxygen has low solubility in water. In blood, it is bound to the protein hemoglobin in red blood cells so it can be transported around the body.	Fats are non-polar substances and are insoluble in water. Cholesterol has a slight charge, but is also water insoluble. Both are transported in blood within lipoprotein complexes, spheres of phospholipids arranged with the hydrophilic heads and proteins facing out and hydrophobic tails facing inside.

1. (a) Describe the difference between a **hydrophilic** and **hydrophobic** molecule: _____

_____

_____

(b) Use an example to describe how a hydrophilic and a hydrophobic molecule are transported in blood: _____

_____

_____

_____

2. How does water act as a coolant during sweating? _____

_____

_____

LINK **236**   LINK **189**   LINK **45**   WEB **40**   ◀ KNOW

41 Sugars

Key Idea: Monosaccharides are the building blocks for larger carbohydrates. They can exist as isomers.

Sugars (monosaccharides and disaccharides) play a central role in cells, providing energy and joining together to form carbohydrate macromolecules, such as starch and glycogen. Monosaccharide polymers form the major component of most plants (as cellulose). Monosaccharides are important as a primary energy source for cellular metabolism. Disaccharides are important in human nutrition and are found in milk (lactose) table sugar (sucrose) and malt (maltose). Carbohydrates have the general formula $(CH_2O)_n$, where $n =$ the number of carbon atoms.

Monosaccharides

Monosaccharides are single-sugar molecules and include glucose (grape sugar and blood sugar) and fructose (honey and fruit juices). They are used as a primary energy source for fuelling cell metabolism. They can be joined together to form disaccharides and polysaccharides.

The most common arrangements found in sugars are hexose (6 sided) or pentose (5 sided) rings. The commonly occurring monosaccharides contain between three and seven carbon atoms in their carbon chains and, of these, the 6C hexose sugars occur most frequently. All monosaccharides are reducing sugars (they can participate in reduction reactions).

Examples of monosaccharide structures

Triose	Pentose	Hexose
C | C | C	(pentagon)	(hexagon)
e.g. glyceraldehyde	e.g. ribose, deoxyribose	e.g. glucose, fructose, galactose

Glucose Isomers

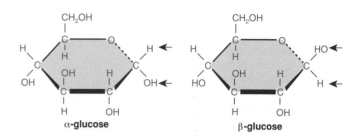

α-glucose   β-glucose

Isomers are compounds with the same chemical formula (same types and numbers of atoms) but they have different arrangements of their atoms. The different arrangement of the atoms means that each isomer has different properties.

In structural isomers (such as α and β glucose, above), the atoms are linked in different sequences. Optical isomers are identical in every way but are mirror images of each other.

Glucose is a versatile molecule. It provides energy to power cellular reactions, can form energy storage molecules such as glycogen, or it can be used to build structural molecules.

Plants make their glucose via the process of photosynthesis. Animals and other heterotrophic organisms obtain their glucose by consuming plants or other organisms.

Fructose, often called fruit sugar, is a simple monosaccharide. It is often derived from sugar cane (above). Both fructose and glucose can be directly absorbed into the bloodstream.

1. Describe the two major functions of monosaccharides:

    (a) _____

    _____

    (b) _____

    _____

2. Define the term **structural isomer**, using glucose molecules as examples: _____

_____

_____

_____

3. Why is glucose such a versatile molecule? _____

_____

_____

© 2012-2014 **BIOZONE** International
ISBN: 978-1-927173-93-0
Photocopying Prohibited

WEB   LINK   LINK

KNOW   41   42   43

42 Condensation and Hydrolysis of Sugars

Key Idea: Condensation reactions join monosaccharides together to form disaccharides and polysaccharides. Hydrolysis reactions split disaccharides and polysaccharides into smaller molecules

Monosaccharide monomers can be linked together by a **condensation reaction**, to produce larger molecules

(disaccharides and polysaccharides). The reverse reaction, **hydrolysis**, breaks compound sugars down into their constituent monosaccharides. **Disaccharides** (double-sugars) are produced when two monosaccharides are joined together. Different disaccharides are formed by joining together different combinations of monosaccharides (below).

Condensation and Hydrolysis Reactions

Monosaccharides can combine to form compound sugars in what is called a condensation reaction. Compound sugars can be broken down by hydrolysis to simple monosaccharides.

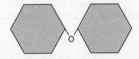
Two mono-saccharides

Condensation Reaction

Two monosaccharides are joined together to form a disaccharide with the release of a water molecule (hence its name). A net energy input is required for the reaction to proceed.

Hydrolysis Reaction

When a disaccharide is split, as in digestion, a water molecule is used as a source of hydrogen and a hydroxyl group. The reaction is catalysed by specific enzymes.

+
H_2O — Glycosidic bond

Disaccharide + water

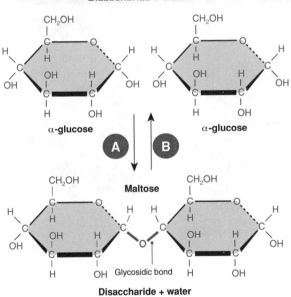

α-glucose   α-glucose

A   **B**

Maltose

Glycosidic bond

Disaccharide + water

Disaccharides

Disaccharides (below) are double-sugar molecules and are used as energy sources and as building blocks for larger molecules. They are important in human nutrition and are found in milk (lactose), table sugar (sucrose), and malt (maltose).

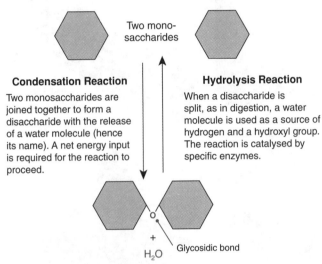

The type of disaccharide formed depends on the monomers involved and whether they are in their α- or β- form. Only a few disaccharides (e.g. lactose) are classified as reducing sugars. Some common disaccharides are described below.

Lactose, a milk sugar, is made up of β-glucose + β-galactose. Milk contains 2-8% lactose by weight. It is the primary carbohydrate source for suckling mammalian infants.

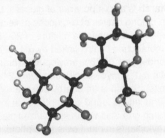

Maltose is composed of two α-glucose molecules. Germinating seeds contain maltose because the plant breaks down their starch stores to use it for food.

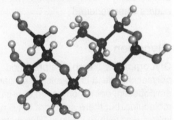

Sucrose (table sugar) is a simple sugar derived from plants such as sugar cane, sugar beet, or maple sap. It is composed of an α-glucose molecule and a β-fructose molecule.

1. Explain briefly how disaccharide sugars are formed and broken down: _____

_____

_____

_____

2. On the diagram above, name the reaction occurring at points **A** and **B** and name the product that is formed:

_____

_____

3. On the lactose, maltose, and sucrose molecules (above right), circle the two monomers on each molecule.

© 2012-2014 **BIOZONE** International
ISBN: 978-1-927173-93-0
Photocopying Prohibited

LINK LINK WEB

46 **45** **42** **KNOW**

43 Polysaccharides

Key Idea: Polysaccharides consist of many monosaccharides joined together through condensation reactions. Their composition and isomerization alter their functional properties. **Polysaccharides** (complex carbohydrates) are straight or branched chains of many monosaccharides joined together. They can consist of one or more types of monosaccharides.

The most common polysaccharides, cellulose, starch, and glycogen contain only glucose, but their properties are very different. These differences are a function of the glucose isomer involved and the types of linkages joining them. Different polysaccharides can thus be a source of readily available glucose or a structural material that resists digestion.

Cellulose

Cellulose is a structural material found in the cell walls of plants. It is made up of unbranched chains of β-glucose molecules held together by β-1,4 glycosidic links. As many as 10,000 glucose molecules may be linked together to form a straight chain. Parallel chains become cross-linked with hydrogen bonds and form bundles of 60-70 molecules called **microfibrils**. Cellulose microfibrils are very strong and are a major structural component of plants, e.g. as the cell wall. Few organisms can break the β-linkages so cellulose is an ideal structural material.

Starch

Starch is also a polymer of glucose, but it is made up of long chains of α-glucose molecules linked together. It contains a mixture of 25-30% amylose (unbranched chains linked by α-1,4 glycosidic bonds) and 70-75% amylopectin (branched chains with α-1, 6 glycosidic bonds every 24-30 glucose units). Starch is an energy storage molecule in plants and is found concentrated in insoluble starch granules within specialized plastids called amyloplasts in plant cells (see photo, right). Starch can be easily hydrolysed by enzymes to soluble sugars when required.

Glycogen

Glycogen, like starch, is a branched polysaccharide. It is chemically similar to amylopectin, being composed of α-glucose molecules, but there are more α-1,6 glycosidic links mixed with α-1,4 links. This makes it more highly branched and more water-soluble than starch. Glycogen is a storage compound in animal tissues and is found mainly in liver and muscle cells (photo, right). It is readily hydrolysed by enzymes to form glucose making it an ideal energy storage molecule for active animals.

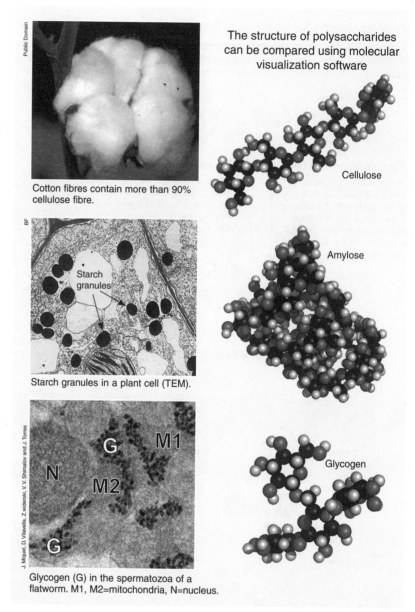

Cotton fibres contain more than 90% cellulose fibre.

Starch granules in a plant cell (TEM).

Glycogen (G) in the spermatozoa of a flatworm. M1, M2=mitochondria, N=nucleus.

The structure of polysaccharides can be compared using molecular visualization software

Cellulose

Amylose

Glycogen

1. (a) Why are polysaccharides such a good source of energy?_____

    _____

    (b) How is the energy stored in polysaccharides mobilized?_____

    _____

2. Contrast the properties of the polysaccharides starch, cellulose, and glycogen and relate these to their roles in the cell:

    _____

    _____

    _____

    _____

    _____

© 2012-2014 **BIOZONE** International
ISBN: 978-1-927173-93-0
Photocopying Prohibited

44 Starch and Cellulose

Key Idea: Starch and cellulose are important polysaccharides in plants. Starch is a storage carbohydrate made up of two α-glucose polymers, amylose and amylopectin. Cellulose is a β-glucose polymer which forms the plant cell wall.

Glucose monomers can be linked in condensation reactions to form large structural and energy storage polysaccharides. The glucose isomer involved and the type of glycosidic linkage determines the properties of the molecule.

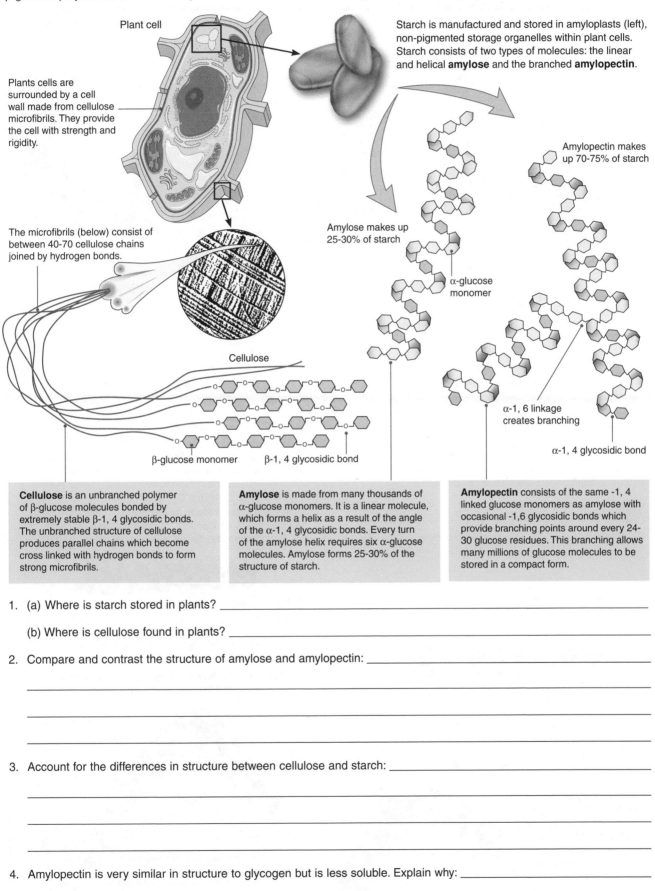

Plant cell

Plants cells are surrounded by a cell wall made from cellulose microfibrils. They provide the cell with strength and rigidity.

Starch is manufactured and stored in amyloplasts (left), non-pigmented storage organelles within plant cells. Starch consists of two types of molecules: the linear and helical **amylose** and the branched **amylopectin**.

Amylopectin makes up 70-75% of starch

The microfibrils (below) consist of between 40-70 cellulose chains joined by hydrogen bonds.

Amylose makes up 25-30% of starch

α-glucose monomer

Cellulose

α-1, 6 linkage creates branching

β-glucose monomer   β-1, 4 glycosidic bond

α-1, 4 glycosidic bond

Cellulose is an unbranched polymer of β-glucose molecules bonded by extremely stable β-1, 4 glycosidic bonds. The unbranched structure of cellulose produces parallel chains which become cross linked with hydrogen bonds to form strong microfibrils.

Amylose is made from many thousands of α-glucose monomers. It is a linear molecule, which forms a helix as a result of the angle of the α-1, 4 glycosidic bonds. Every turn of the amylose helix requires six α-glucose molecules. Amylose forms 25-30% of the structure of starch.

Amylopectin consists of the same -1, 4 linked glucose monomers as amylose with occasional -1,6 glycosidic bonds which provide branching points around every 24-30 glucose residues. This branching allows many millions of glucose molecules to be stored in a compact form.

1. (a) Where is starch stored in plants? _____

   (b) Where is cellulose found in plants? _____

2. Compare and contrast the structure of amylose and amylopectin: _____

   _____

   _____

   _____

3. Account for the differences in structure between cellulose and starch: _____

   _____

   _____

   _____

4. Amylopectin is very similar in structure to glycogen but is less soluble. Explain why: _____

   _____

© 2012-2014 **BIOZONE** International
ISBN: 978-1-927173-93-0
Photocopying Prohibited

LINK 43   LINK 14   WEB 44   APP

45 Lipids

Key Idea: Lipids are non-polar, hydrophobic organic molecules, which have many important biological functions. Fatty acids are a type of lipid.

Lipids are organic compounds which are mostly nonpolar (have no overall charge) and hydrophobic, so they do not readily dissolve in water. Lipids include fats, waxes, sterols,

and phospholipids. Fatty acids are a major component of neutral fats and phospholipids. Fatty acids consist of an even number of carbon atoms, with hydrogen bound along the length of the chain. The presence of a carboxyl group (–COOH) at one end makes them an acid. They are generally classified as saturated or unsaturated fatty acids (below).

Lipids Store Large Amounts of Energy

Energy is required for locomotion

Hibernating squirrel

Fats are an economical way to store energy reserves. A gram of fat yields more than twice as much energy as a gram of carbohydrate because, carbon for carbon, fats require more oxidation to become CO_2 and H_2O than do carbohydrates. The fat-tailed gerbil (above) stores fat in its tail to survive over the winter.

Lipids are concentrated sources of energy and provide fuel for aerobic respiration. Fatty acids undergo β-oxidation in the mitochondrial matrix to release 2-C units which enter the Krebs cycle and are fully oxidized, producing ATP, water, and carbon dioxide. ATP provides the usable energy to drive essential life processes.

Proteins and carbohydrates can be converted into fats and stored. In humans and other mammals, the amount of lipid stored as an energy reserve far exceeds the energy stored as glycogen. During times of plenty, this store is increased, to be used during times of food shortage (e.g. during winter hibernation, above).

Saturated and Unsaturated Fatty Acids

Fatty acids are classed as either saturated or unsaturated. **Saturated fatty acids** contain the maximum number of hydrogen atoms. **Unsaturated fatty acids** contain some double-bonds between carbon atoms and are not fully saturated with hydrogens. A chain with only one double bond is called monounsaturated, whereas a chain with two or more double bonds is called polyunsaturated.

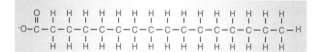

Formula (above) and molecular model (below) for a saturated fatty acid (palmitic acid).

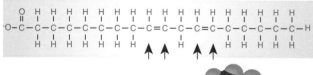

Formula (above) and molecular model (right) for an unsaturated fatty acid (linoleic acid). The arrows indicate double bonded carbon atoms that are not fully saturated with hydrogens.

Trans- and Cis- Fatty Acids

Unsaturated fatty acids can exist as **trans-** and **cis-** isomers. *Trans*-fatty acids (TFA) are rare in nature and have no known benefit to human health. They are produced when vegetable oils are hydrogenated to increase shelf life and make them solid but spreadable. TFA pack together more tightly than *cis*-fatty acids and consequently have higher melting points than the corresponding *cis* forms. TFA have no specific metabolic functions and are metabolized in the liver differently to *cis* isomers, raising levels of low density lipoproteins in the blood and increasing the risk of coronary artery disease.

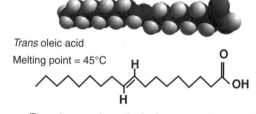

Trans oleic acid

Melting point = 45°C

Trans- isomers have the hydrogens on the opposite side of the double bond, producing a straight molecule.

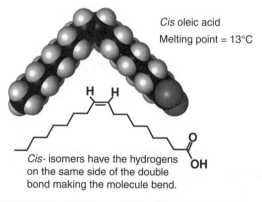

Cis oleic acid

Melting point = 13°C

Cis- isomers have the hydrogens on the same side of the double bond making the molecule bend.

WEB
45

LINK
46

© 2012-2014 **BIOZONE** International
ISBN: 978-1-927173-93-0
Photocopying Prohibited

Triglycerides

Fatty acids

Triglyceride: an example of a neutral fat

The most abundant lipids in living things are neutral fats. They make up the fats and oils found in plants and animals. **Neutral fats** are composed of a glycerol molecule attached to one (monoglyceride), two (diglyceride) or three (triglyceride) fatty acids.

Lipids containing a high proportion of saturated fatty acids tend to be solids at room temperature (e.g. butter). Lipids with a high proportion of unsaturated fatty acids are oils and tend to be liquid at room temperature (e.g. olive oil). This is because the unsaturation causes kinks in the straight chains so that the fatty acid chains do not pack closely together.

Triglycerides are Formed by Condensation Reactions

Triglycerides form when glycerol bonds with three fatty acids. Glycerol is an alcohol containing three carbons. Each of these carbons is bonded to a hydroxyl (-OH) group.

When glycerol bonds with the fatty acid, an ester bond is formed and water is released. Three separate condensation reactions are involved in producing a triglyceride.

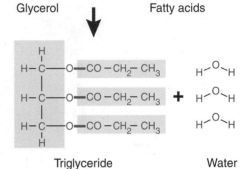

Glycerol   Fatty acids

Triglyceride   Water

1. Why are lipids such a good fuel source for metabolic reactions? _____

_____

2. How are lipids broken down before they are respired? _____

_____

3. Many arid-adapted organisms, including camels and some rodents, obtain all the water they need from the metabolism of fats. How does the metabolism of fats provide an organism with water?

_____

4. (a) Distinguish between saturated and unsaturated fatty acids: _____

_____

   (b) How is the type of fatty acid present in a neutral fat or phospholipid related to that molecule's properties?

_____

_____

   (c) Why are unsaturated fatty acids particularly abundant in the cellular membranes of Antarctic fish?

_____

5. (a) Describe the structural and functional differences between *cis*- and *trans*- fatty acids: _____

_____

_____

_____

   (b) What is the significance of these differences to human health? _____

_____

_____

46 Lipids and Health

Key Idea: Dietary lipids are essential for health, but some types of lipids are associated with disease.

Lipids are an essential macronutrient, but their energy dense nature has made them the target of nutritional concerns over the increasing incidence of obesity and its associated health problems. The dietary intakes of different types of fats, including cholesterol, have been a subject of a great deal of research, some of which is conflicting or inconclusive.

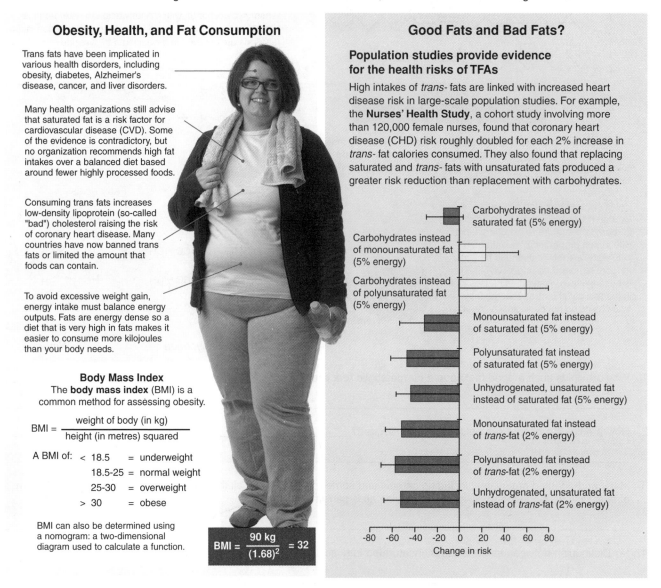

Obesity, Health, and Fat Consumption

Trans fats have been implicated in various health disorders, including obesity, diabetes, Alzheimer's disease, cancer, and liver disorders.

Many health organizations still advise that saturated fat is a risk factor for cardiovascular disease (CVD). Some of the evidence is contradictory, but no organization recommends high fat intakes over a balanced diet based around fewer highly processed foods.

Consuming trans fats increases low-density lipoprotein (so-called "bad") cholesterol raising the risk of coronary heart disease. Many countries have now banned trans fats or limited the amount that foods can contain.

To avoid excessive weight gain, energy intake must balance energy outputs. Fats are energy dense so a diet that is very high in fats makes it easier to consume more kilojoules than your body needs.

Body Mass Index

The **body mass index** (BMI) is a common method for assessing obesity.

$$BMI = \frac{\text{weight of body (in kg)}}{\text{height (in metres) squared}}$$

A BMI of:
- < 18.5 = underweight
- 18.5-25 = normal weight
- 25-30 = overweight
- > 30 = obese

BMI can also be determined using a nomogram: a two-dimensional diagram used to calculate a function.

$$BMI = \frac{90 \text{ kg}}{(1.68)^2} = 32$$

Good Fats and Bad Fats?

Population studies provide evidence for the health risks of TFAs

High intakes of *trans-* fats are linked with increased heart disease risk in large-scale population studies. For example, the **Nurses' Health Study**, a cohort study involving more than 120,000 female nurses, found that coronary heart disease (CHD) risk roughly doubled for each 2% increase in *trans-* fat calories consumed. They also found that replacing saturated and *trans-* fats with unsaturated fats produced a greater risk reduction than replacement with carbohydrates.

Chart (Change in risk, x-axis from -80 to 80):
- Carbohydrates instead of saturated fat (5% energy)
- Carbohydrates instead of monounsaturated fat (5% energy)
- Carbohydrates instead of polyunsaturated fat (5% energy)
- Monounsaturated fat instead of saturated fat (5% energy)
- Polyunsaturated fat instead of saturated fat (5% energy)
- Unhydrogenated, unsaturated fat instead of saturated fat (5% energy)
- Monounsaturated fat instead of *trans*-fat (2% energy)
- Polyunsaturated fat instead of *trans*-fat (2% energy)
- Unhydrogenated, unsaturated fat instead of *trans*-fat (2% energy)

---

1. Analyse the Nurses' Health Study data above. Summarize the findings for the effect of *trans* fat on CHD risk factors:

_____

_____

_____

2. The data regarding the health effects of different types of dietary fat are often conflicting.

(a) What variables might account for difference in the findings of different studies: _____

_____

_____

(b) What are health authorities agreed on? _____

_____

3. Using the BMI, calculate the weight or weight range at which a 1.85 m tall man would be considered:

(a) Underweight:_____ (b) Overweight:_____

© 2012-2014 **BIOZONE** International
ISBN: 978-1-927173-93-0
Photocopying Prohibited

47 Amino Acids

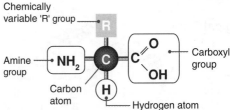

Key Idea: Amino acids can be joined together by condensation reactions to form polypeptides. Proteins are made up of one or more polypeptide molecules.

Amino acids are the basic units from which proteins are made. Twenty amino acids commonly occur in proteins and they can be linked in many different ways by peptide bonds to form a huge variety of polypeptides. Peptide bonds are formed by **condensation reactions** between amino acids.

The Structure and Properties of Amino Acids

Chemically variable 'R' group

Amine group — NH₂
Carbon atom — C
Carboxyl group — C $\begin{smallmatrix}O\\OH\end{smallmatrix}$
Hydrogen atom — H

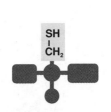

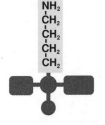

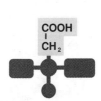

All amino acids have a common structure (above), but the R group is different in each kind of amino acid (right). The property of the R group determines how it will interact with other amino acids and ultimately determines how the amino acid chain folds up into a functional protein. For example, the hydrophobic R groups of soluble proteins are folded into the protein's interior, while the hydrophilic groups are arranged on the outside.

Cysteine

This 'R' group can form **disulfide bridges** with other cysteines to create cross linkages in a polypeptide chain.

Lysine

This 'R' group gives the amino acid an **alkaline** property.

Aspartic acid

This 'R' group gives the amino acid an **acidic** property.

Condensation and Hydrolysis Reactions

Two amino acids

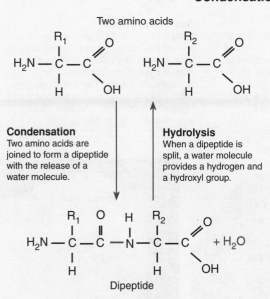

Condensation
Two amino acids are joined to form a dipeptide with the release of a water molecule.

Hydrolysis
When a dipeptide is split, a water molecule provides a hydrogen and a hydroxyl group.

Dipeptide

Amino acids are linked by **peptide bonds** to form long **polypeptide chains** of up to several thousand amino acids. Peptide bonds form between the carboxyl group of one amino acid and the amine group of another (left). Water is formed as a result of this bond formation.

The sequence of amino acids in a polypeptide is called the **primary structure** and is determined by the order of nucleotides in DNA and mRNA. The linking of amino acids to form a polypeptide occurs on ribosomes in the cytoplasm. Once released from the ribosome, a polypeptide will fold into a secondary structure determined by the composition and position of the amino acids making up the chain.

A polypeptide chain

Peptide bond   Peptide bond   Peptide bond   Peptide bond   Peptide bond   Peptide bond

1. (a) What makes each of the amino acids in proteins unique? _____

    (b) What is the primary structure of a protein? _____

    (c) What determines the primary structure? _____

    _____

    (d) How do the sequence and composition of amino acids in a protein influence how a protein folds up? _____

    _____

    _____

2. (a) What type of bond joins neighbouring amino acids together? _____

    (b) How is this bond formed? _____

    (c) Circle this bond in the dipeptide above:

    (d) How are di- and polypeptides broken down? _____

KNOW ▶

48 Proteins

Key Idea: Interactions between amino acid R groups direct a polypeptide chain to fold into its functional shape. When a protein is denatured, it loses its functionality.

A protein may consist of one polypeptide chain, or several polypeptide chains linked together. Hydrogen bonds between amino acids cause it to form its **secondary structure**, either an α-helix or a β-pleated sheet. The interaction between R groups causes a polypeptide to fold into its **tertiary structure**, a three dimensional shape held by ionic bonds and disulfide bridges (bonds formed between sulfur containing amino acids). If bonds are broken (through denaturation), the protein loses its tertiary structure, and its functionality.

The Shape of a Protein Reflects Its Biological Role

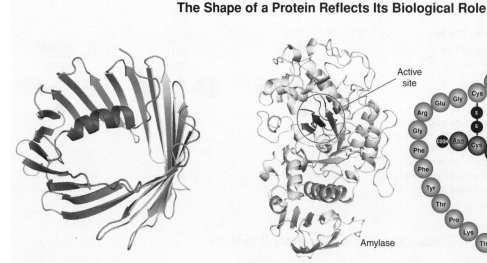

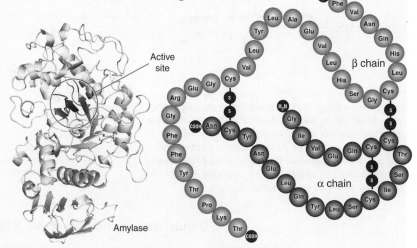

Active site

Amylase

β chain

α chain

Channel Proteins
Proteins that fold to form channels in the plasma membrane present non-polar R groups to the membrane and polar R groups to the inside of the channel. Hydrophilic molecules and ions are then able to pass through these channels into the interior of the cell. Ion channels are found in nearly all cells and many organelles.

Enzymes
Enzymes are globular proteins that catalyse specific reactions. Enzymes that are folded to present polar R groups at the active site will be specific for polar substances. Non-polar active sites will be specific for non-polar substances. Alteration of the active site by extremes of temperature or pH cause a loss of function.

Sub-Unit Proteins
Many proteins, e.g. insulin and hemoglobin, consist of two or more sub-units in a complex quaternary structure, often in association with a metal ion. Active insulin is formed by two polypeptide chains stabilized by disulfide bridges between neighbouring cysteines. Insulin stimulates glucose uptake by cells.

Protein Denaturation

When the chemical bonds holding a protein together become broken the protein can no longer hold its three dimensional shape. This process is called **denaturation**, and the protein usually loses its ability to carry out its biological function.

There are many causes of denaturation including exposure to heat or pH outside of the protein's optimum range. The main protein in egg white is albumin. It has a clear, thick fluid appearance in a raw egg (right). Heat (cooking) denatures the albumin protein and it becomes insoluble, clumping together to form a thick white substance (far right).

Raw (native) egg white

Cooked (denatured) egg white

1. Explain the importance of the amino acid sequence in protein folding: _____

_____

_____

2. Why do channel proteins often fold with non-polar R groups to the channel's exterior and polar R groups to its interior?

_____

_____

3. Why does **denaturation** often result in the loss of protein functionality? _____

_____

_____

_____

© 2012-2014 **BIOZONE** International
ISBN: 978-1-927173-93-0
Photocopying Prohibited

49 The Role of Proteins

Key Idea: Protein structure is related to its biological function. Proteins can be classified according to their structure or their function. Globular proteins are spherical and soluble in water (e.g. enzymes). Fibrous proteins have an elongated structure and are not water soluble. They are often made up of repeating units and provide stiffness and rigidity to the more fluid components of cells and tissues. They have important structural and contractile roles.

Globular Proteins

Properties
- Easily water soluble
- Tertiary structure critical to function
- Polypeptide chains folded into a spherical shape

Function
- Catalytic, *e.g. enzymes*
- Regulatory, *e.g. hormones (insulin)*
- Transport, *e.g. hemoglobin*
- Protective, *e.g. immunoglobulins (antibodies)*
- Signal transduction, *e.g. rhodopsin*

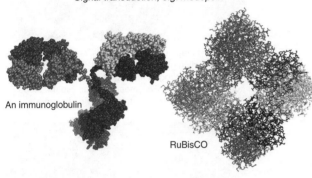

An immunoglobulin

RuBisCO

Immunoglobulins
These are large multi-unit Y-shape plasma proteins that recognize, bind to, and help to destroy bacteria and viruses. The tips of the 'Y' form the binding site for specific antigens.

RuBisCo is a large multi-unit enzyme found in green plants and catalyses the first step of carbon fixation in the Calvin cycle. It consists of 8 large (L) and 8 small (S) subunits arranged as 4 dimers. RuBisCO is the most abundant protein in the world.

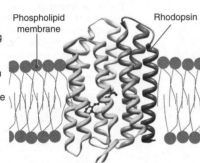

Phospholipid membrane

Rhodopsin

Rhodopsin is a signalling protein involved in photoreception in the retina. It consists of seven transmembrane helices and a bound photoreactive chromophore, which is a cofactor essential to rhodopsin's function.

Fibrous Proteins

Properties
- Water insoluble
- Very tough physically; may be supple or stretchy
- Parallel polypeptide chains in long fibres or sheets

Function
- Structural role in cells and organisms e.g. *collagen found in connective tissue, cartilage, bones, tendons, and blood vessel walls.*
- Contractile e.g. *myosin, actin*

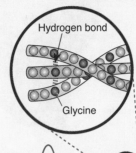

Hydrogen bond

Glycine

The collagen molecule consists of three polypeptides wound together to form a helical 'rope'. Every third amino acid in each polypeptide is a glycine (Gly) where hydrogen bonding holds the three strands together. Collagen molecules self assemble into **fibrils** held together by covalent cross linkages (below). Bundles of fibrils form fibres.

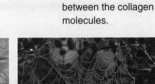

Many collagen molecules form fibrils and the fibrils group together to form larger fibres.

Covalent cross links between the collagen molecules.

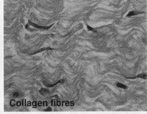

Collagen fibres

Spider silk

Collagen is the main component of connective tissue, and is mostly found in fibrous tissues (e.g. tendons, ligaments, and skin). **Spider silk** is a protein spun into a web by spiders to capture prey. Like all fibrous proteins, it is very strong.

1. How are proteins involved in the following roles? Give examples to help illustrate your answer:

   (a) Structural tissues of the body: _____

   _____

   _____

   (b) Catalysing metabolic reactions in cells: _____

   _____

   _____

2. How does the shape of a fibrous protein relate to its functional role? _____

   _____

3. How does the shape of a catalytic protein (enzyme) relate to its functional role? _____

   _____

© 2012-2014 **BIOZONE** International
ISBN: 978-1-927173-93-0
Photocopying Prohibited

LINK
50

WEB
49

KNOW

50 Enzymes

Key Idea: Enzymes are biological catalysts. They speed up biological reactions by increasing the number of successful collisions between reactants.

Most enzymes are proteins. They are called biological catalysts because they speed up biochemical reactions, but the enzyme itself remains unchanged. They may break down a single substrate molecule into simpler substances, or join two or more substrate molecules together.

The Active Site

Enzymes have an **active site** to which specific substrates bind. The shape and chemistry of the active site is specific to a enzyme, and is a function of the polypeptide's complex tertiary structure.

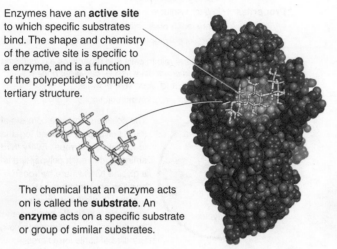

The chemical that an enzyme acts on is called the **substrate**. An **enzyme** acts on a specific substrate or group of similar substrates.

Enzymes Catalyse Metabolic Reactions

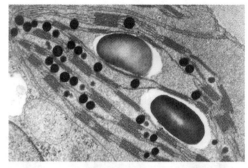

In the first stage of photosynthesis, enzymes catalyse the production of ATP and NADPH. These provide the energy and hydrogen molecules for the second stage of photosynthesis. Enzymes also catalyse the steps that fix carbon from CO_2 to produce carbohydrates.

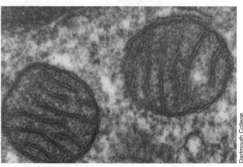

Dartmouth College

The breakdown of glucose is catalysed by several enzymes. Glycolysis uses ten different enzymes, one for each step of the process. The Krebs cycle involves another eight enzymes. The electron transport chain moves H^+ across a membrane to create the proton gradient required to produce ATP using the enzyme ATP synthase.

Substrates Collide with an Enzyme's Active Site

For a reaction to occur reactants must collide with sufficient speed and with the correct orientation. Enzymes enhance reaction rates by providing a site for reactants to come together in such a way that a reaction will occur. They do this by orientating the reactants so that the reactive regions are brought together. They may also destabilize the bonds within the reactants making it easier for a reaction to occur.

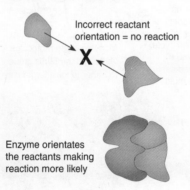

Incorrect reactant orientation = no reaction

Enzyme orientates the reactants making reaction more likely

1. (a) What is meant by the **active site** of an enzyme and relate it to the enzyme's tertiary structure: _____

_____

_____

_____

   (b) Why do enzymes usually only work on one substrate (or group of closely related substrates)? _____

_____

_____

2. How do substrate molecules come into contact with an enzyme's active site? _____

_____

3. Using examples, explain the role of enzymes in metabolic processes: _____

_____

_____

_____

_____

© 2012-2014 **BIOZONE** International
ISBN: 978-1-927173-93-0
Photocopying Prohibited

51 Enzyme Reaction Rates

Key Idea: Enzymes operate most effectively within a narrow range of conditions. The rate of enzyme catalysed reactions is influenced by both enzyme and substrate concentration. Enzymes usually have an optimum set of conditions (e.g. of pH and temperature) under which their activity is greatest. Many plant and animal enzymes show little activity at low temperatures. Enzyme activity increases with increasing temperature, but falls off after the optimum temperature is exceeded and the enzyme is denatured. Extremes in pH can also cause denaturation. Within their normal operating conditions, enzyme reaction rates are influenced by enzyme and substrate concentration in a predictable way (below).

In the graphs below, the rate of reaction or degree of enzyme activity is plotted against each of four factors that affect enzyme performance. Answer the questions relating to each graph:

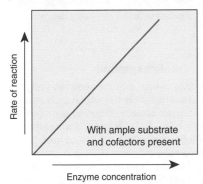

With ample substrate and cofactors present

Rate of reaction / Enzyme concentration

1. Enzyme Concentration

(a) Describe the change in the rate of reaction when the enzyme concentration is increased (assuming substrate is not limiting):

_____

_____

(b) Suggest how a cell may vary the amount of enzyme present in a cell:

_____

_____

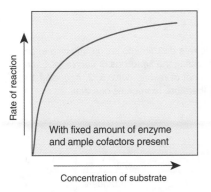

With fixed amount of enzyme and ample cofactors present

Rate of reaction / Concentration of substrate

2. Substrate Concentration

(a) Describe the change in the rate of reaction when the substrate concentration is **increased** (assuming a fixed amount of enzyme):

_____

_____

(b) Explain why the rate changes the way it does: _____

_____

_____

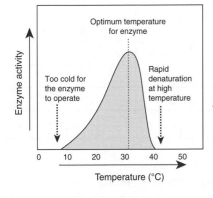

Optimum temperature for enzyme

Too cold for the enzyme to operate

Rapid denaturation at high temperature

Enzyme activity / Temperature (°C)

0   10   20   30   40   50

3. Temperature

Higher temperatures speed up all reactions, but few enzymes can tolerate temperatures higher than 50–60°C. The rate at which enzymes are **denatured** (change their shape and become inactive) increases with higher temperatures.

(a) Describe what is meant by an *optimum temperature* for enzyme activity:

_____

_____

(b) Explain why most enzymes perform poorly at low temperatures:

_____

_____

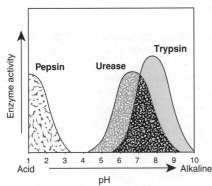

Pepsin   Urease   Trypsin

Enzyme activity

1   2   3   4   5   6   7   8   9   10
Acid ————————————→ Alkaline
pH

4. Acidity or Alkalinity (pH)

Like all proteins, enzymes are **denatured** by *extremes* of **pH** (very acid or alkaline). Within these extremes, most enzymes are still influenced by pH. Each enzyme has a preferred pH range for optimum activity.

(a) State the optimum pH for each of the enzymes:

Pepsin: _____ Trypsin: _____ Urease: _____

(b) Pepsin acts on proteins in the stomach. Explain how its optimum pH is suited to its working environment:

_____

_____

LINK   WEB
52   51   SKILL

52 Investigating Catalase Activity

Key Idea: Catalase activity in germinating seeds changes with time.

Catalase is an enzyme that catalyses the breakdown of hydrogen peroxide to water and oxygen. This activity describes an experiment to investigate the effects of germination age on the level of catalase activity in mung beans. Completing this activity will help to design and evaluate your own experiment into enzyme activity.

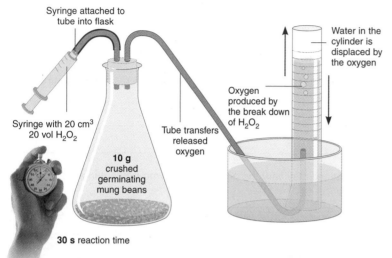

Syringe attached to tube into flask

Syringe with 20 cm^3 20 vol H$_2$O$_2$

10 g crushed germinating mung beans

30 s reaction time

Tube transfers released oxygen

Oxygen produced by the break down of H$_2$O$_2$

Water in the cylinder is displaced by the oxygen

The Aim

To investigate the effect of germination age on the level of catalase activity in mung beans.

Background

Germinating seeds are metabolically very active and this metabolism inevitably produces reactive oxygen species, including hydrogen peroxide (H$_2$O$_2$). H$_2$O$_2$ is helps germination by breaking dormancy, but it is also toxic. To counter the toxic effects of H$_2$O$_2$ and prevent cellular damage, germinating seeds also produce **catalase**, an enzyme that catalyses the breakdown of H$_2$O$_2$ to water and oxygen.

The Apparatus

In this experiment, 10 g germinating mung bean seeds (0.5, 2, 4, 6, or 10 days old) were ground by hand with a mortar and pestle and placed in a conical flask as above. There were six trials at each of the five seedling ages. With each trial, 20 cm^3 of 20 vol H$_2$O$_2$ was added to the flask at time 0 and the reaction was allowed to run for 30 seconds. The oxygen released by the decomposition of the H$_2$O$_2$ by catalase in the seedlings was collected via a tube into an inverted measuring cylinder. The volume of oxygen produced is measured by the amount of water displaced from the cylinder. The results from all trials are tabulated below:

A class was divided into six groups with each group testing the seedlings of each age. Each group's set of results (for 0.5, 2, 4, 6, and 10 days) therefore represents one trial.

Stage of germination (days) \ Trial #	Volume of oxygen collected after 30s (cm^3)						Mean	Standard deviation	Mean rate (cm^3 s^{-1} g^{-1})
	1	2	3	4	5	6			
0.5	9.5	10	10.7	9.5	10.2	10.5			
2	36.2	30	31.5	37.5	34	40			
4	59	66	69	60.5	66.5	72			
6	39	31.5	32.5	41	40.3	36			
10	20	18.6	24.3	23.2	23.5	25.5			

1. Write the equation for the catalase reaction with hydrogen peroxide: _____

2. Complete the table above to summarize the data from the six trials:

   (a) Calculate the mean volume of oxygen for each stage of germination and enter the values in the table.

   (b) Calculate the standard deviation for each mean and enter the values in the table (you may use a spreadsheet).

   (c) Calculate the mean rate of oxygen production in cm^3 per second per gram. For the purposes of this exercise, assume that the weight of germinating seed in every case was 10.0 g.

3. In another scenario, group (trial) #2 obtained the following measurements for volume of oxygen produced: 0.5 d: 4.8 cm^3, 2 d: 29.0 cm^3, 4 d: 70 cm^3, 6 d: 30.0 cm^3, 10 d: 8.8 cm^3 (pencil these values in beside the other group 2 data set).

   (a) Describe how group 2's new data accords with the measurements obtained from the other groups: _____

   _____

   (b) Describe how you would approach a reanalysis of the data set incorporating group 2's new data: _____

   _____

   _____

© 2012-2014 **BIOZONE** International
ISBN: 978-1-927173-93-0
Photocopying Prohibited

(c) Explain the rationale for your approach _____

_____

4. Use the tabulated data to plot an appropriate graph of the results on the grid provided below:

5. (a) Describe the trend in the data: _____

_____

_____

(b) Explain the relationship between stage of germination and catalase activity shown in the data: _____

_____

_____

6. Describe any potential sources of errors in the apparatus or the procedure: _____

_____

_____

7. Describe two things that might affect the validity of findings in this experimental design: _____

_____

_____

8. Describe one improvement you could make to the experiment in order to generate more reliable data: _____

_____

_____

_____

53 Applications of Enzymes

Key Idea: Immobilized lactase is used to remove lactose from milk, making it suitable for people with lactose intolerance. Milk is a high quality food containing protein, fat, carbohydrate, minerals, and vitamins. Many people become lactose intolerant (cannot digest lactose) as they grow older. They avoid milk products, and lose out on the benefits of milk. Removing the lactose from milk allows lactose intolerant people to gain the nutritional benefits from milk.

Removing Lactose from Milk

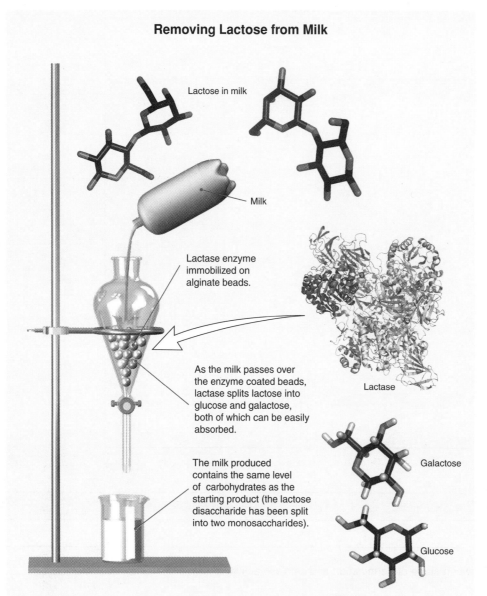

Lactose in milk

Milk

Lactase enzyme immobilized on alginate beads.

As the milk passes over the enzyme coated beads, lactase splits lactose into glucose and galactose, both of which can be easily absorbed.

Lactase

The milk produced contains the same level of carbohydrates as the starting product (the lactose disaccharide has been split into two monosaccharides).

Galactose

Glucose

Lactase and Humans

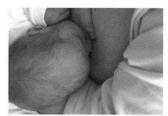

Lactose is a disaccharide found in milk. It is less sweet than glucose. All infant humans produce the enzyme **lactase**, which hydrolyses lactose into glucose and galactose.

As humans become older, their production of lactase gradually declines and they lose their ability to hydrolyse lactose. As adults, they are **lactose intolerant**, and feel bloated after drinking milk.

In humans of mainly European, East African, or Indian descent, lactase production continues into adulthood. But people of mainly Asian descent cease production early in life and become lactose intolerant.

1. Explain why being able to continue to drink milk throughout life is of benefit to humans: _____

_____

_____

_____

2. How is lactase used to produce lactose-free milk? _____

_____

_____

_____

3. Why does lactose-free milk often have a slightly sweeter taste that ordinary milk? _____

_____

_____

© 2012-2014 **BIOZONE** International
ISBN: 978-1-927173-93-0
Photocopying Prohibited

54 Nucleotides and Nucleic Acids

Key Idea: Nucleotides are the building blocks of DNA and RNA. Nucleic acids are long chains of nucleotides which store and transmit genetic information.

A nucleotide has three components: a base, a sugar, and a phosphate group. They are the building blocks of nucleic acids (DNA and RNA), which are involved in the transmission of inherited information. Nucleic acids have the capacity to store the information that controls cellular activity. The central nucleic acid is called deoxyribonucleic acid (DNA). Ribonucleic acids (RNA) are involved in the 'reading' of the DNA information. All nucleic acids are made up of nucleotides linked together to form chains or strands. The strands vary in the sequence of the bases found on each nucleotide. It is this sequence which provides the 'genetic instructions' for the cell.

Chemical Structure of a Nucleotide

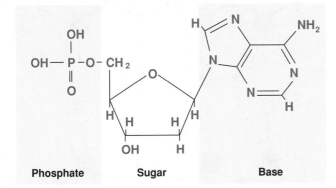

Phosphate   Sugar   Base

Symbolic Form of a Nucleotide

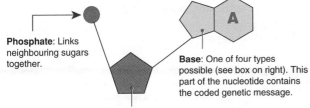

Phosphate: Links neighbouring sugars together.

Base: One of four types possible (see box on right). This part of the nucleotide contains the coded genetic message.

Sugar: One of two types possible: ribose in RNA and deoxyribose in DNA.

Nucleotides are the building blocks of DNA. Their precise sequence in a DNA molecule provides the genetic instructions for the organism to which it governs. Accidental changes in nucleotide sequences are a cause of mutations, usually harming the organism, but occasionally providing benefits.

Bases

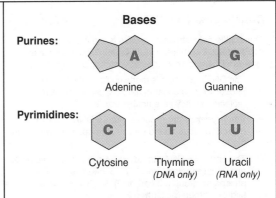

Purines: Adenine   Guanine

Pyrimidines: Cytosine   Thymine *(DNA only)*   Uracil *(RNA only)*

The two-ringed bases above are **purines**. The single-ringed bases are **pyrimidines**. Although only one of four kinds of base can be used in a nucleotide, **uracil** is found only in RNA, replacing **thymine**. DNA contains A, T, G, and C, while RNA contains A, U, G, and C.

Sugars

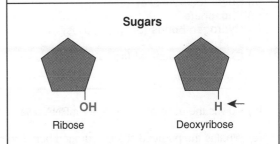

Ribose   Deoxyribose

Deoxyribose sugar is found only in DNA. It differs from **ribose** sugar, found in RNA, by the lack of a single oxygen atom (arrowed).

RNA Molecule

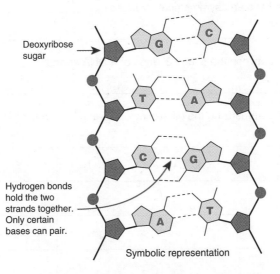

In RNA, uracil replaces thymine in the code.

Ribose sugar

DNA Molecule

Deoxyribose sugar

Hydrogen bonds hold the two strands together. Only certain bases can pair.

Symbolic representation

DNA Molecule

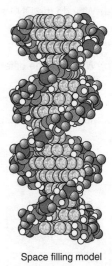

Space filling model

Ribonucleic acid (RNA) comprises a *single strand* of nucleotides linked together.

Deoxyribonucleic acid (DNA) comprises a *double strand* of nucleotides linked together. It is shown unwound in the symbolic representation (left). The DNA molecule takes on a twisted, double-helix shape as shown in the space filling model on the right.

LINK 59   LINK 56   LINK 55   WEB 54

KNOW

Formation of a nucleotide

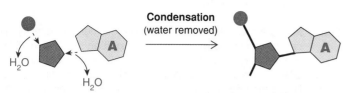

Condensation
(water removed)

A nucleotide is formed when phosphoric acid and a base are chemically bonded to a sugar molecule. In both cases, water is given off, and they are therefore condensation reactions. In the reverse reaction, a nucleotide is broken apart by the addition of water (**hydrolysis**).

Formation of a dinucleotide

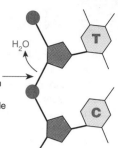

Two nucleotides are linked together by a condensation reaction between the phosphate of one nucleotide and the sugar of another.

Double-Stranded DNA

The double-helix structure of DNA is like a ladder twisted into a corkscrew shape around its longitudinal axis. It is 'unwound' here to show the relationships between the bases.

- The DNA backbone is made up from alternating phosphate and sugar molecules, giving the DNA molecule an asymmetrical structure.

- The asymmetrical structure gives a DNA strand a **direction**. Each strand runs in the opposite direction to the other.

- The ends of a DNA strand are labelled the 5' (five prime) and 3' (three prime) ends. The **5'** end has a terminal phosphate group (off carbon 5), the **3'** end has a terminal hydroxyl group (off carbon 3).

- The way the pairs of bases come together to form hydrogen bonds is determined by the number of bonds they can form and the configuration of the bases.

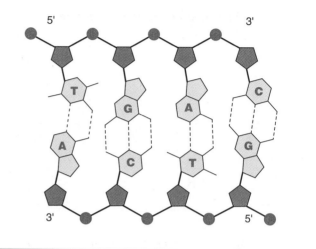

---

1. The diagram above depicts a double-stranded DNA molecule. Label the following parts on the diagram:

   (a) **Sugar** (deoxyribose)
   (b) **Phosphate**
   (c) **Hydrogen bonds** (between bases)
   (d) **Purine** bases
   (e) **Pyrimidine** bases

2. (a) Explain the **base-pairing rule** that applies in double-stranded DNA: _____

   _____

   (b) How is the base-pairing rule for mRNA different? _____

   (c) What is the purpose of the hydrogen bonds in double-stranded DNA? _____

   _____

3. Describe the functional role of nucleotides: _____

   _____

4. (a) Why do the DNA strands have an asymmetrical structure? _____

   _____

   (b) What are the differences between the 5' and 3' ends of a DNA strand? _____

   _____

5. Complete the following table summarizing the differences between DNA and RNA molecules:

	DNA	RNA
Sugar present		
Bases present		
Number of strands		
Relative length		

© 2012-2014 **BIOZONE** International
ISBN: 978-1-927173-93-0

55 Creating a DNA Model

Key Idea: Nucleotides pair together in a specific way called the base pairing rule. In DNA, adenine always pairs with thymine, and cytosine always pairs with guanine. DNA molecules are double stranded. Each strand is made up of nucleotides. The chemical properties of each nucleotide mean it can only bind with a one other type of nucleotide. This is called the **base pairing rule** and is explained in the table below. This exercise will help you to learn this rule.

DNA Base Pairing Rule			
Adenine	is always attracted to	**Thymine**	A ⟷ T
Thymine	is always attracted to	**Adenine**	T ⟷ A
Cytosine	is always attracted to	**Guanine**	C ⟷ G
Guanine	is always attracted to	**Cytosine**	G ⟷ C

1. Cut out page 73 and separate each of the 24 nucleotides by cutting along the columns and rows (see arrows indicating cutting points). Although drawn as geometric shapes, these symbols represent chemical structures.

2. Place one of each of the four kinds of nucleotide on their correct spaces below:

Place a cut-out symbol for **thymine** here

Thymine

Place a cut-out symbol for **cytosine** here

Cytosine

Place a cut-out symbol for **adenine** here

Adenine

Place a cut-out symbol for **guanine** here

Guanine

3. Identify and **label** each of the following features on the *adenine* nucleotide immediately above: **phosphate, sugar, base, hydrogen bonds**

4. Create one strand of the DNA molecule by placing the 9 correct 'cut out' nucleotides in the labelled spaces on the following page (DNA molecule). Make sure these are the right way up (with the **P** on the left) and are aligned with the left hand edge of each box. Begin with thymine and end with guanine.

5. Create the complementary strand of DNA by using the base pairing rule above. Note that the nucleotides have to be arranged upside down.

6. Under normal circumstances, it is not possible for adenine to pair up with guanine or cytosine, nor for any other mismatches to occur. Describe the two factors that prevent a mismatch from occurring:

(a) Factor 1: _____

_____

(b) Factor 2: _____

_____

7. Once you have checked that the arrangement is correct, you may glue, paste or tape these nucleotides in place.

NOTE:	There may be some value in keeping these pieces loose in order to practice the base pairing rule. For this purpose, *removable tape* would be best.

SKILL

DNA Molecule

Put the named nucleotides on the left hand side to create the template strand

Put the matching **complementary** nucleotides opposite the template strand

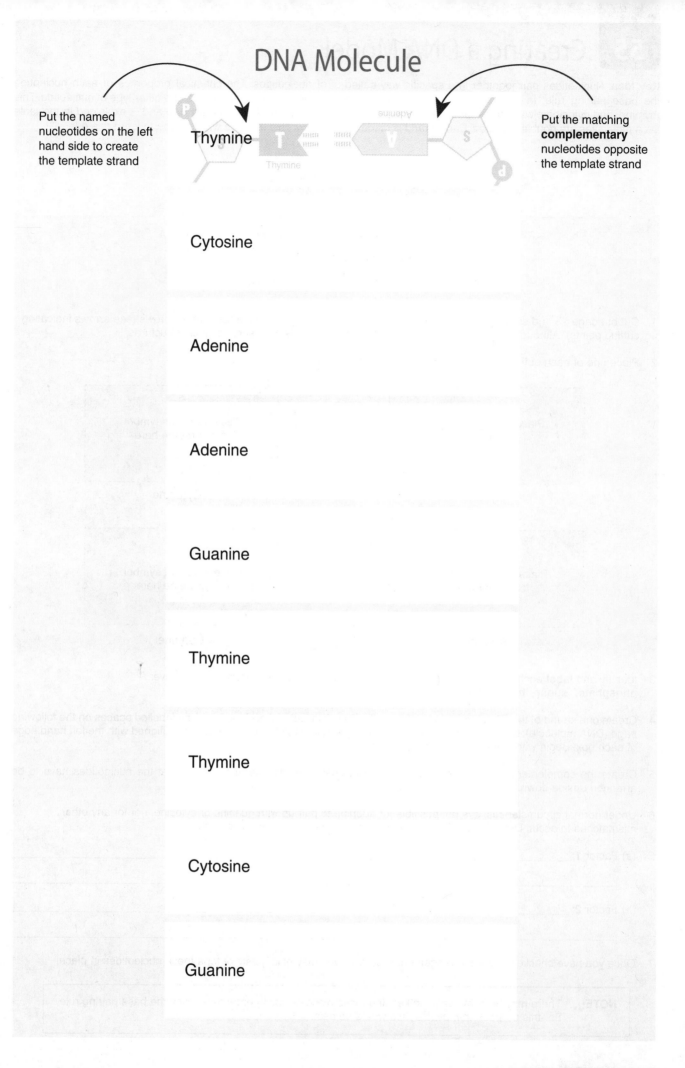

Thymine

Cytosine

Adenine

Adenine

Guanine

Thymine

Thymine

Cytosine

Guanine

Nucleotides

Tear out this page and separate each of the 24 nucleotides
by cutting along the columns and rows (see arrows indicating the cutting points).

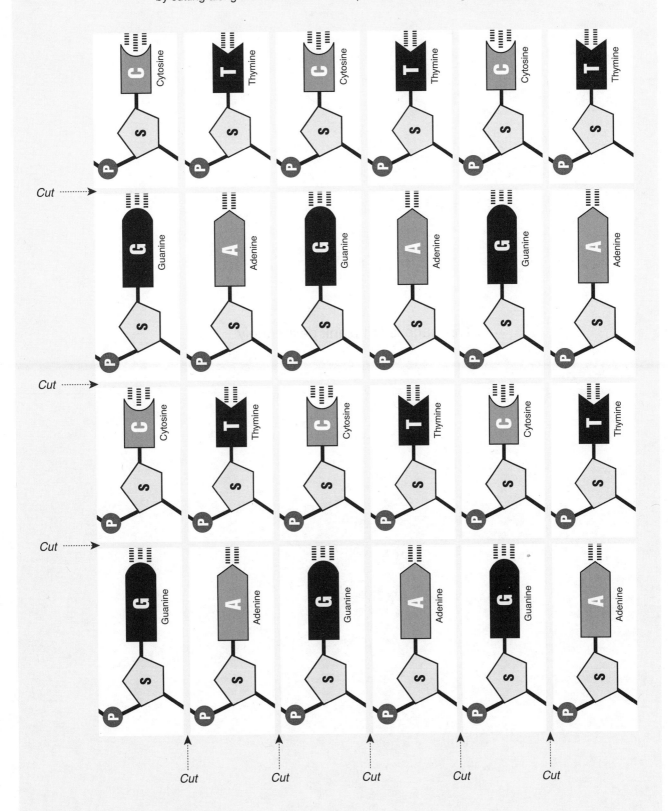

Cut ⟶

Cut ⟶

Cut ⟶

Cut   *Cut*   *Cut*   *Cut*   *Cut*

This page is left blank deliberately

56 DNA Replication

Key Idea: Semi conservative DNA replication produces two identical copies of DNA, each containing half original material and half new material. Before a cell can divide, it must double its DNA. It does this by a process called DNA replication. This process ensures that each resulting cell receives a complete set of genes from the original cell. After the DNA has replicated, each chromosome is made up of two chromatids, joined at the centromere. The two chromatids will become separated during cell division to form two separate chromosomes. During DNA replication, nucleotides are added at the replication fork. Enzymes are responsible for all of the key events.

Step 1
Unwinding the DNA molecule

A normal chromosome consists of an unreplicated DNA molecule. Before cell division, this long molecule of double stranded DNA must be replicated.

For this to happen, it is first untwisted and separated (unzipped) at high speed at its replication fork by an enzyme called helicase. Another enzyme relieves the strain that this generates by cutting, winding and rejoining the DNA strands.

Step 2
Making new DNA strands

The formation of new DNA is carried out mostly by an enzyme complex called **DNA polymerase**.

DNA polymerase works in a 5' to 3' direction so nucleotides are assembled in a continuous fashion on one strand but in short fragments on the other strand. These fragments are later joined by an enzyme to form one continuous length.

Step 3
Rewinding the DNA molecule

Each of the two new double-helix DNA molecules has one strand of the original DNA (dark gray and white) and one strand that is newly synthesized (blue). The two DNA molecules rewind into their double-helix shape again.

DNA replication is semi-conservative, with each new double helix containing one old (parent) strand and one newly synthesized (daughter) strand. The new chromosome has twice as much DNA as a non-replicated chromosome. The two chromatids will become separated in the cell division process to form two separate chromosomes.

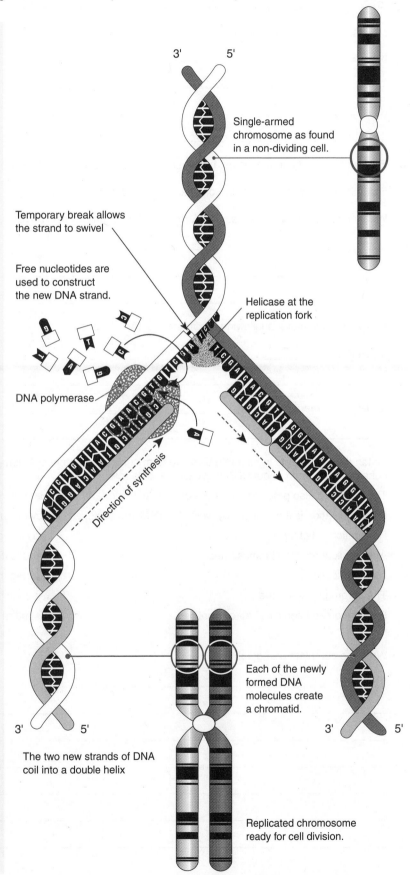

Single-armed chromosome as found in a non-dividing cell.

Temporary break allows the strand to swivel

Free nucleotides are used to construct the new DNA strand.

Helicase at the replication fork

DNA polymerase

Direction of synthesis

Each of the newly formed DNA molecules create a chromatid.

The two new strands of DNA coil into a double helix

Replicated chromosome ready for cell division.

1. What is the purpose of DNA replication? _____

_____

_____

2. Summarize the three main steps involved in DNA replication:

(a) _____

_____

(b) _____

_____

(c) _____

_____

3. For a cell with 22 chromosomes, how many chromatids would exist following DNA replication? _____

4. State the percentage of new and original DNA in each daughter cell: _____

5. What does it mean when we say DNA replication is **semi-conservative**? _____

_____

_____

_____

6. Explain the roles of the following enzymes in DNA replication:

(a) Helicase: _____

_____

(b) DNA polymerase: _____

_____

7. Match the statements in the table below to form complete sentences, then put the sentences in order to make a coherent paragraph about DNA replication and its role:

The enzymes also proofread the DNA during replication...	...is required before mitosis can occur.
DNA replication is the process by which the DNA molecule...	...by enzymes.
Replication is tightly controlled...	...to correct any mistakes.
After replication, the chromosome...	...and half new DNA.
DNA replication...	...during mitosis.
The chromatids separate...	...is copied to produce two identical DNA strands.
Each chromatid contains half original ...	...is made up of two chromatids.

Write the complete paragraph here:

_____

_____

_____

_____

_____

_____

© 2012-2014 **BIOZONE** International
ISBN: 978-1-927173-93-0
Photocopying Prohibited

57 Meselson and Stahl's Experiment

Key Idea: Meselson and Stahl devised an experiment that showed DNA replication is semi-conservative.

Three models were proposed to explain how DNA replicated. Watson and Crick proposed the **semi-conservative model** in which each DNA strand served as a template, forming a new DNA molecule that was half old and half new DNA. The conservative model proposed that the original DNA served as a complete template so that the resulting DNA was completely new. The dispersive model proposed that the two new DNA molecules had part new and part old DNA interspersed throughout them. Meselson and Stahl's experiment confirmed that DNA replication is semi-conservative.

Meselson and Stahl's Experiment

E. coli were grown for several generations in a medium containing a **heavy nitrogen isotope** (^{15}N) and transferred to a medium containing a **light nitrogen isotope** (^{14}N) once all the bacterial DNA contained ^{15}N. Newly synthesized DNA would contain ^{14}N, old DNA would contain ^{15}N.

1

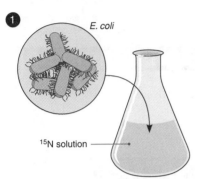

E. coli were grown in a nutrient solution containing ^{15}N. After 14 generations, all the bacterial DNA contained ^{15}N. A sample is removed. This is **generation 0**.

2

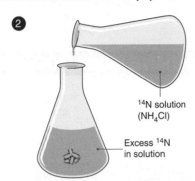

Generation 0 is added to a solution containing excess ^{14}N (as NH_4Cl). During replication, new DNA will incorporate ^{14}N and be 'lighter' than the original DNA (which contains only ^{15}N).

3

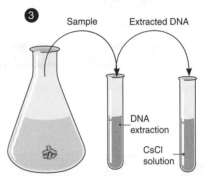

Every generation (~ 20 minutes), a sample is taken and treated to release the DNA. The DNA is placed in a CsCl solution which provides a density gradient for separation of the DNA.

4

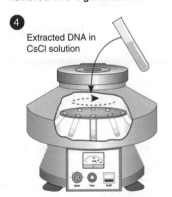

Samples are spun in a high speed ultracentrifuge at 140,000 *g* for 20 hours. Heavier ^{15}N DNA moves closer to the bottom of the test tube than light ^{14}N DNA or $^{14}N/^{15}N$ intermediate DNA.

5

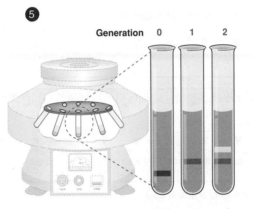

All the DNA in the generation 0 sample moved to the bottom of the test tube. All the DNA in the generation 1 sample moved to an intermediate position. At generation 2 half the DNA was at the intermediate position and half was near the top of the test tube. In subsequent generations, more DNA was near the top and less was in the intermediate position.

6

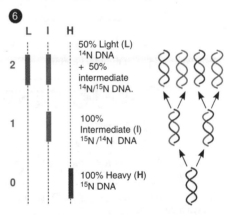

All of the generation 1 DNA contained one light strand (^{14}N) and one heavy (^{15}N) strand to produce an intermediate density. At generation 2, 50% of the DNA was light and 50% was intermediate DNA. This combination of light and intermediate (hybrid) DNA confirmed the semi conservative replication model.

1. Explain why Meselson and Stahl's experiment supports the semi-conservative replication model: _____

_____

_____

_____

_____

2. Identify the replication model that fits the following hypothetical data:

(a) 100% of generation 0 is "heavy DNA", 50% of generation 1 is "heavy" and 50% is "light", and 25% of generation 2 is "heavy" and 75% is "light":

_____

(b) 100% of generation 0 is "heavy DNA", 100% of generation 1 is "intermediate DNA", and 100% generation 2 lies between the "intermediate" and "light" DNA regions:

_____

58 Genes to Proteins

Key Idea: Genes are sections of DNA that code for proteins. Genes are expressed when they are transcribed into messenger RNA (mRNA) and then translated into a protein. **Gene expression** is the process of rewriting a gene into a protein. It involves **transcription** of the DNA into mRNA and **translation** of the mRNA into protein. A gene is bounded by a start (promoter) region, upstream of the gene, and a terminator region, downstream of the gene. These regions control transcription by telling RNA polymerase where to start and stop transcription of the gene. The information flow for gene to protein is shown below. Nucleotides are read in groups of three called triplets. The equivalent on the mRNA molecule is the codon. Some codons have a special control functions (start and stop) in the making of a protein.

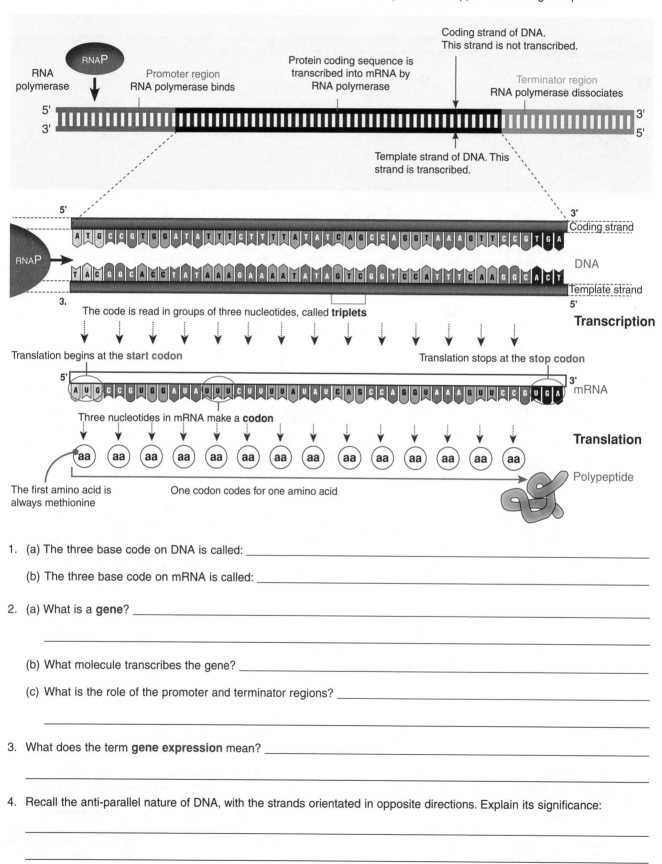

1. (a) The three base code on DNA is called: _____

    (b) The three base code on mRNA is called: _____

2. (a) What is a **gene**? _____

    _____

    (b) What molecule transcribes the gene? _____

    (c) What is the role of the promoter and terminator regions? _____

    _____

3. What does the term **gene expression** mean? _____

    _____

4. Recall the anti-parallel nature of DNA, with the strands orientated in opposite directions. Explain its significance:

    _____

    _____

© 2012-2014 **BIOZONE** International
ISBN: 978-1-927173-93-0
Photocopying Prohibited

59 The Genetic Code

Key Idea: The genetic code is the set of rules by which the genetic information in DNA or mRNA is translated into proteins. The genetic information for the assembly of amino acids is stored as three-base sequence. These three letter codes on mRNA are called **codons**. Each codon represents one of 20 amino acids used to make proteins. The code is effectively universal, being the same in all living things (with a few minor exceptions). The genetic code is summarized in a mRNA-amino acid table, which identifies the amino acid encoded by each mRNA codon. The code is degenerate, meaning there may be more than one codon for each amino acid. Most of this degeneracy is in the third nucleotide of a codon.

1. (a) Use the base-pairing rule for to create the complementary strand for the DNA template strand shown.

   (b) For the same DNA strand, determine the mRNA sequence and then use the mRNA–amino acid table to determine the corresponding amino acid sequence. Note that in mRNA, uracil (U) replaces thymine (T) and pairs with adenine.

Template strand

DNA | T A C | C C A | A T G | G A C | T C C | C A T | T A T | G C C | C G T | G A A | A T C |

Complementary strand

Gene expression

Template strand

DNA | T A C | C C A | A T G | G A C | T C C | C A T | T A T | G C C | C G T | G A A | A T C |

Transcription

mRNA

Translation

Amino acids

mRNA - Amino Acid Table

The table on the right is used to 'decode' the genetic code. It shows which amino acid each mRNA codon codes for. There are 64 different codons possible, 61 code for amino acids, and three are stop codons.

Amino acid names are written as three letter abbreviations (e.g. Ser = serine). To work out which amino acid a codon codes for, carry out the following steps:

i   Find the first letter of the codon in the row on the left hand side of the table. AUG is the start codon.

ii Find the column that intersects that row from the top, second letter, row.

iii Locate the third base in the codon by looking along the row on the right hand side that matches your codon.

E.g. **GAU** codes for Asp (aspartic acid)

Read second letter here
Read first letter here
Read third letter here

		Second Letter				
		U	**C**	**A**	**G**	
First Letter	**U**	UUU Phe UUC Phe UUA Leu UUG Leu	UCU Ser UCC Ser UCA Ser UCG Ser	UAU Tyr UAC Tyr UAA STOP UAG STOP	UGU Cys UGC Cys UGA STOP UGG Trp	U C A G
	C	CUU Leu CUC Leu CUA Leu CUG Leu	CCU Pro CCC Pro CCA Pro CCG Pro	CAU His CAC His CAA Gln CAG Gln	CGU Arg CGC Arg CGA Arg CGG Arg	U C A G
	A	AUU Ile AUC Ile AUA Ile AUG Met	ACU Thr ACC Thr ACA Thr ACG Thr	AAU Asn AAC Asn AAA Lys AAG Lys	AGU Ser AGC Ser AGA Arg AGG Arg	U C A G
	G	GUU Val GUC Val GUA Val GUG Val	GCU Ala GCC Ala GCA Ala GCG Ala	GAU Asp GAC Asp GAA Glu GAG Glu	GGU Gly GGC Gly GGA Gly GGG Gly	U C A G

Third Letter

2. (a) State the mRNA START and STOP codons: _____

   (b) Give an example to illustrate the degeneracy of the genetic code: _____

   _____

LINK 210 LINK 60 **SKILL**

60 Transcription and Translation

Key Idea: In eukaryotic cells, transcription occurs in the nucleus and translation occurs in the cytoplasm. The translation phase is carried out by ribosomes.
As we have seen, in order for a gene in DNA to be expressed, its base sequence must be transcribed into mRNA and then translated into a functional product such as a protein.

Ribosomes are responsible for catalysing the synthesis of polypeptides, linking amino acids in the order specified by the mRNA. Transfer RNA (tRNA) molecules, which have an anticodon region complementary to the codons on mRNA, bring amino acids into position within the ribosome-mRNA complex, where they are joined to form a polypeptide chain.

Gene Expression in Eukaryotes

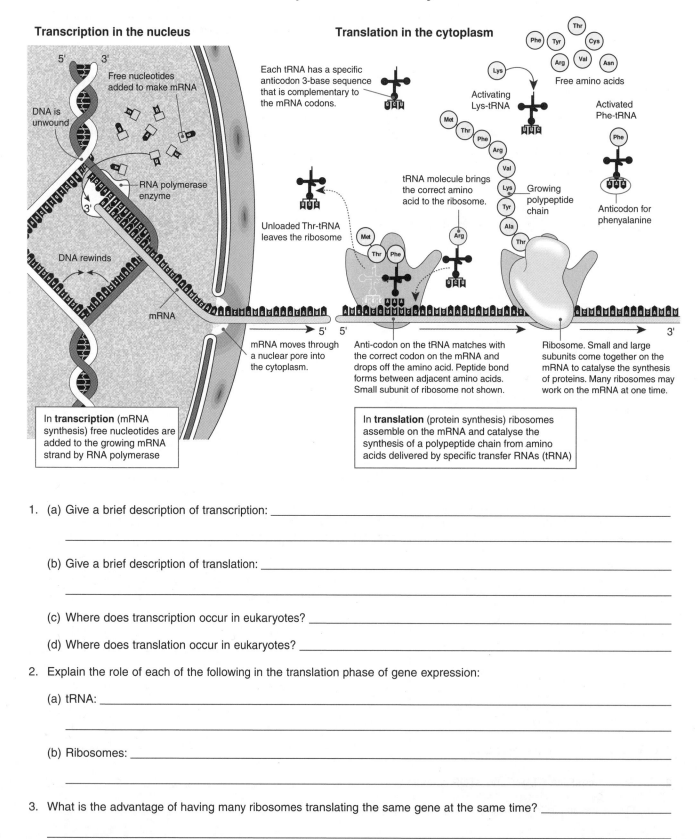

Transcription in the nucleus

5' 3'
Free nucleotides added to make mRNA
DNA is unwound
RNA polymerase enzyme
3'
DNA rewinds
mRNA

In **transcription** (mRNA synthesis) free nucleotides are added to the growing mRNA strand by RNA polymerase

Translation in the cytoplasm

Each tRNA has a specific anticodon 3-base sequence that is complementary to the mRNA codons.

Unloaded Thr-tRNA leaves the ribosome

tRNA molecule brings the correct amino acid to the ribosome.

Free amino acids

Activating Lys-tRNA

Activated Phe-tRNA

Growing polypeptide chain

Anticodon for phenyalanine

mRNA moves through a nuclear pore into the cytoplasm.

Anti-codon on the tRNA matches with the correct codon on the mRNA and drops off the amino acid. Peptide bond forms between adjacent amino acids. Small subunit of ribosome not shown.

Ribosome. Small and large subunits come together on the mRNA to catalyse the synthesis of proteins. Many ribosomes may work on the mRNA at one time.

In **translation** (protein synthesis) ribosomes assemble on the mRNA and catalyse the synthesis of a polypeptide chain from amino acids delivered by specific transfer RNAs (tRNA)

1. (a) Give a brief description of transcription: _____

_____

(b) Give a brief description of translation: _____

_____

(c) Where does transcription occur in eukaryotes? _____

(d) Where does translation occur in eukaryotes? _____

2. Explain the role of each of the following in the translation phase of gene expression:

(a) tRNA: _____

_____

(b) Ribosomes: _____

_____

3. What is the advantage of having many ribosomes translating the same gene at the same time? _____

_____

© 2012-2014 **BIOZONE** International
ISBN: 978-1-927173-93-0
Photocopying Prohibited

61 The Role of ATP in Cells

Key Idea: ATP transports chemical energy within the cell for use in metabolic processes.

All organisms require energy to be able to perform the metabolic processes required for them to function and reproduce. This energy is obtained by cell respiration, a set of metabolic reactions which ultimately convert biochemical energy from 'food' into the nucleotide adenosine triphosphate (ATP). ATP is considered to be a universal energy carrier, transferring chemical energy within the cell for use in metabolic processes such as biosynthesis, cell division, cell signalling, thermoregulation, cell mobility, and active transport of substances across membranes.

The Structure of ATP

The ATP molecule is a nucleotide derivative. It consists of three components; a purine base (adenine), a pentose sugar (ribose), and three phosphate groups, which attach to the 5' carbon of the pentose sugar. The structure of ATP is shown as a schematic below and as a three dimensional structure (right).

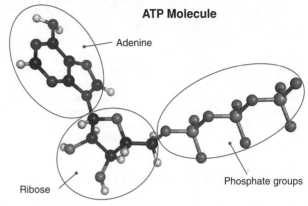

ATP Molecule

Adenine + ribose = adenosine

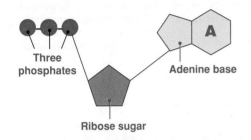

Three phosphates

Adenine base

Ribose sugar

How does ATP Provide Energy?

ATP releases its energy during hydrolysis. Water is split and added to the terminal phosphate group resulting in ADP and Pi. For every mole of ATP hydrolysed **30.7 kJ** of energy is released. Note that energy is released during the formation of chemical bonds not from the breaking of chemical bonds.

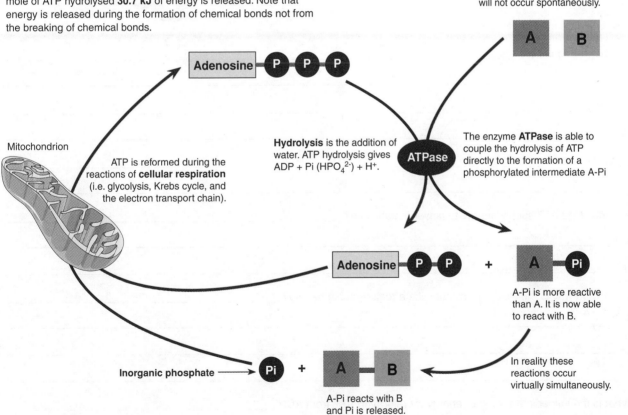

The reaction of A + B is endergonic. It requires energy to proceed and will not occur spontaneously.

Adenosine — P P P

Mitochondrion

ATP is reformed during the reactions of **cellular respiration** (i.e. glycolysis, Krebs cycle, and the electron transport chain).

Hydrolysis is the addition of water. ATP hydrolysis gives $ADP + Pi (HPO_4^{2-}) + H^+$.

ATPase

The enzyme **ATPase** is able to couple the hydrolysis of ATP directly to the formation of a phosphorylated intermediate A-Pi

Adenosine — P P + A — Pi

A-Pi is more reactive than A. It is now able to react with B.

In reality these reactions occur virtually simultaneously.

Inorganic phosphate → Pi + A — B

A-Pi reacts with B and Pi is released.

Note! The phosphate bonds in ATP are often referred to as high energy bonds. This can be misleading. The bonds contain *electrons in a high energy state* (making the bonds themselves relatively weak). A small amount of energy is required to break the bonds, but when the intermediates recombine and form new chemical bonds a large amount of energy is released. The final product is less reactive than the original reactants.

In many textbooks the reaction series above is simplified and the intermediates are left out:

$$A + B \longrightarrow AB$$

ATP   ADP + Pi

LINK 66   LINK 62   WEB 61   KNOW

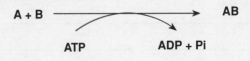

ATP Powers Metabolism

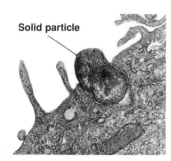

Solid particle

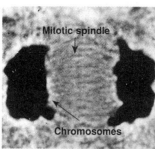

Mitotic spindle

Chromosomes

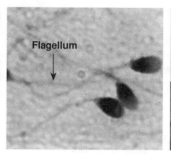

Flagellum

The energy released from the hydrolysis of ATP is used to actively transport molecules and substances across the cellular membrane. Phagocytosis (above), (the engulfment of solid particles) is one such example.

Cell division (mitosis), as observed in this onion cell, requires ATP to proceed. Formation of the mitotic spindle and chromosome separation are two aspects of cell division which require energy from ATP hydrolysis to occur.

The hydrolysis of ATP provides the energy for motile cells to achieve movement via a tail-like structure called a flagellum. For example, mammalian sperm must be able to move to the ovum to fertilize it.

Not all of the energy released in the oxidation of glucose is captured in ATP. The rest is lost as heat. This heat energy is used to maintain body temperature. Thermoregulatory mechanisms such as shivering and sweating also involve energy expenditure.

1. In which organelle is ATP produced in the cell? _____

2. Which enzyme catalyses the hydrolysis of ATP? _____

3. On the space filling model of ATP to the right:

    (a) Label the three components of an ATP molecule:

    (b) Show which phosphate bond is hydrolysed to provide the energy for cellular work:

4. How is ATP is involved in:

    (a) Thermoregulation: _____

    _____

    (b) Cell motility: _____

    _____

    (c) Cell division: _____

    _____

5. (a) How does ATP supply energy to power metabolism? _____

    _____

    _____

    (b) In what way is the ADP/ATP system like a rechargeable battery? _____

    _____

    _____

    _____

6. What is the immediate source of energy for reforming ATP from ADP? _____

7. Explain why metabolically active cells (e.g. sperm, secretory cells, muscle fibres) have large numbers of mitochondria:

_____

_____

_____

_____

62 Cell Respiration

Key Idea: Cell respiration is the process by which the energy in glucose is transferred to ATP.

Cellular respiration can be **aerobic** (requires oxygen) or **anaerobic** (does not require oxygen). Some plants and animals can generate ATP anaerobically for short periods of time. Other organisms (anaerobic bacteria) use only anaerobic respiration and live in oxygen-free environments. Cell respiration occurs in the cytoplasm and mitochondria.

An Overview of Cell Respiration

Respiration involves three metabolic stages (plus a link reaction) summarized below. The first two stages are the catabolic pathways that decompose glucose and other organic fuels. In the third stage, the electron transport chain accepts electrons from the first two stages and passes these from one electron acceptor to another. The energy released at each stepwise transfer is used to make ATP. The final electron acceptor in this process is molecular oxygen.

> **Cellular respiration equation**
>
> Glucose + Oxygen $\longrightarrow$ Carbon dioxide + Water + Energy
>
> $C_6H_{12}O_6 + 6O_2 \longrightarrow 6CO_2 + 6H_2O + \text{Energy}$

1 **Glycolysis**. In the cytoplasm, glucose is broken down into two molecules of pyruvate.

2 **The link reaction**. Pyruvate is split and added to coenzyme A ready to enter the Krebs cycle.

3 **Krebs cycle**. In the mitochondrial matrix, a derivative of pyruvate is decomposed to CO_2.

4 **Electron transport chain**. This occurs in the inner membranes of the mitochondrion and accounts for almost 90% of the ATP generated by respiration.

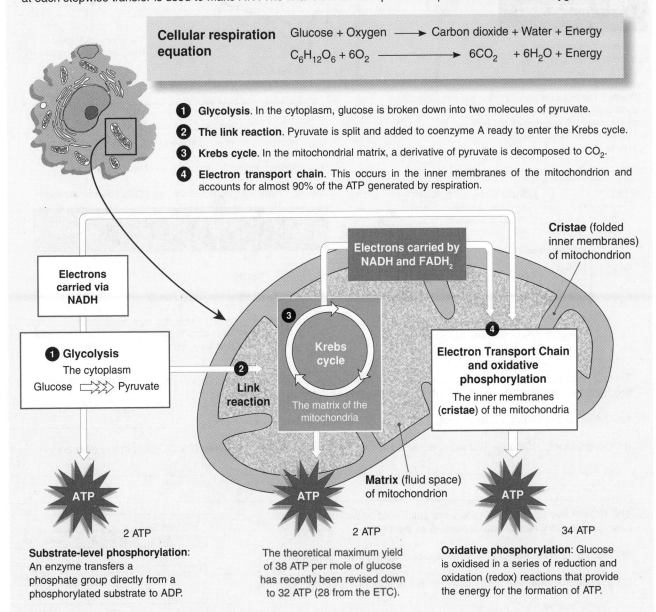

Cristae (folded inner membranes) of mitochondrion

Electrons carried by NADH and FADH$_2$

Electrons carried via NADH

1 **Glycolysis**
The cytoplasm
Glucose $\Longrightarrow$ Pyruvate

2 **Link reaction**

3 **Krebs cycle**
The matrix of the mitochondria

4 **Electron Transport Chain and oxidative phosphorylation**
The inner membranes (**cristae**) of the mitochondria

Matrix (fluid space) of mitochondrion

ATP — 2 ATP

ATP — 2 ATP

ATP — 34 ATP

Substrate-level phosphorylation: An enzyme transfers a phosphate group directly from a phosphorylated substrate to ADP.

The theoretical maximum yield of 38 ATP per mole of glucose has recently been revised down to 32 ATP (28 from the ETC).

Oxidative phosphorylation: Glucose is oxidised in a series of reduction and oxidation (redox) reactions that provide the energy for the formation of ATP.

1. What is the purpose of cell respiration? _____

2. Describe precisely in which part of the cell the following take place:

   (a) Glycolysis: _____

   (b) Krebs cycle reactions: _____

   (c) Electron transport chain: _____

3. How many ATP molecules **per molecule of glucose** are generated during the following stages of respiration?

   (a) Glycolysis: _____ (b) Krebs cycle: _____ (c) Electron transport chain: _____ (d) Total: _____

© 2012-2014 **BIOZONE** International
ISBN: 978-1-927173-93-0
Photocopying Prohibited

LINK 227   LINK 226   LINK 64   WEB 62   **KNOW**

63 Measuring Respiration

Key Idea: The respiratory quotient (RQ) provides a useful indication of the respiratory substrate being used.

In small animals or germinating seeds, the rate of cell respiration can be measured using a simple respirometer: a sealed unit where the a dioxide (CO_2) produced by the respiring tissues is absorbed by soda lime and the volume of oxygen (O_2) consumed is detected by fluid displacement in a manometer. Germinating seeds are often used to calculate the respiratory quotient (**RQ**): the ratio of the amount of CO_2 produced in cell respiration to the amount of O_2 consumed.

Respiratory Substrates and RQ

The respiratory quotient (RQ) can be expressed simply as:

$$RQ = \frac{CO_2 \text{ produced}}{O_2 \text{ consumed}}$$

When pure carbohydrate is oxidized in cellular respiration, the RQ is 1.0; more oxygen is required to oxidize fatty acids (RQ = 0.7). The RQ for protein is about 0.9. Organisms usually respire a mix of substrates, giving RQ values of between 0.8 and 0.9 (see table 1, below).

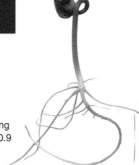

Table 1: RQ values for a range of substrates.

RQ	Substrate
> 1.0	Carbohydrate with some anaerobic respiration
1.0	Carbohydrates e.g. glucose
0.9	Protein
0.7	Fat
0.5	Fat with associated carbohydrate synthesis
0.3	Carbohydrate with associated organic acid synthesis

Using RQ to determine respiratory substrate

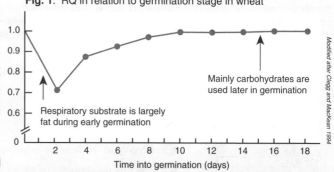

Fig. 1: RQ in relation to germination stage in wheat

Respiratory substrate is largely fat during early germination

Mainly carbohydrates are used later in germination

Time into germination (days)

Respiratory quotient

Modified after Clegg and MacKean 1994

Fig. 1, above, shows how experimental RQ values have been used to determine the respiratory substrate utilized by germinating wheat seeds (*Triticum sativum*) over the period of their germination.

Table 2: Rates of O_2 consumption and CO_2 production in crickets

Time after last fed (h)	Temperature (°C)	Rate of O_2 consumption (mL g^{-1} h^{-1})	Rate of CO_2 production (mL g^{-1} h^{-1})
1	20	2.82	2.82
48	20	2.82	1.97
1	30	5.12	5.12
48	30	5.12	3.57

Table 2 shows the rates of oxygen consumption and carbon dioxide production of crickets kept under different experimental conditions.

1. Table 2 above shows the results of an experiment to measure the rates of oxygen consumption and carbon dioxide production of crickets 1 hour and 48 hours after feeding at different temperatures:

   (a) Calculate the RQ of a cricket kept at 20°C, 48 hours after feeding (show working): _____

   (b) Compare this RQ to the RQ value obtained for the cricket 1 hour after being fed (20°C). Explain the difference:

   _____

   _____

2. The RQs of two species of seeds were calculated at two day intervals after germination. Results are tabulated to the right:

   (a) Plot the change in RQ of the two species during early germination:

   (b) Explain the values in terms of the possible substrates being respired:

   _____

   _____

   _____

   _____

   _____

   _____

   _____

   _____

Days after germination	RQ	
	Seedling A	Seedling B
2	0.65	0.70
4	0.35	0.91
6	0.48	0.98
8	0.68	1.00
10	0.70	1.00

64 Anaerobic Metabolism

Key Idea: Glucose can be metabolized aerobically and anaerobically to produce ATP. The ATP yield from aerobic processes is higher than from anaerobic processes. Aerobic respiration occurs in the presence of oxygen. Organisms can also generate ATP when oxygen is absent by using a molecule other than oxygen as the terminal electron acceptor for the pathway. In alcoholic fermentation in yeasts, the electron acceptor is ethanal. In lactic acid fermentation, which occurs in mammalian muscle even when oxygen is present, the electron acceptor is pyruvate itself.

Alcoholic Fermentation

In alcoholic fermentation, the H^+ acceptor is ethanal which is reduced to ethanol with the release of carbon dioxide (CO_2). Yeasts respire aerobically when oxygen is available but can use alcoholic fermentation when it is not. At ethanol levels above 12-15%, the ethanol produced by alcoholic fermentation is toxic and this limits their ability to use this pathway indefinitely. The root cells of plants also use fermentation as a pathway when oxygen is unavailable but the ethanol must be converted back to respiratory intermediates and respired aerobically.

Lactic Acid Fermentation

Skeletal muscles produce ATP in the absence of oxygen using lactic acid fermentation. In this pathway, pyruvate is reduced to lactic acid, which dissociates to form lactate and H^+. The conversion of pyruvate to lactate is reversible and this pathway operates alongside the aerobic system all the time to enable greater intensity and duration of activity. Lactate can be metabolized in the muscle itself or it can enter the circulation and be taken up by the liver to replenish carbohydrate stores. This 'lactate shuttle' is an important mechanism for balancing the distribution of substrates and waste products.

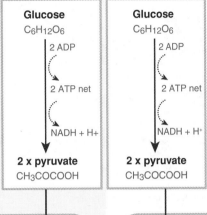

Glucose
$C_6H_{12}O_6$

2 ADP
2 ATP net
$NADH + H^+$

2 x pyruvate
$CH_3COCOOH$

Glucose
$C_6H_{12}O_6$

2 ADP
2 ATP net
$NADH + H^+$

2 x pyruvate
$CH_3COCOOH$

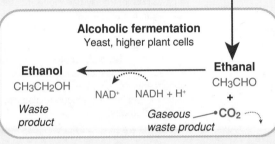

Alcoholic fermentation
Yeast, higher plant cells

Ethanol
CH_3CH_2OH

NAD^+ $NADH + H^+$

Waste product

Ethanal
CH_3CHO
+
Gaseous — CO_2
waste product

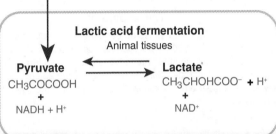

Lactic acid fermentation
Animal tissues

Pyruvate
$CH_3COCOOH$
+
$NADH + H^+$

Lactate
$CH_3CHOHCOO^- + H^+$
+
NAD^+

The alcohol and CO_2 produced from alcoholic fermentation form the basis of the brewing and baking industries. In baking, the dough is left to ferment and the yeast metabolizes sugars to produce ethanol and CO_2. The CO_2 causes the dough to rise.

Yeasts are used to produce almost all alcoholic beverages (e.g. wine and beers). The yeast used in the process breaks down the sugars into ethanol (alcohol) and CO_2. The alcohol produced is a metabolic by-product of fermentation by the yeast.

The lactate shuttle in vertebrate skeletal muscle works alongside the aerobic system to enable maximal muscle activity. Lactate moves from its site of production to regions within and outside the muscle (e.g. liver) where it can be respired aerobically.

1. (a) In aerobic respiration, a **theoretical maximum** of 38 ATP can be generated (achieved yield is closer to 32 ATP). Only 2 ATP are generated in fermentation. Calculate the percentage efficiency of fermentation compared to aerobic respiration:

_____

(b) Why is the efficiency of fermentation so low? _____

_____

2. Why can't yeasts produce ATP anaerobically indefinitely? _____

_____

3. Describe an advantage of the lactate shuttle in working muscle: _____

_____

LINK 65   WEB 64   APP

65 Investigating Alcoholic Fermentation in Yeast

Key Idea: Fermentation can be studied in brewer's yeast. Brewer's yeast is a facultative anaerobe (meaning it can respire aerobically or use fermentation). It will preferentially use alcoholic fermentation when sugars are in excess. One would expect glucose to be the preferred substrate, as it is the starting molecule in cell respiration, but brewer's yeast is capable of utilizing a variety of sugars, including disaccharides that can be broken down into single units. Completing this activity will involve a critical evaluation of the second-hand data provided.

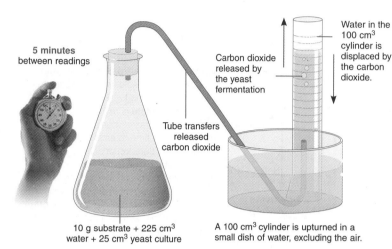

5 minutes between readings

Carbon dioxide released by the yeast fermentation

Water in the 100 cm³ cylinder is displaced by the carbon dioxide.

Tube transfers released carbon dioxide

10 g substrate + 225 cm³ water + 25 cm³ yeast culture

A 100 cm³ cylinder is upturned in a small dish of water, excluding the air.

The Apparatus

In this experiment, all substrates tested used the same source culture of 30 g active yeast dissolved in 150 cm³ of room temperature (24°C) tap water. For each substrate, 25 g of the substrate to be tested was added to 225 cm³ room temperature (24°C) tap water buffered to pH 4.5. Then 25 cm³ of source culture was added to the test solution. The control contained yeast solution but no substrate:

The Aim

To investigate the suitability of different mono- and disaccharide sugars as substrates for alcoholic fermentation in yeast.

Background

The rate at which brewer's or baker's yeast (*Saccharomyces cerevisiae*) metabolizes carbohydrate substrates is influenced by factors such as temperature, solution pH, and type of carbohydrate available. The literature describes yeast metabolism as optimal in warm, acidic environments. High levels of sugars suppress aerobic respiration, so yeast will preferentially use the fermentation pathway in the presence of excess substrate.

The substrates

Glucose is a monosaccharide. Maltose (glucose-glucose), sucrose (glucose-fructose), and lactose (glucose-galactose) are disaccharides.

Substrate Time (min)	Group 1: Volume of carbon dioxide collected (cm³)				
	None	**Glucose**	**Maltose**	**Sucrose**	**Lactose**
0	0	0	0	0	0
5	0	0	0.8	0	0
10	0	0	0.8	0	0
15	0	0	0.8	0.1	0
20	0	0.5	2.0	0.8	0
25	0	1.2	3.0	1.8	0
30	0	2.8	3.6	3.0	0.5
35	0	4.2	5.4	4.8	0.5
40	0	4.6	5.6	4.8	0.5
45	0	7.4	8.0	7.2	1.0
50	0	10.8	8.9	7.6	1.3
55	0	13.6	9.6	7.7	1.3
60	0	16.1	10.4	9.6	1.3
65	0	22.0	12.1	10.2	1.8
70	0	23.8	14.4	12.0	1.8
75	0	26.7	15.2	12.6	2.0
80	0	32.5	17.3	14.3	2.1
85	0	37.0	18.7	14.9	2.4
90	0	39.9	21.6	17.2	2.6

Substrate Time (min)	Group 2: Volume of carbon dioxide collected (cm³)				
	None	**Glucose**	**Maltose**	**Sucrose**	**Lactose**
90	0	24.4	19.0	17.5	0

1. Write the equation for the fermentation of glucose by yeast:

   _____

2. Calculate the rate of carbon dioxide production per minute for each substrate in group 1's results:

   (a) None: _____

   (b) Glucose: _____

   (c) Maltose: _____

   (d) Sucrose: _____

   (e) Lactose: _____

3. A second group of students performed the same experiment. Their results are summarized, below left. Calculate the rate of carbon dioxide production per minute for each substrate in group 2's results:

   (a) None: _____

   (b) Glucose: _____

   (c) Maltose: _____

   (d) Sucrose: _____

   (e) Lactose: _____

Experimental design and results adapted from Tom Schuster, Rosalie Van Zyl, & Harold Coller , California State University Northridge 2005

© 2012-2014 **BIOZONE** International
ISBN: 978-1-927173-93-0

4. What assumptions are being made in this experimental design and do you think they were reasonable?

_____

_____

5. Use the tabulated data to plot an appropriate graph of group 1's results on the grid provided:

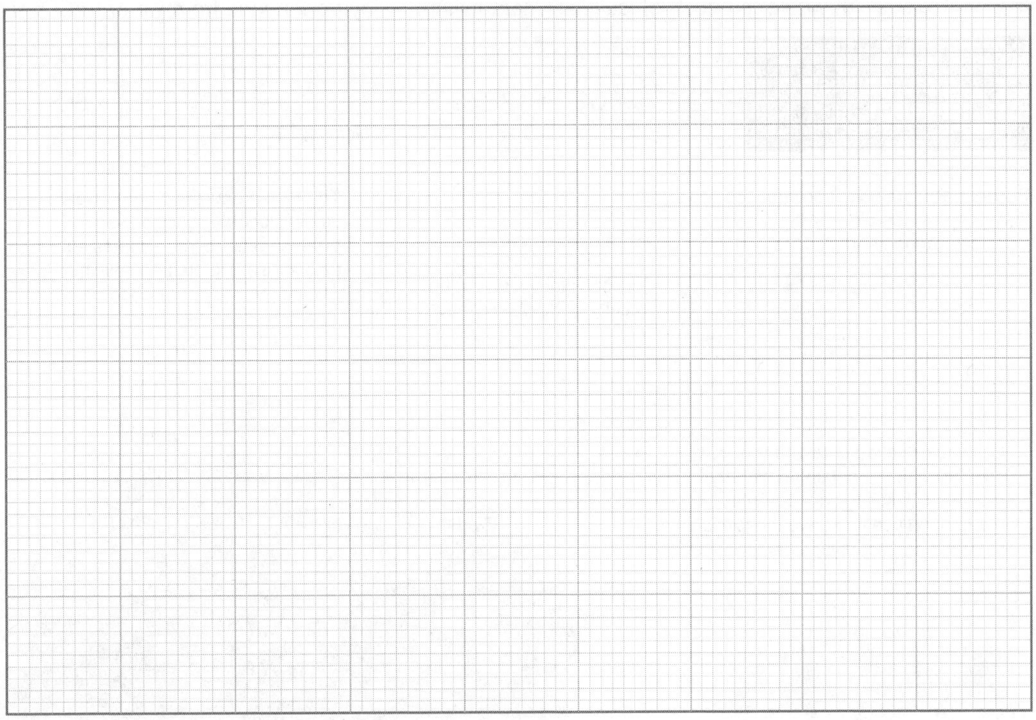

6. (a) Summarize the results of group 1's fermentation experiment: _____

_____

_____

(b) Explain the findings based on your understanding of cell respiration and carbohydrate chemistry:

_____

_____

_____

_____

7. (a) Plot a column chart to compare the results of the two groups in the volume of CO_2 collected after 90 minutes for each substrate (axes have been completed):

(b) Compare the results of the two groups:

_____

_____

_____

_____

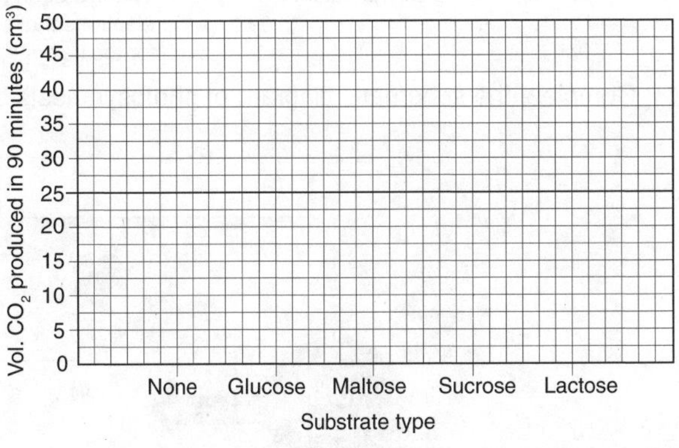

(c) Provide a probable explanation for any differences in the results: _____

_____

(d) As a group, discuss how you could improve this experiment. Staple a summary of your suggestions here.

66 Photosynthesis

Key Idea: Photosynthesis is the process of converting sunlight, carbon dioxide, and water into glucose and oxygen. **Photosynthesis** is of fundamental importance to living things because it transforms sunlight energy into chemical energy stored in molecules, releases free oxygen gas, and absorbs carbon dioxide (a waste product of cellular metabolism).

Photosynthetic organisms use special pigments, called **chlorophylls**, to absorb light of specific wavelengths and capture the light energy. Like cellular respiration, photosynthesis is a redox process, but in photosynthesis, water is split and electrons are transferred together with hydrogen ions from water to CO_2, reducing it to sugar.

Photosynthesis Equation	
$6CO_2 + 12H_2O \xrightarrow[\text{Chlorophyll}]{\text{Light}} C_6H_{12}O_6 + 6O_2 + 6H_2O$	

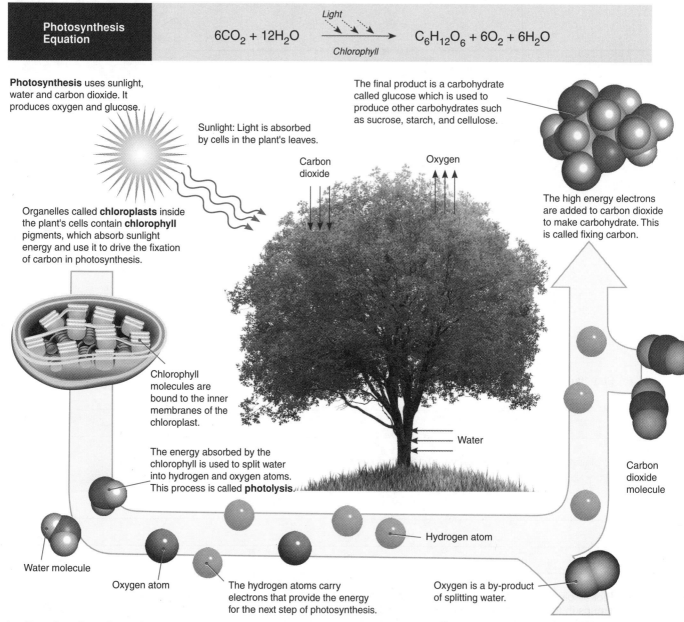

Photosynthesis uses sunlight, water and carbon dioxide. It produces oxygen and glucose.

Sunlight: Light is absorbed by cells in the plant's leaves.

Carbon dioxide

Oxygen

The final product is a carbohydrate called glucose which is used to produce other carbohydrates such as sucrose, starch, and cellulose.

Organelles called **chloroplasts** inside the plant's cells contain **chlorophyll** pigments, which absorb sunlight energy and use it to drive the fixation of carbon in photosynthesis.

The high energy electrons are added to carbon dioxide to make carbohydrate. This is called fixing carbon.

Chlorophyll molecules are bound to the inner membranes of the chloroplast.

Water

The energy absorbed by the chlorophyll is used to split water into hydrogen and oxygen atoms. This process is called **photolysis**.

Carbon dioxide molecule

Hydrogen atom

Water molecule

Oxygen atom

The hydrogen atoms carry electrons that provide the energy for the next step of photosynthesis.

Oxygen is a by-product of splitting water.

1. Complete the schematic diagram of photosynthesis below:

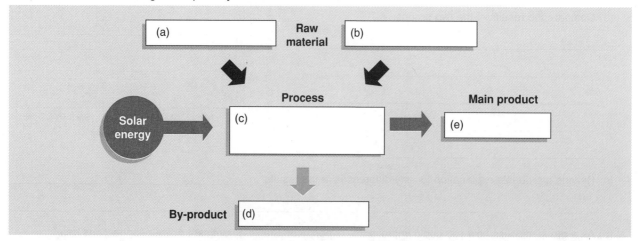

(a) Raw material (b)

Solar energy → Process (c) → Main product (e)

By-product (d)

67 Pigments and Light Absorption

Key Idea: Chlorophyll pigments absorb light of specific wavelengths and capture light energy for photosynthesis. Substances that absorb visible light are called **pigments**, and different pigments absorb light of different wavelengths. The ability of a pigment to absorb particular wavelengths of light can be measured with a spectrophotometer. The light absorption vs the wavelength is called the **absorption**

spectrum of that pigment. The absorption spectrum of different photosynthetic pigments provides clues to their role in photosynthesis, since light can only perform work if it is absorbed. An **action spectrum** profiles the effectiveness of different wavelengths of light in fuelling photosynthesis. It is obtained by plotting wavelength against a measure of photosynthetic rate (e.g. oxygen production).

The Electromagnetic Spectrum

Light is a form of energy known as electromagnetic radiation. The segment of the electromagnetic spectrum most important to life is the narrow band between about 380 nm and 750 nm. This radiation is known as **visible light** because it is detected as colours by the human eye (although some other animals, such as insects, can see in the UV range). It is the visible light that drives photosynthesis.

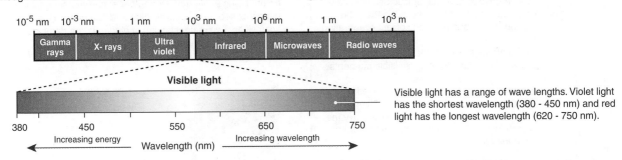

Visible light has a range of wave lengths. Violet light has the shortest wavelength (380 - 450 nm) and red light has the longest wavelength (620 - 750 nm).

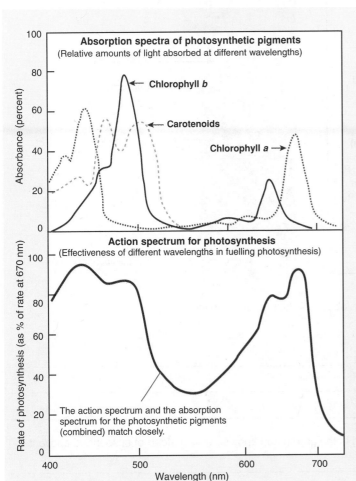

The photosynthetic pigments of plants

The photosynthetic pigments of plants fall into two categories: chlorophylls (which absorb red and blue-violet light) and carotenoids (which absorb strongly in the blue-violet and appear orange, yellow, or red). The pigments are located on the chloroplast membranes (the thylakoids) and are associated with membrane transport systems.

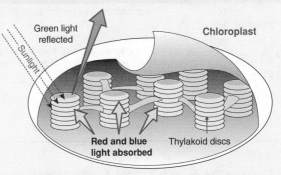

The pigments of chloroplasts in higher plants (above) absorb blue and red light. The leaves appear green because the green light is reflected. Each photosynthetic pigment has its own characteristic absorption spectrum (left, top graph). Although only chlorophyll a can participate directly in the light reactions of photosynthesis, the accessory pigments (chlorophyll b and carotenoids) can absorb wavelengths of light that chlorophyll a cannot. The accessory pigments pass the energy (photons) to chlorophyll a, thus broadening the spectrum that can effectively drive photosynthesis.

Left: Graphs comparing absorption spectra of photosynthetic pigments compared with the action spectrum for photosynthesis.

1. What is meant by the absorption spectrum of a pigment? _____

_____

2. Why doesn't the **action spectrum** for photosynthesis exactly match the absorption spectrum of chlorophyll *a*?

_____

_____

_____

© 2012-2014 **BIOZONE** International
ISBN: 978-1-927173-93-0
Photocopying Prohibited

LINK 228 · LINK 68 · WEB 67 APP

68 Separation of Pigments by Chromatography

Key Idea: Photosynthetic pigments can be separated using paper chromatography.

Chromatography is a technique used to separate a mixture of molecules and can be used on small samples. Chromatography is based on passing a mixture dissolved in a mobile phase (a solvent) through a stationary phase, which separates the molecules according to their specific characteristics. Paper chromatography is a simple technique in which porous paper serves as the stationary phase, and a solvent, either water or ethanol, serves as the mobile phase.

Paper chromatography

Set up and procedure

The chromatography paper is folded so it can be secured by the bung inside the test tube. The bung also prevents the solvent evaporating.

Chromatography paper may be treated with chemicals to stain normally invisible pigments.

A spot of concentrated sample is added using a pipette and suspended above the solvent. As the solvent travels up the paper it will carry the sample with it. The distance the sample travels depends on its solubility.

A pencil line is used to show the starting point.

Solvent

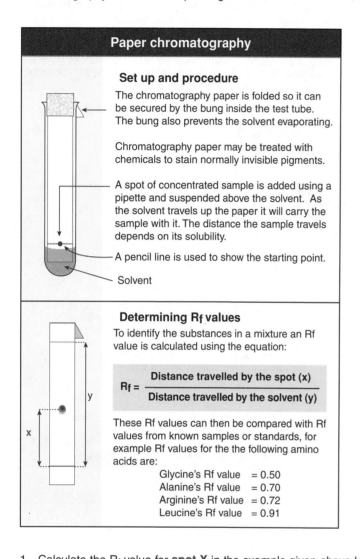

Determining R$_f$ values

To identify the substances in a mixture an Rf value is calculated using the equation:

$$R_f = \frac{\text{Distance travelled by the spot (x)}}{\text{Distance travelled by the solvent (y)}}$$

These Rf values can then be compared with Rf values from known samples or standards, for example Rf values for the the following amino acids are:

Glycine's Rf value = 0.50
Alanine's Rf value = 0.70
Arginine's Rf value = 0.72
Leucine's Rf value = 0.91

Separation of Photosynthetic Pigments

The four primary pigments of green plants can easily be separated and identified using paper chromatography. The pigments from the leaves are first extracted with acetone before being separated. During paper chromatography the pigments separate out according to differences in their relative solubilities. Two major classes of pigments are detected: the two greenish chlorophyll pigments and two yellowish carotenoid pigments.

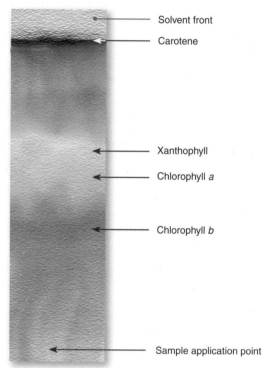

Solvent front
Carotene
Xanthophyll
Chlorophyll *a*
Chlorophyll *b*
Sample application point

1. Calculate the R$_f$ value for **spot X** in the example given above left (show your working): _____

_____

2. Why is the R$_f$ value of a substance always less than 1? _____

_____

_____

3. Predict what would happen if a sample was immersed in the chromatography solvent, instead of suspended above it:

_____

_____

4. With reference to their R$_f$ values, rank the four amino acids (listed above) in terms of their solubility: _____

_____

5. Compare the solubility of chlorophyll pigments to that of the other pigments separated by paper chromatography:

_____

_____

69 Factors Affecting Photosynthetic Rate

Key Idea: Environmental factors, such as CO_2 availability and light intensity, affect photosynthesis rate.

The photosynthetic rate is the rate at which plants make carbohydrate. It is dependent on environmental factors, particularly the availability of light and carbon dioxide (CO_2). Temperature is important, but its influence is less clear because it depends on the availability of the other two limiting factors (CO_2 and light) and the temperature tolerance of the plant. The relative importance of these factors can be tested experimentally by altering one of the factors while holding the others constant. The results for such an experiment are shown below.

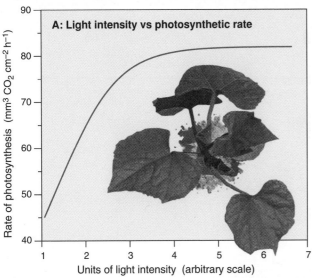

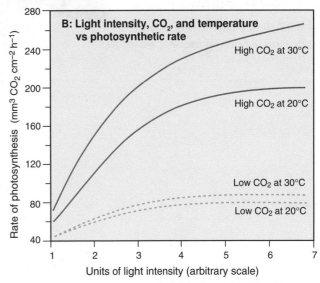

These figures illustrate the effect of different limiting factors on the rate of photosynthesis in cucumber plants. Figure A shows the effect of different light intensities when the temperature and carbon dioxide (CO_2) level are kept constant. Figure B shows the effect of different light intensities at two temperatures and two CO_2 concentrations. In each of these experiments, either CO_2 level or temperature was changed at each light intensity in turn.

1. Based on the figures above, summarize and explain the effect of each of the following factors on photosynthetic rate:

    (a) CO_2 concentration: _____

    _____

    (b) Light intensity: _____

    _____

    (c) Temperature: _____

    _____

2. Explain why photosynthetic rate declines when the CO_2 level is reduced: _____

    _____

    _____

3. (a) In figure B, explain how the effects of CO_2 concentration were distinguished from the effects of temperature:

    _____

    (b) Identify which factor (CO_2 or temperature) had the greatest effect on photosynthetic rate: _____

    (c) Explain how you can tell this from the graph: _____

    _____

4. Explain how glasshouses can be used to create an environment in which photosynthetic rates are maximized:

    _____

    _____

    _____

5. Design an experiment to demonstrate the effect of temperature on photosynthetic rate. You should include a hypothesis, list of equipment, and methods. Staple your experiment to this page.

© 2012-2014 **BIOZONE** International
ISBN: 978-1-927173-93-0
Photocopying Prohibited

70 Chapter Review

Summarize what you know about this topic under the headings provided. You can draw diagrams or mind maps, or write short notes to organize your thoughts. Use the images and hints included to help you:

Water
HINT: Describe how water can form bonds with other molecules, and why this is important.

Molecules to metabolism

HINT: Describe the importance of carbon in forming organic molecules.

ATP

Carbohydrates and lipids:
HINT: How does the structure of carbohydrates and lipids influence their function?

Proteins:
HINT: Describe the structure and function of proteins. How can the structure be disrupted?

Enzymes:
HINT: Describe the role of enzymes. What is the active site?

DNA and RNA:
HINT: Describe the structure and function of nucleotides in DNA and RNA. How does DNA replication occur?

Cell respiration
HINT: Compare cell respiration and fermentation.
Include differences in ATP yield.

WMU

Mitochondrion

Photosynthesis
HINT: Identify the substrates and products formed in photosynthesis using an equation. What role does chlorophyll play?

71 KEY TERMS: Did You Get It?

1. Test your vocabulary by matching each term to its definition, as identified by its preceding letter code.

amino acid _____

ATP _____

catabolism _____

catalyst _____

condensation _____

denaturation _____

disaccharide _____

DNA _____

enzyme _____

fermentation _____

hydrolysis _____

lipid _____

monosaccharide _____

nucleic acid _____

polypeptides _____

polysaccharide _____

semi-conservative _____

translation _____

A Chemical reaction that combines two molecules. Water is produced as a by-product.

B Chemical reaction in which a molecule is split by water (as H^+ and OH^-).

C A double sugar molecule used as an energy source and a building block of larger molecules. Examples are sucrose and lactose.

D The process by which mRNA is decoded to produce a specific polypeptide.

E A nucleotide comprising a purine base, a pentose sugar, and three phosphate groups, which acts as the cell's energy carrier.

F A model for DNA replication which proposes each DNA strand serves as a template, forming a new DNA molecule with half old and half new DNA.

G A globular protein which acts as a catalyst to speed up a specific biological reaction.

H A complex carbohydrate with a structural and energy storage role in cells. Examples include cellulose, starch, and glycogen.

I A substance or molecule that lowers the activation energy of a reaction but is itself not used up during the reaction. In biological systems, this function is carried out by enzymes.

J Also called fat or oil. A biological compound made up of glycerol and fatty acid components.

K The loss of a protein's three-dimensional functional structure.

L Macromolecules that form from the joining of multiple amino acids together.

M A building block of proteins.

N A carbohydrate monomer. Examples include fructose and glucose.

O A polynucleotide molecule that occurs in two forms, DNA and RNA.

P Metabolic process in which complex molecules are broken down into simpler ones.

Q Process which provides an alternative way to produce energy if oxygen is temporarily unavailable. Does not involve a terminal electron acceptor.

R Universally found macromolecules composed of chains of nucleotides. These molecules carry genetic information within cells.

2. (a) Name the process described in the equation (right): $C_6H_{12}O_6 + 6O_2 \longrightarrow 6CO_2 + 6H_2O + Energy$

_____

(b) Where does this process occur? _____

(c) Name the type of energy molecule that is produced in this process: _____

(d) How is energy released by this molecule? _____

3. (a) Write the process of photosynthesis as:

A word equation: _____

A chemical equation: _____

(b) Where does photosynthesis occur? _____

TEST

Topic 3

Genetics

3.1 Genes

Understandings, applications, skills

		Activity number
☐	1 Define gene, locus, and allele. Explain how alleles differ from each other.	72
☐	2 Explain how new alleles are formed by mutation. Identify the role of copying errors and mutagens in causing mutations.	73 74
☐	3 Describe the causes of sickle cell anemia (sickle cell disease) with respect to the DNA mutation, the consequent change to the mRNA, and the change to the amino acid sequence in the polypeptide.	76

> **TOK** How do we determine if the link between sickle cell disease and the prevalence of malaria is a correlation or a cause and effect?

		Activity number
☐	4 Explain what is meant by the genome.	77
☐	5 Describe the aims and results of the Human Genome Project. Compare the number of genes in humans with other species.	77 85
☐	6 Use a database to determine differences in the base sequence of a gene (e.g. cytochrome C gene) in two species.	

3.2 Chromosomes

Understandings, applications, skills

		Activity number
☐	1 Describe the nature of the bacterial chromosome and extra-chromosomal plasmids. Explain why bacteria are haploid for most genes.	78
☐	2 Describe the nature of eukaryotic chromosomes, including the role of histone proteins in packaging DNA in the nucleus. Explain how the length of DNA molecules is measured.	79
☐	3 Compare genome size in T2 phage, *E. coli*, *Drosophila melanogaster*, *Homo sapiens*, and *Paris japonica*.	72 80
☐	4 Chromosome number is a species specific characteristic. Compare diploid chromosome number in *Homo sapiens*, *Pan troglodytes*, *Canis familiaris*, *Oryza sativa*, *Parascaris equorum*.	81
☐	5 Distinguish between a karyotype and a karyogram. Distinguish sex chromosomes and autosomes. Use a karyogram to deduce sex and diagnose Down syndrome in humans.	81 82
☐	6 Use a database to identify the locus of a human gene and its polypeptide product.	83

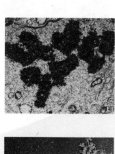

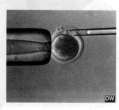

3.3 Meiosis

Understandings, applications, skills

☐ 1 Describe the purpose and genetic outcome of meiosis. Outline the important stages in meiosis and their significance, including condensation of the chromosomes, pairing of homologues and crossing over, random orientation of homologues prior to separation, and reduction division. Comment on the contribution of meiosis to genetic variation in the offspring. Draw diagrams to show the stages in meiosis resulting in the formation of four haploid cells.

80 81

☐ 2 Explain how non-disjunction in meiosis can result in chromosome abnormalities, e.g. Down syndrome. Explain the effect of parental age on the incidence of chromosomal abnormalities arising as a result of non-disjunction.

82

☐ 3 Explain how chromosome abnormalities are detected using karyotype analysis and describe the methods used to obtain the cells for analysis. Describe the risks associated with these techniques.

83

3.4 Inheritance

Activity number

Understandings, applications, skills

☐ 1 Explain Mendel's principles of inheritance and how they were established.

86

> **TOK** *Mendel's ideas were not immediately accepted. What factors encourage acceptance of new ideas by the scientific community?*

☐ 2 Describe the production of haploid gametes by meiosis. Describe how formation of the diploid zygote in fertilization can result in an individual that is homozygous or heterozygous for a gene. Distinguish between dominant and recessive alleles.

72 80

☐ 3 Using examples, describe the effect of radiation and mutagenic chemicals on the mutation rate and the incidence of genetic diseases and cancer.

74

☐ 4 Describe how genetic diseases in humans can result from recessive, dominant, or codominant alleles on autosomes. Describe the inheritance of cystic fibrosis (recessive allele) and Huntington disease (dominant allele). Know that many genetic diseases in humans are recognized, but most are rare.

73 75 76

☐ 5 Use Punnett squares to predict the outcome of monohybrid genetic crosses.

88-90 95

☐ 6 Compare predicted and actual outcomes of genetic crosses.

95 265

☐ 7 Describe the effect of codominant alleles with reference to the inheritance of ABO blood groups in humans.

91 92

☐ 8 Describe and explain the pattern of inheritance in sex linked genes. Describe the inheritance patterns of the sex-linked disorders red-green colour blindness and hemophilia.

93 94

☐ 9 Analyse pedigree charts to determine the pattern of inheritance of genetic diseases.

96

3.5 Genetic modification and biotechnology

Activity number

Understandings, applications, skills

☐ 1 Explain how and why PCR is used to amplify small amounts of DNA.

97 98

☐ 2 Explain how gel electrophoresis separates proteins or DNA fragments for analysis.

99

☐ 3 Describe the use of DNA profiling to compare DNA samples, e.g. in paternity cases or forensic investigations. Analyse examples of DNA profiles.

100 101

☐ 4 Describe the genetic modification of organisms by gene transfer between species, Using examples, explain the production and role of recombinant plasmids in gene transfers.

102-109

☐ 5 With reference to specific examples, evaluate the potential risks and benefits associated with GM crops.

110

☐ 6 Explain what is meant by a clone and give examples of natural clones in plants (e.g. cuttings) and animals (e.g. embryo splitting or twinning). With reference to plant cloning, identify and assess experimentally the factors affecting the rooting of stem cuttings.

111 112

☐ 7 Describe the production of cloned embryos by somatic cell nuclear transfer (also called SCNT).

113

> **TOK** *DNA evidence is now widely used to secure convictions for crimes, but what criteria are necessary to assess the reliability of evidence?*

72 Alleles

Key Idea: Eukaryotes generally have paired chromosomes. Each chromosome contains many genes and each gene may have a number of versions called alleles.

Sexually reproducing organisms usually have paired sets of chromosomes, one set from each parent. The equivalent chromosomes that form a pair are termed **homologues**. They carry equivalent sets of genes, but there is the potential for different versions of a gene (**alleles**) to exist in a population.

Homologous Chromosomes

In sexually reproducing organisms, most cells have a homologous pair of chromosomes (one coming from each parent). This diagram shows the position of three different genes on the same chromosome that control three different traits (A, B and C).

Chromosomes are formed from DNA and proteins. DNA tightly winds around special proteins to form the chromosome.

Having two different versions (alleles) of gene A is a **heterozygous** condition. Only the dominant allele (A) will be expressed. Alleles differ by only a few bases.

When both chromosomes have identical copies of the dominant allele for gene B the organism is **homozygous dominant** for that gene.

When both chromosomes have identical copies of the recessive allele for gene C the organism is said to be **homozygous recessive** for that gene.

Maternal chromosome originating from the egg of the female parent.

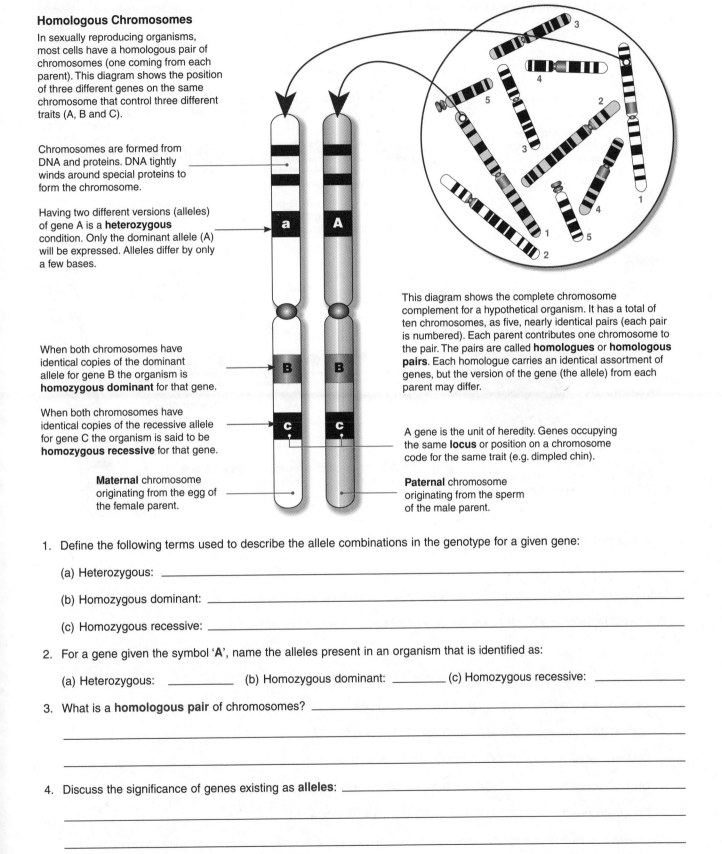

This diagram shows the complete chromosome complement for a hypothetical organism. It has a total of ten chromosomes, as five, nearly identical pairs (each pair is numbered). Each parent contributes one chromosome to the pair. The pairs are called **homologues** or **homologous pairs**. Each homologue carries an identical assortment of genes, but the version of the gene (the allele) from each parent may differ.

A gene is the unit of heredity. Genes occupying the same **locus** or position on a chromosome code for the same trait (e.g. dimpled chin).

Paternal chromosome originating from the sperm of the male parent.

1. Define the following terms used to describe the allele combinations in the genotype for a given gene:

   (a) Heterozygous: _____

   (b) Homozygous dominant: _____

   (c) Homozygous recessive: _____

2. For a gene given the symbol '**A**', name the alleles present in an organism that is identified as:

   (a) Heterozygous: _____   (b) Homozygous dominant: _____   (c) Homozygous recessive: _____

3. What is a **homologous pair** of chromosomes? _____

   _____

   _____

4. Discuss the significance of genes existing as **alleles**: _____

   _____

   _____

   _____

   _____

© 2012-2014 **BIOZONE** International
ISBN: 978-1-927173-93-0
Photocopying Prohibited

LINK
88

LINK
83

LINK
73

KNOW

73 Changes to the DNA Sequence

Key Idea: Changes to the DNA sequence are called mutations. Mutations are the ultimate source of new genetic information, i.e. new alleles.

Mutations are changes to the DNA sequence. They arise through errors in DNA copying and involve alterations in the DNA from a single base pair to large parts of chromosomes. Bases may be inserted into, substituted, or deleted from the DNA. Mutations are the source of all new alleles. Most often mutations are harmful, but occasionally they can be beneficial. Sometimes they produce no phenotypic change and are silent. An example of a mutation producing a new allele is described below. This mutation causes a form of genetic hearing loss (called NSRD). It occurs in the gene coding for the protein connexin 26.

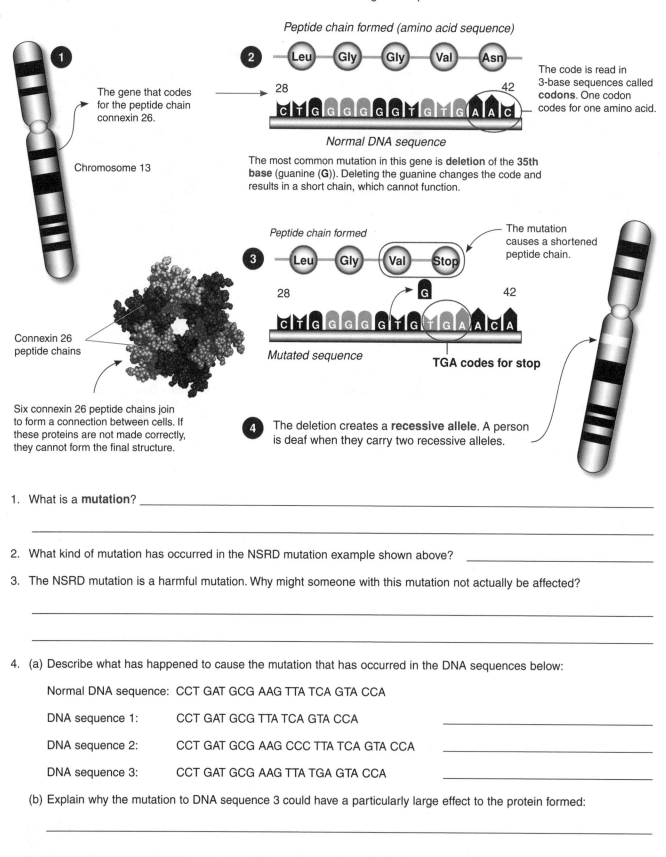

Peptide chain formed (amino acid sequence)

The gene that codes for the peptide chain connexin 26.

Chromosome 13

The code is read in 3-base sequences called **codons**. One codon codes for one amino acid.

Normal DNA sequence

The most common mutation in this gene is **deletion** of the **35th base** (guanine (**G**)). Deleting the guanine changes the code and results in a short chain, which cannot function.

Peptide chain formed

The mutation causes a shortened peptide chain.

Connexin 26 peptide chains

Mutated sequence

TGA codes for stop

Six connexin 26 peptide chains join to form a connection between cells. If these proteins are not made correctly, they cannot form the final structure.

4. The deletion creates a **recessive allele**. A person is deaf when they carry two recessive alleles.

1. What is a **mutation**? _____

_____

2. What kind of mutation has occurred in the NSRD mutation example shown above? _____

3. The NSRD mutation is a harmful mutation. Why might someone with this mutation not actually be affected?

_____

_____

4. (a) Describe what has happened to cause the mutation that has occurred in the DNA sequences below:

Normal DNA sequence: CCT GAT GCG AAG TTA TCA GTA CCA

DNA sequence 1:   CCT GAT GCG TTA TCA GTA CCA   _____

DNA sequence 2:   CCT GAT GCG AAG CCC TTA TCA GTA CCA _____

DNA sequence 3:   CCT GAT GCG AAG TTA TGA GTA CCA   _____

(b) Explain why the mutation to DNA sequence 3 could have a particularly large effect to the protein formed:

_____

_____

© 2012-2014 **BIOZONE** International
ISBN: 978-1-927173-93-0
Photocopying Prohibited

74 Mutagens

Key Idea: Mutagens are chemical or physical agents that cause a change in the DNA sequence.

Mutations occur spontaneously in all organisms. The natural rate at which a gene will undergo change is normally very low, but this rate can be increased by environmental factors such as ionizing radiation and mutagenic chemicals (e.g. benzene). Only mutations in cells producing gametes (**gametic mutations**) will be inherited. If they occur in a body cell after the organism has begun to develop beyond the zygote stage, they are called **somatic mutations**. In some cases, these may disrupt the normal controls over gene regulation and expression and trigger the onset of cancer.

Mutagen and Effect

Ionizing Radiation

Radiation in the form of gamma rays and particle emission from radioactive isotopes can cause a wide range of mutations. Rates of thyroid cancer increased in areas near Chernobyl after the explosion of the No. 4 reactor there. Skin cancer (from high exposure to ultraviolet) is increasingly common. Fair skinned people at low latitudes are at risk from ultraviolet radiation. Safer equipment has considerably reduced the risks to those working with ionizing radiation (e.g. radiographers).

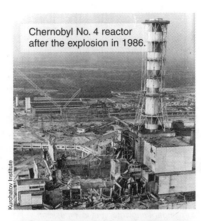
Chernobyl No. 4 reactor after the explosion in 1986.

Viruses and Microorganisms

Some viruses integrate into the human chromosome, upsetting genes and triggering cancers. Examples include hepatitis B virus (liver cancer), HIV (Kaposi's sarcoma), Epstein-Barr virus (Burkitt's lymphoma, Hodgkin's disease), and HPV (right) which is implicated in cervical cancer. Aflatoxins produced by the fungus *Aspergillus flavus* are potent inducers of liver cancer. Those at higher risk of viral infections include intravenous drug users and those with unsafe sex practices

Poisons and Irritants

Many chemicals are mutagenic. Synthetic and natural examples include organic solvents such as benzene, asbestos, formaldehyde, tobacco tar, vinyl chlorides, coal tars, some dyes, and nitrites. Those most at risk include workers in the chemicals industries, including the glue, paint, rubber, resin, and leather industries, petrol pump attendants, and those in the coal and other mining industries.

Photo right: Firefighters and those involved in environmental clean-up of toxic spills are at high risk of exposure to mutagens.

Diet, Alcohol and Tobacco Smoke

Diets high in fat, especially those containing burned or fatty, highly preserved meat, slow the passage of food through the gut giving time for mutagenic irritants to form in the lower bowel.
High alcohol intake increases the risk of some cancers and increases susceptibility to tobacco-smoking related cancers.
Tobacco tar is one of the most damaging constituents of tobacco smoke. Tobacco tars contain at least 17 known carcinogens (cancer inducing mutagens) that cause chronic irritation of the gas exchange system and cause cancer in smokers.

1. Describe examples of environmental factors that induce mutations under the following headings:

    (a) Radiation: _____

    _____

    _____

    _____

    _____

    (b) Chemical agents: _____

    _____

    _____

    _____

    _____

    _____

2. Explain how **mutagens** cause mutations:

    _____

    _____

    _____

    _____

    _____

    _____

    _____

    _____

    _____

    _____

3. Distinguish between **gametic** and **somatic mutations** and comment on the significance of the difference:

    _____

    _____

    _____

    _____

    _____

    _____

LINK 35   WEB 74   **KNOW**

75 Gene Mutations and Genetic Diseases

Key Idea: Many genetic diseases in humans are the result of mutations to recessive alleles, but some are also caused by dominant or codominant alleles.

There are more than 6000 human diseases attributed to mutations in single genes, although most are uncommon. The three genetic diseases described below occur with relatively high frequency and are the result of recessive, dominant, and codominant allele mutations respectively.

Cystic Fibrosis (CF)	Huntington Disease (HD)	Sickle Cell Anemia

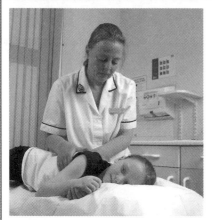

Cystic fibrosis is traditionally treated with physical therapy to clear mucus from the airways.

American singer-songwriter and folk musician Woody Guthrie died from complications of HD

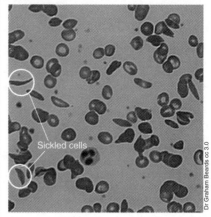

Sickled cells

Dr Graham Beards cc 3.0

In a person heterozygous for the sickle cell allele, only some of the red blood cells are deformed.

Incidence: Varies with populations:
United States: 1 in 1000 (0.1%).
Asians in England: 1 in 10,000
European descent: 1 in 20-28 are carriers.

Incidence: An uncommon disease affecting 3-7 per 100,000 people of European descent. Less common in other ethnicities, including people of Japanese, Chinese, and African descent.

Incidence: Occurs most commonly in people of African descent. West Africans: 1% (10-45% are carriers). West Indians: 0.5%.

Gene type: Autosomal recessive. The most common mutation is ΔF508, which accounts for around 70% of all defective CF genes. The mutation is a deletion of the 508th triplet in the DNA code for the chloride transport protein CFTR. As a result, the amino acid phenylalanine is missing and the CFTR protein cannot carry out its function of regulating chloride ion balance in the cell.

Gene type: Autosomal dominant mutation of the HTT gene caused by a trinucleotide repeat expansion on the short arm of chromosome 4. In the mutation (**mHTT**), the number of CAG repeats increases from the normal 6-30 to 36-125. The severity of the disease increases with the number of repeats. The repeats result in the production of an abnormally long version of the huntingtin protein.

Gene type: Autosomal mutation involving substitution of a single nucleotide in the HBB gene that codes for the beta chain of hemoglobin. The allele is codominant. The substitution causes a change in a single amino acid. The mutated hemoglobin behaves differently when deprived of oxygen, causing distortion of the red blood cells, anemia, and circulatory problems.

Gene location: Chromosome 7

Gene location: Short arm of chromosome 4

Gene location: Short arm of chromosome 11

CFTR

p   *q*

HTT

p   *q*

HBB

p   *q*

Symptoms: Disruption of all glands: the pancreas, intestinal glands, biliary tree (biliary cirrhosis), bronchial glands (chronic lung infections), and sweat glands (high salt content of which becomes depleted). Infertility occurs in both sexes.

Symptoms: The long huntingtin protein is cut into smaller toxic fragments, which accumulate in nerve cells and eventually kill them. The disease becomes apparent in mid-adulthood, with jerky, involuntary movements and loss of memory, reasoning, and personality.

Symptoms: Sickling of the red blood cells, which are removed from circulation, anemia, pain, damage to tissues and organs.

Inheritance: Autosomal recessive pattern. Affected people are homozygous recessive for the mutation. Heterozygotes are carriers.

Inheritance: Autosomal dominance pattern. Affected people may be homozygous or heterozygous for the mutant allele.

Inheritance: Autosomal codominance pattern. People who are homozygous for the mutant allele have sickle cell, heterozygotes are only mildly affected and are carriers. The allele show heterozygote advantage in malaria-prone regions (see opposite).

1. For each of genetic disorder below, indicate the following:

   (a) Sickle cell anemia: Gene name: _____   Chromosome: _____ Mutation type: _____

   (b) Cystic fibrosis: Gene name: _____   Chromosome: _____ Mutation type: _____

   (c) Huntington disease: Gene name: _____   Chromosome: _____ Mutation type: _____

2. Describe the inheritance pattern characterizing each of the following genetic diseases:

   (a) Cystic fibrosis: _____

   (b) Huntington's disease: _____

   (c) Sickle cell anemia: _____

3. Explain why mHTT, which is dominant and lethal, does not disappear from the population: _____
   _____

© 2012-2014 **BIOZONE** International
ISBN: 978-1-927173-93-0
Photocopying Prohibited

76 The Sickle Cell Mutation

Key Idea: Sickle cell anemia is caused by a mutation that affects the beta chain of the hemoglobin (Hb) molecule.
Sickle cell anemia is an inherited disorder caused by a gene mutation that codes for a faulty beta (β) chain Hb protein.

This in turn causes the red blood cells to deform, resulting in a range of medical problems. The allele is codominant and the mutation is lethal in homozygotes, but individuals with only one mutated allele show resistance to malaria.

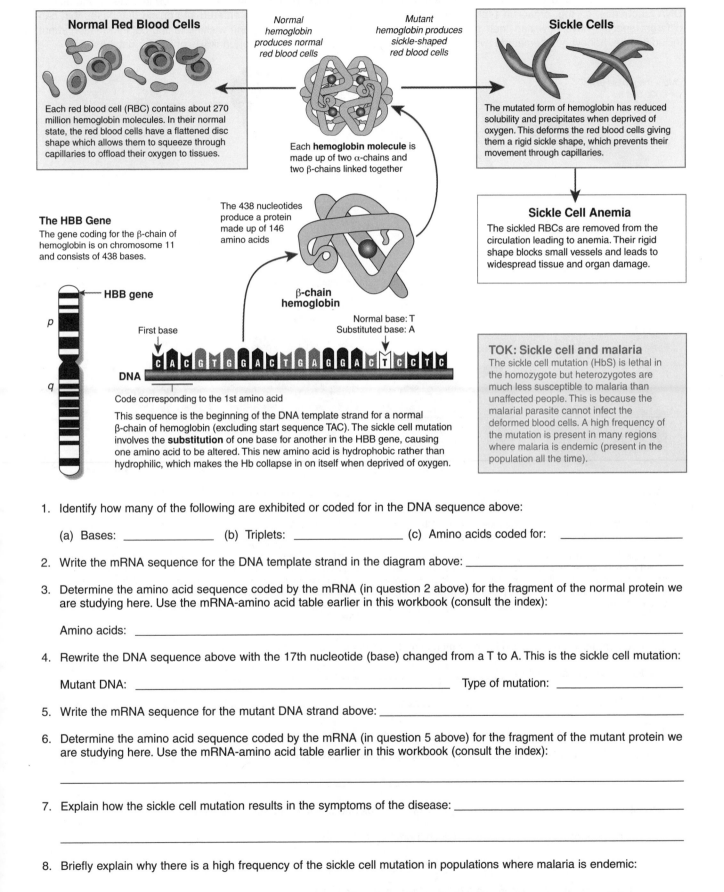

Normal Red Blood Cells

Each red blood cell (RBC) contains about 270 million hemoglobin molecules. In their normal state, the red blood cells have a flattened disc shape which allows them to squeeze through capillaries to offload their oxygen to tissues.

Normal hemoglobin produces normal red blood cells

Mutant hemoglobin produces sickle-shaped red blood cells

Each **hemoglobin molecule** is made up of two α-chains and two β-chains linked together

Sickle Cells

The mutated form of hemoglobin has reduced solubility and precipitates when deprived of oxygen. This deforms the red blood cells giving them a rigid sickle shape, which prevents their movement through capillaries.

Sickle Cell Anemia

The sickled RBCs are removed from the circulation leading to anemia. Their rigid shape blocks small vessels and leads to widespread tissue and organ damage.

The HBB Gene
The gene coding for the β-chain of hemoglobin is on chromosome 11 and consists of 438 bases.

The 438 nucleotides produce a protein made up of 146 amino acids

β-chain hemoglobin

Normal base: T
Substituted base: A

HBB gene

First base

C A C G T G G A C T G A G G A C T C C T C

DNA

Code corresponding to the 1st amino acid

This sequence is the beginning of the DNA template strand for a normal β-chain of hemoglobin (excluding start sequence TAC). The sickle cell mutation involves the **substitution** of one base for another in the HBB gene, causing one amino acid to be altered. This new amino acid is hydrophobic rather than hydrophilic, which makes the Hb collapse in on itself when deprived of oxygen.

TOK: Sickle cell and malaria
The sickle cell mutation (HbS) is lethal in the homozygote but heterozygotes are much less susceptible to malaria than unaffected people. This is because the malarial parasite cannot infect the deformed blood cells. A high frequency of the mutation is present in many regions where malaria is endemic (present in the population all the time).

1. Identify how many of the following are exhibited or coded for in the DNA sequence above:

   (a) Bases: _____   (b) Triplets: _____   (c) Amino acids coded for: _____

2. Write the mRNA sequence for the DNA template strand in the diagram above: _____

3. Determine the amino acid sequence coded by the mRNA (in question 2 above) for the fragment of the normal protein we are studying here. Use the mRNA-amino acid table earlier in this workbook (consult the index):

   Amino acids: _____

4. Rewrite the DNA sequence above with the 17th nucleotide (base) changed from a T to A. This is the sickle cell mutation:

   Mutant DNA: _____   Type of mutation: _____

5. Write the mRNA sequence for the mutant DNA strand above: _____

6. Determine the amino acid sequence coded by the mRNA (in question 5 above) for the fragment of the mutant protein we are studying here. Use the mRNA-amino acid table earlier in this workbook (consult the index):

   _____

7. Explain how the sickle cell mutation results in the symptoms of the disease: _____

   _____

8. Briefly explain why there is a high frequency of the sickle cell mutation in populations where malaria is endemic:

   _____

© 2012-2014 **BIOZONE** International
ISBN: 978-1-927173-93-0
Photocopying Prohibited

LINK 142   LINK 75   LINK 59   WEB 76   **APP**

77 Genomes

Key Idea: A genome is an organism's complete set of genetic material, including all of its genes. The genomes of many organisms have been sequenced, allowing genes to be compared. These can be searched for on gene databases.

The aim of most genome projects is to determine the DNA sequence of the organism's entire genome. Many different species have now had their genomes sequenced including the honeybee, nematode worm, African clawed frog, pufferfish, zebra fish, rice, cow, dog, and rat. Genome sizes and the number of genes per genome vary, and are not necessarily correlated with the size and structural complexity of the organism. Once completed, genome sequences are analysed by computer to identify genes. Gene sequences and details are often entered into online databases that can be searched by anyone wishing to find out information about a particular gene. Genbank® is one such database.

*Mb = megabase pairs or 1,000,000 bp

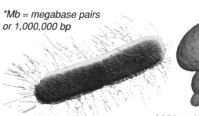

Bacterium
(*Escherichia coli*)

Genome size: 4.6 Mb*
Number of genes: 4403

E. coli has been used as a laboratory organism for over 70 years. Various strains of *E. coli* are responsible for several human diseases.

Artist's impression

Yeast
(*Saccharomyces cerevisiae*)

Genome size: 13 Mb
Number of genes: 6000

The first eukaryotic genome to be completely sequenced. Yeast is used as a model organism to study human cancer.

Human
(*Homo sapiens*)

Genome size: 3000 Mb
Number of genes: < 22,500

The completion of the human genome has allowed advances in medical research, especially in cancer research.

Rice
(*Oryza sativa*)

Genome size: 466 Mb (indica) and 420 Mb (japonica)
Number of genes: 46,000

A food staple for much of the world's population. The importance of rice as a world food crop made sequencing it a high priority.

Mouse
(*Mus musculus*)

Genome size: 2500 Mb
Number of genes: 30,000

New drugs destined for human use are often tested on mice because more than 90% of their proteins show similarities to human proteins.

Fruit fly
(*Drosophila melanogaster*)

Genome size: 150 Mb
Number of genes: 14,000

Drosophila has been used extensively for genetic studies for many years. About 50% of all fly proteins show similarities to mammalian proteins.

alpsdake PD

Japanese canopy plant
(*Paris japonica*)

Genome size: 149,000 Mb

This rare native Japanese plant has the largest genome sequenced so far (15% larger than any previous estimate for a eukaryote). Plants with very large genomes reproduce and grow slowly.

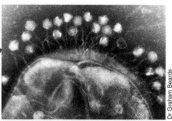

Dr Graham Beards

T2 phage

Genome size: 160,000 bp
Number of genes: Approx. 300

T2 phage is one of a group of related T-even phages that infect bacteria. Analysis of these phages indicates a small core genome with variations being the result of genetic transfers during evolution.

1. For each organism below, calculate how much smaller or larger the genome is than the human genome:

   (a) Japanese canopy plant: _____

   (b) *E. coli*: _____

   (c) T2 phage: _____

2. Plants with very large genome sizes are at higher risk of extinction. Can you suggest why? _____

3. Access Genbank® through the Weblinks page for this book and use it to answer the following:

   (a) The number of base pairs in human mitochondrial DNA: _____

   (b) The number of bases on human chromosome 1: _____

   (c) The guanine/cytosine percentage (GC%) in human chromosome 1: _____

78 Prokaryotic Chromosome Structure

Key Idea: Prokaryote DNA is packaged as one single chromosome that is not associated with protein.

DNA is a universal carrier of genetic information but it is packaged differently in prokaryotic and eukaryotic cells. Unlike eukaryotic chromosomes, the prokaryotic chromosome is not enclosed in a nuclear membrane and is not associated with protein. It is a single circular (rather than linear) molecule of double stranded DNA, attached to the plasma membrane and located in a region called the nucleoid, which is in direct contact with the cytoplasm. As well as the bacterial

chromosome, bacteria often contain small circular, double-stranded DNA molecules called plasmids. Plasmids are not connected to the main bacterial chromosome and usually contain 5-100 genes that are not crucial to cell survival under normal conditions. However, in certain environments, they may provide a selective advantage as they may carry genes for properties such as antibiotic resistance and heavy metal tolerance. Horizontal gene transfer of plasmid DNA by bacterial conjugation is a major factor in the spread drug resistance and in rapid bacterial evolution.

The Prokaryotic Chromosome

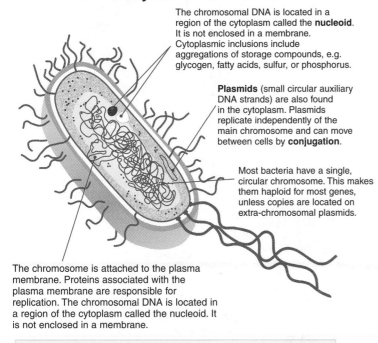

The chromosomal DNA is located in a region of the cytoplasm called the **nucleoid**. It is not enclosed in a membrane. Cytoplasmic inclusions include aggregations of storage compounds, e.g. glycogen, fatty acids, sulfur, or phosphorus.

Plasmids (small circular auxiliary DNA strands) are also found in the cytoplasm. Plasmids replicate independently of the main chromosome and can move between cells by **conjugation**.

Most bacteria have a single, circular chromosome. This makes them haploid for most genes, unless copies are located on extra-chromosomal plasmids.

The chromosome is attached to the plasma membrane. Proteins associated with the plasma membrane are responsible for replication. The chromosomal DNA is located in a region of the cytoplasm called the nucleoid. It is not enclosed in a membrane.

One Gene-one Protein?

In contrast to eukaryotes, prokaryotic DNA consists almost entirely of protein coding genes and their regulatory sequences. It was the study of prokaryotic genomes that gave rise to the one gene-one protein hypothesis, which still largely holds true for bacteria.

The Bacterial Plasmid

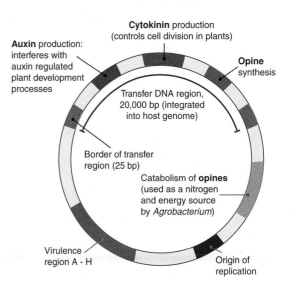

Cytokinin production (controls cell division in plants)

Auxin production: interferes with auxin regulated plant development processes

Transfer DNA region, 20,000 bp (integrated into host genome)

Opine synthesis

Border of transfer region (25 bp)

Catabolism of **opines** (used as a nitrogen and energy source by *Agrobacterium*)

Virulence region A - H

Origin of replication

The bacterium *Agrobacterium tumefaciens* often contains the *Ti* (tumour inducing) plasmid. This plasmid is able to transfer genetic material into plant cells and causes crown gall disease. Several regions on the plasmid (identified above) help it to infect plants. The plasmid is just over 200,000 bp long and contains 196 genes. The mapping of its genes has made it of great importance in the creation of transgenic plants.

1. Describe two important ways in which prokaryote and eukaryote chromosomes differ:

   (a) _____

   _____

   (b) _____

   _____

2. Explain the consequences to protein synthesis of the prokaryotic chromosome being free in the cytoplasm:

   _____

   _____

3. Most of the bacterial genome comprises protein coding genes and their regulatory sequences. Explain the consequence of this to the relative sizes of bacterial and eukaryotic chromosomes:

   _____

   _____

   _____

79 Eukaryotic Chromosome Structure

Key Idea: Eukaryotic DNA is located in the cell nucleus. A DNA molecule is very long. It must be wound up to fit into the cell's nucleus.

Eukaryotes package their DNA as discrete linear chromosomes. The number of chromosomes varies from species to species. The way the DNA is packaged changes during the life cycle of the cell, but classic chromosome structures (below) appear during metaphase of mitosis.

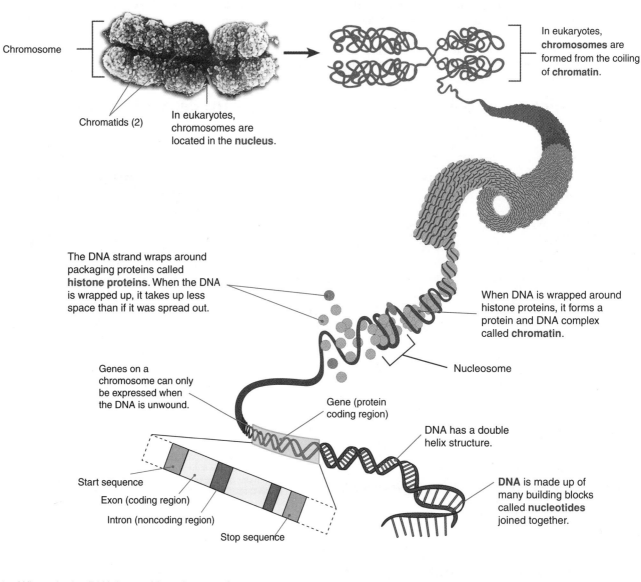

Chromosome

Chromatids (2)

In eukaryotes, chromosomes are located in the **nucleus**.

In eukaryotes, **chromosomes** are formed from the coiling of **chromatin**.

The DNA strand wraps around packaging proteins called **histone proteins**. When the DNA is wrapped up, it takes up less space than if it was spread out.

When DNA is wrapped around histone proteins, it forms a protein and DNA complex called **chromatin**.

Nucleosome

Genes on a chromosome can only be expressed when the DNA is unwound.

Gene (protein coding region)

DNA has a double helix structure.

Start sequence

Exon (coding region)

Intron (noncoding region)

Stop sequence

DNA is made up of many building blocks called **nucleotides** joined together.

1. Where is the DNA located in eukaryotes? _____

2. Why does DNA need to be packaged up to fit inside a cell nucleus? _____
   _____

3. How do histone proteins help in the coiling of DNA? _____
   _____
   _____
   _____
   _____

4. What is the difference between an exon and an intron? _____
   _____
   _____

80 Stages in Meiosis

Key Idea: Meiosis is a special type of cell division. It produces sex cells (gametes) for the purpose of sexual reproduction. Meiosis involves a single chromosomal duplication followed by two successive nuclear divisions, and results in a halving of the diploid chromosome number. Meiosis occurs in the sex organs of animals and the sporangia of plants. If genetic mistakes (**gene** and **chromosome mutations**) occur here, they will be passed on to the offspring (they will be inherited).

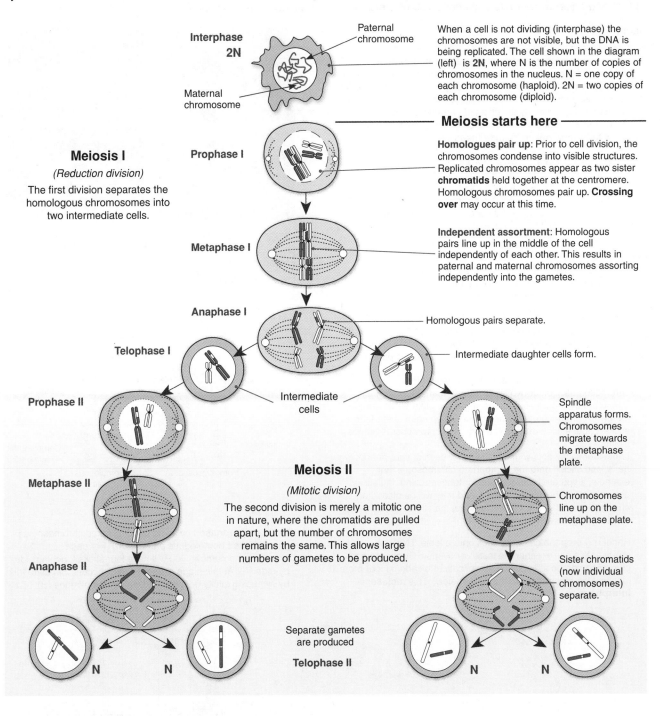

Interphase 2N

Paternal chromosome

Maternal chromosome

When a cell is not dividing (interphase) the chromosomes are not visible, but the DNA is being replicated. The cell shown in the diagram (left) is **2N**, where N is the number of copies of chromosomes in the nucleus. N = one copy of each chromosome (haploid). 2N = two copies of each chromosome (diploid).

——————— **Meiosis starts here** ———————

Meiosis I
(Reduction division)

The first division separates the homologous chromosomes into two intermediate cells.

Prophase I

Homologues pair up: Prior to cell division, the chromosomes condense into visible structures. Replicated chromosomes appear as two sister **chromatids** held together at the centromere. Homologous chromosomes pair up. **Crossing over** may occur at this time.

Metaphase I

Independent assortment: Homologous pairs line up in the middle of the cell independently of each other. This results in paternal and maternal chromosomes assorting independently into the gametes.

Anaphase I

Homologous pairs separate.

Telophase I

Intermediate daughter cells form.

Intermediate cells

Prophase II

Spindle apparatus forms. Chromosomes migrate towards the metaphase plate.

Metaphase II

Meiosis II
(Mitotic division)

The second division is merely a mitotic one in nature, where the chromatids are pulled apart, but the number of chromosomes remains the same. This allows large numbers of gametes to be produced.

Chromosomes line up on the metaphase plate.

Anaphase II

Sister chromatids (now individual chromosomes) separate.

Separate gametes are produced

Telophase II

N   **N**   **N**   **N**

1. Identify two ways meiosis produces variation in the gametes: _____

_____

_____

_____

2. How is meiosis I different from meiosis II? _____

_____

_____

LINK **82**   LINK **81**   WEB **80**   **KNOW**

81 Modelling Meiosis

Key Idea: We can simulate crossing over, gamete production, and the inheritance of alleles during meiosis using ice-block sticks to represent chromosomes.

This practical activity simulates the production of gametes (sperm and eggs) by **meiosis** and shows you how **crossing** **over** increases genetic variability. This is demonstrated by studying how two of your own alleles are inherited by the child produced at the completion of the activity. Completing this activity will help you to visualize and understand meiosis. It will take 25-45 minutes.

Background

Each of your somatic cells contain 46 chromosomes. You received 23 chromosomes from your mother (**maternal chromosomes**), and 23 chromosomes from your father (**paternal chromosomes**). Therefore, you have 23 homologous (same) pairs. For simplicity, the number of chromosomes studied in this exercise has been reduced to four (two homologous pairs). To study the effect of crossing over on genetic variability, you will look at the inheritance of two of your own traits: the ability to **tongue roll** and **handedness**.

Chromosome #	Phenotype	Genotype
10	Tongue roller	TT, Tt
10	Non-tongue roller	tt
2	Right handed	RR, Rr
2	Left handed	rr

Record your phenotype and genotype for each trait in the table (right). **NOTE:** If you have a dominant trait, you will not know if you are heterozygous or homozygous for that trait, so you can choose either genotype for this activity.

BEFORE YOU START THE SIMULATION: Partner up with a classmate. Your gametes will combine with theirs (fertilization) at the end of the activity to produce a child. Decide who will be the female, and who will be the male. You will need to work with this person again at step 6.

1. Collect four ice-blocks sticks. These represent four chromosomes. Colour two sticks blue or mark them with a P. These are the paternal chromosomes. The plain sticks are the maternal chromosomes. Write your initial on each of the four sticks. Label each chromosome with their chromosome number (right).
   Label four sticky dots with the alleles for each of your phenotypic traits, and stick it onto the appropriate chromosome. For example, if you are heterozygous for tongue rolling, the sticky dots with have the alleles **T** and **t**, and they will be placed on chromosome 10. If you are left handed, the alleles will be **r** and **r** and be placed on chromosome 2 (right).

2. Randomly drop the chromosomes onto a table. This represents a cell in either the testes or ovaries. **Duplicate** your chromosomes (to simulate DNA replication) by adding four more identical ice-block sticks to the table (below). This represents **interphase**.

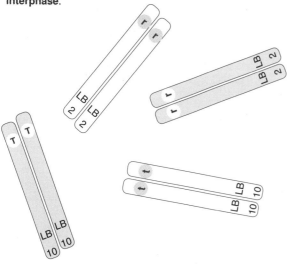

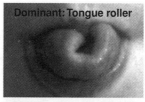

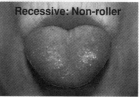

Trait	Phenotype	Genotype
Handedness		
Tongue rolling		

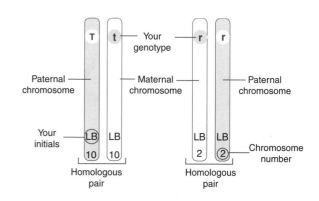

3. Simulate **prophase I** by lining the duplicated chromosome pair with their homologous pair (below). For each chromosome number, you will have four sticks touching side-by-side (A). At this stage **crossing over** occurs. Simulate this by swapping sticky dots from adjoining homologues (B).

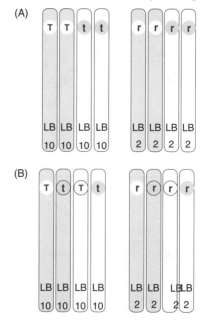

© 2012-2014 **BIOZONE** International
ISBN: 978-1-927173-93-0
Photocopying Prohibited

4. Randomly align the homologous chromosome pairs to simulate alignment on the metaphase plate (as occurs in **metaphase I**). Simulate **anaphase I** by separating chromosome pairs. For each group of four sticks, two are pulled to each pole.

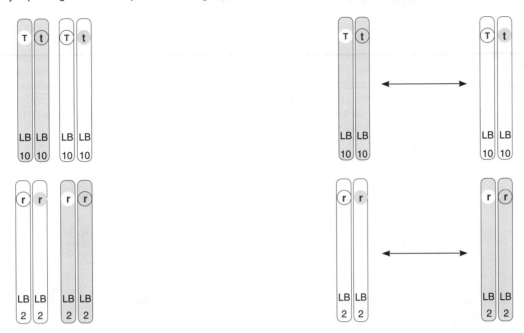

5. **Telophase I:** Two intermediate cells are formed. If you have been random in the previous step, each intermediate cell will contain a mixture of maternal and paternal chromosomes. This is the end of **meiosis 1**.

Now that meiosis 1 is completed, your cells need to undergo **meiosis 2.** Carry out prophase II, metaphase II, anaphase II, and telophase II. Remember, there is no crossing over in meiosis II. At the end of the process each intermediate cell will have produced two haploid gametes (below).

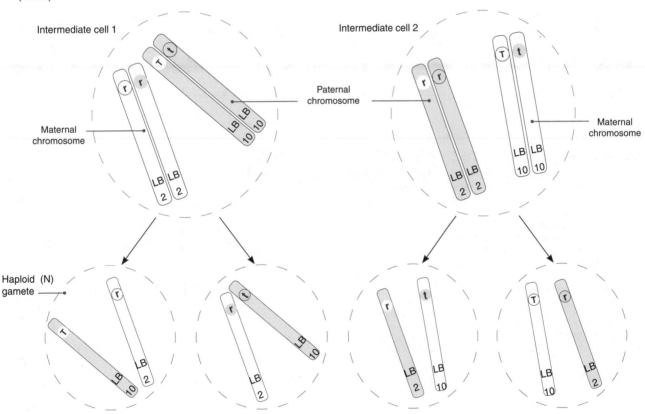

6. Pair up with the partner you chose at the beginning of the exercise to carry out **fertilization**. Randomly select one sperm and one egg cell. The unsuccessful gametes can be removed from the table. Combine the chromosomes of the successful gametes. You have created a child! Fill in the following chart to describe your child's genotype and phenotype for tongue rolling and handedness.

Trait	Phenotype	Genotype
Handedness		
Tongue rolling		

82 Non-Disjunction in Meiosis

Key Idea: Non-disjunction during meiosis results in incorrect apportioning of chromosomes to the gametes. Down syndrome is caused by non-disjunction of chromosome 21.

The meiotic spindle normally distributes chromosomes to daughter cells without error. However, mistakes can occur in which the homologous chromosomes fail to separate properly at anaphase during meiosis I, or sister chromatids fail to separate during meiosis II. In these cases, one gamete receives two of the same type of chromosome and the other gamete receives no copy. This mishap is called **non-disjunction** and it results in abnormal numbers of chromosomes passing to the gametes. If either aberrant gamete unites with a normal one at fertilization, the offspring will have an abnormal chromosome number, known as an **aneuploidy**. Down syndrome (trisomy 21) results from non-disjunction in this way.

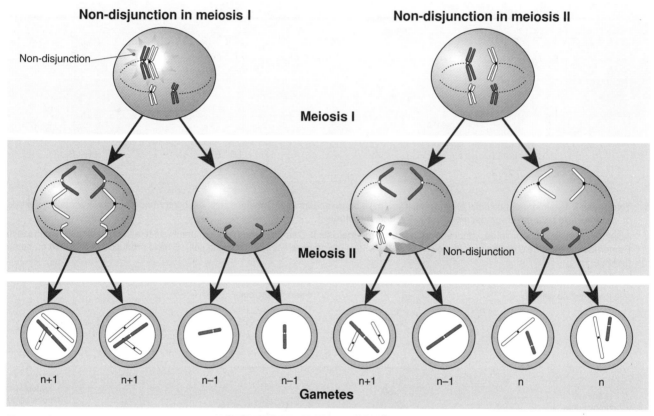

Down Syndrome (Trisomy 21)

Down syndrome is the most common of the human aneuploidies. The incidence rate is subject to a **maternal age effect** (the rate increases rapidly with maternal age). Nearly all cases (approximately 95%) result from **non-disjunction** of chromosome 21 during **meiosis**. When this happens, a gamete (most commonly the oocyte) ends up with 24 rather than 23 chromosomes, and fertilization produces a trisomic offspring.

Right: The karyotype of an individual with trisomy 21. The chromosomes are circled.

Incidence of Down syndrome related to maternal age	
Maternal age (years)	Incidence per 1000 live births
< 30	<1
30 - 34	1 - 2
35 - 39	2 - 5
40 - 44	5 - 10
> 44	10 -20

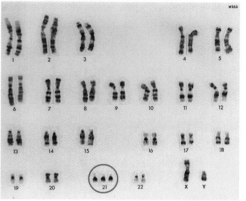

Photo: Waikato Hospital

1. Describe the consequences of non-disjunction during meiosis: _____

_____

_____

2. Explain why non-disjunction in meiosis I results in a higher proportion of faulty gametes than non-disjunction in meiosis II:

_____

_____

3. What is the maternal age effect and what are its consequences? _____

_____

LINK

APP 83

83 Karyotypes

Key Idea: The karyotype is the number and appearance of chromosomes in the nucleus of a eukaryotic cell. The karyotype can be pictured in a standard format, called a karyogram, in which the chromosomes are ordered by size. The diagram below shows a **karyogram** for a normal human. Karyotyping begins with 'freezing' the nuclei of cultured white blood cells in metaphase of mitosis. A photograph of the chromosomes is then cut up and the chromosomes are rearranged on a grid to produce the karyogram. The homologous pairs are placed together, identified by their general shape, length, and banding pattern after staining. In humans, the **male karyotype** has 44 autosomes, an X chromosome, and a Y chromosome (44 + XY). The **female karyotype** has two X chromosomes (44 + XX).

Typical Layout of a Human Karyogram

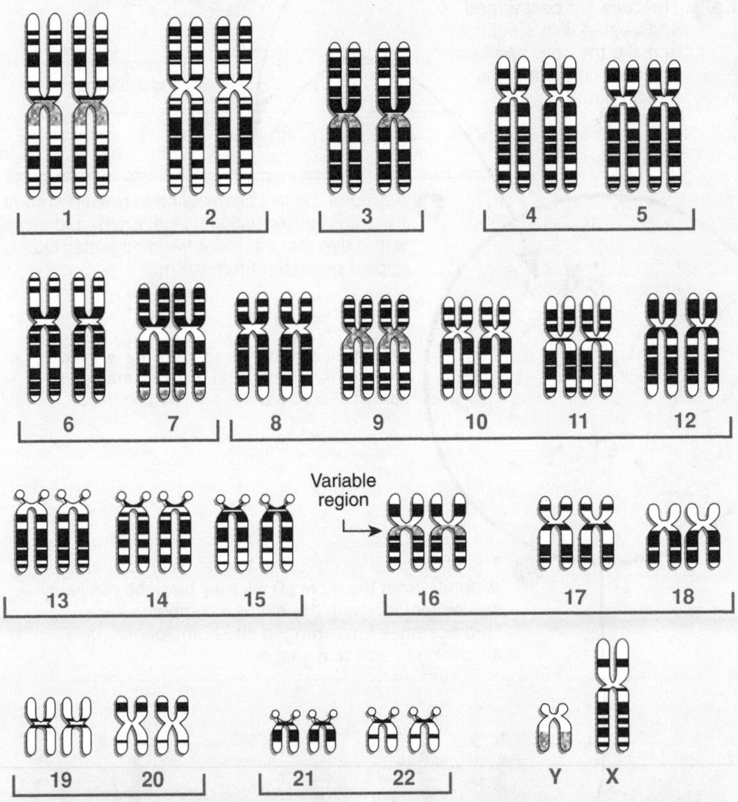

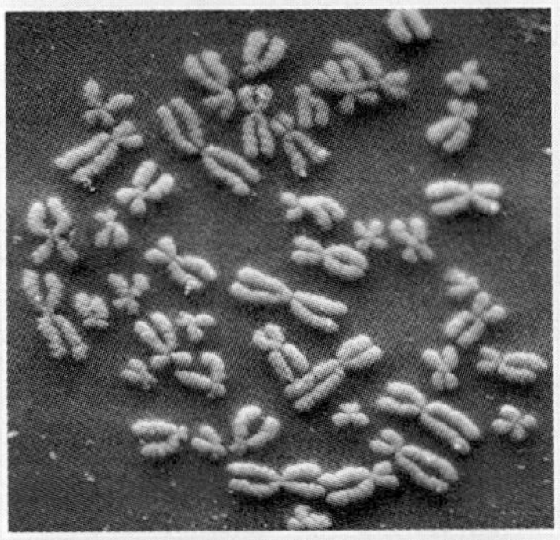

A scanning electron micrograph (SEM) of human chromosomes clearly showing their double chromatids.

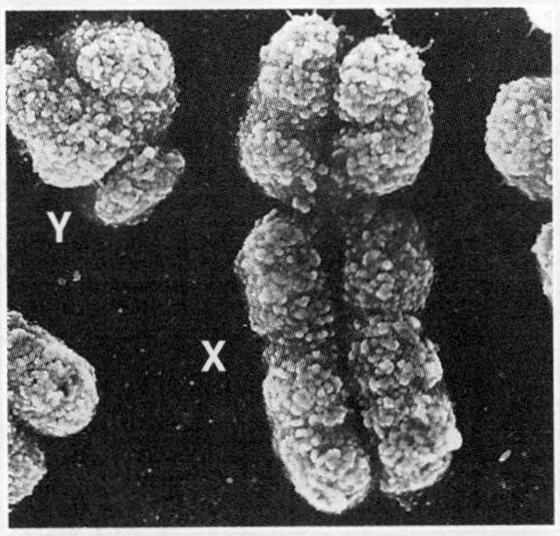

This SEM shows the human X and Y chromosomes. Although these two are the sex chromosomes, they are not homologous.

Karyotypes for different species

The term **karyotype** refers to the chromosome complement of a cell or a whole organism. In particular, it shows the number, size, and shape of the chromosomes as seen during metaphase of mitosis. The diagram on the left depicts the human karyotype. Chromosome numbers vary considerably among organisms and may differ markedly between closely related species:

Organism	Chromosome number (2N)
Vertebrates	
cat	38
rat	42
rabbit	44
human	46
chimpanzee	48
gorilla	48
cattle	60
horse	64
dog	78
turkey	82
goldfish	94
Invertebrates	
horse roundworm	2
fruit fly *Drosophila*	8
housefly	12
honey bee	32 or 16
Hydra	32
Plants	
broad bean	12
cabbage	18
garden pea	14
rice	24
Ponderosa pine	24
orange	18, 27 or 36
potato	48

NOTE: The number of chromosomes is not a measure of the quantity of genetic information.

1. (a) What is a **karyogram**? _____

   _____

   (b) What information can it provide? _____

   _____

2. Distinguish between **autosomes** and **sex chromosomes**: _____

   _____

   _____

LINK 84   WEB 83   APP

Preparing a Karyogram

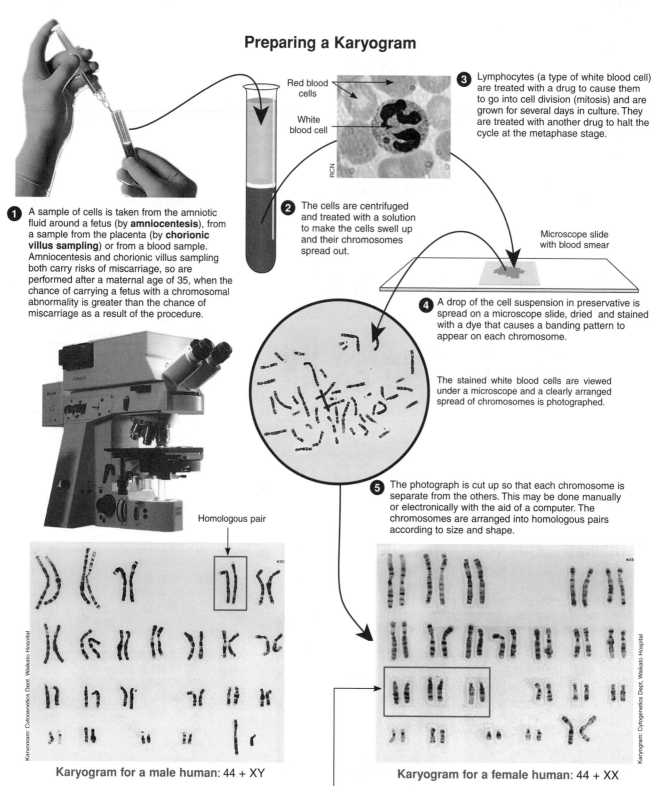

1 A sample of cells is taken from the amniotic fluid around a fetus (by **amniocentesis**), from a sample from the placenta (by **chorionic villus sampling**) or from a blood sample. Amniocentesis and chorionic villus sampling both carry risks of miscarriage, so are performed after a maternal age of 35, when the chance of carrying a fetus with a chromosomal abnormality is greater than the chance of miscarriage as a result of the procedure.

Red blood cells

White blood cell

RCN

3 Lymphocytes (a type of white blood cell) are treated with a drug to cause them to go into cell division (mitosis) and are grown for several days in culture. They are treated with another drug to halt the cycle at the metaphase stage.

2 The cells are centrifuged and treated with a solution to make the cells swell up and their chromosomes spread out.

Microscope slide with blood smear

4 A drop of the cell suspension in preservative is spread on a microscope slide, dried and stained with a dye that causes a banding pattern to appear on each chromosome.

The stained white blood cells are viewed under a microscope and a clearly arranged spread of chromosomes is photographed.

5 The photograph is cut up so that each chromosome is separate from the others. This may be done manually or electronically with the aid of a computer. The chromosomes are arranged into homologous pairs according to size and shape.

Homologous pair

Karyogram: Cytogenetics Dept, Waikato Hospital

Karyogram for a male human: 44 + XY

Karyogram: Cytogenetics Dept, Waikato Hospital

Karyogram for a female human: 44 + XX

Chromosomes are arranged in groups according to size, shape and banding pattern.

3. On the male and female karyograms above, **number** each homologous pair of chromosomes using the diagram on the previous page as a guide.

4. **Circle** the sex chromosomes (**X** and **Y**) in the karyogram of the female and the male.

5. Write down the number of *autosomes* and the arrangement of *sex chromosomes* for each sex:

    (a) **Female**:   No. of autosomes: _____   Sex chromosomes: _____

    (b) **Male**:   No. of autosomes: _____   Sex chromosomes: _____

6. State how many chromosomes are found in a:

    (a) Normal human (somatic) body cell: _____   (b) Normal human sperm or egg cell: _____

84 Human Karyotype Exercise

Key Idea: A karyogram can be created by matching the size and banding pattern of individual chromosomes.

Each chromosome has specific distinguishing features. Chromosomes are stained in a special technique that gives them a banded appearance in which the banding pattern represents regions containing up to many hundreds of genes.

Cut out the chromosomes below and arrange them on the record sheet in order to determine the sex and chromosome condition of the individual whose karyotype is shown. The karyograms presented on the previous pages and the hints on how to recognize chromosome pairs can be used to help you complete this activity.

Distinguishing Characteristics of Chromosomes

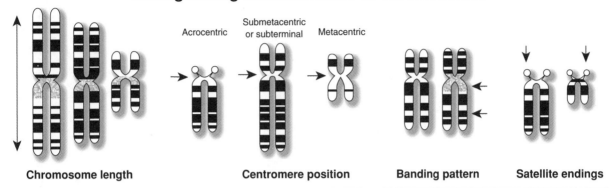

| Chromosome length | Centromere position | Banding pattern | Satellite endings |

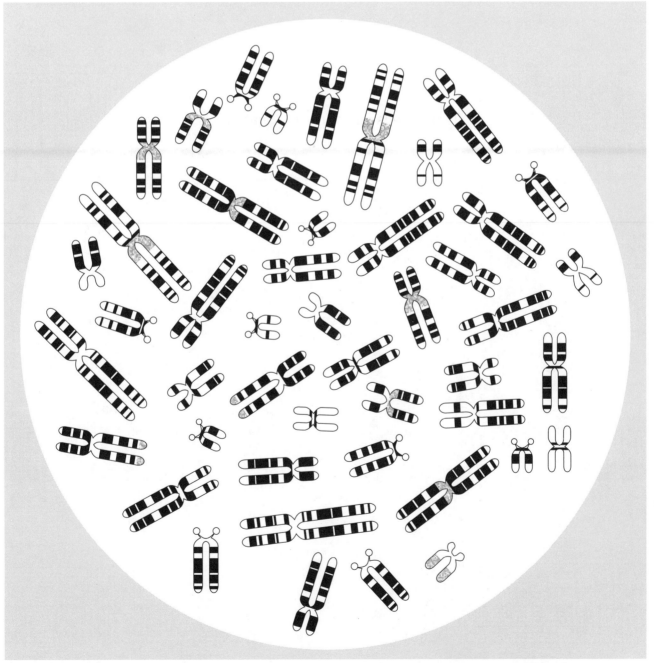

LINK **83**   LINK **82**   WEB **84**   **APP**

This page is left blank deliberately

1. Cut out the chromosomes on page 111 and arrange them on the record sheet below in their homologous pairs.

2. (a) Determine the sex of this individual: **male** or **female** (circle one)

    (b) State whether the individual's *chromosome arrangement* is: **normal** or **abnormal** (circle one)

    (c) If the arrangement is abnormal, state in what way and name the syndrome displayed: _____

    _____

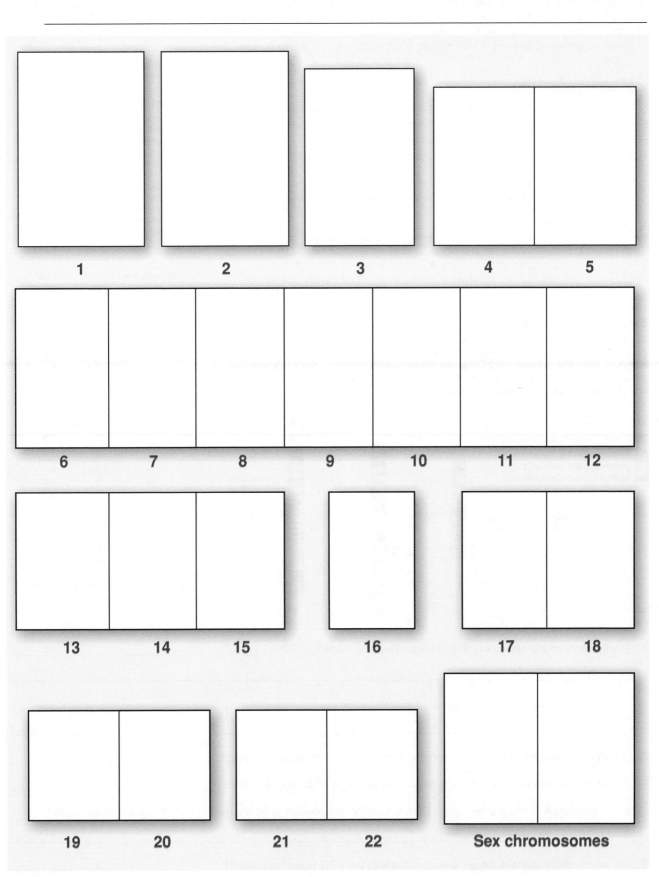

85 The Human Genome Project

Key Idea: The Human Genome Project (HGP) was a publicly funded global venture to determine the sequence of bases in the human genome and identify and map the genes.

The HGP was completed in 2003, ahead of schedule, although analysis of it continues. Other large scale sequencing projects have arisen as a result of the initiative to sequence the human genome. In 2002, for example, the International HapMap Project was started with the aim of describing the common patterns of human genetic variation. The HGP has provided an immense amount of information, but it is not the whole story. The next task is to find out what the identified genes do. The identification and study of the protein products of genes (**proteomics**) is an important development of the HGP and will give a better understanding of the functioning of the genome. There is also increasing research in the area of the human **epigenome**, which is the record of chemical changes to the DNA and histone proteins that do not involve changes in the DNA base sequence itself. The epigenome is important in regulating genome function. Understanding how it works is important to understanding disease processes.

Key results of the HGP

- There are perhaps only 20,000-25,000 protein-coding genes in our human genome.
- It covers 99% of the gene containing parts of the genome and is 99.999% accurate.
- The new sequence correctly identifies almost all known genes (99.74%).
- Its accuracy and completeness allows systematic searches for causes of disease.

Count of Mapped Genes

The aim of the HGP was to produce a continuous block of sequence information for each chromosome. Initially the sequence information was obtained to draft quality, with an error rate of 1 in 1000 bases. The **Gold Standard** sequence, with an error rate of <1 per 100,000 bases, was completed in October 2004. This table shows the length and number of mapped genes for each chromosome.

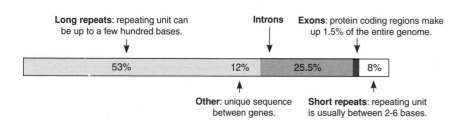

Long repeats: repeating unit can be up to a few hundred bases.

Introns

Exons: protein coding regions make up 1.5% of the entire genome.

| 53% | 12% | 25.5% | 8% |

Other: unique sequence between genes.

Short repeats: repeating unit is usually between 2-6 bases.

Examples of Mapped Genes

The positions of an increasing number of genes have been mapped onto human chromosomes (see below). Sequence variations can cause or contribute to identifiable disorders. Note that chromosome 21 (the smallest human chromosome) has a relatively low gene density, while others are gene rich. This is possibly why trisomy 21 (Down syndrome) is one of the few viable human autosomal trisomies.

Key

- Variable regions (heterochromatin)
- Regions reflecting the unique patterns of light and dark bands seen on stained chromosomes

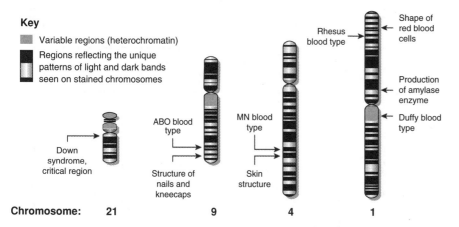

Down syndrome, critical region

ABO blood type

Structure of nails and kneecaps

MN blood type

Skin structure

Rhesus blood type

Shape of red blood cells

Production of amylase enzyme

Duffy blood type

Chromosome: 21   9   4   1

Chromosome	Length (Mb)	No. of Mapped Genes
1	263	1873
2	255	1113
3	214	965
4	203	614
5	194	782
6	183	1217
7	171	995
8	155	591
9	145	804
10	144	872
11	144	1162
12	143	894
13	114	290
14	109	1013
15	106	510
16	98	658
17	92	1034
18	85	302
19	67	1129
20	72	599
21	50	386
22	56	501
X	164	1021
Y	59	122
Total:		**19,447**

1. Briefly describe the objectives of the Human Genome Project (HGP): _____

_____

_____

_____

2. (a) What percentage of the human genome is made up of long repeating units? _____

(b) What percentage of the human genome is made up of short repeating units? _____

(c) How would knowing the full human DNA sequence and position of genes help in identifying certain diseases?

_____

3. On which chromosome is the gene associated with the ABO blood type found? _____

© 2012-2014 **BIOZONE** International
ISBN: 978-1-927173-93-0

86 Mendel's Pea Plant Experiments

Key Idea: Many genes produce phenotypic traits that are inherited in predictable ratios, as shown by Mendel's pea experiments.

Gregor Mendel (1822-84), right, was an Austrian monk who carried out the pioneering studies of inheritance. Mendel bred pea plants to study the inheritance patterns of a number of **traits** (specific characteristics). He showed that characters could be masked in one generation but could reappear in later generations and proposed that inheritance involved the transmission of discrete units of inheritance from one generation to the next. At the time the mechanism of inheritance was unknown, but further research has provided an accepted mechanism and we know these units of inheritance are **genes**. The entire genetic makeup of an organism is its **genotype**.

Mendel examined six traits and found that they were inherited in predictable ratios, depending on the **phenotypes** (the physical appearance) of the parents. Some of his results from crossing heterozygous plants are tabulated below. The numbers in the results column represent how many offspring had those traits.

1. Study the results for each of the six experiments below. Determine which of the two phenotypes is dominant, and which is the recessive. Place your answers in the spaces in the **dominance** column in the table below.
2. Calculate the ratio of dominant phenotypes to recessive phenotypes (to two decimal places). The first one has been done for you (5474 ÷ 1850 = 2.96). Place your answers in the spaces provided in the table below:

Trait	Possible Phenotypes		Results		Dominance	Ratio
Seed shape	Wrinkled	Round	Wrinkled	1850	Dominant: Round	2.96 : 1
			Round	5474	Recessive: Wrinkled	
			TOTAL	**7324**		
Seed colour	Green	Yellow	Green	2001	Dominant:	
			Yellow	6022		
			TOTAL	**8023**	Recessive	
Pod colour	Green	Yellow	Green	428	Dominant:	
			Yellow	152		
			TOTAL	**580**	Recessive	
Flower position	Axial	Terminal	Axial	651	Dominant:	
			Terminal	207		
			TOTAL	**858**	Recessive	
Pod shape	Constricted	Inflated	Constricted	299	Dominant:	
			Inflated	882		
			TOTAL	**1181**	Recessive	
Stem length	Tall	Dwarf	Tall	787	Dominant:	
			Dwarf	277		
			TOTAL	**1064**	Recessive	

3. Mendel's experiments identified that two heterozygous parents should produce offspring in the ratio of three times as many dominant offspring to those showing the recessive phenotype.

(a) Which three of Mendel's experiments provided ratios closest to the theoretical 3:1 ratio? _____

_____

(b) Suggest why these results deviated less from the theoretical ratio than the others: _____

_____

_____

© 2012-2014 **BIOZONE** International
ISBN: 978-1-927173-93-0
Photocopying Prohibited

LINK
87
WEB
86
KNOW

87 Mendel's Laws of Inheritance

Key Idea: Genetic information is inherited from parents in discrete units called genes. Mendel's laws of inheritance govern how these genes are passed to the offspring.

Mendel's laws account for the inheritance patterns he observed in his experiments.

Particulate Inheritance

Characteristics of both parents are passed on to the next generation as discrete entities (genes).

This model explained many observations that could not be explained by the idea of blending inheritance, which was universally accepted prior to this theory. The trait for flower colour (right) appears to take on the appearance of only one parent plant in the first generation, but reappears in later generations.

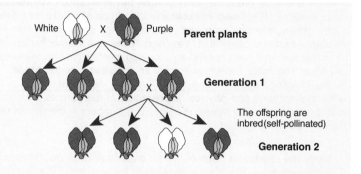

White X Purple **Parent plants**

Generation 1

The offspring are inbred (self-pollinated)

Generation 2

Law of Segregation

During gametic meiosis, the two members of any pair of alleles segregate unchanged and are passed into different gametes.

These gametes are eggs (ova) and sperm cells. The allele in the gamete will be passed on to the offspring.

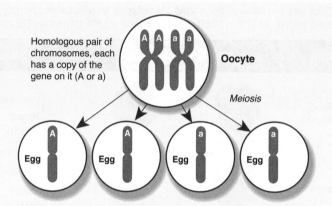

Homologous pair of chromosomes, each has a copy of the gene on it (A or a)

Oocyte

Meiosis

Egg Egg Egg Egg

Law of Independent Assortment

Allele pairs separate independently during gamete formation, and traits are passed on to offspring independently of one another (this is only true for unlinked genes).

This diagram shows two genes (A and B) that code for different traits. Each of these genes is represented twice, one copy (allele) on each of two homologous chromosomes. The genes A and B are located on different chromosomes and, because of this, they will be inherited independently of each other i.e. the gametes may contain any combination of the parental alleles.

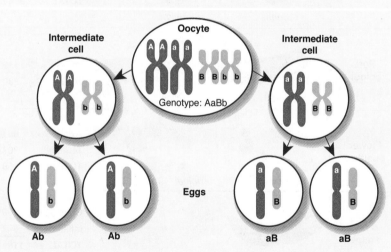

Intermediate cell **Oocyte** **Intermediate cell**

Genotype: AaBb

Eggs

Ab Ab aB aB

---

1. State the **property of genetic inheritance** that allows parent pea plants of different flower colour to give rise to flowers of a single colour in the first generation, with both parental flower colours reappearing in the following generation:

   _____

2. The oocyte is the egg producing cell in the ovary of an animal. In the diagram illustrating the **law of segregation** above:

   (a) State the genotype for the oocyte (adult organism): _____

   (b) State the genotype of each of the **four** gametes: _____

   (c) State how many different kinds of gamete can be produced by this oocyte: _____

3. The diagram illustrating the **law of independent assortment** (above) shows only one possible result of the random sorting of the chromosomes to produce: Ab and aB in the gametes.
   (a) List another possible combination of genes (on the chromosomes) ending up in gametes from the same oocyte:

   _____

   (b) How many different gene combinations are possible for the oocyte? _____

© 2012-2014 **BIOZONE** International
ISBN: 978-1-927173-93-0
Photocopying Prohibited

88 Basic Genetic Crosses

Key Idea: The outcome of a cross depends on the parental genotypes. A true breeding parent is homozygous for the gene involved.

Examine the diagrams depicting monohybrid (single gene) inheritance. The F₁ generation by definition describes the offspring of a cross between distinctly different, **true-breeding** (homozygous) parents. A **back cross** refers to any cross between an offspring and one of its parents. If the back cross is to a homozygous recessive, it is diagnostic, and is therefore called a test cross.

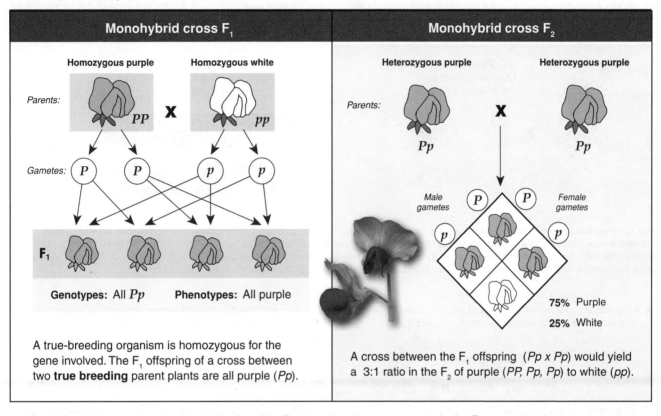

Monohybrid cross F₁

Homozygous purple **PP** X Homozygous white **pp**

Gametes: P, P, p, p

F₁ — **Genotypes:** All *Pp*   **Phenotypes:** All purple

A true-breeding organism is homozygous for the gene involved. The F₁ offspring of a cross between two **true breeding** parent plants are all purple (*Pp*).

Monohybrid cross F₂

Heterozygous purple **Pp** X Heterozygous purple **Pp**

Male gametes P, P / Female gametes p, p

75% Purple   **25%** White

A cross between the F₁ offspring (*Pp x Pp*) would yield a 3:1 ratio in the F₂ of purple (*PP, Pp, Pp*) to white (*pp*).

1. Study the diagrams above and explain why white flower colour does not appear in the F₁ generation but reappears in the F₂ generation:

_____

_____

_____

_____

2. Complete the crosses below:

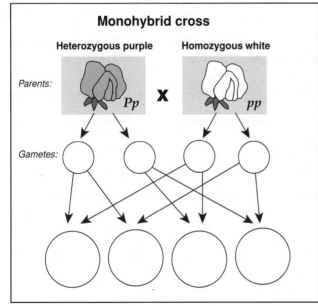

Monohybrid cross

Parents: Heterozygous purple *Pp* X Homozygous white *pp*

Gametes:

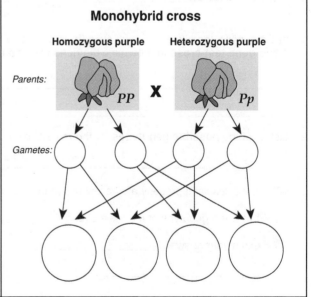

Monohybrid cross

Parents: Homozygous purple *PP* X Heterozygous purple *Pp*

Gametes:

LINK **95** LINK **90** LINK **89** WEB **88** **KNOW**

89 The Test Cross

Key Idea: An unknown genotype of an individual can be determined by crossing the individual with another that is homozygous recessive for the same trait and observing the phenotypes of the offspring.

It is not always possible to determine an organism's genotype by its appearance because gene expression is complicated by patterns of dominance and by gene interactions. The **test cross** was developed by Mendel as a way to establish the

genotype of an organism with the dominant phenotype for a particular trait. The principle is simple. The individual with the unknown genotype is bred with a homozygous recessive individual for the trait(s) of interest. The homozygous recessive can produce only one type of allele (recessive), so the phenotypes of the offspring will reveal the genotype of the unknown parent. The test cross can be used to determine genotypes involving single genes or multiple genes.

Parent 1
Unknown genotype
(but with dominant traits)

Parent 2
Homozygous recessive genotype
(no dominant traits)

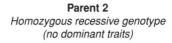

X

The common fruit fly (*Drosophila melanogaster*) is often used to illustrate basic principles of inheritance because it has several genetic markers whose phenotypes are easily identified. Once such phenotype is body colour. Wild type (normal) *Drosophila* have yellow-brown bodies. The allele for yellow-brown body colour (E) is dominant. The allele for an ebony coloured body (e) is recessive. The test crosses below show the possible outcomes for an individual with homozygous and heterozygous alleles for ebony body colour.

A. A homozygous recessive female (ee) with an ebony body is crossed with a homozyogous dominant male (EE).

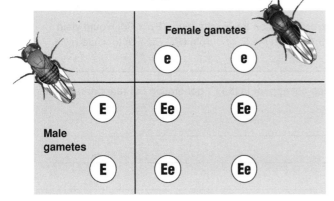

Cross A:
(a) Genotype frequency: _100% Ee_

(b) Phenotype frequency: _100% yellow-brown_

B. A homozygous recessive female (ee) with an ebony body is crossed with a heterozygous male (Ee).

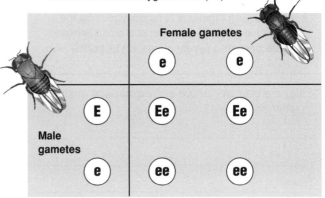

Cross B:
(a) Genotype frequency: _50% Ee, 50% ee_

(b) Phenotype frequency: _50% yellow-brown, 50% ebony_

1. In *Drosophila*, the allele for brown eyes (**b**) is recessive, while the red eye allele (**B**) is dominant. How would you set up a **two gene test cross** to determine the genotype of a male who has a normal body colour and red eyes?

_____

_____

2. List all of the **possible genotypes** for the male *Drosophila*: _____

_____

3. 50% of the resulting progeny are yellow-brown bodies with red eyes, and 50% have ebony bodies with red eyes.

(a) What is the genotype of the male *Drosophila*? _____

(b) Explain your answer: _____

_____

© 2012-2014 **BIOZONE** International
ISBN: 978-1-927173-93-0
Photocopying Prohibited

90 Monohybrid Cross

Key Idea: A monohybrid cross studies the inheritance pattern of one gene. The offspring of these crosses occur in predictable ratios.

In this activity, you will examine six types of matings possible for a pair of alleles governing coat colour in guinea pigs. A dominant allele (**B**) produces **black** hair and its recessive allele (**b**), produces white. Each parent can produce two types of gamete by meiosis. Determine the **genotype** and **phenotype frequencies** for the crosses below. For crosses 3 to 6, also determine the gametes produced by each parent (write these in the circles) and offspring genotypes and phenotypes (write these inside the offspring shapes).

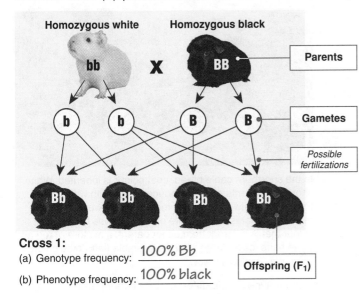

Homozygous white — Homozygous black — **Parents** — **Gametes** — *Possible fertilizations* — **Offspring (F₁)**

Cross 1:
(a) Genotype frequency: 100% Bb

(b) Phenotype frequency: 100% black

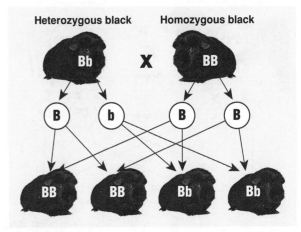

Heterozygous black — Homozygous black

Cross 2:
(a) Genotype frequency: _____

(b) Phenotype frequency: _____

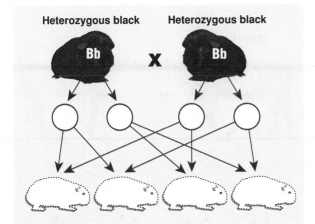

Heterozygous black — Heterozygous black

Cross 3:
(a) Genotype frequency: _____

(b) Phenotype frequency: _____

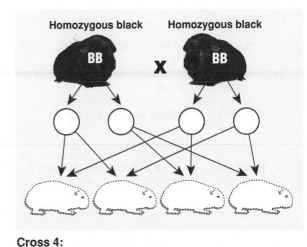

Homozygous black — Homozygous black

Cross 4:
(a) Genotype frequency: _____

(b) Phenotype frequency: _____

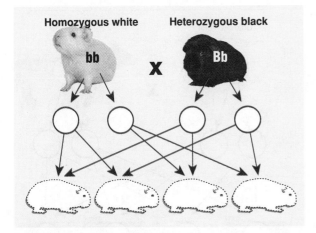

Homozygous white — Heterozygous black

Cross 5:
(a) Genotype frequency: _____

(b) Phenotype frequency: _____

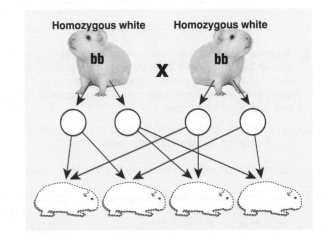

Homozygous white — Homozygous white

Cross 6:
(a) Genotype frequency: _____

(b) Phenotype frequency: _____

© 2012-2014 **BIOZONE** International
ISBN: 978-1-927173-93-0
Photocopying Prohibited

LINK **95** WEB **90** **KNOW**

91 Codominance of Alleles

Key Idea: In codominance, neither allele is recessive and both alleles are equally and independently expressed in the heterozygote.

Codominance is an inheritance pattern in which both alleles in a heterozygote contribute to the phenotype and both alleles are **independently** and **equally expressed**. Examples include the human blood group AB and certain coat colours in horses and cattle. Reddish coat colour is equally dominant with white. Animals that have both alleles have coats that are roan (both red and white hairs are present).

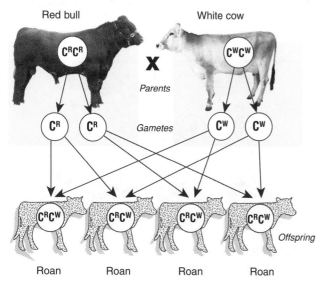

Red bull — $C^R C^R$
White cow — $C^W C^W$

X

Parents

Gametes: C^R C^R C^W C^W

Offspring: $C^R C^W$ $C^R C^W$ $C^R C^W$ $C^R C^W$

Roan   Roan   Roan   Roan

A roan shorthorn heifer

In the shorthorn cattle breed, coat colour is inherited. White shorthorn parents always produce calves with white coats. Red parents always produce red calves. However, when a red parent mates with a white one, the calves have a coat colour that is different from either parent; a mixture of red and white hairs, called roan. Use the example (left) to help you to solve the problems below.

1. Explain how codominance of alleles can result in offspring with a phenotype that is different from either parent:

_____

2. A white bull is mated with a roan cow (right):

   (a) Fill in the spaces to show the genotypes and phenotypes for parents and calves:

   (b) What is the phenotype ratio for this cross?

   _____

   (c) How could a cattle farmer control the breeding so that the herd ultimately consisted of only red cattle?

   _____

   _____

   _____

White bull   Roan cow

X

3. A farmer has only roan cattle on his farm. He suspects that one of the neighbours' bulls may have jumped the fence to mate with his cows earlier in the year because half the calves born were red and half were roan. One neighbour has a red bull, the other has a roan.

   (a) Fill in the spaces (right) to show the genotype and phenotype for parents and calves.

   (b) Which bull serviced the cows? **red** or **roan** (*delete one*)

4. Describe the classical phenotypic ratio for a codominant gene resulting from the cross of two heterozygous parents (e.g. a cross between two roan cattle):

   _____

   _____

Unknown bull ?   Roan cow

X

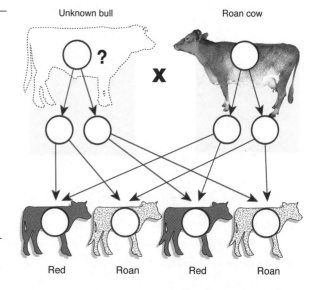

Red   Roan   Red   Roan

92 Codominance in Multiple Allele Systems

Key Idea: The human ABO blood group system is a multiple allele system involving the codominant alleles I^A and I^B and the recessive allele i.

The four common blood groups of the human 'ABO blood group system', determined by the alleles: I^A, I^B, and i, are A, B, AB, and O. The ABO antigens consist of sugars attached to the surface of red blood cells. The alleles code for enzymes (proteins) that join these sugars together. The allele i produces a non-functioning enzyme that is unable to make any changes to the basic antigen (sugar) molecule. The other two alleles (I^A, I^B) are **codominant** and are expressed equally. They each produce a different functional enzyme that adds a different, specific sugar to the basic sugar molecule. The blood group A and B antigens are able to react with antibodies present in the blood of other people so blood must always be matched for transfusion.

Recessive allele: **i**	produces a non-functioning protein	
Dominant allele: **I^A**	produces an enzyme which forms **A antigen**	
Dominant allele: **I^B**	produces an enzyme which forms **B antigen**	

Blood group (phenotype)	Possible genotypes	Frequency*		
		White	Black	Native American
O	ii	45%	49%	79%
A	$I^A I^A$ $I^A i$	40%	27%	16%
B		11%	20%	4%
AB		4%	4%	1%

* Frequency is based on North American population
Source: www.kcom.edu/faculty/chamberlain/Website/MSTUART/Lect13.htm

If a person has the $I^A i$ allele combination then their blood group will be group **A**. The presence of the recessive allele has no effect on the blood group in the presence of a dominant allele. Another possible allele combination that can create the same blood group is $I^A I^A$.

1. Use the information above to complete the table (right) for the possible genotypes for blood group B and group AB.

2. Below are six crosses possible between couples of various blood group types. The first example has been completed for you. Complete the genotype and phenotype for the other five crosses below:

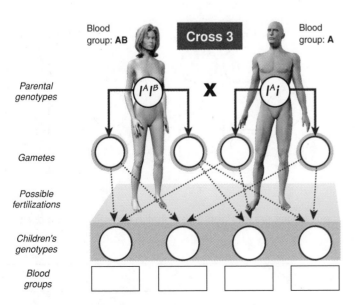

Cross 1 — Blood group: AB × Blood group: AB
Parental genotypes: $I^A I^B$ × $I^A I^B$
Gametes: I^A, I^B, I^A, I^B
Children's genotypes: $I^A I^A$, $I^A I^B$, $I^A I^B$, $I^B I^B$
Blood groups: A, AB, AB, B

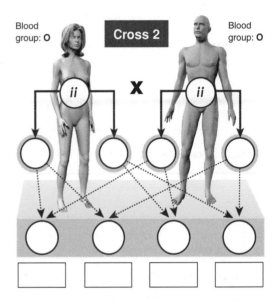

Cross 2 — Blood group: O × Blood group: O
Parental genotypes: ii × ii

Cross 3 — Blood group: AB × Blood group: A
Parental genotypes: $I^A I^B$ × $I^A i$

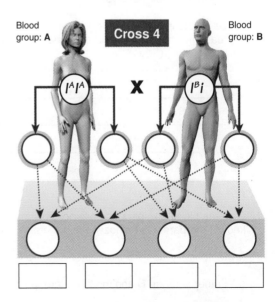

Cross 4 — Blood group: A × Blood group: B
Parental genotypes: $I^A I^A$ × $I^B i$

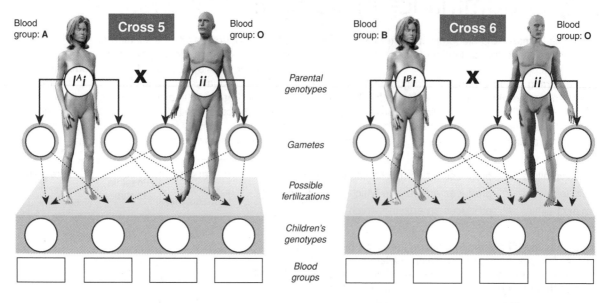

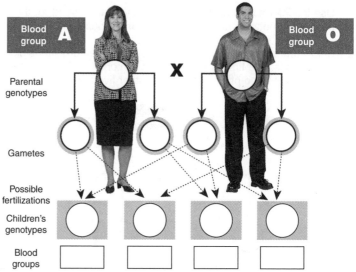

3. A wife is heterozygous for blood group **A** and the husband has blood group **O**.

   (a) Give the genotypes of each parent (fill in spaces on the diagram on the right).

   Determine the probability of:

   (b) One child having blood group **O**:

   _____

   (c) One child having blood group **A**:

   _____

   (d) One child having blood group **AB**:

   _____

4. In a court case involving a paternity dispute, a man claims that the child (blood group **B**) born to a woman is his son and wants custody. The woman claims that he is not the father.

   (a) List the possible genotypes for the child: _____

   (b) The man has a blood group **O** and the woman has a blood group **A**. Use the Punnett squares below to help you determine their genotypes.

   (c) Using the information in (a) and (b), state with reasons whether the child could be the man's son:

   _____

   _____

   _____

   _____

   _____

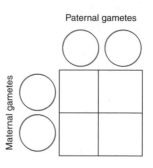

5. Give the blood groups which are possible for children of the following parents (remember that in some cases you don't know if the parent is homozygous or heterozygous).

   (a) Mother is group **AB** and father is group **O**: _____

   (b) Father is group **B** and mother is group **A**: _____

93 Sex Linkage

Key Idea: Many genes on the X chromosome do not have a match on the Y chromosome. In males, a recessive allele cannot therefore be masked by a dominant allele.

Sex linkage is a special case of linkage occurring when a gene is located on a sex chromosome (usually the X). The result is that the character encoded by the gene is usually seen only in the heterogametic sex (XY) and occurs rarely in the homogametic sex (XX). In humans, recessive sex linked genes are responsible for a number of heritable disorders in males, e.g. hemophilia. Women who have the recessive alleles on their chromosomes are said to be **carriers**.

Hemophilia is an inherited genetic disorder linked to the X-chromosome that results in ineffective blood clotting when a blood vessel is damaged. The most common type, hemophilia A, occurs in 1 in 5000 male births. Any male who carries the gene will express the phenotype. Hemophilia is extremely rare in women.

1. A couple wish to have children. The woman knows she a carrier for hemophilia. The man is not a hemophiliac. Use the notation X^h for hemophilia and X^H for the dominant allele to complete the diagram on the right including the parent genotypes, gametes and possible fertilizations. Write the genotypes and phenotypes in the table below.

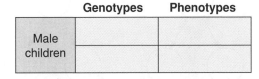

	Genotypes	Phenotypes
Male children		

Female children		

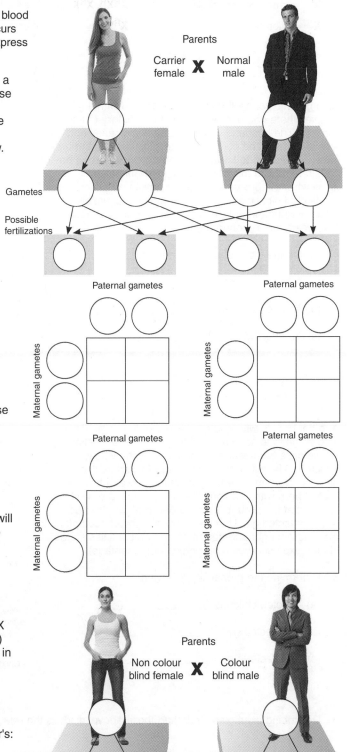

Parents

Carrier female **X** Normal male

Gametes

Possible fertilizations

Paternal gametes

Maternal gametes

Paternal gametes

Maternal gametes

Paternal gametes

Maternal gametes

Paternal gametes

Maternal gametes

2. (a) A second couple also wish to have children. The woman knows her maternal grandfather was a hemophiliac, but neither her mother or father were. Determine the probability she is a carrier (X^HX^h) Use the Punnett squares, right, to help you:

_____

_____

(b) The man is a normal non-hemophiliac male. Determine the probability that their first male child will have hemophilia. Use the Punnett squares, right, to help you:

_____

_____

3. The gene for red-green colour vision is carried on the X chromosome. If the gene is faulty, colour blindness (X^b) will occur in males. Red-green colour blindness occurs in about 8% of males but in less than 1% of females.

A colour blind man has children with a woman who is not colour blind. The couple have four children. Their phenotypes are: 1 non colour blind son, 1 colour blind son, 2 non colour blind daughters. Describe the mother's:

(a) Genotype: _____

(b) Phenotype: _____

(c) Identify the genotype not possessed by any of the children:

_____

Parents

Non colour blind female **X** Colour blind male

Gametes

Possible fertilizations

Dominant allele in humans

A rare form of rickets in humans is determined by a **dominant** allele of a gene on the **X chromosome** (it is not found on the Y chromosome). This condition is not successfully treated with vitamin D therapy. The allele types, genotypes, and phenotypes are as follows:

Allele types		Genotypes	Phenotypes
X^R = affected by rickets		$X^R X^R$, $X^R X$ =	Affected female
X = normal		$X^R Y$ =	Affected male
		XX, XY =	Normal female, male

As a genetic counsellor you are presented with a married couple where one of them has a family history of this disease. The husband is affected by this disease and the wife is normal. The couple, who are thinking of starting a family, would like to know what their chances are of having a child born with this condition. They would also like to know what the probabilities are of having an affected boy or affected girl. Use the symbols above to complete the diagram right and determine the probabilities stated below (expressed as a proportion or percentage).

4. Determine the probability of having:

    (a) Affected children: _____

    (b) An affected girl: _____

    (c) An affected boy: _____

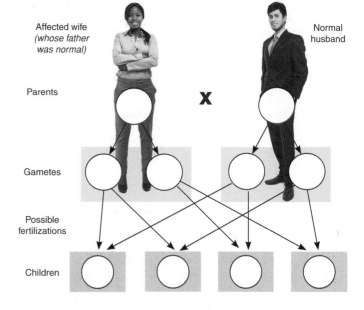

Another couple with a family history of the same disease also come in to see you to obtain genetic counseling. In this case the husband is normal and the wife is affected. The wife's father was not affected by this disease. Determine what their chances are of having a child born with this condition. They would also like to know what the probabilities are of having an affected boy or affected girl. Use the symbols above to complete the diagram right and determine the probabilities stated below (expressed as a proportion or percentage).

5. Determine the probability of having:

    (a) Affected children: _____

    (b) An affected girl: _____

    (c) An affected boy: _____

6. Describing examples other than those above, discuss the role of **sex linkage** in the inheritance of genetic disorders:

_____

_____

_____

_____

© 2012-2014 **BIOZONE** International
ISBN: 978-1-927173-93-0

94 Inheritance Patterns

Key Idea: Sex-linked traits and autosomal traits have different inheritance patterns.

Complete the following monohybrid crosses for different types of inheritance patterns in humans: autosomal recessive, autosomal dominant, sex linked recessive, and sex linked dominant inheritance.

1. **Inheritance of autosomal recessive traits**
   Example: *Albinism*

   Albinism (lack of pigment in hair, eyes and skin) is inherited as an autosomal recessive allele (not sex-linked).

   Using the codes: **PP** (normal) **Pp** (carrier)
   pp (albino)

   (a) Enter the parent phenotypes and complete the Punnett square for a cross between two carrier genotypes.

   (b) Give the ratios for the phenotypes from this cross.

   Phenotype ratios: _____

   _____

2. **Inheritance of autosomal dominant traits**
   Example: *Woolly hair*

   Woolly hair is inherited as an autosomal dominant allele. Each affected individual will have at least one affected parent.

   Using the codes: **WW** (woolly hair)
   Ww (woolly hair, heterozygous)
   ww (normal hair)

   (a) Enter the parent phenotypes and complete the Punnett square for a cross between two heterozygous individuals.

   (b) Give the ratios for the phenotypes from this cross.

   Phenotype ratios: _____

   _____

3. **Inheritance of sex linked recessive traits**
   Example: *Hemophilia*

   Inheritance of hemophilia is sex linked. Males with the recessive (hemophilia) allele, are affected. Females can be carriers.

   Using the codes: **XX** (normal female)
   XXʰ (carrier female)
   XʰXʰ (hemophiliac female)
   XY (normal male)
   XʰY (hemophiliac male)

   (a) Enter the parent phenotypes and complete the Punnett square for a cross between a normal male and a carrier female.

   (b) Give the ratios for the phenotypes from this cross.

   Phenotype ratios: _____

   _____

4. **Inheritance of sex linked dominant traits**
   Example: *Sex linked form of rickets*

   A rare form of rickets is inherited on the X chromosome.

   Using the codes: **XX** (normal female); **XY** (normal male)
   XᴿX (affected heterozygote female)
   XᴿXᴿ (affected female)
   XᴿY (affected male)

   (a) Enter the parent phenotypes and complete the Punnett square for a cross between an affected male and heterozygous female.

   (b) Give the ratios for the phenotypes from this cross.

   Phenotype ratios: _____

   _____

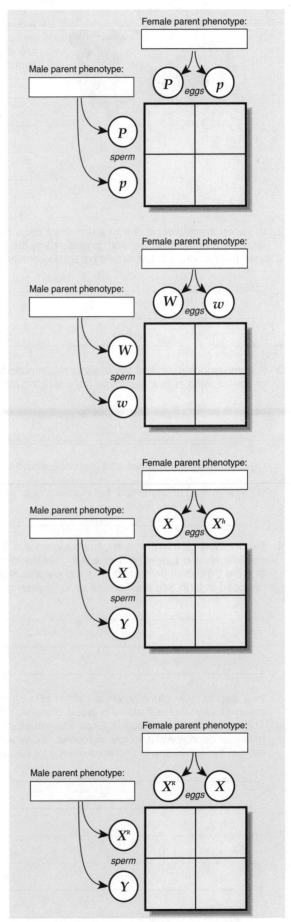

© 2012-2014 **BIOZONE** International
ISBN: 978-1-927173-93-0
Photocopying Prohibited

95 Problems Involving Monohybrid Inheritance

Key Idea: Monohybrid crosses involve only a single gene. The following problems involve Mendelian crosses. The alleles involved are associated with various phenotypic traits controlled by a single gene. The problems are to give you practice in problem solving using Mendelian genetics.

1. A dominant gene (**W**) produces wire-haired texture in dogs; its recessive allele (**w**) produces smooth hair. A group of heterozygous wire-haired individuals are crossed and their F_1 progeny are then test-crossed. Determine the expected genotypic and phenotypic ratios among the **test cross** progeny:

_____

_____

_____

_____

_____

2. In sheep, black wool is due to a recessive allele (**b**) and white wool to its dominant allele (**B**). A white ram is crossed to a white ewe. Both animals carry the black allele (**b**). They produce a white ram lamb, which is then back crossed to the female parent. Determine the probability of the **back cross** offspring being black:

_____

_____

_____

3. A homozygous recessive allele, aa, is responsible for albinism. Humans can exhibit this phenotype. In each of the following cases, determine the possible genotypes of the mother and father, and of their children:

(a) Both parents have normal phenotypes; some of their children are albino and others are unaffected: _____

_____

(b) Both parents are albino and have only albino children: _____

(c) The woman is unaffected, the man is albino, and they have one albino child and three unaffected children:

_____

4. Two mothers give birth to sons at a busy hospital. The son of the first couple has hemophilia, a recessive, X-linked disease. Neither parent from couple #1 has the disease. The second couple has an unaffected son, despite the fact that the father has hemophilia. The two couples challenge the hospital in court, claiming their babies must have been swapped at birth. You must advise as to whether or not the sons could have been swapped. What would you say?

_____

_____

_____

5. In a dispute over parentage, the mother of a child with blood group O identifies a male with blood group A as the father. The mother is blood group B. Draw Punnett squares to show possible genotype/phenotype outcomes to determine if the male is the father and the reasons (if any) for further dispute:

_____

_____

_____

_____

_____

LINK   LINK

© 2012-2014 **BIOZONE** International
ISBN: 978-1-927173-93-0
Photocopying Prohibited

96 Pedigree Analysis

Key Idea: Pedigree charts are a way of graphically illustrating inheritance patterns over a number of generations. They are used to study the inheritance of genetic disorders and make it possible to follow the genetic history of an individual.

Sample Pedigree Chart

Pedigree charts use various symbols to indicate an individuals particular traits. The key (right) should be consulted to make sense of the various symbols. Particular individuals are identified by their generation number and their order number in that generation. For example, **II-6** is the sixth person in the second row. The arrow indicates the **propositus**; the person through whom the pedigree was discovered (i.e. who reported the condition).

If the chart on the right were illustrating a human family tree, it would represent three generations: grandparents (I-1 and I-2) with three sons and one daughter. Two of the sons (II-3 and II-4) are identical twins, but did not marry or have any children. The other son (II-1) married and had a daughter and another child (sex unknown). The daughter (II-5) married and had two sons and two daughters (plus a child that died in infancy).

For the particular trait being studied, the grandfather was expressing the phenotype (showing the trait) and the grandmother was a carrier. One of their sons and one of their daughters also show the trait, together with one of their granddaughters.

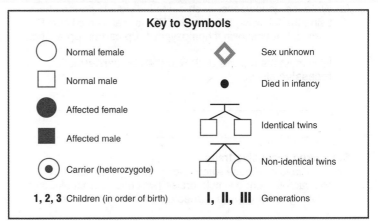

Key to Symbols

○ Normal female
□ Normal male
● Affected female
■ Affected male
◉ Carrier (heterozygote)
1, 2, 3 Children (in order of birth)
◇ Sex unknown
● Died in infancy
Identical twins
Non-identical twins
I, II, III Generations

1. Pedigree chart of your family

Using the symbols in the key above and the example illustrated as a guide, construct a pedigree chart of your own family (or one that you know of) starting with the parents of your mother and/or father on the first line. Your parents will appear on the second line (II) and you will appear on the third line (III). There may be a fourth generation line (IV) if one of your brothers or sisters has had a child. Use a ruler to draw up the chart carefully.

2. Autosomal recessive traits

Albinos lack pigment in the hair, skin and eyes. This trait is inherited as an autosomal recessive allele (i.e. it is not carried on the sex chromosome).

(a) Write the genotype for each of the individuals on the chart using the following letter codes: **PP** normal skin colour; **P-** normal, but unknown if homozygous; **Pp** carrier; **pp** albino.

(b) Why must the parents (II-3) and (II-4) be **carriers** of a **recessive** allele:

_____

_____

Albinism in humans

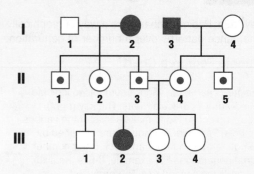

3. Sex linked recessive traits

Hemophilia is a disease where blood clotting is affected. A person can die from a simple bruise (which is internal bleeding). The clotting factor gene is carried on the X chromosome.

(a) Write the genotype for each of the individuals on the chart using the codes: **XY** normal male; **X^hY** affected male; **XX** normal female; **X^hX** female carrier; **X^hX^h** affected female:

(b) Why can males never be carriers?

_____

_____

_____

_____

Hemophilia in humans

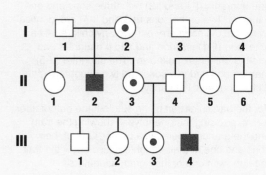

4. Autosomal dominant traits

An unusual trait found in some humans is woolly hair (not to be confused with curly hair). Each affected individual will have at least one affected parent.

(a) Write the genotype for each of the individuals on the chart using the following letter codes:
WW woolly hair; **Ww** woolly hair (heterozygous); **W-** woolly hair, but unknown if homozygous; **ww** normal hair

(b) Describe a feature of this inheritance pattern that suggests the trait is the result of a **dominant** allele:

_____

_____

_____

Woolly hair in humans

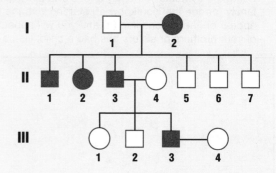

5. Sex linked dominant traits

A rare form of rickets is inherited on the X chromosome. All daughters of affected males will be affected. More females than males will show the trait.

(a) Write the genotype for each of the individuals on the chart using the following letter codes:
XY normal male; **X^RY** affected male; **XX** normal female; **X^R-** female (unknown if homozygous); **X^RX^R** affected female.

(b) Why will more females than males be affected?

_____

_____

_____

_____

A rare form of rickets in humans

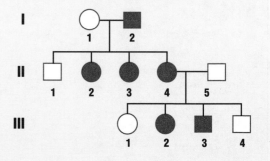

© 2012-2014 **BIOZONE** International
ISBN: 978-1-927173-93-0
Photocopying Prohibited

97 Amazing Organisms, Amazing Enzymes

Key Idea: The substances produced by organisms to survive in their environment can provide solutions to biotechnological problems.

Before the 1980s scientists knew of only a few organisms that could survive in extreme conditions. Indeed, many scientists believed that life in highly saline or high temperature and pressure environments was impossible. That view changed with the discovery of bacteria inhabiting the deep sea hydrothermal vents. They tolerate temperatures over 110°C and pressures of over 200 atmospheres. Bacteria were also found in volcanic hot pools on land, some surviving at temperatures in excess of 80°C. Most enzymes are denatured at temperatures above 40°C, but these **thermophilic** bacteria have enzymes that are fully functional at high temperatures. This discovery led to the development of one of the most important techniques in biotechnology, the **polymerase chain reaction** (PCR).

PCR is a technique, first described in the 1970s, that allows scientists to copy and multiply a piece of DNA millions of times. The DNA is heated to 98°C so that it separates into single strands and polymerase enzyme is added to synthesize new DNA strands from supplied free nucleotides. This earlier technique was labor intensive and expensive because the polymerase denatured at the high temperatures and had to be replaced every cycle. In 1985, a thermophilic polymerase (*Taq* **polymerase**) was isolated from the bacterium *Thermophilus aquaticus,* which inhabited the hot springs of Yellowstone National Park. Isolating this enzyme enabled automation of the PCR process, because the polymerase was stable throughout multiple cycles of synthesis. This led to a rapid growth in biotechnology, and gene technology in particular, because DNA samples could be easily copied for sequencing.

Searching for novel compounds in organisms from extreme environments is important in the development of new biotechnologies. Organisms must have compounds that can work in their specific environment, and the identification and extraction of these may allow them to be adapted for human use. For example, the Antarctic sea sponge *Kirkpatrickia variolosa* produces an alkaloid excreted as a toxic defence to prevent other organisms growing nearby. Tests indicate that this same chemical may have biological activity against cancer cells. Compounds from other sponge species are currently being assessed to treat a range of diseases including cancer, AIDS, tuberculosis and other bacterial infections, and cystic fibrosis.

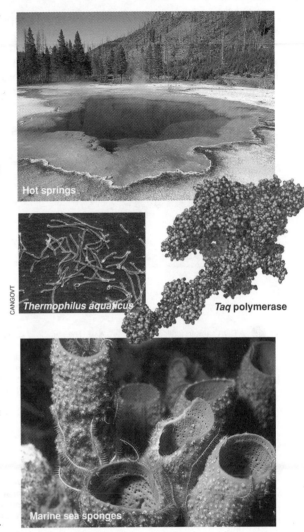

Hot springs

Thermophilus aquaticus

Taq polymerase

Marine sea sponges

1. Why was PCR not a viable technique until the mid 1980s? _____

_____

_____

_____

2. Explain why *Taq* polymerase was so important in the development of PCR: _____

_____

_____

_____

_____

3. Explain how investigating the lifestyles of other organisms can lead to advances in unrelated areas of science:

_____

_____

_____

_____

_____

_____

LINK
98

LINK
56

KNOW

98 Polymerase Chain Reaction

Key Idea: PCR uses a polymerase enzyme to copy a DNA sample, producing billions of copies in a few hours.

Many procedures in DNA technology, e.g. DNA sequencing and profiling, require substantial amounts of DNA yet, very often, only small amounts are obtainable (e.g. DNA from a crime scene or from an extinct organism). **PCR (polymerase**

chain reaction) is a technique for reproducing large quantities of DNA in the laboratory from an original sample. For this reason, it is often called **DNA amplification**. The technique is outlined below for a single cycle of replication. Subsequent cycles replicate DNA at an exponential rate, so PCR can produce billions of copies of DNA in only a few hours.

A Single Cycle of the Polymerase Chain Reaction

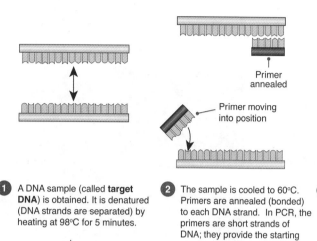

DNA polymerase: A thermally stable form of the enzyme is used (e.g. *Taq polymerase*). This is extracted from thermophilic bacteria.

Primer annealed

Primer moving into position

Nucleotides

Direction of synthesis

1. A DNA sample (called **target DNA**) is obtained. It is denatured (DNA strands are separated) by heating at 98°C for 5 minutes.

2. The sample is cooled to 60°C. Primers are annealed (bonded) to each DNA strand. In PCR, the primers are short strands of DNA; they provide the starting sequence for DNA extension.

3. DNA polymerase binds to the primers and synthesizes complementary strands of DNA using the free nucleotides.

4. After one cycle, there are now two copies of the original DNA.

Repeat for about 25 cycles

Repeat cycle of heating and cooling until enough copies of the target DNA have been produced

---

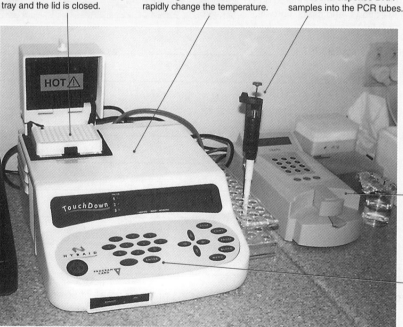

Loading Tray
Prepared samples in tiny PCR tubes are placed in the loading tray and the lid is closed.

Temperature Control
Inside the machine are heating and refrigeration mechanisms to rapidly change the temperature.

Dispensing Pipette
Pipettes with disposable tips are used to dispense DNA samples into the PCR tubes.

Thermal Cycler

Amplification of DNA can be carried out with simple-to-use machines called thermal cyclers. Once a DNA sample has been prepared, in just a few hours the amount of DNA can be increased billions of times. Thermal cyclers are in common use in the biology departments of universities, as well as other kinds of research and analytical laboratories. The one pictured on the left is typical of this modern piece of equipment.

DNA Quantitation
The amount of DNA in a sample can be determined by placing a known volume in this quantitation machine. For many genetic engineering processes, a minimum amount of DNA is required.

Controls
The control panel allows a number of different PCR programmes to be stored in the machine's memory. Carrying out a PCR run usually just involves starting one of the stored programmes.

1. Explain the purpose of PCR: _____

_____

_____

_____

© 2012-2014 **BIOZONE** International
ISBN: 978-1-927173-93-0

2. Describe how the **polymerase chain reaction** works: _____

_____

_____

_____

_____

_____

_____

3. Describe three situations where only very small DNA samples may be available for sampling and PCR could be used:

(a) _____

_____

(b) _____

_____

(c) _____

_____

4. After only two cycles of replication, four copies of the double-stranded DNA exist. Calculate how much a DNA sample will have increased after:

(a) 10 cycles: _____   (b) 25 cycles: _____

5. The risk of contamination in the preparation for PCR is considerable.

(a) Describe the effect of having a single molecule of unwanted DNA in the sample prior to PCR:

_____

_____

_____

(b) Describe two possible sources of DNA contamination in preparing a PCR sample:

Source 1: _____

Source 2: _____

(c) Describe two precautions that could be taken to reduce the risk of DNA contamination:

Precaution 1: _____

_____

Precaution 2: _____

_____

6. Describe two other genetic engineering/genetic manipulation procedures that require PCR amplification of DNA:

(a) _____

_____

_____

(b) _____

_____

_____

99 Gel Electrophoresis

Key Idea: Gel electrophoresis is used to separate DNA fragments on the basis of size.

DNA can be loaded onto an electrophoresis gel and separated by size. DNA has an overall negative charge, so when an electrical current is run through a gel, the DNA moves towards the positive electrode. The rate at which the DNA molecules move through the gel depends primarily on their size and the strength of the electric field. The gel they move through is full of pores (holes). Smaller DNA molecules move through the pores more easily (and more quickly) than larger ones. At the end of the process, the DNA molecules can be stained and visualized as a series of bands. Each band contains DNA molecules of a particular size. The bands furthest from the start of the gel contain the smallest DNA fragments. The bands closest to the start of the gel contain the largest DNA fragments.

Analyzing DNA using Gel Electrophoresis

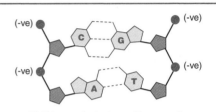

DNA is negatively charged because the phosphates (blue) that form part of the backbone of a DNA molecule have a negative charge.

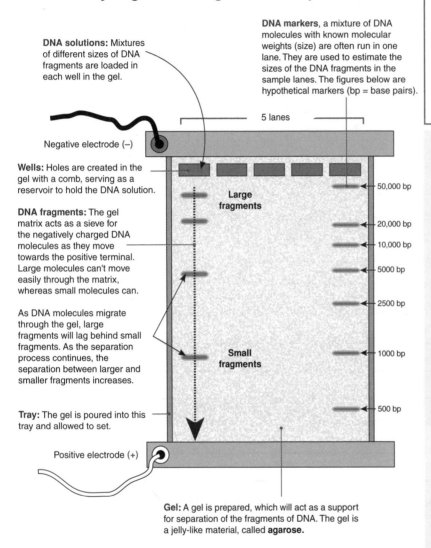

DNA solutions: Mixtures of different sizes of DNA fragments are loaded in each well in the gel.

DNA markers, a mixture of DNA molecules with known molecular weights (size) are often run in one lane. They are used to estimate the sizes of the DNA fragments in the sample lanes. The figures below are hypothetical markers (bp = base pairs).

5 lanes

Negative electrode (–)

Wells: Holes are created in the gel with a comb, serving as a reservoir to hold the DNA solution.

DNA fragments: The gel matrix acts as a sieve for the negatively charged DNA molecules as they move towards the positive terminal. Large molecules can't move easily through the matrix, whereas small molecules can.

As DNA molecules migrate through the gel, large fragments will lag behind small fragments. As the separation process continues, the separation between larger and smaller fragments increases.

Large fragments

Small fragments

50,000 bp
20,000 bp
10,000 bp
5000 bp
2500 bp
1000 bp
500 bp

Tray: The gel is poured into this tray and allowed to set.

Positive electrode (+)

Gel: A gel is prepared, which will act as a support for separation of the fragments of DNA. The gel is a jelly-like material, called **agarose.**

Steps in the Process of Gel Electrophoresis of DNA

1. A tray is prepared to hold the gel matrix.

2. A gel comb is used to create holes in the gel. The gel comb is placed in the tray.

3. Agarose gel powder is mixed with a buffer solution (this carries the DNA in a stable form). The solution is heated until dissolved and poured into the tray and allowed to cool.

4. The gel tray is placed in an electrophoresis chamber and the chamber is filled with buffer, covering the gel. This allows the electric current from electrodes at either end of the gel to flow through the gel.

5. DNA samples are mixed with a "loading dye" to make the DNA sample visible. The dye also contains glycerol or sucrose to make the DNA sample heavy so that it will sink to the bottom of the well and not disperse into the buffer solution.

6. A safety cover is placed over the gel, electrodes are attached to a power supply and turned on.

7. When the dye marker has moved through the gel, the current is turned off and the gel is removed from the tray.

8. DNA molecules are made visible by staining the gel dye (e.g. **methylene blue** or ethidium bromide) which binds to DNA and will fluoresce in UV light.

1. What is the purpose of gel electrophoresis? _____

_____

_____

2. Name the two forces that control the speed at which fragments pass through the gel:

   (a) _____

   (b) _____

3. Why do the smallest fragments travel through the gel the fastest? _____

_____

_____

© 2012-2014 **BIOZONE** International
ISBN: 978-1-927173-93-0
Photocopying Prohibited

100 DNA Profiling Using PCR

Key Idea: Short units of DNA that repeat a different number of times in different people can be used to produce individual genetic profiles.

In chromosomes, some of the DNA contains simple, repetitive sequences. These non-coding nucleotide sequences repeat over and over again and are found scattered throughout the genome. Some repeating sequences, called **microsatellites** or **short tandem repeats** (STRs), are very short (2-6 base pairs) and can repeat up to 100 times. The human genome has many different microsatellites. Equivalent sequences in different people vary considerably in the numbers of the repeating unit. This phenomenon has been used to develop

DNA profiling, which identifies the natural variations found in every person's DNA. Identifying these DNA differences is a useful tool for forensic investigations. In the USA, there are many laboratories approved for forensic DNA testing. Increasingly, these are targeting the 13 core STR loci; enough to guarantee that the odds of someone else sharing the same result are extremely unlikely (less than one in a billion). DNA profiling has been used to help solve previously unsolved crimes and to assist in current or future investigations. DNA profiling can also be used to establish genetic relatedness (e.g. in paternity disputes or pedigree disputes), or when searching for a specific gene (e.g. screening for disease).

Microsatellites (Short Tandem Repeats)

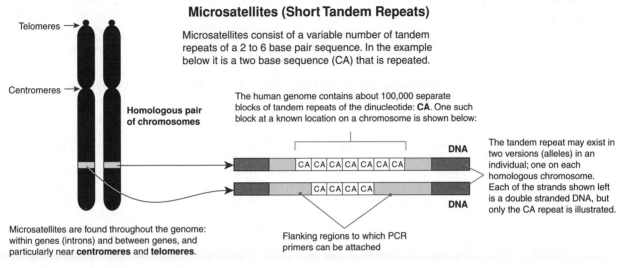

Microsatellites consist of a variable number of tandem repeats of a 2 to 6 base pair sequence. In the example below it is a two base sequence (CA) that is repeated.

Telomeres

Centromeres

Homologous pair of chromosomes

The human genome contains about 100,000 separate blocks of tandem repeats of the dinucleotide: **CA**. One such block at a known location on a chromosome is shown below:

DNA

DNA

The tandem repeat may exist in two versions (alleles) in an individual; one on each homologous chromosome. Each of the strands shown left is a double stranded DNA, but only the CA repeat is illustrated.

Microsatellites are found throughout the genome: within genes (introns) and between genes, and particularly near **centromeres** and **telomeres**.

Flanking regions to which PCR primers can be attached

How short tandem repeats are used in DNA profiling

This diagram shows how three people can have quite different microsatellite arrangements at the same point (locus) in their DNA. Each will produce a different DNA profile using gel electrophoresis:

1 **Extract DNA from sample**

A sample collected from the tissue of a living or dead organism is treated with chemicals and enzymes to extract the DNA, which is separated and purified.

2 **Amplify microsatellite using PCR**

Specific primers (arrowed) that attach to the flanking regions (light gray) either side of the microsatellite are used to make large quantities of the micro-satellite and flanking regions sequence only (no other part of the DNA is amplified/replicated).

3 **Visualize fragments on a gel**

The fragments are separated by length, using **gel electrophoresis**. DNA, which is negatively charged, moves toward the positive terminal. The smaller fragments travel faster than larger ones.

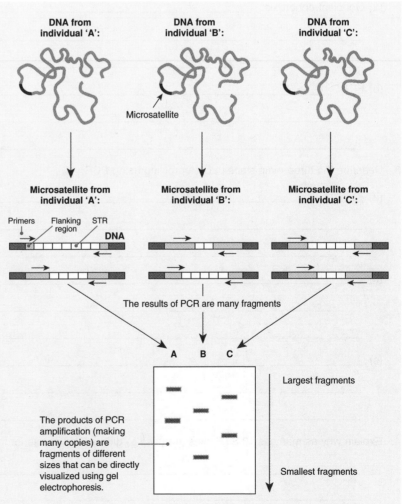

DNA from individual 'A':

DNA from individual 'B':

DNA from individual 'C':

Microsatellite

Microsatellite from individual 'A':

Microsatellite from individual 'B':

Microsatellite from individual 'C':

Primers Flanking region STR DNA

The results of PCR are many fragments

A B C

Largest fragments

The products of PCR amplification (making many copies) are fragments of different sizes that can be directly visualized using gel electrophoresis.

Smallest fragments

134

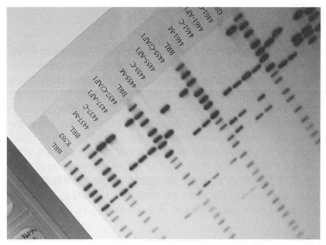

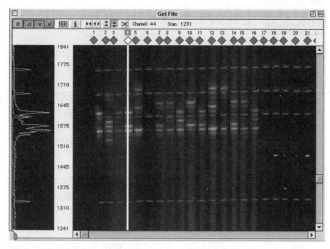

The photo above shows a film output from a DNA profiling procedure. The lanes with many regular bands are used for calibration; they contain DNA fragment sizes of known length. These calibration lanes can be used to determine the length of fragments in the unknown samples.

DNA profiling can be automated in the same way as DNA sequencing. Powerful computer software is able to display the results of many samples that are run at the same time. In the photo above, the sample in lane 4 has been selected and displays fragments of different length on the left of the screen.

1. Describe the properties of **short tandem repeats** that are important to the application of **DNA profiling** technology:

_____

_____

_____

_____

_____

2. Explain the role of each of the following techniques in the process of DNA profiling:

(a) Gel electrophoresis: _____

_____

_____

(b) PCR: _____

_____

_____

3. Describe the three main steps in DNA profiling using PCR:

(a) _____

_____

_____

(b) _____

_____

_____

(c) _____

_____

_____

4. Explain why as many as 10 STR sites are used to gain a DNA profile for forensic evidence: _____

_____

_____

101 Forensic Applications of DNA Profiling

Key Idea: DNA profiling has many forensic applications. The use of DNA as a tool for solving crimes such as homicide is well known, but it can also has several other applications. DNA evidence has been used to identify body parts, solve cases of industrial sabotage and contamination, for paternity testing, and even in identifying animal products illegally made from endangered species.

①
Offender was wearing a cap but lost it when disturbed. DNA can be retrieved from flakes of skin and hair.

DNA left behind when offender drunk from a cup in the kitchen.

Bloodstain. DNA can be extracted from white blood cells in the sample

Hair. DNA can be recovered from cells at the base of the strand of hair.

During the initial investigation, samples of material that may contain DNA are taken for analysis. At a crime scene, this may include blood and body fluids as well as samples of clothing or objects that the offender might have touched. Samples from the victim are also taken to eliminate them as a possible source of contamination.

② DNA is isolated and profiles are made from all samples and compared to known DNA profiles such as that of the victim.

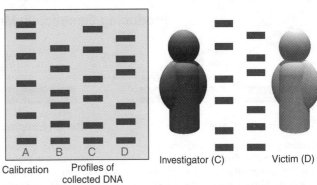

Calibration Profiles of collected DNA Investigator (C) Victim (D)

③ Unknown DNA samples are compared to DNA databases of convicted offenders and to the DNA of the alleged offender.

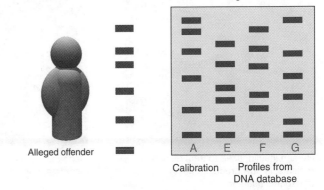

Alleged offender

Calibration Profiles from DNA database

④ Although it does not make a complete case, DNA profiling, in conjunction with other evidence, is one of the most powerful tools in identifying offenders or unknown tissues.

1. Why are DNA profiles obtained for both the victim and investigator?

_____

_____

_____

2. Use the evidence to decide if the alleged offender is innocent or guilty and explain your decision:

_____

_____

_____

3. Why is a father-child match between CSF1PO allele 9 more significant that a match between allele 12?

_____

_____

_____

_____

Paternity Testing

DNA profiling can be used to determine paternity (and maternity) by looking for matches in alleles between parents and children. This can be used in cases such as child support or inheritance. DNA profiling can establish the certainty of paternity (and maternity) to a 99.99% probability of parentage.

Every person has two copies of each chromosome and therefore two copies (alleles) of every testable DNA marker. In a DNA profile each marker's alleles are given a number: for the mother it may 1, 2 and the father 3, 4. The child will have a combination of these. The table below illustrates this:

DNA Marker	Mother's Alleles	Child's Alleles	Father's Alleles
CSF1PO	7, 8	8, 9	9, 12
D10S1248	14, 15	11, 14	10, 11
D12S391	16, 17	17, 17	17, 18
D13S317	10, 11	9, 10	8, 9

The frequency of the each allele occurring in the population is important when determining paternity (or maternity). For example DNA marker CSF1PO allele 9 has a frequency of 0.0294 making the match between father and child very significant (whereas allele 12 has a frequency of 0.3446, making a match less significant). For each allele a paternity index (PI) is calculated. These indicate the significance of the match. The PIs are combined to produce a percentage probability of parentage.

LINK
100

WEB
101

APP

102 What is Genetic Modification?

Key Idea: Genetically modified organisms (GMOs) are organisms with artificially altered DNA. GMOs may provide solutions to many human health and food supply issues.

The genetic modification of organisms has many potential applications including medical cures, increasing crop yields, and potentially helping to solve the world's pollution and resource problems. Organisms with artificially altered DNA are referred to as **genetically modified organisms** or **GMOs**. They may be modified in one of three ways (see below). Some of the current and proposed applications of gene technology raise complex ethical and safety issues. The benefits of their use must be carefully weighed against the risks to human health, and the health and well-being of other organisms and the environment.

Producing Genetically Modified Organisms (GMOs)

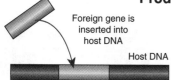

Foreign gene is inserted into host DNA

Host DNA

Existing gene is altered

Host DNA

Gene is deleted or deactivated

Host DNA

Add a Foreign Gene

A novel (foreign) gene is inserted from another species. This will enable the GMO to express the trait coded by the new gene. Organisms genetically altered in this way are referred to as **transgenic**.

Alter an Existing Gene

An existing gene may be edited to alter or correct its expression. New gene editing technologies (such as CRISPR) are making successful gene therapy more feasible than ever before.

Delete or 'Turn off' a Gene

An existing gene may be deleted or deactivated (switched off) to prevent the expression of a trait (e.g. the deactivation of the ripening gene in tomatoes produced the Flavr-Savr tomato).

Human insulin, used to treat diabetic patients, is now produced using transgenic bacteria.

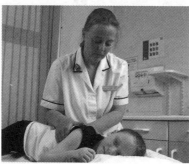

Gene therapy could be used treat genetic disorders, such as cystic fibrosis.

Manipulating gene action is one way in which to control processes such as ripening in fruit.

1. Using examples, discuss the ways in which an organism may be genetically modified (to produce a GMO):

_____

_____

_____

_____

_____

_____

2. Explain how human needs or desires have provided a stimulus for the development of the following biotechnologies:

(a) Gene therapy: _____

_____

_____

(b) The production and use of transgenic organisms: _____

_____

_____

(c) The extension of shelf life in stored produce: _____

_____

_____

© 2012-2014 **BIOZONE** International
ISBN: 978-1-927173-93-0
Photocopying Prohibited

103 Making Recombinant DNA

Key Idea: Recombinant DNA (rDNA) is produced by first isolating a DNA sequence, then inserting it into the DNA of a different organism.

The production of rDNA is possible because the DNA of every organism is made of the same building blocks (**nucleotides**).

rDNA allows a gene from one organism to be moved into, and expressed in, a different organism. Two important tools used to create rDNA are restriction digestion (chopping up the DNA) using **restriction enzymes** and DNA ligation (joining of sections of DNA) using the enzyme **DNA ligase**.

Information about Restriction Enzymes

▶ A **restriction enzyme** is an enzyme that cuts a double-stranded DNA molecule at a specific **recognition site** (a specific DNA sequence). There are many different types of restriction enzymes, each has a unique recognition site.

▶ Some restriction enzymes produce DNA fragments with two **sticky ends** (right). A sticky end has exposed nucleotide bases at each end. DNA cut in such a way is able to be joined to other DNA with matching sticky ends. Such joins are specific to their recognition sites.

▶ Some restriction enzymes produce a DNA fragment with two **blunt ends** (ends with no exposed nucleotide bases). The piece it is removed from is also left with blunt ends. DNA cut in such a way can be joined to any other blunt end fragment. Unlike sticky ends, blunt end joins are non-specific because there are no sticky ends to act as specific recognition sites.

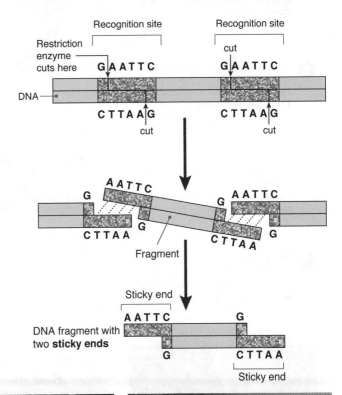

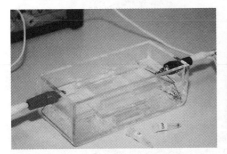

The fragments of DNA produced by the restriction enzymes are mixed with ethidium bromide, a molecule that fluoresces under UV light. The DNA fragments are then placed on an electrophoresis gel to separate the different lengths of DNA.

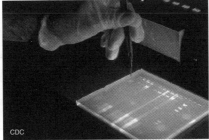

Once the DNA fragments are separated, the gel is placed on a UV viewing platform. The area of the gel containing the DNA fragments of the correct length is cut out and placed in a solution that dissolves the gel. This releases the DNA into the solution.

The solution containing the DNA is centrifuged at high speed to separate out the DNA. Centrifugation works by separating molecules of different densities. Once isolated, the DNA can be spliced into another DNA molecule.

1. What is the purpose of restriction enzymes in making recombinant DNA? _____

_____

_____

2. Distinguish between sticky end and blunt end fragments: _____

_____

_____

_____

3. Why is it useful to have many different kinds of restriction enzymes? _____

_____

_____

© 2012-2014 **BIOZONE** International
ISBN: 978-1-927173-93-0
Photocopying Prohibited

LINK 105 WEB 103 **KNOW**

Creating a Recombinant DNA Plasmid

1 Two pieces of DNA are cut by the same restriction enzyme (they will produce fragments with matching sticky ends).

2 Fragments with matching sticky ends can be joined by base-pairing. This process is called **annealing**. This allows DNA fragments from different sources to be joined.

The fragments of DNA are joined together by the enzyme **DNA ligase**, producing a molecule of **recombinant DNA**.

3 The joined fragments will usually form either a linear or a circular molecule, as shown here (right) as recombinant **plasmid** DNA.

pGLO is a plasmid engineered to contain Green Fluorescent Protein (*gfp*). pGLO has been used to create fluorescent organisms, including the bacteria above (bright patches on agar plates).

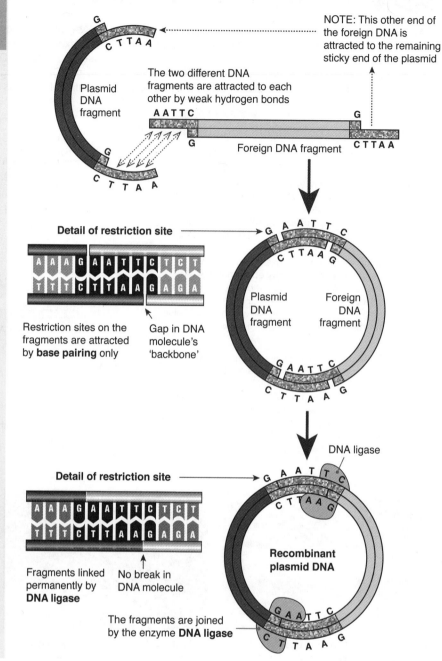

NOTE: This other end of the foreign DNA is attracted to the remaining sticky end of the plasmid

The two different DNA fragments are attracted to each other by weak hydrogen bonds

Plasmid DNA fragment

Foreign DNA fragment

Detail of restriction site

Restriction sites on the fragments are attracted by **base pairing** only

Gap in DNA molecule's 'backbone'

Plasmid DNA fragment

Foreign DNA fragment

Detail of restriction site

Fragments linked permanently by **DNA ligase**

No break in DNA molecule

DNA ligase

Recombinant plasmid DNA

The fragments are joined by the enzyme **DNA ligase**

4. Explain in your own words the two main steps in the process of joining two DNA fragments together:

(a) Annealing: _____

_____

(b) DNA ligase: _____

_____

5. Why can **ligation** be considered the reverse of the **restriction digestion** process? _____

_____

6. Why can recombinant DNA be expressed in any kind of organism, even if it contains DNA from another species?

_____

_____

_____

_____

© 2012-2014 **BIOZONE** International
ISBN: 978-1-927173-93-0

104 Applications of GMOs

Key Idea: Techniques for genetic manipulation are widely applied throughout modern biotechnology: in food and enzyme technology, in industry and medicine, and in agriculture and horticulture.

Microorganisms are among the most widely used GMOs, with applications ranging from pharmaceutical production and vaccine development to environmental clean-up. Crop plants are also popular candidates for genetic modification although their use, as with much of genetic engineering of higher organisms, is controversial and sometimes problematic.

Application of GMOs

Extending Shelf Life
Some fresh produce (e.g. tomatoes) have been engineered to have an extended shelf life. In the case of tomatoes, the gene for ripening has been switched off, delaying the process of softening in the fruit.

Pest or Herbicide Resistance
Plants can be engineered to produce their own insecticide and become pest resistant. Genetically engineered herbicide resistance is also common. In this case, chemical weed killers can be used freely without damaging the crop.

Crop Improvement
Gene technology is now an integral part of the development of new crop varieties. Crops can be engineered to produce higher protein levels or to grow in inhospitable conditions (e.g. salty or dry conditions).

Environmental Clean-Up
Some bacteria have been engineered to grow on waste products, such as liquefied newspaper pulp or oil. As well as degrading pollutants and wastes, the bacteria may be harvested as a commercial protein source.

Biofactories
Transgenic bacteria are widely used to produce desirable products: often hormones or proteins. Large quantities of a product can be produced using bioreactors (above). For example, insulin production by recombinant yeast, growth hormone production.

Vaccine Development
The potential exists for multipurpose vaccines to be made using gene technology. Genes coding for vaccine components (e.g. viral protein coat) are inserted into an unrelated live vaccine (e.g. polio vaccine), and deliver proteins to stimulate an immune response.

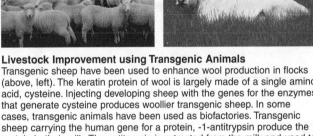

Livestock Improvement using Transgenic Animals
Transgenic sheep have been used to enhance wool production in flocks (above, left). The keratin protein of wool is largely made of a single amino acid, cysteine. Injecting developing sheep with the genes for the enzymes that generate cysteine produces woollier transgenic sheep. In some cases, transgenic animals have been used as biofactories. Transgenic sheep carrying the human gene for a protein, -1-antitrypsin produce the protein in their milk. The antitrypsin is extracted from the milk and used to treat hereditary emphysema (a type of lung disease).

1. Research the potential benefits and disadvantages of using GMOs for one of the applications described above, and discuss these in the space provided below:

_____

_____

_____

_____

_____

_____

_____

_____

_____

_____

_____

LINK 108   LINK 107   LINK 106   **KNOW**

105 *In Vivo* Gene Cloning

Key Idea: *In vivo* cloning describes the insertion of a gene into an organism and using the replication machinery of that organism to multiply the gene or produce its protein product. Recombinant DNA techniques (restriction digestion and ligation) are used to insert a gene of interest into the DNA of a vector (e.g. plasmid or viral DNA). This produces a recombinant DNA molecule called a **molecular clone** that can transmit the gene of interest to another organism. To be useful, all vectors must be able to replicate inside their host organism, they must have one or more sites at which a restriction enzyme can cut, and they must have some kind of **genetic marker** that allows them to be identified. Bacterial plasmids are commonly used vectors because they are easy to manipulate, their restriction sites are well known, and they are readily taken up by cells in culture. Once the molecular clone has been taken up by bacterial cells, and those cells are identified, the gene can be replicated (cloned) many times as the bacteria grow and divide in culture.

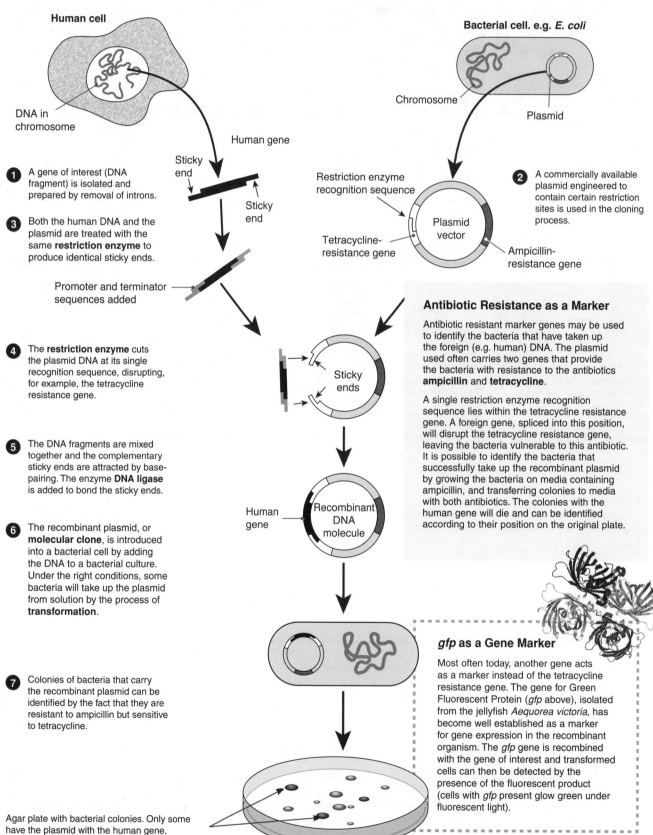

Human cell

DNA in chromosome

Human gene

Sticky end

Sticky end

Bacterial cell. e.g. *E. coli*

Chromosome

Plasmid

Restriction enzyme recognition sequence

Plasmid vector

Tetracycline-resistance gene

Ampicillin-resistance gene

1 A gene of interest (DNA fragment) is isolated and prepared by removal of introns.

2 A commercially available plasmid engineered to contain certain restriction sites is used in the cloning process.

3 Both the human DNA and the plasmid are treated with the same **restriction enzyme** to produce identical sticky ends.

Promoter and terminator sequences added

4 The **restriction enzyme** cuts the plasmid DNA at its single recognition sequence, disrupting, for example, the tetracycline resistance gene.

Sticky ends

5 The DNA fragments are mixed together and the complementary sticky ends are attracted by base-pairing. The enzyme **DNA ligase** is added to bond the sticky ends.

Human gene

Recombinant DNA molecule

6 The recombinant plasmid, or **molecular clone**, is introduced into a bacterial cell by adding the DNA to a bacterial culture. Under the right conditions, some bacteria will take up the plasmid from solution by the process of **transformation**.

Antibiotic Resistance as a Marker

Antibiotic resistant marker genes may be used to identify the bacteria that have taken up the foreign (e.g. human) DNA. The plasmid used often carries two genes that provide the bacteria with resistance to the antibiotics **ampicillin** and **tetracycline**.

A single restriction enzyme recognition sequence lies within the tetracycline resistance gene. A foreign gene, spliced into this position, will disrupt the tetracycline resistance gene, leaving the bacteria vulnerable to this antibiotic. It is possible to identify the bacteria that successfully take up the recombinant plasmid by growing the bacteria on media containing ampicillin, and transferring colonies to media with both antibiotics. The colonies with the human gene will die and can be identified according to their position on the original plate.

7 Colonies of bacteria that carry the recombinant plasmid can be identified by the fact that they are resistant to ampicillin but sensitive to tetracycline.

gfp as a Gene Marker

Most often today, another gene acts as a marker instead of the tetracycline resistance gene. The gene for Green Fluorescent Protein (*gfp* above), isolated from the jellyfish *Aequorea victoria*, has become well established as a marker for gene expression in the recombinant organism. The *gfp* gene is recombined with the gene of interest and transformed cells can then be detected by the presence of the fluorescent product (cells with *gfp* present glow green under fluorescent light).

Agar plate with bacterial colonies. Only some have the plasmid with the human gene.

© 2012-2014 **BIOZONE** International
ISBN: 978-1-927173-93-0
Photocopying Prohibited

1. Why might it be desirable to use *in vivo* methods to clone genes rather than PCR? _____

_____

_____

2. Explain when it may not be desirable to use bacteria to clone genes: _____

_____

_____

3. Explain how a human gene is removed from a chromosome and placed into a plasmid. _____

_____

_____

_____

_____

4. A bacterial plasmid replicates at the same rate as the bacteria. If a bacteria containing a recombinant plasmid replicates and divides once every thirty minutes, calculate the number of plasmid copies there will be after twenty four hours:

_____

_____

5. (a) Where are the genes for antibiotic resistance located? _____

   (b) What will happen to the colonies when they are plated on to a medium with both antibiotics? _____

   _____

   (c) How does placing the colonies on media with both antibiotics help to identify colonies with the human gene? _____

   _____

   _____

6. Explain why the *gfp* marker is a more desirable gene marker than genes for antibiotic resistance:

   _____

   _____

   _____

   _____

7. Bacteriophages are viruses that infect bacteria:

   (a) What feature of bacteriophages make them useful for genetic engineering? _____

   _____

   (b) How could a bacteriophage be used to clone a gene? _____

   _____

   _____

   _____

   _____

106 Using Recombinant Bacteria

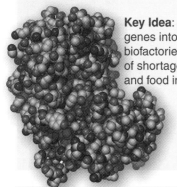

Key Idea: Inserting useful genes into bacteria to produce biofactories can solve the problem of shortages in the manufacturing and food industries.

The Issue

▶ **Chymosin** (also known as **rennin**) is an enzyme that digests milk proteins. It is the active ingredient in rennet, a substance used by cheesemakers to clot milk into curds.

▶ Traditionally rennin is extracted from "chyme", i.e. the stomach secretions of suckling calves (hence its name of chymosin).

▶ By the 1960s, a shortage of chymosin was limiting the volume of cheese produced.

▶ Enzymes from fungi were used as an alternative but were unsuitable because they caused variations in the cheese flavour.

Concept 1
Enzymes are proteins made up of amino acids. The amino acid sequence of chymosin can be determined and the mRNA coding sequence for its translation identified.

Concept 2
Reverse transcriptase can be used to synthesize a DNA strand from the mRNA. This process produces DNA without the introns, which cannot be processed by bacteria.

Concept 3
DNA can be cut at specific sites using **restriction enzymes** and rejoined using **DNA ligase**. New genes can be inserted into self-replicating bacterial **plasmids**.

Concept 4
Under certain conditions, bacteria are able to lose or take up plasmids from their environment. Bacteria are readily grown in vat cultures at little expense.

Concept 5
The protein in made by the bacteria in large quantities.

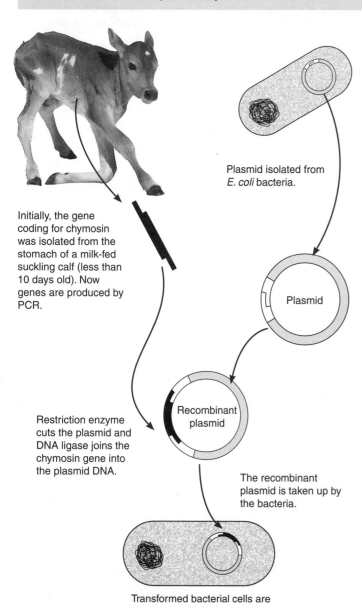

Initially, the gene coding for chymosin was isolated from the stomach of a milk-fed suckling calf (less than 10 days old). Now genes are produced by PCR.

Plasmid isolated from *E. coli* bacteria.

Plasmid

Restriction enzyme cuts the plasmid and DNA ligase joins the chymosin gene into the plasmid DNA.

Recombinant plasmid

The recombinant plasmid is taken up by the bacteria.

Transformed bacterial cells are grown in a vat culture

Techniques

The amino acid sequence of chymosin is first determined and the RNA codons for each amino acid identified.

mRNA matching the identified sequence is isolated from the stomach of young calves. **Reverse transcriptase** is used to transcribe mRNA into DNA. The DNA sequence can also be made synthetically once the sequence is determined.

The DNA is amplified using PCR.

Plasmids from *E. coli* bacteria are isolated and cut using **restriction enzymes.** The DNA sequence for chymosin is inserted using **DNA ligase**.

Plasmids are returned to *E. coli* by placing the bacteria under conditions that induce them to take up plasmids.

Outcomes

The transformed bacteria are grown in vat culture. Chymosin is produced by *E. coli* in packets within the cell that are separated during the processing and refining stage.

Recombinant chymosin entered the marketplace in 1990. It established a significant market share because cheesemakers found it to be cost effective, of high quality, and in consistent supply. Most cheese is now produced using recombinant chymosin such as CHY-MAX.

Further Applications

A large amount of processing is required to extract chymosin from *E.coli*. There are now other bacteria and fungi that have been engineered to produce the enzyme. Most chymosin is now produced in a similar way using the fungi *Aspergillus niger* and *Kluyveromyces lactis*. Both fungi have greater capacity to produce chymosin because their cells are larger than *E.coli*. Their secretory pathways are also more similar to humans than those of *E.coli*.

Enzymes from GMOs are widely used in the baking industry. Maltogenic alpha amylase from *Bacillus subtilis* bacteria is used as an anti-staling agent to prolong shelf life. Hemicellulases from *B. subtilis* and xylanase from the fungus *Aspergillus oryzae* are used for improvement of dough, crumb structure, and volume during the baking process.

Lipase from *Aspergillus oryzae* is used in processing of palm oil to produce low cost cocoa butter substitutes (above), which have a similar 'mouth feel' to cocoa butter.

Acetolactate decarboxylase from *B. subtilis* is one of several enzymes used in the brewing industry. It reduces maturation time of the beer by by-passing a rate-limiting step.

1. Describe the main use of chymosin: _____

   _____

2. What was the traditional source of chymosin? _____

   _____

3. Summarize the key concepts that led to the development of the technique for producing chymosin:

   (a) Concept 1: _____

   (b) Concept 2: _____

   (c) Concept 3: _____

   (d) Concept 4: _____

   (e) Concept 5: _____

4. Explain how the gene for chymosin was isolated and how the technique could be applied to isolating other genes:

   _____

   _____

   _____

   _____

   _____

   _____

5. Describe three advantages of using chymosin produced by GE bacteria over chymosin from traditional sources:

   (a) _____

   _____

   (b) _____

   _____

   (c) _____

   _____

6. Explain why the fungus *Aspergillus niger* is now more commonly used to produce chymosin instead of *E. coli*:

   _____

   _____

   _____

107 Golden Rice

Key Idea: The use of recombinant DNA to build a new metabolic pathway has greatly increased the nutritional value of a rice.

The Issue

▶ **Beta-carotene** (β-carotene) is a precursor to **vitamin A** which is involved in many functions including vision, immunity, fetal development, and skin health.

▶ Vitamin A deficiency is common in developing countries where up to 500,000 children suffer from night blindness, and death rates due to infections are high due to a lowered immune response.

▶ Providing enough food containing useful quantities of β-carotene is difficult and expensive in many countries.

Concept 1
Rice is a staple food in many developing countries. It is grown in large quantities and is available to most of the population, but it lacks many of the essential nutrients required by the human body for healthy development. It is low in β-carotene.

Concept 2
Rice plants produce β-carotene but not in the edible rice **endosperm**. Engineering a new biosynthetic pathway would allow β-carotene to be produced in the endosperm. Genes expressing enzymes for carotene synthesis can be inserted into the rice genome.

Concept 3
The enzyme **carotene desaturase (CRT1)** in the soil bacterium *Erwinia uredovora,* catalyses multiple steps in carotenoid biosynthesis. **Phytoene synthase (PSY)** overexpresses a colourless carotene in the daffodil plant *Narcissus pseudonarcissus.*

Concept 4
DNA can be inserted into an organism's genome using a suitable **vector**. *Agrobacterium tumefaciens* is a tumour-forming bacterial plant pathogen that is commonly used to insert novel DNA into plants.

The Development of Golden Rice

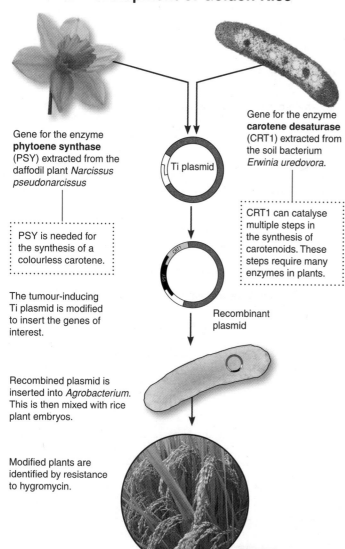

Gene for the enzyme **phytoene synthase** (PSY) extracted from the daffodil plant *Narcissus pseudonarcissus*

Gene for the enzyme **carotene desaturase** (CRT1) extracted from the soil bacterium *Erwinia uredovora.*

Ti plasmid

> PSY is needed for the synthesis of a colourless carotene.

> CRT1 can catalyse multiple steps in the synthesis of carotenoids. These steps require many enzymes in plants.

The tumour-inducing Ti plasmid is modified to insert the genes of interest.

Recombinant plasmid

Recombined plasmid is inserted into *Agrobacterium*. This is then mixed with rice plant embryos.

Modified plants are identified by resistance to hygromycin.

SGR1

Techniques

The **PSY** gene from daffodils and the **CRT1** gene from *Erwinia uredovora* are sequenced.

DNA sequences are synthesized into packages containing the CRT1 or PSY gene, terminator sequences, and **endosperm specific promoters** (these ensure expression of the gene only in the edible portion of the rice).

The *Ti* plasmid from *Agrobacterium* is modified using restriction enzymes and DNA ligase to delete the tumour-forming gene and insert the synthesized DNA packages. A gene for resistance to the antibiotic **hygromycin** is also inserted so that transformed plants can be identified later. The parts of the *Ti* plasmid required for plant transformation are retained.

Modified *Ti* plasmid is inserted into the bacterium.

Agrobacterium is incubated with rice plant embryo. Transformed embryos are identified by their resistance to hygromycin.

Outcomes

The rice produced had endosperm with a distinctive yellow colour. Under greenhouse conditions golden rice (**SGR1**) contained 1.6 μg per g of carotenoids. Levels up to five times higher were produced in the field, probably due to improved growing conditions.

Further Applications

Further research on the action of the PSY gene identified more efficient methods for the production of β-carotene. The second generation of golden rice now contains up to 37 μg per g of carotenoids. Golden rice was the first instance where a complete biosynthetic pathway was engineered. The procedures could be applied to other food plants to increase their nutrient levels.

The ability of *Agrobacterium* to transfer genes to plants is exploited for crop improvement. The tumour-inducing *Ti* plasmid is modified to delete the tumour-forming gene and insert a gene coding for a desirable trait. The parts of the *Ti* plasmid required for plant transformation are retained.

Soybeans are one of the many food crops that have been genetically modified for broad spectrum herbicide resistance. The first GM soybeans were planted in the US in 1996. By 2007, nearly 60% of the global soybean crop was genetically modified; the highest of any other crop plant.

GM cotton was produced by inserting the gene for the BT toxin into its genome. The bacterium *Bacillus thuringiensis* naturally produces BT toxin, which is harmful to a range of insects, including the larvae that eat cotton. The BT gene causes cotton to produce this insecticide in its tissues.

1. Describe the basic methodology used to create golden rice: _____

2. Explain how scientists ensured β-carotene was produced in the endosperm: _____

3. What property of *Agrobacterium tumefaciens* makes it an ideal vector for introducing new genes into plants?

4. (a) How could this new variety of rice reduce disease in developing countries? _____

   (b) Absorption of vitamin A requires sufficient dietary fat. Explain how this could be problematic for the targeted use of golden rice in developing countries:

5. As well as increasing nutrient content as in golden rice, other traits of crop plants are also desirable. For each of the following traits, suggest features that could be desirable in terms of increasing yield:

   (a) Grain size or number: _____

   (b) Maturation rate: _____

   (c) Pest resistance: _____

108 Production of Insulin

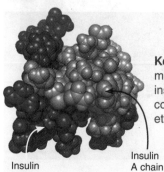

Insulin B chain

Insulin A chain

Key Idea: By using microorganisms to make human insulin, problematic issues of cost, allergic reactions, and ethics have been addressed.

The Issue

▶ **Type I diabetes mellitus** is a metabolic disease caused by a lack of **insulin**. Around 25 people in every 100,000 suffer from type I diabetes.

▶ It is treatable only with injections of insulin.

▶ In the past, insulin was taken from the pancreases of cows and pigs and purified for human use. The method was expensive and some patients had severe allergic reactions to the foreign insulin or its contaminants.

Concept 1

DNA can be cut at specific sites using **restriction enzymes** and joined together using **DNA ligase**. New genes can be inserted into self-replicating bacterial **plasmids** at the point where the cuts are made.

Concept 2

Plasmids are small, circular pieces of DNA found in some bacteria. They usually carry genes useful to the bacterium. *E. coli* plasmids can carry promoters required for the transcription of genes.

Concept 3

Under certain conditions, Bacteria are able to lose or pick up plasmids from their environment. Bacteria can be readily grown in vat cultures at little expense.

Concept 4

The DNA sequences coding for the production of the two polypeptide chains (A and B) that form human insulin can be isolated from the human genome.

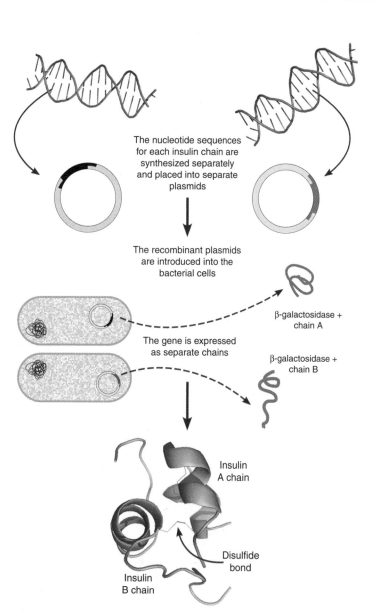

The nucleotide sequences for each insulin chain are synthesized separately and placed into separate plasmids

The recombinant plasmids are introduced into the bacterial cells

The gene is expressed as separate chains

β-galactosidase + chain A

β-galactosidase + chain B

Insulin A chain

Disulfide bond

Insulin B chain

Techniques

The **gene** is **chemically synthesized** as two nucleotide sequences, one for the **insulin A chain** and one for the **insulin B chain**. The two sequences are small enough to be inserted into a plasmid.

Plasmids are extracted from *Escherichia coli*. The gene for the bacterial enzyme **β-galactosidase** is located on the plasmid. To make the bacteria produce insulin, the insulin gene must be linked to the β-galactosidase gene, which carries a promoter for transcription.

Restriction enzymes are used to cut plasmids at the appropriate site and the A and B insulin sequences are inserted. The sequences are joined with the plasmid DNA using **DNA ligase**.

The **recombinant plasmids** are inserted back into the bacteria by placing them together in a culture that favors plasmid uptake by bacteria.

The bacteria are then grown and multiplied in vats under carefully controlled growth conditions.

Outcomes

The product consists partly of β-galactosidase, joined with either the A or B chain of insulin. The chains are extracted, purified, and mixed together. The A and B insulin chains connect via **disulfide cross linkages** to form the functional insulin protein. The insulin can then be made ready for injection in various formulations.

Further Applications

The techniques involved in producing human insulin from genetically modified bacteria can be applied to a range of human proteins and hormones. Proteins currently being produced include human growth hormone, interferon, and factor VIII.

© 2012-2014 **BIOZONE** International
ISBN: 978-1-927173-93-0

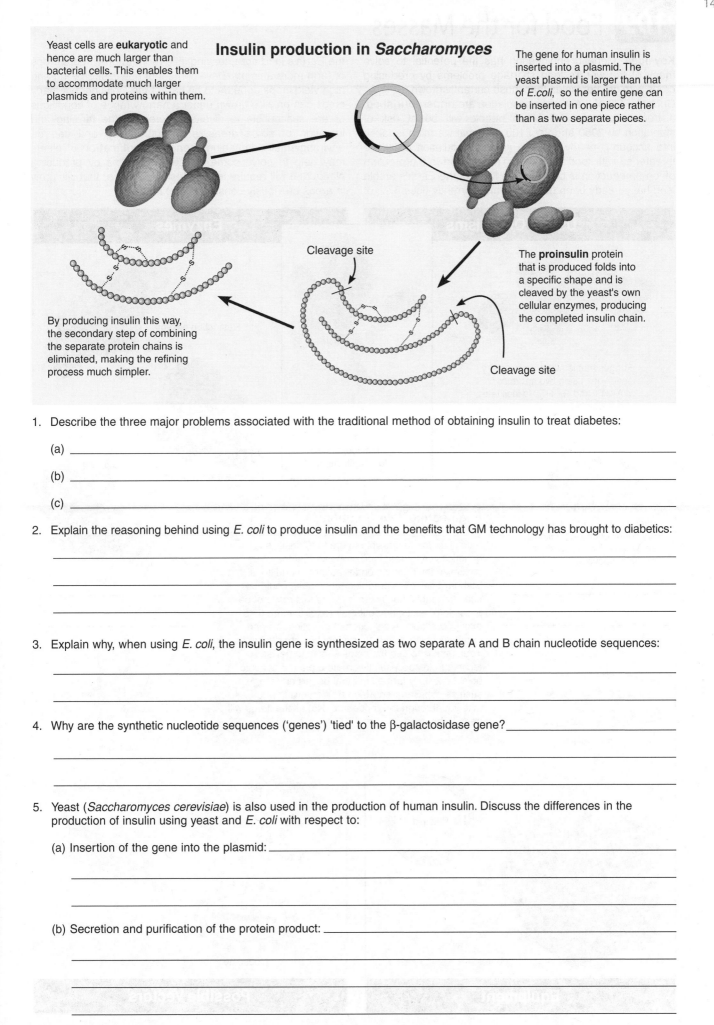

Yeast cells are **eukaryotic** and hence are much larger than bacterial cells. This enables them to accommodate much larger plasmids and proteins within them.

Insulin production in *Saccharomyces*

The gene for human insulin is inserted into a plasmid. The yeast plasmid is larger than that of *E.coli*, so the entire gene can be inserted in one piece rather than as two separate pieces.

Cleavage site

The **proinsulin** protein that is produced folds into a specific shape and is cleaved by the yeast's own cellular enzymes, producing the completed insulin chain.

By producing insulin this way, the secondary step of combining the separate protein chains is eliminated, making the refining process much simpler.

Cleavage site

1. Describe the three major problems associated with the traditional method of obtaining insulin to treat diabetes:

   (a) _____

   (b) _____

   (c) _____

2. Explain the reasoning behind using *E. coli* to produce insulin and the benefits that GM technology has brought to diabetics:

   _____

   _____

   _____

3. Explain why, when using *E. coli*, the insulin gene is synthesized as two separate A and B chain nucleotide sequences:

   _____

   _____

4. Why are the synthetic nucleotide sequences ('genes') 'tied' to the β-galactosidase gene?_____

   _____

   _____

5. Yeast (*Saccharomyces cerevisiae*) is also used in the production of human insulin. Discuss the differences in the production of insulin using yeast and *E. coli* with respect to:

   (a) Insertion of the gene into the plasmid: _____

   _____

   _____

   (b) Secretion and purification of the protein product: _____

   _____

   _____

© 2012-2014 **BIOZONE** International
ISBN: 978-1-927173-93-0

109 Food for the Masses

Key Idea: Genetic engineering has the potential to solve many of the world's food shortage problems by producing crops with greater yields than those currently grown.

Currently 1/6 of the world's population are **undernourished**. If trends continue, 1.5 billion people will be at risk of starvation by 2050 and, by 2100 (if global warming is taken into account), nearly half the world's population could be threatened with food shortages. The solution to the problem of food production is complicated. Most of the Earth's arable land has already been developed and currently uses 37% of

the Earth's land area, leaving little room to grow more crops or farm more animals. Development of new fast growing and high yield crops appears to be part of the solution, but many crops can only be grown under a narrow range of conditions or are susceptible to disease. Moreover, the farming and irrigation of some areas is difficult, costly, and can be environmentally damaging. **Genetic modification** of plants may help to solve some of these problems by producing plants that will require less intensive culture or that will grow in areas previously considered not arable.

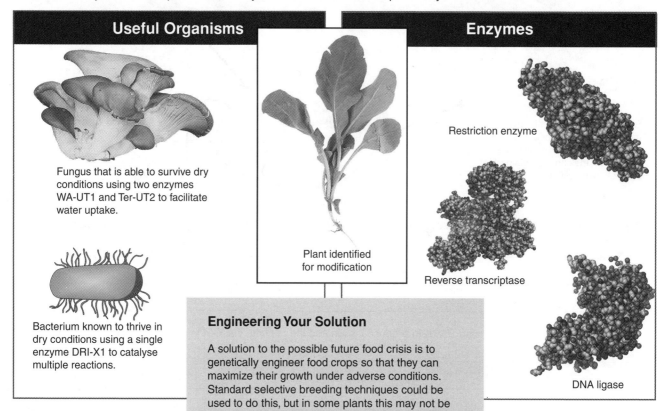

Useful Organisms

Fungus that is able to survive dry conditions using two enzymes WA-UT1 and Ter-UT2 to facilitate water uptake.

Bacterium known to thrive in dry conditions using a single enzyme DRI-X1 to catalyse multiple reactions.

Plant identified for modification

Enzymes

Restriction enzyme

Reverse transcriptase

DNA ligase

Engineering Your Solution

A solution to the possible future food crisis is to genetically engineer food crops so that they can maximize their growth under adverse conditions. Standard selective breeding techniques could be used to do this, but in some plants this may not be possible or feasible and it may require more time than is available. A selection of genetic tools and organisms with useful characteristics are described. **Your task** is to use the items shown to devise a technique to successfully create a plant that could be successfully farmed in semi-desert environments such as sub-Saharan Africa. The following page will take you through the procedure. Not all the items will need to be used.

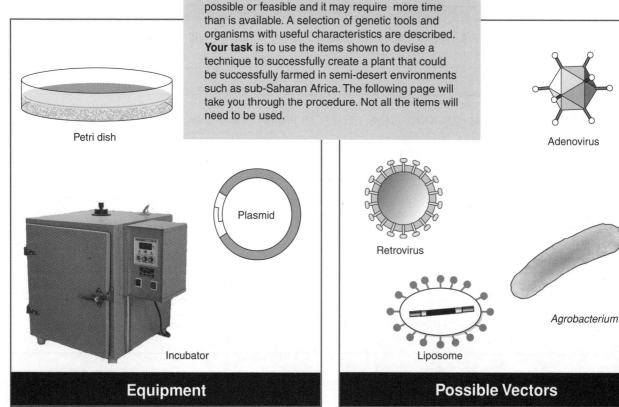

Equipment

Petri dish

Incubator

Plasmid

Possible Vectors

Adenovirus

Retrovirus

Liposome

Agrobacterium

© 2012-2014 **BIOZONE** International
ISBN: 978-1-927173-93-0
Photocopying Prohibited

1. Identify the organism you would chose as a 'donor' of drought survival genes and explain your choice:

   _____

2. Describe a process to identify and isolate the required gene(s) and identify the tools to be used:_____

   _____

   _____

   _____

   _____

   _____

3. Identify a vector for the transfer of the isolated gene(s) into the crop plant and explain your decision: _____

   _____

4. Explain how the isolated gene(s) would be integrated into the vector's genome:_____

   _____

   _____

   _____

   _____

   _____

5. (a) Explain how the vector will transform the identified plant: _____

   _____

   _____

   (b) Identify the stage of development at which the plant would most easily be transformed. Explain your choice:

   _____

   _____

   _____

6. Explain how the transformed plants could be identified:_____

   _____

   _____

   _____

   _____

   _____

7. Explain how a large number of plants can be grown from the few samples that have taken up the new DNA:

   _____

   _____

   _____

   _____

   _____

   _____

110 The Ethics of GMO Technology

Key Idea: There are many potential benefits and risks in using genetically modified organisms.

Genetically modified organisms (GMOs) have many potential benefits, but their use raises a number of biological and ethical concerns. Some of these include risk to human health, animal welfare issues, and environmental safety. Currently a matter of concern to consumers is the adequacy of government regulations for the labelling of food products with GMO content. In some countries GM products must be clearly labelled, while other countries have no requirements for GM labelling. This can take away consumer choice about the types of products they buy. The use of GM may also have trade implications for countries exporting and importing GMO produce.

Potential Benefits of GMOs

1. Increase in crop yields, including crops with more nutritional value and that store for longer.
2. Decrease in use of pesticides, herbicides and animal remedies.
3. Production of crops that are drought tolerant or salt tolerant.
4. Improvement in the health of the human population and the medicines used to achieve it.
5. Development of animal factories for the production of proteins used in manufacturing, the food industry, and health.

Potential Risks of GMOs

1. Possible (uncontrollable) spread of transgenes into other species of plants, or animals.
2. Concerns that the release of GMOs into the environment may be irreversible.
3. Animal welfare and ethical issues: GM animals may suffer poor health and reduced life span.
4. GMOs may cause the emergence of pest, insect, or microbial resistance to traditional control methods.
5. May create a monopoly and dependence of developing countries on companies who are seeking to control the world's commercial seed supply.

What's Killing the Monarchs? It's not the Corn

Bt corn is a corn variety genetically engineered to contain a gene from the soil bacterium *Bacillus thuringiensis*. The gene allows the corn to produce a toxin, which acts as a pesticide against butterfly and moth larvae but does not affect other insects such as beetles or bees (or indeed any other animal). The target insect pest for Bt corn is the larval stage of the European corn borer, which causes hundreds of millions of dollars worth of damage to crops annually. The Bt endotoxin had been used since the 1960s as a microbial insecticide and is considered safe because of its selectivity. Bt corn was developed by the company Monsanto and sales began in 1996. There are many different types of Bt corn, each one engineered to produce the toxin in slightly different ways. One of the first produced was Bt 176.

By 1999 monarch butterfly populations in the American Midwest began declining. During that year, Cornell University published a paper showing that the Bt toxin could be dispersed to other plants by the corn's pollen. Pollen landing on milkweed near corn crops could potentially kill the monarch caterpillars that fed exclusively on the milkweed. This resulted in a backlash against Bt corn by environmental activists. However, in 2001 a study was released that argued the toxin in pollen was not causing monarch decline. The toxicity in pollen was due mainly to the Bt 176 variety which was used in less than 2% of the corn grown and was in the process of being phased out. Other Bt corn varieties did not develop enough toxin, or their pollen density was too low to affect monarch caterpillars.

It now appears that there is a related but quite different reason for the Monarch butterfly decline. In 1996, Monsanto also began selling "Roundup Ready" corn, engineered to withstand glyphosate herbicide. Corn crops could be sprayed with herbicide and while the weeds die the corn would keep on growing, allowing less targeted spraying applications. As a result milkweed, which often grew in or near corn crops, was also killed, leaving no food for monarch caterpillars.

So... What's killing the monarchs?

David R. Tibble

Scott McDougall

Above: North American populations of monarchs migrate (above) to overwintering sites in Mexico and California.
Right: Monarch caterpillars feed exclusively on milkweed.

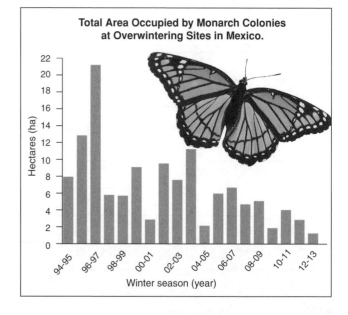

Total Area Occupied by Monarch Colonies at Overwintering Sites in Mexico.

Hectares (ha) vs Winter season (year)

© 2012-2014 **BIOZONE** International
ISBN: 978-1-927173-93-0
Photocopying Prohibited

1. Identify three benefits of GMOs and explain why they are of benefit:

    (a) _____
    _____

    (b) _____
    _____

    (c) _____
    _____

2. Identify three possible risks of GMOs and explain the significance of these risks:

    (a) _____
    _____
    _____

    (b) _____
    _____
    _____

    (c) _____
    _____
    _____

3. As a group, discuss the ethical issues surrounding GM corn and monarch declines. Who is to blame for the decline of monarchs and what can be done to help the population recover? Summarize the main points of your discussion below:

    _____
    _____
    _____
    _____
    _____
    _____
    _____
    _____

4. Some of the public's fears and concerns about genetically modified food arise from moral or religious beliefs, while others have a biological basis and are related to the potential biological threat posed by GMOs.

    (a) Conduct a class discussion or debate to identify any concerns, and list them below:

    _____
    _____
    _____
    _____
    _____

    (b) Identify which of those you have listed above pose a real biological threat: _____
    _____
    _____

111 Natural Clones

Key Idea: Many plants and some animals have the ability to produce clones naturally.

Clones are organisms genetically identical to the parent organism. Clones can be produced very rapidly, but their lack of genetic variability makes their populations vulnerable if the environment changes. Plants can reproduce asexually and produce clones through the production of vegetative structures such as tubers, rhizomes, and bulbs. They can also be propagated from cuttings or may develop roots wherever a part of the plant is touching the ground or wounded. Animals are less able to produce clones, although it does happen in a relatively small number of taxa.

Natural Vegetative Structures in Plants

Tubers are the swollen part of an underground stem or root, usually modified for storing food. Tubers can be cut into pieces that will grow into new plants provided a lateral bud is included in the piece.

Potato stem tuber

'Eye' (lateral bud)

A true **bulb** is simply a typical shoot compressed into a shortened form. They act as a storage organ for the plant. The bulbs can be sectioned to induce the production of new bulbs for propagation.

Garlic bulbs

In **rhizomes**, food is stored in the horizontal, underground stem. Rhizomes tend to be thick, fleshy or woody, and bear nodes with scale or foliage leaves and buds. Growth occurs at the buds on the ends of the rhizome or nearby nodes. Rhizomes can be cut into pieces to produce more plants.

Underground stem containing stored food

Iris rhizome

Shoot

Corm

Adventitious roots

Corms also store food in stem tissue. Like bulbs they can divided to produce more identical plants Cyclamen, gladiolus, and crocus (above) are corms.

Growth in crocus, a typical corm

Natural Cloning in Animals

Full natural cloning in animals is rare and usually only occurs in simple animals such as *Planaria*, starfish and *Hydra*.

Planarians are well known for their ability to form clones. The flatworm can be cut into up to three pieces, all of which will regenerate into new individuals. In nature some species reproduce by pinching off a short section of tail which develops into a new organism.

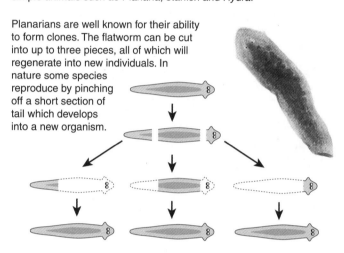

Sponges and most cnidarians (e.g. *Hydra*) can reproduce by **budding**. A small part of the parent body separates from the rest and develops into a new individual. This new individual may remain attached as part of the colony, or the bud may constrict at its point of attachment and be released as an independent organism.

This photo (right) shows *Hydra* budding. The new individuals are budding from the main body of the parent animal. The photograph shows the bulge and constriction where each new offspring will separate to form an independent individual.

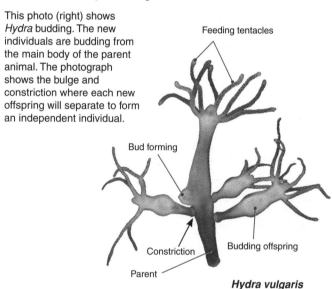

Feeding tentacles

Bud forming

Constriction

Budding offspring

Parent

Hydra vulgaris

1. (a) What is the genetic relationship between a population of clones? _____

_____

_____

(b) Explain the effect of this in relation to the population adapting to environment change: _____

_____

_____

_____

© 2012-2014 **BIOZONE** International
ISBN: 978-1-927173-93-0
Photocopying Prohibited

112 Cloning by Embryo Splitting

Key Idea: Cloning by embryo splitting replicates the natural twinning process, but enables multiple clones to be produced from just one high-value individual.

Livestock often produce only one or two offspring per year, so building a herd with desirable traits by selective breeding alone is a lengthy process. Cloning makes it possible to produce animals with desirable characteristics (e.g. high milk yield) more quickly than otherwise. Embryo splitting, or artificial twinning, is the simplest way to create a clone. It replicates

the natural twinning process *in-vitro*, and the genetically identical embryos are implanted into surrogates to complete development. The individuals produced by embryo splitting will have many of the same characteristics as the parents, although their exact phenotype is not known until after birth. Cloning provides genetically identical animals for studying disease processes. It can also be used (controversially) to produce embryos from which undifferentiated stem cells can be isolated for use in therapeutic medicine.

Livestock are selected on the basis of desirable qualities such as wool, meat, or milk production. Multiple eggs are taken from chosen individuals. These are then fertilized and grown *in-vitro* to produce multiple embryos for implantation into surrogates.

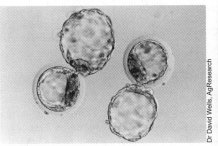

Cloned embryos immediately prior to implantation into a surrogate. These are at the blastocyst stage (50 - 150 cells). A single livestock animal may provide numerous eggs and therefore many blastocysts for implantation.

Embryo splitting produces multiple clones, but the clones are derived from an embryo whose physical characteristics are not completely known. This represents a limitation for practical applications when the purpose of the procedure is to produce high value livestock.

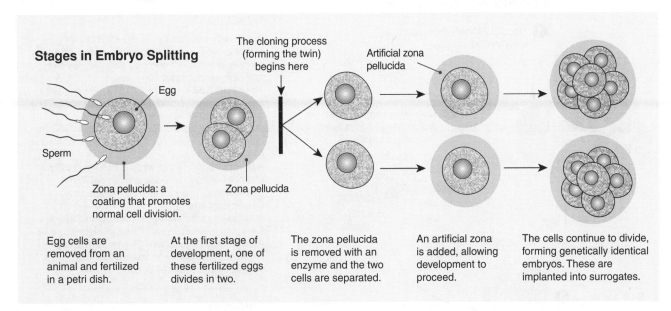

Stages in Embryo Splitting

The cloning process (forming the twin) begins here

Artificial zona pellucida

Egg

Sperm

Zona pellucida: a coating that promotes normal cell division.

Zona pellucida

Egg cells are removed from an animal and fertilized in a petri dish.

At the first stage of development, one of these fertilized eggs divides in two.

The zona pellucida is removed with an enzyme and the two cells are separated.

An artificial zona is added, allowing development to proceed.

The cells continue to divide, forming genetically identical embryos. These are implanted into surrogates.

1. With respect to animals, explain what is meant by **cloning**: _____

_____

2. How does embryo splitting enable breeders to produce multiple clones from a single high value animal? _____

_____

_____

_____

3. Describe the possible benefits to be gained from cloning high milk yielding cows: _____

_____

_____

4. Why would it be undesirable to produce all livestock using embryo splitting? _____

_____

113 Cloning by Somatic Cell Nuclear Transfer

Blackface ewe

Dolly

PHOTO: Courtesy Roslin Institute ©

Key Idea: Clones can be made by fusing the empty egg cell with a cell from the organism to be cloned.

The Issue

▶ Individuals vary in characteristics, even within specific breeds of animal, such as sheep.

▶ Clones remove the variability and produce livestock that would develop in a predictable way or produce a consistent quality of product such as wool or milk.

▶ Clones produced using traditional embryo-splitting are derived from an embryo whose physical characteristics are not completely known. Scientists wanted to speed up the process and produce clones from a proven phenotype.

Concept 1
Somatic cells can be made to return to a dormant or embryonic state so that their genes will not be expressed.

Concept 2
The nucleus of a cell can be removed and replaced with the nucleus of an unrelated cell. Cells can be made to fuse together.

Concept 3
Fertilized egg cells produce embryos. Egg cells that contain the nucleus of a donor cell will produce embryos with DNA identical to the donor cell.

Concept 4
Embryos can be implanted into surrogate mothers and develop to full term with seemingly no ill effects.

Somatic Cell Nuclear Transfer (SCNT)

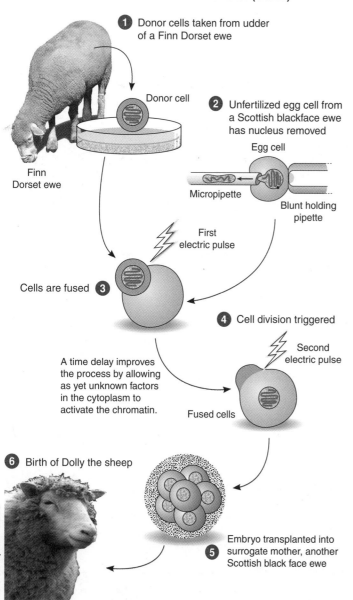

① Donor cells taken from udder of a Finn Dorset ewe

Donor cell

Finn Dorset ewe

② Unfertilized egg cell from a Scottish blackface ewe has nucleus removed

Egg cell

Micropipette

Blunt holding pipette

First electric pulse

Cells are fused ③

④ Cell division triggered

A time delay improves the process by allowing as yet unknown factors in the cytoplasm to activate the chromatin.

Second electric pulse

Fused cells

⑥ Birth of Dolly the sheep

PHOTO: Courtesy Roslin Institute ©

Embryo transplanted into surrogate mother, another Scottish black face ewe ⑤

Techniques

Donor cells from the udder of a Finn Dorset ewe are taken and cultured in a low nutrient media for a week. The nutrient deprived cells stop dividing and become **dormant**.

An **unfertilized egg** from a Scottish blackface ewe has the nucleus removed using a micropipette. The rest of the cell contents are left intact.

The dormant udder cell and the recipient denucleated egg cell are fused using a mild electric pulse.

A second electric pulse triggers cellular activity and cell division, jump starting the cell into development. This can also be triggered by chemical means.

After six days the embryo is transplanted into a surrogate mother, another Scottish blackface ewe. After a 148 day gestation 'Dolly' is born. DNA profiling shows she is genetically identical to the original Finn Dorset cell donor.

Outcomes

Dolly, a Finn Dorset lamb, was born at the Roslin Institute (near Edinburgh) in July 1996. She was the first mammal to be cloned from **non-embryonic** cells, i.e. cells that had already differentiated into their final form. Dolly's birth showed that the process leading to cell specialization is not irreversible and that cells can be 'reprogrammed' into an embryonic state. Although cloning seems relatively easy there are many problems that occur. Of the hundreds of eggs that were reconstructed only 29 formed embryos and only Dolly survived to birth.

Further Applications

In animal reproductive technology, cloning has facilitated the rapid production of genetically superior stock. These animals may then be dispersed among commercial herds. The **primary focus** of the new cloning technologies is to provide an economically viable way to rapidly produce transgenic animals with very precise genetic modifications.

© 2012-2014 **BIOZONE** International
ISBN: 978-1-927173-93-0
Photocopying Prohibited

Adult cloning heralds a new chapter in the breeding of livestock. Traditional breeding methods are slow, unpredictable, and suffer from a time delay in waiting to see what the phenotype is like before breeding the next generation. Adult cloning methods now allow a rapid spread of valuable livestock into commercial use among farmers. It will also allow the livestock industry to respond rapidly to market changes in the demand for certain traits in livestock products. In New Zealand, 10 healthy clones were produced from a single cow (the differences in coat colour patterns arise from the random migration of pigment cells in early embryonic development).

10 cloned calves

Dr David Wells, AgResearch

Saving the Enderby Island Cattle

Lady

Elsie ("L-C" Lady clone)

Enderby Island lies about 320 km south of New Zealand. During the late 1800s and early 1900s, there was an attempt to settle the island and farm cattle (the original breed is not known). The attempt failed and the cattle were left behind when the settlers left. For almost 90 years, the cattle population survived on a poor diet of scrub and seaweed, becoming a unique breed. In 1991, New Zealand's Department of Conservation determined that the remaining 53 cattle were interfering with the island's ecological recovery and began an eradication programme. During the first expedition, 47 of the cattle were killed. Semen and egg cells were taken from the cattle and stored.

Lady and her clones
In 1992, it was discovered that two cattle remained on the island, a cow (later named Lady) and her calf. They were captured and moved to a research centre in New Zealand where the calf later died for unknown reasons, leaving Lady the last of her breed. Attempts were made to use the collected sperm and eggs to produce embryos. Lady was moved to a second research centre. There, a bull (Derby) was produced by *in vitro* fertilization (IVF) and implantation in a surrogate. The low success rate of IVF prompted an attempt to clone Lady using SCNT. Of the 74 cloned embryos that were implanted into 37 cows, five survived to term. One died soon after birth and two died in the following year. The surviving two female calves were later bred with Derby and produced four calves. In 2006, the total population for the breed was 6, although further *in vitro* fertilization and implantation to surrogates has produced a third generation of cattle and the population is now slowly growing. Enderby Island cattle remain the only rare breed to be saved from extinction using SCNT.

1. What is **adult cloning** (as it relates to somatic cell nuclear transfer or **SCNT**)?:

   _____

   _____

2. Explain how each of the following events is controlled in the **SCNT** process:

   (a) The switching off of all genes in the donor cell: _____

   _____

   (b) The fusion (combining) of donor cell with enucleated egg cell: _____

   _____

   (c) The activation of the cloned cell into producing an embryo: _____

   _____

3. Describe two potential applications of nuclear transfer technology for the cloning of animals:

   (a) _____

   _____

   (b) _____

   _____

114 Chapter Review

Summarize what you know about this topic under the headings provided. You can draw diagrams or mind maps, or write short notes to organize your thoughts. Use the images and hints, included to help you:

Genes

HINT: Define genes and alleles. What is the effect of mutation on alleles? How do different organisms compare in their number of genes?

Chromosomes

HINT: Compare eukaryotic and prokaryotic DNA. Define homologous chromosomes, autosomes and sex chromosomes.

Meiosis

HINT: Define diploid and haploid. Draw the stages of meiosis.

REVISE

Inheritance

HINT: Mendel's laws of inheritance, Punnett squares and monohybrid crosses.

Genetic Modification

HINT: Review DNA profiling, its procedure and its use. Define the term clone and review natural and artificial methods for cloning. Review Genetic modification and its uses, risks, and ethics.

115 Key Terms: Did You Get It?

1. Test your vocabulary by matching each term to its definition, as identified by its preceding letter code.

alleles

autosome

diploid

DNA profiling

dominant

gel electrophoresis

genetic modification

genotype

karyotype

meiosis

monohybrid cross

mutation

non-disjunction

PCR

phenotype

Punnett square

recessive

A A process that is used to separate different lengths of DNA by placing them in a gel matrix placed in a buffered solution through which an electric current is passed.

B Allele that will only express its trait in the absence of the dominant allele.

C Genetic cross between two individuals that differ in one trait of particular interest.

D The process of altering the genetic makeup of cells or organisms by the selective removal, insertion, or modification of DNA.

E Sequences of DNA occupying the same gene locus (position) on different, but homologous, chromosomes.

F The process of double nuclear division (reduction division) to produce four nuclei, each containing half the original number of chromosomes (haploid).

G A change to the DNA sequence of an organism. This may be a deletion, insertion, duplication, inversion or translocation of DNA in a gene or chromosome.

H A graphical way of illustrating the outcome of a cross.

I Allele that expresses its trait irrespective of the other allele.

J A non-sex chromosome.

K The allele combination of an organism.

L The number and appearance of chromosomes in the nucleus of a eukaryotic cell.

M The process of identifying regions of a DNA sequence that are variable between individuals in order to distinguish between them.

N Observable characteristics in an organism.

O The failure of chromosome pairs to separate properly during meiosis I or failure of sister chromatids to separate properly during meiosis II (or mitosis).

P A reaction that is used to amplify fragments of DNA using cycles of heating and cooling.

Q Having two homologous copies of each chromosome (2N), usually one from the mother and one from the father.

2. Use lines to match the statements in the table below to form complete sentences:

Mutations are the ultimate...

Alleles are variations...

A person carrying two of the same alleles (one on each homologous chromosome)...

If the person carries two different alleles...

Alleles may be...

A dominant allele...

A recessive allele only...

... of a gene.

... dominant or recessive.

... for the gene, they are heterozygous.

... is said to be homozygous.

... expresses its trait if it is in the homozygous condition.

... source of new alleles.

... always expresses its trait whether it is in the homozygous or the heterozygous condition.

TEST

Topic 4

Ecology

4.1 Species, communities, and ecosystems

Understandings, applications, skills

Activity number

☐ 1 Define the terms: species, population, and community. Explain how the community and the abiotic (physical) environment interact to form an ecosystem. — 116

☐ 2 Discuss the influence of human activity on the potential of ecosystems to be sustainable. Explain how ecosystems can be investigated using mesocosms. — 117

☐ 3 Distinguish between autotrophs and heterotrophs. Classify species as autotrophs, consumers, detritivores, or saprotrophs based on their mode of nutrition. — 122

☐ 4 Describe the reliance of autotrophs on inorganic nutrients. Describe the role of nutrient cycles in maintaining the supply of these nutrients to autotrophs. — 127

☐ 5 Use chi-squared to test the statistical significance of associations between species in ecosystems sampled using quadrats. — 118-121

4.2 Energy flow

Understandings, applications, skills

Activity number

☐ 1 Describe the dependence of most ecosystems on sunlight energy. Explain how sunlight energy is converted by autotrophs into chemical energy in carbon compounds. Explain how energy flows through food chains and webs. Describe the efficiency of these energy transfers and explain the consequences of this to the length of food chains. How is energy lost from the system? — 122-125

☐ 2 Make quantitative representations of energy flow using pyramids of energy. — 126

4.3 Carbon cycling

Understandings, applications, skills

Activity number

☐ 1 Construct and annotate a diagram of the carbon cycle to show carbon sinks and processes involved in carbon fluxes. Estimate carbon fluxes due to processes in the carbon cycle including photosynthesis, respiration, decomposition, fossilization, and combustion. Explain the role of methane in the carbon cycle, including its production from organic matter and its oxidation in the atmosphere. — 128

☐ 2 Analyse data from air monitoring stations to explain annual fluctuations in CO_2. — 128

4.4 Climate change

Understandings, applications, skills

Activity number

☐ 1 Identify greenhouse gases and describe their role in retaining heat in the atmosphere and sustaining life on Earth. Identify the factors determining the impact of a greenhouse gas and explain the role of carbon dioxide (CO_2) and water vapour as the most significant greenhouse gases. — 129

☐ 2 Describe how global temperatures and climate patterns are influenced by greenhouse gas concentrations. Discuss the correlation between rising atmospheric CO_2 since the start of the Industrial Revolution and average global temperatures. Describe the primary cause of recent increases in atmospheric CO_2. — 130

☐ 3 Evaluate claims by climate change sceptics that human activities are not causing climate change. — 130

☐ 4 Explain what is meant by ocean acidification. Discuss the threat to coral reefs posed by increasing concentrations of dissolved CO_2 in the oceans. — 133

☐ 5 Comment on the global impact of climate change and explain why a reduction in greenhouse gas emissions requires an international cooperative approach. — 131 132

☐ 6 **TOK** *The precautionary principle aims to guide decision making where there is uncertainty. Is certainty possible in the natural sciences?* — 134

116 Components of an Ecosystem

Key Idea: An ecosystem consists of all the organisms living in a particular area and their physical environment.

An **ecosystem** is a community of living organisms and the physical (non-living) components of their environment. The community (living component of the ecosystem) is in turn made up of a number of **populations**, these being organisms of the same species living in the same geographical area. The structure and function of an ecosystem is determined by the physical (abiotic) and the living (biotic) factors, which determine species distribution and survival.

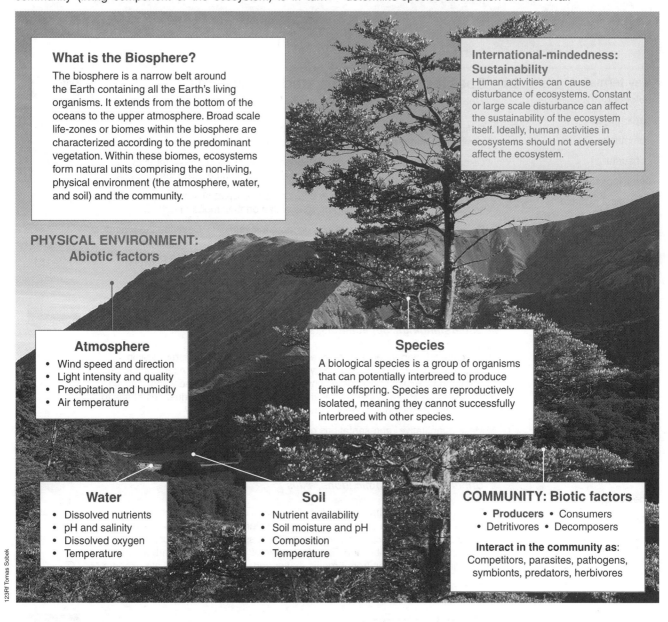

What is the Biosphere?

The biosphere is a narrow belt around the Earth containing all the Earth's living organisms. It extends from the bottom of the oceans to the upper atmosphere. Broad scale life-zones or biomes within the biosphere are characterized according to the predominant vegetation. Within these biomes, ecosystems form natural units comprising the non-living, physical environment (the atmosphere, water, and soil) and the community.

International-mindedness: Sustainability

Human activities can cause disturbance of ecosystems. Constant or large scale disturbance can affect the sustainability of the ecosystem itself. Ideally, human activities in ecosystems should not adversely affect the ecosystem.

PHYSICAL ENVIRONMENT: Abiotic factors

Atmosphere
- Wind speed and direction
- Light intensity and quality
- Precipitation and humidity
- Air temperature

Species
A biological species is a group of organisms that can potentially interbreed to produce fertile offspring. Species are reproductively isolated, meaning they cannot successfully interbreed with other species.

Water
- Dissolved nutrients
- pH and salinity
- Dissolved oxygen
- Temperature

Soil
- Nutrient availability
- Soil moisture and pH
- Composition
- Temperature

COMMUNITY: Biotic factors
- **Producers** • Consumers
- Detritivores • Decomposers

Interact in the community as:
Competitors, parasites, pathogens, symbionts, predators, herbivores

123RF Tomas Sobek

1. Distinguish clearly between a community and an ecosystem: _____

_____

_____

_____

2. Distinguish between biotic and abiotic factors: _____

_____

_____

3. Use one or more of the following terms to describe each of the features of a beech community listed below:
   Terms: *population, community, ecosystem, physical factor.*

   (a) All the beech trees present: _____   (c) All the organisms present: _____

   (b) The entire forest: _____   (d) The humidity: _____

WEB LINK LINK
116 117 118

© 2012-2014 **BIOZONE** International
ISBN: 978-1-927173-93-0
Photocopying Prohibited

KNOW

Measuring the Diversity of Ecosystems

In most field studies, it is not possible to measure or count every member of a population. Instead, the population is sampled in a way that provides a fair (unbiased) representation of the organisms present and their distribution. This is usually achieved through **random sampling**, a technique in which every possible sample of a given size has the same chance of selection.

Students sampling an intertidal community

The methods you use to sample must be appropriate to the community being studied and the information you want to obtain. You must also think about the time and equipment available, the organisms involved, and the impact your study might have on the environment. Communities in which the populations are at low density and have a random or clumped distribution will require a different sampling strategy to those in which the populations are uniformly distributed and at higher density. There are many sampling options, each with advantages and drawbacks for particular communities. Quadrats are often used to sample communities of plants and invertebrates and can be placed randomly or along a transect (below).

Random point sampling	Line and belt transects	Point sampling: systematic grid	Random quadrats
A	**B**	**C**	**D**

Random sampling is achieved using random number tables which provide points on a grid (A) or pairs of coordinates, which are joined to form a line (B). Random sampling produces an unbiased result and can be used to sample large populations, but it can provide a poor representation of the area if not enough samples are taken. Systematic sampling (C) has more bias but provides good coverage of the sample area. Quadrats (D) can be used for point sampling or transects.

Types of Ecosystems

Ecosystems vary greatly in their features, from dry, inhospitable environments with very little vegetation, to lush tropical rainforests containing many different types of vegetation. Each ecosystem has a unique combination of abiotic factors, which collectively influence its community structure.

Desert | Tropical rainforest

Alpine | Temperate forest

Geographical Barriers can Isolate Species

Species found on almost every continent of the world are called cosmopolitan species. Examples include wild pigeons, house sparrows, brown (or Norway) rats, and the housefly. Although they are the same species, if the populations are separated by a significant barrier (e.g. mountains, oceans) or on different continents, they will not be able to interbreed. Long term isolation can eventually lead to differences between populations. In some cases, these differences may result in reproductive isolation and the formation of a new species. The distribution of the brown rat is shown in dark grey below.

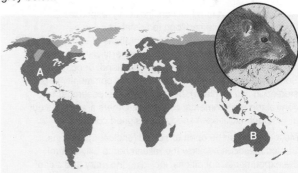

4. (a) What is a biological species? _____

_____

(b) Why are the two brown rat populations (A and B) still considered to be the same species?

_____

5. (a) What is meant by random sampling? _____

_____

(b) Why is it usually recommended that population sampling is random? _____

_____

_____

© 2012-2014 **BIOZONE** International
ISBN: 978-1-927173-93-0

117 The Stability of Ecosystems

Key Idea: Ecosystems have the potential to remain stable, i.e. in a relatively unchanged state, over long periods of time. Although the biotic and abiotic components of ecosystems are constantly responding to environmental changes, ecosystems as a whole are potentially stable (unchanging) for long periods of time. The long term stability of an ecosystem depends partly on its ability to resist change and recover from disturbance (its resilience). Human activity can alter the long term unchanging nature of ecosystems by interfering with important aspects of ecosystem function, such as nutrient cycling. These effects can be investigated on a small scale using experimental systems called mesocosms.

An ecosystem may remain stable for many hundreds or thousands of years provided that the biotic and abiotic components interacting within it remain stable.

Small scale disturbances usually have a minor ecosystem effect. Fire or flood may destroy some parts, but enough is left for the ecosystem to return to its original state.

Large scale disturbances such as volcanic eruptions, sea level rise, or large scale open cast mining remove all components of the ecosystem, changing it forever.

Experimental Systems Can Model Ecosystem Functions

A mesocosm set up in a laboratory

Aspects of ecosystem function, including responses to changes in inputs and long term stability, can be investigated using physical representations of ecosystems called **mesocosms**. Examples include artificial ponds and streams, or enclosed areas of land, wetland, or ocean. Some mesocosm studies allow a natural community to be studied *in situ* (in place), but still allow the researcher to control the environmental conditions. Others are carried out at research facilities in specially designed containers.

Mesocosms can be open or sealed (enclosed) systems. Sealed mesocosms allow the researcher to fully control the experimental conditions, including the entry and exit of matter. Mesocosms, especially small ones, are generally not stable in the long term, and change over time as a result of their smaller scale and isolated nature.

A Mesocosm Study

Small, closed ecological chambers were used by researchers at the University of Washington to test system responses to changes in environment and inputs. One aspect of the study is described here.

Researchers altered the levels of algal growth nutrients added to the mesocosm chambers and measured the effect of the algal response on the population growth of a marine copepod (*Tigriopus californicus*), an algal grazer.

Algal growth-promoting medium was added at 2, 10, or 20% to seawater, together with 0.1 mL of an algal mix. Two days after adding the growth medium and algae, six copepods were added to each chamber. The chambers were sealed and the population size in each mesocosm was measured over time (results below).

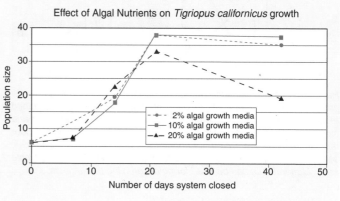

Effect of Algal Nutrients on *Tigriopus californicus* growth

Y-axis: Population size (0 to 40)
X-axis: Number of days system closed (0 to 50)

- ● 2% algal growth media
- ■ 10% algal growth media
- ▲ 20% algal growth media

Adapted from Armentrout, B & Kappes, H; University of Washington with corrections

1. Analyse the data in the graph (above right). Describe the results and comment on the stability of each chamber: _____

_____

_____

_____

2. What assumptions are made in this experiment? _____

_____

_____

_____

© 2012-2014 **BIOZONE** International
ISBN: 978-1-927173-93-0
Photocopying Prohibited

118 Quadrat Sampling

Key Idea: Quadrat sampling involves a series of random placements of a frame of known size over an area of habitat to assess the abundance or diversity of organisms.

Quadrat sampling is a method by which organisms in a certain proportion (sample) of the habitat are counted directly. It is used when the organisms are too numerous to count in total. It can be used to estimate population **abundance** (number), **density, frequency of occurrence**, and **distribution**. Quadrats may be used without a transect when studying a relatively uniform habitat. In this case, the quadrat positions are chosen randomly using a random number table.

The general procedure is to count all the individuals (or estimate their percentage cover) in a number of quadrats of known size and to use this information to work out the abundance or percentage cover value for the whole area.

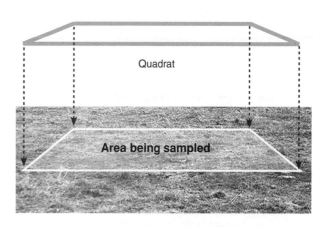

Quadrat

Area being sampled

$$\text{Estimated average density} = \frac{\text{Total number of individuals counted}}{\text{Number of quadrats X area of each quadrat}}$$

Guidelines for Quadrat Use:

1. The **area of each quadrat** must be known exactly and ideally quadrats should be the same shape. The quadrat does not have to be square (it may be rectangular, hexagonal etc.).

2. **Enough quadrat samples** must be taken to provide results that are representative of the total population.

3. The **population of each quadrat** must be known exactly. Species must be distinguishable from each other, even if they have to be identified at a later date. It has to be decided beforehand what the count procedure will be and how organisms over the quadrat boundary will be counted.

4. The size of the quadrat should be appropriate to the organisms and habitat, e.g. a large size quadrat for trees.

5. The quadrats must be **representative of the whole area.** This is usually achieved by **random sampling** (right).

The area to be sampled is divided up into a grid pattern with indexed coordinates

Quadrats are applied to the predetermined grid on a random basis. This can be achieved by using a random number table.

Sampling a centipede population

A researcher by the name of Lloyd (1967) sampled centipedes in Wytham Woods, near Oxford in England. A total of 37 hexagon–shaped quadrats were used, each with a diameter of 30 cm (see diagram on right). These were arranged in a pattern so that they were all touching each other. Use the data in the diagram to answer the following questions.

1. Determine the average number of centipedes captured per quadrat:

    _____

2. Calculate the estimated average density of centipedes per square metre (remember that each quadrat is 0.08 square metres in area):

    _____

3. Looking at the data for individual quadrats, describe in general terms the distribution of the centipedes in the sample area:

    _____

    _____

4. Describe one factor that might account for the distribution pattern:

    _____

    _____

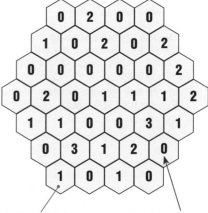

Each quadrat was a hexagon with a diameter of 30 cm and an area of 0.08 square metres.

The number in each hexagon indicates how many centipedes were caught in that quadrat.

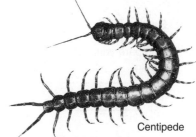

Centipede

119 Sampling a Rocky Shore Community

Key Idea: The estimates of a population gained from using a quadrat may vary depending on where the quadrats are placed. Larger samples can account for variation.
The diagram (opposite page) represents an area of seashore with its resident organisms. The distribution of coralline algae and four animal species are shown. This exercise is designed to prepare you for planning and carrying out a similar procedure to practically investigate a natural community.

1. **Decide on the sampling method**
   For the purpose of this exercise, it has been decided that the populations to be investigated are too large to be counted directly and a quadrat sampling method is to be used to estimate the average density of the four animal species as well as that of the algae.

2. **Mark out a grid pattern**
   Use a ruler to mark out 3 cm intervals along each side of the sampling area (area of quadrat = 0.03 x 0.03 m). **Draw lines** between these marks to create a 6 x 6 grid pattern (total area = 0.18 x 0.18 m). This will provide a total of 36 quadrats that can be investigated.

3. **Number the axes of the grid**
   Only a small proportion of the possible quadrat positions are going to be sampled. It is necessary to select the quadrats in a random manner. It is not sufficient to simply guess or choose your own on a 'gut feeling'. The best way to choose the quadrats randomly is to create a numbering system for the grid pattern and then select the quadrats from a random number table. Starting at the *top left hand corner*, **number the columns** and **rows** from 1 to 6 on each axis.

4. **Choose quadrats randomly**
   To select the required number of quadrats randomly, use random numbers from a random number table. The random numbers are used as an index to the grid coordinates. Choose 6 quadrats from the total of 36 using table of random numbers provided for you at the bottom of the next page. Make a note of which column of random numbers you choose. Each member of your group should choose a different set of random numbers (i.e. different column: A–D) so that you can compare the effectiveness of the sampling method.

   Column of random numbers chosen: _____

   NOTE: Highlight the boundary of each selected quadrat with coloured pen/highlighter.

5. **Decide on the counting criteria**
   Before the counting of the individuals for each species is carried out, the criteria for counting need to be established.

There may be some problems here. You must decide before sampling begins as to what to do about individuals that are only partly inside the quadrat. Possible answers include:

(a) Only counting individuals that are completely inside the quadrat.
(b) Only counting individuals with a clearly defined part of their body inside the quadrat (such as the head).
(c) Allowing for 'half individuals' (e.g. 3.5 barnacles).
(d) Counting an individual that is inside the quadrat by half or more as one complete individual.

Discuss the merits and problems of the suggestions above with other members of the class (or group). You may even have counting criteria of your own. Think about other factors that could cause problems with your counting.

6. **Carry out the sampling**
   Carefully examine each selected quadrat and **count the number of individuals** of each species present. Record your data in the spaces provided on the next page.

7. **Calculate the population density**
   Use the combined data TOTALS for the sampled quadrats to estimate the average density for each species by using the formula:

$$\text{Density} = \frac{\text{Total number in all quadrats sampled}}{\text{Number of quadrats sampled} \times \text{area of a quadrat}}$$

Remember that a total of 6 quadrats are sampled and each has an area of 0.0009 m^2. The density should be expressed as the number of individuals *per square metre* (no. m^{-2}).

Plicate barnacle: [ ]   Snakeskin chiton: [ ]

Oyster borer: [ ]   Coralline algae: [ ]

Limpet: [ ]

8. (a) In this example the animals are not moving. Describe the problems associated with sampling moving organisms. Explain how you would cope with sampling these same animals if they were really alive and very active:

_____

_____

_____

(b) Carry out a direct count of all 4 animal species and the algae for the whole sample area (all 36 quadrats). Apply the data from your direct count to the equation given in (7) above to calculate the actual population density (remember that the number of quadrats in this case = 36):

Barnacle: [ ]   Oyster borer: [ ]   Chiton: [ ]   Limpet: [ ]   Algae: [ ]

Compare your estimated population density to the actual population density for each species:

_____

_____

_____

LINK
118

SKILL

© 2012-2014 **BIOZONE** International
ISBN: 978-1-927173-93-0
Photocopying Prohibited

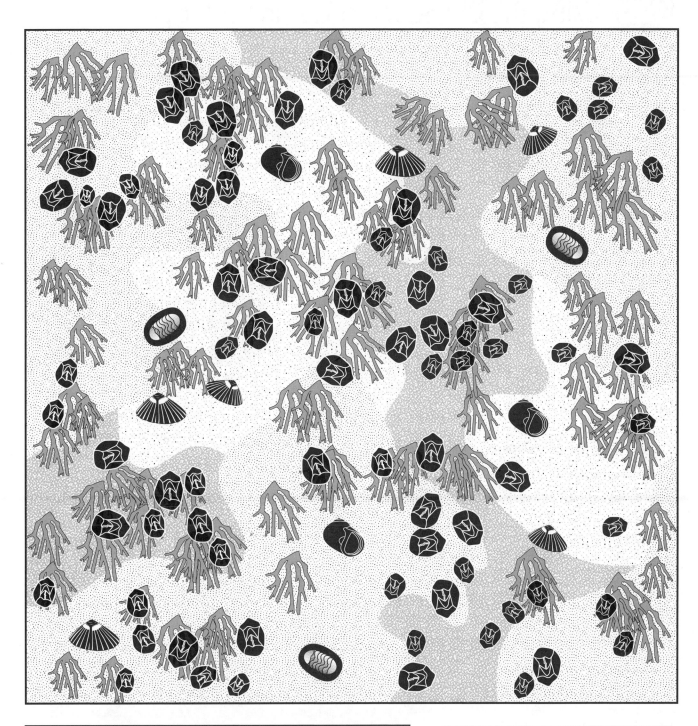

Coordinates for each quadrat	Plicate barnacle	Oyster borer	Snakeskin chiton	Limpet	Coralline algae
1:					
2:					
3:					
4:					
5:					
6:					
TOTAL					

Table of random numbers

A	B	C	D
2 2	3 1	6 2	2 2
3 2	1 5	6 3	4 3
3 1	5 6	3 6	6 4
4 6	3 6	1 3	4 5
4 3	4 2	4 5	3 5
5 6	1 4	3 1	1 4

The table above has been adapted from a table of random numbers from a statistics book. Use this table to select quadrats randomly from the grid above. Choose one of the columns (A to D) and use the numbers in that column as an index to the grid. The first digit refers to the row number and the second digit refers to the column number. To locate each of the 6 quadrats, find where the row and column intersect, as shown below:

Example: 5 2   refers to the 5th row and the 2nd column

120 Using the Chi-Squared Test in Ecology

Key Idea: The chi-squared test is used to compare sets of categorical data and evaluate if differences between them are statistically significant or due to chance.

The chi-squared test (χ^2) is used to determine differences between categorical data sets when working with frequencies (counts). For the test to be valid, the data recorded for each categorical variable (e.g. species) must be raw counts (not measurements or derived data). The chi-squared test is used for two types of comparison: test for goodness of fit and tests of independence (association). A test for goodness of fit is used to compare an experimental result with an expected theoretical outcome. You will perform this test later to compare the outcome of genetic crosses to an expected theoretical ratio. A test for independence evaluates whether two variables are associated. The chi squared test is not valid when sample sizes are small (<20). Like all statistical tests, it aims to test the null hypothesis; the hypothesis of no difference (or no association) between groups of data. The worked example below uses the chi-squared test for association in a study of habitat preference in mudfish.

Using the Chi-Squared Test for Independence

Black mudfish (*Neochanna diversus*) is a small fish species native to New Zealand and found in wetlands and swampy streams. Researchers were interested in finding environmental indicators of favourable mudfish habitat. They sampled 80 wetland sites for the presence or absence of mudfish and recorded if there was emergent vegetation present or absent. Emergent vegetation, defined as vegetation rooted in water but emerging above the water surface, is an indicator of a relatively undisturbed environment. A chi-squared for association was used to test if mudfish were found more often at sites with emergent vegetation than by chance alone. The null hypothesis was that there is no association. The worked example is below. The table of observed values records the number of sites with or without mudfish and with or without emergent vegetation.

Black mudfish are able to air-breathe and can survive through seasonal drying of their wetland habitat.

Photo and data: Rhys Barrier, University of Waikato

Step 1: Enter the observed values (O) in a contingency table

A χ^2 test for association requires that the data (counts or frequencies) are entered in a **contingency table** (a matrix format to analyse and record the relationship between two or more categorical variables). Marginal totals are calculated for each row and column and a grand total is recorded in the bottom right hand corner (right).

	Mudfish absent (0)	Mudfish present (1)	Total
Emergent vegetation absent (0)	15	0	15
Emergent vegetation present (1)	26	39	65
Total	41	39	80

Step 2: Calculate the expected values (E)

Calculating the expected values for a contingency table is simple. For each category, divide the row total by the grand total and multiply by the column total. You can enter these in a separate table or as separate columns next to the observed values (right).

	Mudfish absent (0)	Mudfish present (1)	Total
Emergent vegetation absent (0)	7.69	7.31	15
Emergent vegetation present (1)	33.31	31.69	65
Total	41	39	80

Step 3: Calculate the value of chi-squared (χ^2) of $(O - E)^2 \div (E)$

The difference between the observed (O) and expected (E) values is calculated as a measure of the deviation from a predicted result. Since some deviations are negative, they are all squared to give positive values. This step is best done as a tabulation to obtain a value for $(O - E)^2 \div (E)$ for each category. The sum of all these values is the value of chi squared (blue table right).

$$\chi^2 = \sum \frac{(O - E)^2}{E}$$

Where: O = the observed result
E = the expected result
Σ = sum of

Category	O	E	O–E	$(O–E)^2$	$\dfrac{(O–E)^2}{E}$
Mudfish 0/EmVeg 0	15	7.69	7.31	53.44	6.95
Mudfish 1/EmVeg 0	0	7.31	-7.31	53.44	7.31
Mudfish 0/EmVeg 1	26	33.31	-7.31	53.44	1.60
Mudfish 1/EmVeg 1	39	31.69	7.31	53.44	1.69

Total = 80 $\chi^2 \longrightarrow \Sigma = 17.55$

Step 4: Calculate the degrees of freedom (df)

The degrees of freedom for a contingency table is given by the formula: (rows-1) x (columns-1). For this example, degrees of freedom (df) is therefore (2-1) x (2-1) = 1.

Critical values of χ^2 at different levels of probability. By convention, the critical probability for rejecting the null hypothesis (H_0) is 5%. If the test statistic is greater than the tabulated value for P = 0.05 we reject H_0 in favour of the alternative hypothesis.

Step 5: Using the chi squared table

On the χ^2 table (relevant part reproduced in the table right) with 1 degree of freedom, the calculated value for χ^2 of 17.55 corresponds to a probability of less than 0.001 (see arrow). *This means that by chance alone a χ^2 value of 17.55 could be expected less than 0.1% of the time.* This probability is much lower than the 0.05 value which is generally regarded as significant. The null hypothesis can be rejected and we have reason to believe that black mudfish are associated with sites with emergent vegetation more than expected by chance alone.

	Level of Probability (P)				
df	0.05	0.025	0.01	0.005	0.001
1	3.84	5.02	6.63	7.88	10.83
2	5.99	7.38	9.21	10.60	13.82
3	7.81	9.35	11.34	12.84	16.27

LINK
REFER 266

© 2012-2014 **BIOZONE** International
ISBN: 978-1-927173-93-0
Photocopying Prohibited

121 Chi-Squared Exercise in Ecology

Key Idea: Chi-squared can be used to determine if an association between two species is statistically significant.
In ecological studies, it is often found that two or more species are found in association. This is usually because of similar environmental requirements or because one species depends on the other. The following hypothetical example outlines a study in which the presence or absence of two plant species was recorded in a marked area. The two species are sometimes, but not always, found together. The chi squared test is used to test the significance of the association.

Using Chi Square to Test Species Associations in a Successful Marsh-Meadow Community

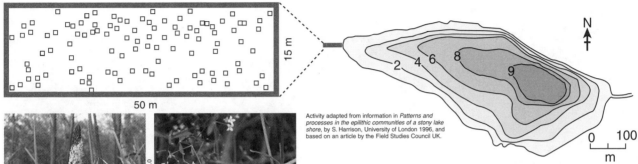

Activity adapted from information in *Patterns and processes in the epilithic communities of a stony lake shore*, by S. Harrison, University of London 1996, and based on an article by the Field Studies Council UK.

Lesser pond sedge (*Carex acutiformis*) is a swamp plant

Marsh bedstraw (*Galium palustre*) grows in ditches and wet meadows

Lake Crosemere (above) is a one of a series of kettle hole lakes in England, formed by glacial retreat at the end of the last glacial period. In a natural process of succession, the lake is gradually infilling from its western edge, and wet meadow and marsh species are replacing the species of the open water. Students investigated the association between two plants previously recorded in studies of the area: the lesser pond sedge (LPS) and marsh bedstraw (MBS). They recorded species presence or absence in 100 quadrats (0.5 m²) placed in an area 15 X 50 m using coordinates generated using a random number function on a spreadsheet. The results are summarized in table 1 below. Follow the steps to complete the analysis.

1. State the null hypothesis (H$_0$) for this investigation:

_____

_____

	LPS present (1)	LPS absent (0)	Total
MBS present (1)	11	3	14
MBS absent (0)	31	55	86
Total	42	58	100

Table 1: Observed results for presence/absence of lesser pond-sedge (LPS) and marsh bedstraw (MBS).

2. In words, summarize the observed results in table 1:

_____

_____

_____

_____

	LPS present (1)	LPS absent (0)	Total
MBS present (1)			
MBS absent (0)			
Total			

Table 2: Expected results for presence/absence of lesser pond-sedge (LPS) and marsh bedstraw (MBS).

3. Calculate the expected values for presence/absence of LPS and MBS. Enter the figures in table 2:

4. Complete the table to calculate the χ^2 value: $\chi^2 =$ _____

5. Calculate the degrees of freedom: _____

6. Using the χ^2, state the *P* value corresponding to your calculated χ^2 value (use the χ^2 table opposite):

_____

7. State whether or not you reject your null hypothesis:

reject H$_0$ / do not reject H$_0$ (*circle one*)

8. What could you conclude about this plant community:

_____

_____

_____

Category	O	E	O–E	(O–E)²	$\frac{(O-E)^2}{E}$
LPS 1/MBS 1					
LPS 0/MBS 1					
LPS 1/MBS 0					
LPS 0/MBS 0					
Total = 100					Σ =

LINK

120 SKILL

122 Food Chains

Key Idea: A food chain is a model to illustrate the feeding relationships between organisms.

Organisms in ecosystems interact by way of their feeding (trophic) relationships. These interactions can be shown in a **food chain**, which is a simple model to illustrate how energy, in the form of food, passes from one organism to the next. Each organism in the chain is a food source for the next. The levels of a food chain are called **trophic levels**. An organism is assigned to a trophic level based on its position in the food chain. Organisms may occupy different trophic levels in different food chains or during different stages of their life. Arrows link the organisms in a food chain. The direction of the arrow shows the flow of energy through the trophic levels. Most food chains begin with a producer, which is eaten by a primary consumer (**herbivore**). Higher level consumers (**carnivores** and **omnivores**) eat other consumers.

Millipede

Producers (autotrophs) e.g. plants, algae, and autotrophic bacteria, make their own food from simple inorganic substances, often by photosynthesis using energy from the sun. Inorganic nutrients are obtained from the abiotic environment, such as the soil and atmosphere.

Consumers (heterotrophs) e.g. animals, get their energy from other organisms. Consumers are ranked according to the trophic level they occupy, i.e. 1st order, 2nd order, and classified according to diet (e.g. carnivores eat animal tissue, omnivores eat plant and animal tissue).

Detritivores and **saprotrophs** both are consumers that gain nutrients from digesting dead organic matter (DOM). Detritivores consume DOM (e.g. earthworms, millipedes) whereas saprotrophs secrete enzymes to digest the DOM extracellularly and absorb the products of digestion (e.g. fungi, soil bacteria).

The diagram at the bottom of the page represents the basic elements of a food chain. In the questions below, you are asked to add to the diagram the features that indicate the flow of energy through the community of organisms.

1. (a) What is the original energy source for this food chain? _____

    (b) Draw arrows on the diagram above to show how the energy flows through the organisms in the food chain. Label each arrow with the process involved in the energy transfer. Draw arrows to show how energy is lost by respiration.

2. Describe how the following obtain their energy:

    (a) Producers: _____

    (b) Consumers: _____

    (c) Detritivores: _____

    (d) Saprotrophs: _____

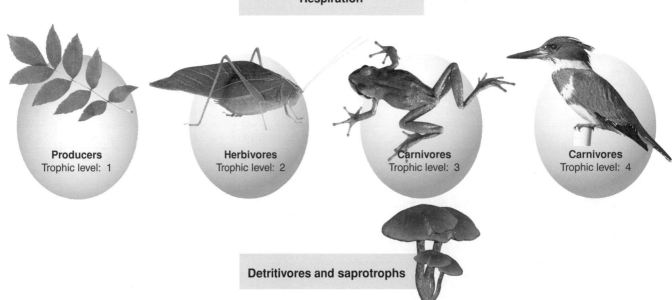

Respiration

Producers
Trophic level: 1

Herbivores
Trophic level: 2

Carnivores
Trophic level: 3

Carnivores
Trophic level: 4

Detritivores and saprotrophs

© 2012-2014 **BIOZONE** International
ISBN: 978-1-927173-93-0
Photocopying Prohibited

123 Food Webs

Key Idea: A food web depicts all the interconnected food chains in an ecosystem. Sunlight is the energy source for most ecosystems. Some energy is lost at each trophic level.

The different food chains in an ecosystem are interconnected to form a complex web of feeding interactions called a **food web**. Sunlight is the initial energy source for almost all ecosystems. Sunlight provides a continuous, but variable, energy supply, which is fixed in carbon compounds by photosynthesis. Energy flows through ecosystems in the chemical bonds within organic matter (food) and, in accordance with the second law of thermodynamics, is dissipated as heat as it is transferred through trophic levels. This loss of energy from the system limits how many links can be made in each food chain, as living organisms cannot convert heat to other forms of energy. Two simplified food webs showing the transfer of energy are depicted below.

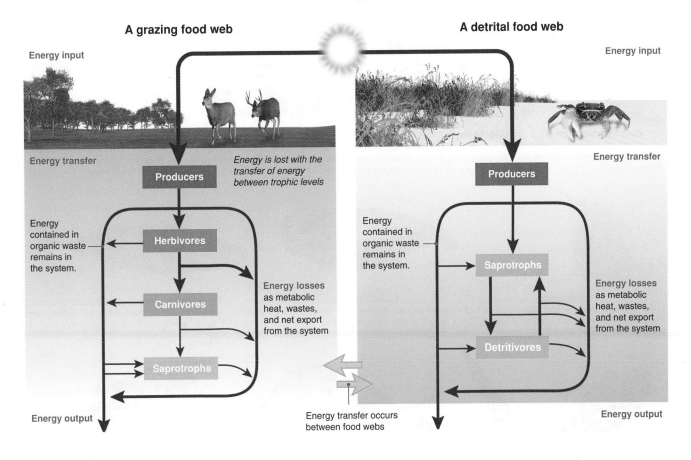

1. Describe how energy is transferred through ecosystems: _____
_____
_____

2. (a) Describe what happens to the **amount** of energy available to each successive trophic level in a food chain:
_____

(b) Explain why this is the case: _____
_____
_____
_____

3. With respect to energy flow, describe a major difference between a detrital and a grazing food web: _____
_____
_____
_____

LINK 125   LINK 124   WEB 123   **KNOW**

124 Constructing a Food Web

Key Idea: Food chains can be put together to form food webs. The complexity of a food web depends on the number of foods chains and trophic levels involved.

Species are assigned to trophic levels on the basis of their sources of nutrition, with the first trophic level (the producers), ultimately supporting all other (consumer) levels.

Consumers are ranked according to the trophic level they occupy, although some consumers may feed at several different trophic levels. In the example of a lake ecosystem below, your task is assemble the organisms into a food web in a way that illustrates their trophic status and their relative trophic position(s).

Feeding Requirements of Lake Organisms

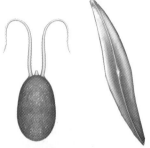

Autotrophic protists
Chlamydomonas (above), *Euglena* Two of the many genera that form the phytoplankton.

Macrophytes (various species)
A variety of flowering aquatic plants are adapted for being submerged, free-floating, or growing at the lake margin.

Detritus
Decaying organic matter from within the lake itself or it may be washed in from the lake margins.

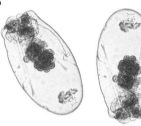

Asplanchna (planktonic rotifer)
A large, carnivorous rotifer that feeds on protozoa and young zooplankton (e.g. small *Daphnia*).

Daphnia
Small freshwater crustacean that forms part of the zooplankton. It feeds on planktonic algae by filtering them from the water with its limbs.

Leech (*Glossiphonia*)
Leeches are fluid feeding predators of smaller invertebrates, including rotifers, small pond snails and worms.

Three-spined stickleback (*Gasterosteus*)
A common fish of freshwater ponds and lakes. It feeds mainly on small invertebrates such as *Daphnia* and insect larvae.

Diving beetle (*Dytiscus*)
Diving beetles feed on aquatic insect larvae and adult insects blown into the lake community. They will also scavenge organic detritus. Adults will also take fish fry. Adults (left) and larvae (right) are voracious top predators in small ponds.

Carp (*Cyprinus*)
A heavy bodied freshwater fish that feeds mainly on bottom living insect larvae and snails, but will also take some plant material (not algae).

Dragonfly larva
Large aquatic insect larvae that are voracious predators of small invertebrates including *Hydra*, *Daphnia*, other insect larvae, and leeches.

Great pond snail (*Limnaea*)
Omnivorous pond snail, eating both plant and animal material, living or dead, although the main diet is aquatic macrophytes.

Herbivorous water beetles (e.g. *Hydrophilus*)
Feed on water plants, although the young beetle larvae are carnivorous, feeding primarily on small pond snails.

Protozan (e.g. *Paramecium*)
Ciliated protozoa such as *Paramecium* feed primarily on bacteria and microscopic green algae such as *Chlamydomonas*.

Pike (*Esox lucius*)
A top ambush predator of all smaller fish and amphibians. They are also opportunistic predators of rodents and small birds.

Mosquito larva (*Culex* spp.)
The larvae of most mosquito species, e.g. *Culex*, feed on planktonic algae and small protozoans before passing through a pupal stage and undergoing metamorphosis into adult mosquitoes.

Hydra
A small carnivorous cnidarian that captures small prey items, e.g. small *Daphnia* and insect larvae, using its stinging cells on the tentacles.

© 2012-2014 **BIOZONE** International
ISBN: 978-1-927173-93-0
Photocopying Prohibited

1. From the information provided for the lake food web components on the previous page, construct **ten** different **food chains** to show the feeding relationships between the organisms. Some food chains may be shorter than others and most species will appear in more than one food chain. An example has been completed for you.

Example 1:   Macrophyte ⟶ Herbivorous water beetle ⟶ Carp ⟶ Pike

(a) _____

(b) _____

(c) _____

(d) _____

(e) _____

(f) _____

(g) _____

(h) _____

(i) _____

(j) _____

2. (a) Use the food chains created above to help you to draw up a **food web** for this community. Use the information supplied to draw arrows showing the flow of **energy** between species (only energy **from** the detritus is required).

   (b) Label each species to indicate its position in the food web, i.e. its trophic level (**T1, T2, T3, T4, T5**). Where a species occupies more than one trophic level, indicate this, e.g. **T2/3**:

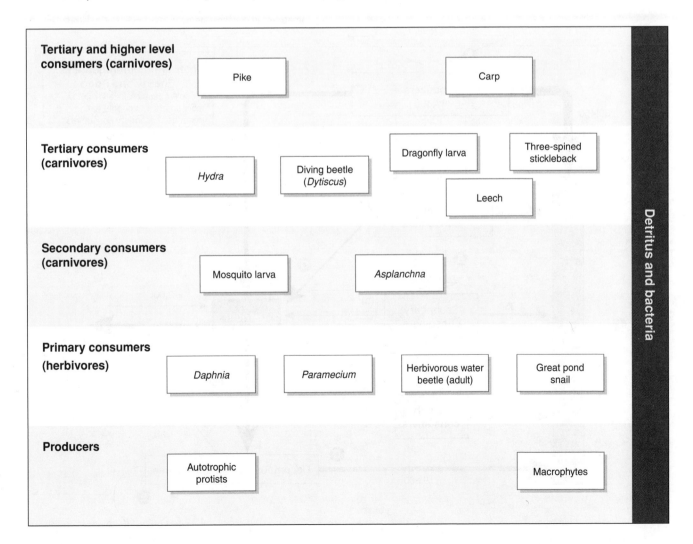

125 Energy Flow in an Ecosystem

Key Idea: Energy flows through an ecosystem between trophic levels. Only 5-20% of energy is transferred from one trophic level to the next.

Energy cannot be created or destroyed, only transformed from one form (e.g. light energy) to another (e.g. chemical energy in the bonds of molecules). This means that the flow of energy through an ecosystem can be measured. Each time energy is transferred from one trophic level to the next (by eating, defecation, etc), some energy is given out as heat to the environment, usually during cellular respiration. Living

organisms cannot convert heat to other forms of energy, so the amount of energy available to one trophic level is always less than the amount at the previous level. Potentially, we can account for the transfer of energy from its input (as solar radiation) to its release as heat from organisms, because energy is conserved. The percentage of energy transferred from one trophic level to the next is the **trophic efficiency**. It varies between 5% and 20% and measures the efficiency of energy transfer. An average figure of 10% trophic efficiency is often used. This is called the **ten percent rule** (below).

Energy Flow Through an Ecosystem

NOTE

Numbers represent **kilojoules** of energy per square metre per year (kJ m^{-2} yr $^{-1}$)

Sunlight falling on plant surfaces
7,000,000

Light absorbed by plants
1,700,000

A

Energy absorbed from the previous trophic level

100

Energy lost as heat **65** | **Trophic level** | **15** Energy lost to detritus

20

Energy passed on to the next trophic level

The energy available to each trophic level will always equal the amount entering that trophic level, minus total losses to that level (due to metabolic activity, death, excretion etc). Energy lost as heat will be lost from the ecosystem. Other losses become part of the detritus and may be utilized by other organisms in the ecosystem

International-mindedness: Energy and Food
People eating crops are primary consumers, but people eating meat are secondary consumers. Crops can feed more people per hectare than livestock can because there is one less step in the food chain.

Producers
87,400

50,450

(a)

7,800 ← Primary consumers

B

1600

Secondary consumers

(b)

4,600

G

22,950

Detritus

2,000 ←

10,465

D

1,330 ←

90

(c)

19,300

Heat loss in metabolic activity

F

55 ← Tertiary consumers

Decomposers and detritivores

(d)

C

19,200

E

© 2012-2014 **BIOZONE** International
ISBN: 978-1-927173-93-0

1. Study the diagram on the previous page illustrating energy flow through a hypothetical ecosystem. Use the example at the top of the page as a guide to calculate the missing values (a)–(d) in the diagram. Note that the sum of the energy inputs always equals the sum of the energy outputs. Write your answers in the spaces provided on the diagram.

2. Identify the processes occurring at the points labelled **A – G** on the diagram:

A. _____   E. _____

B. _____   F. _____

C. _____   G. _____

D. _____

3. (a) Calculate the percentage of light energy falling on the plants that is absorbed at point **A**:

Light absorbed by plants ÷ sunlight falling on plant surfaces x 100 = _____

(b) What happens to the light energy that is not absorbed? _____

4. (a) Calculate the percentage of light energy absorbed that is actually converted (fixed) into producer energy:

Producers ÷ light absorbed by plants x 100 = _____

(b) How much light energy is absorbed but not fixed: _____

(c) Account for the difference between the amount of energy absorbed and the amount actually fixed by producers:

_____

_____

5. Of the total amount of energy **fixed** by producers in this ecosystem (at point **A**) calculate:

(a) The total amount that ended up as metabolic waste heat (in kJ): _____

(b) The percentage of the energy fixed that ended up as waste heat: _____

6. (a) State the groups for which detritus is an energy source: _____

(b) How could detritus be removed or added to an ecosystem? _____

_____

7. Under certain conditions, decomposition rates can be very low or even zero, allowing detritus to accumulate:

(a) From your knowledge of biological processes, what conditions might slow decomposition rates?

_____

(b) What are the consequences of this lack of decomposer activity to the energy flow? _____

_____

(c) Add an additional arrow to the diagram on the previous page to illustrate your answer: _____

(d) Describe three examples of materials that have resulted from a lack of decomposer activity on detrital material:

_____

_____

8. The **ten percent rule** states that the total energy content of a trophic level in an ecosystem is only about one-tenth (or 10%) that of the preceding level. For each of the trophic levels in the diagram on the preceding page, determine the amount of energy passed on to the next trophic level as a percentage:

(a) Producer to primary consumer: _____

(b) Primary consumer to secondary consumer: _____

(c) Secondary consumer to tertiary consumer: _____

126 Ecological Pyramids

Key Idea: Ecological pyramids can be used to illustrate the amount of energy at each trophic level in an ecosystem.
The energy, biomass, or numbers of organisms at each trophic level in any ecosystem can be represented by an ecological pyramid. The first trophic level is placed at the bottom of the pyramid and subsequent trophic levels are stacked on top in their 'feeding sequence'. Ecological pyramids provide a convenient model to illustrate the relationship between different trophic levels in an ecosystem. Pyramids of energy shows the energy contained within each trophic level.

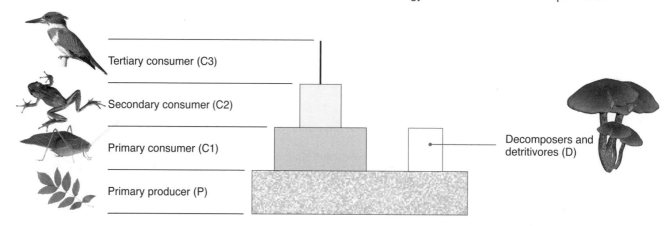

Tertiary consumer (C3)

Secondary consumer (C2)

Primary consumer (C1)

Primary producer (P)

Decomposers and detritivores (D)

The generalized ecological pyramid pictured above shows a conventional pyramid shape, with a large base at the primary producer level, and increasingly smaller blocks at subsequent levels. Not all pyramids have this appearance. Decomposers are placed at the level of the primary consumers and off to the side because they may obtain energy from many different trophic levels and so do not fit into the conventional pyramid structure. Pyramid of biomass measures the mass of the biological material at each trophic level. They are usually similar in appearance to pyramids of energy (biomass diminishes along food chains as the energy retained in the food chain diminishes).

Pyramid of Energy for a Plankton Community

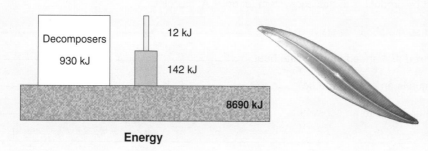

Decomposers

930 kJ

12 kJ

142 kJ

8690 kJ

Energy

Phytoplankton (such as diatoms, left) are producers, transforming sunlight energy into the energy in the chemical bonds within organic matter (food) via photosynthesis. There are many different species of phytoplankton. They form the first trophic level in an ecological pyramid for an aquatic plankton community.

The pyramid illustrated above relates to a hypothetical plankton community. The energy at each trophic level is reduced with each progressive stage in the food chain. As a general rule, a maximum of 10% of the energy is passed on to the next level in the food chain. The remaining energy is lost due to respiration, waste, and heat.

1. Determine the **energy transfer** between trophic levels in the plankton community example in the above diagram:

   (a) Between producers and the primary consumers: _____

   (b) Between the primary consumers and the secondary consumers: _____

   (c) Why is the amount of energy transferred from the producer level to primary consumers considerably less than the expected 10% that occurs in many other communities?

   _____

   _____

   (d) After the producers, which trophic group has the greatest energy content? _____

   (e) Give a likely explanation for this: _____

   _____

   _____

   _____

© 2012-2014 **BIOZONE** International
ISBN: 978-1-927173-93-0
Photocopying Prohibited

SKILL

127 Nutrient Cycles

Key Idea: Matter cycles through the biotic and abiotic compartments of Earth's ecosystems in nutrient cycles.

Nutrient cycles move and transfer chemical elements (e.g. carbon, hydrogen, nitrogen, and oxygen) through the abiotic and biotic components of an ecosystem. Commonly, nutrients must be in an ionic (rather than elemental) form in order for plants and animals to have access to them. The supply of nutrients in an ecosystem is finite and limited.

Essential Nutrients

Macronutrient	Common form	Function
Carbon (C)	CO_2	Organic molecules
Oxygen (O)	O_2	Respiration
Hydrogen (H)	H_2O	Cellular hydration
Nitrogen (N)	N_2, NO_3^-, NH_4^+	Proteins, nucleic acids
Potassium (K)	K^+	Principal ion in cells
Phosphorus (P)	$H_2PO_4^-$, HPO_4^{2-}	Nucleic acids, lipids
Calcium (Ca)	Ca^{2+}	Membrane permeability
Magnesium (Mg)	Mg^{2+}	Chlorophyll
Sulfur (S)	SO_4^{2-}	Proteins

Micronutrient	Common form	Function
Iron (Fe)	Fe^{2+}, Fe^{3+}	Chlorophyll, blood
Manganese (Mn)	Mn^{2+}	Enzyme activation
Molybdenum (Mo)	MoO_4^-	Nitrogen metabolism
Copper (Cu)	Cu^{2+}	Enzyme activation
Sodium (Na)	Na^+	Ion in cells
Silicon (Si)	$Si(OH)_4$	Support tissues

Tropical Rainforest

Nutrients in plants and animals

High speed return of nutrients to plants and animals leaving soil nutrient deficient

Nutrients in soil

Temperate Woodland

Nutrients in plants and animals

Slow return of nutrients allowing greater build up of nutrients in the soil

Nutrients in soil

The speed of nutrient cycling can vary. Some nutrients are cycled slowly, others quickly. The environment and diversity of an ecosystem can also have a large effect on the speed at which nutrients are recycled.

The Role of Organisms in Nutrient Cycling

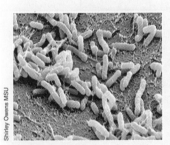

Bacteria
Bacteria play an essential role in nutrient cycles. They act as decomposers, but can also convert nutrients into forms accessible to plants and animals.

Fungi
Fungi are saprophytes and are important decomposers, returning nutrients to the soil or converting them into forms accessible to plants and animals.

Plants
Plants have a role in absorbing nutrients from the soil and making them directly available to browsing animals. They also add their own decaying matter to soils.

Animals
Animals utilize and break down materials from bacteria, plants and fungi and return the nutrients to soils and water via their wastes and when they die.

1. How do the movement of energy and nutrients in an ecosystem differ? _____

_____

_____

_____

2. Describe the role of each of the following in nutrient cycling:

(a) Bacteria: _____

(b) Fungi: _____

(c) Plants: _____

(d) Animals: _____

LINK
128

WEB
127

KNOW

128 The Carbon Cycle

Key Idea: The continued availability of carbon in ecosystems depends on carbon cycling through the abiotic and biotic elements of an ecosystem.

Carbon is an essential element of life and is incorporated into the organic molecules that make up living organisms. Large quantities of carbon are stored in **sinks**, which include the atmosphere as carbon dioxide gas (CO_2), the ocean as carbonate and bicarbonate, and rocks such as coal and

limestone. Carbon cycles between the biotic and abiotic environment. Carbon dioxide is converted by autotrophs into carbohydrates via photosynthesis and returned to the atmosphere as CO_2 through respiration (**fluxes**). These fluxes can be measured. Some of the sinks and processes involved in the carbon cycle, together with the carbon fluxes, are shown below.

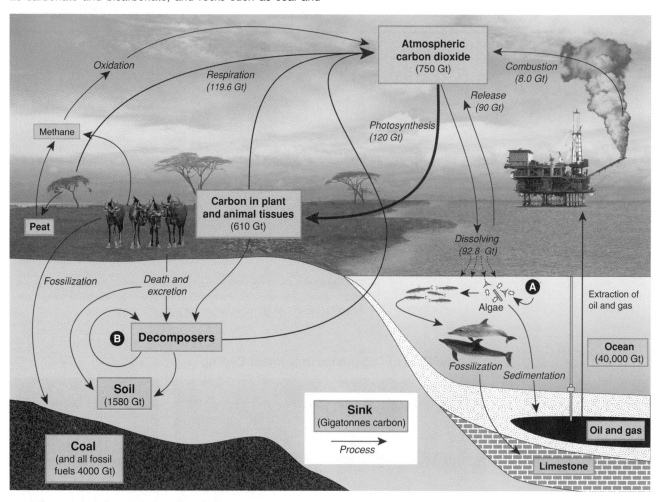

1. Add **arrows** and **labels** to the diagram above to show:

   (a) Dissolving of limestone by acid rain   (c) Mining and burning of coal

   (b) Release of carbon from the marine food chain   (d) Burning of plant material.

2. (a) Name the processes that release carbon into the atmosphere: _____

   _____

   (b) In what form is the carbon released? _____

3. Name the four geological reservoirs (sinks), in the diagram above, that can act as a source of carbon:

   (a) _____   (c) _____

   (b) _____   (d) _____

4. (a) Identify the process carried out by algae at point [**A**]: _____

   (b) Identify the process carried out by decomposers at [**B**]: _____

5. What would be the effect on carbon cycling if there were no decomposers present in an ecosystem? _____

   _____

   _____

© 2012-2014 **BIOZONE** International
ISBN: 978-1-927173-93-0
Photocopying Prohibited

Carbon may be locked up in biotic or abiotic systems for long periods of time, e.g. in the wood of trees or in fossil fuels such as coal or oil. Human activity, e.g. extraction and combustion of fossil fuels, has disturbed the balance of the carbon cycle.

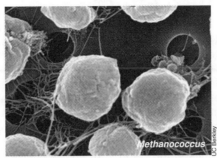

Methanococcus

Methanogenic archaea produce methane when organic material is metabolized in anaerobic conditions. Some methane diffuses into the atmosphere where it is converted in to CO_2 and H_2O and some accumulates in the ground.

Coal mine in Wyoming

Coal is formed from the remains of terrestrial plant material buried in shallow swamps and subsequently compacted under sediments to form a hard black material. Coal is composed primarily of carbon and is a widely used fuel source.

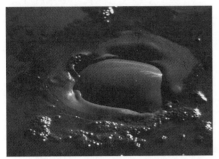

Oil and **natural gas** formed when dead algae and zooplankton settled to the bottom of shallow seas and lakes. These remains were buried and compressed under layers of non-porous sediment.

Limestone is a type of sedimentary rock containing mostly calcium carbonate. It forms when the shells of molluscs and other marine organisms with $CaCO_3$ skeletons become fossilized.

Peat (partly decayed vegetation or organic material) forms when plant material is not fully decomposed due to acidic or anaerobic conditions. Peatlands are a very efficient carbon sink.

6. Describe the **biological origin** of the following geological deposits:

(a) Coal: _____

_____

(b) Oil: _____

_____

(c) Limestone: _____

_____

(d) Peat: _____

_____

7. (a) Explain the role of methanogenic (methane producing) archaea in the carbon cycle: _____

_____

(b) Suggest another biological source of methane: _____

8. In natural circumstances, accumulated reserves of carbon such as peat, coal and oil represent a sink or natural diversion from the cycle. Eventually, the carbon in these sinks returns to the cycle through the action of geological processes which return deposits to the surface for oxidation.

(a) What is the effect of human activity on the amount of carbon stored in sinks? _____

_____

(b) Describe two **global effects** resulting from this activity: _____

_____

(c) What could be done to prevent or alleviate these effects? _____

_____

_____

129 The Greenhouse Effect

Key Idea: The greenhouse effect is the natural effect of having an atmosphere that retains heat received from the Sun. The Earth's atmosphere comprises a mix of gases including nitrogen, oxygen, and water vapour. Small quantities of carbon dioxide and methane are also present. These gases are called **greenhouse gases**. A natural process called the **greenhouse effect** describes how the atmosphere lets in sunlight, but traps the heat that would normally radiate back into space. This natural process results in the Earth having a mean surface temperature of about 15°C (33°C warmer than it would have without an atmosphere). Water vapour contributes the largest greenhouse effect, followed by CO_2. Methane has less effect as it is not as abundant in the atmosphere. Water is often ignored as a greenhouse gas because the amount of water in the atmosphere is related to temperature and therefore to the effect of other greenhouse gases. It is also subject to a positive feedback loop. More water vapour causes temperatures to increase and produce more water vapour. It is likely that the amount of anthropogenic (human generated) water vapour has not increased as much as other human-generated greenhouse gases.

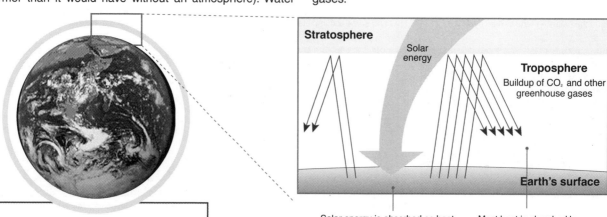

Solar energy is absorbed as heat by Earth, where it is radiated back into the atmosphere

Most heat is absorbed by CO_2 in the troposphere and radiated back to Earth

Sources of 'Greenhouse Gases'

Carbon dioxide
- Exhaust from cars
- Combustion of coal, wood, oil
- Burning rainforests

Methane
- Plant debris and growing vegetation
- Belching and flatus of cows

Chloro-fluoro-carbons (CFCs)
- Leaking coolant from refrigerators
- Leaking coolant from air conditioners

Nitrous oxide
- Car exhaust

Tropospheric ozone*
- Triggered by car exhaust (smog)

*Tropospheric ozone is found in the lower atmosphere (not to be confused with ozone in the stratosphere)

Greenhouse gas (excluding water vapour)	Tropospheric conc.		Global warming potential (compared to CO_2)¶	Atmospheric lifetime (years)§
	Pre-industrial 1750	Present day (2012*)		
Carbon dioxide	280 ppm	395.4 ppm	1	120
Methane	700 ppb	1796 ppb	25	12
Nitrous oxide	270 ppb	324 ppb	310	120
CFCs	0 ppb	0.39 ppbb	4000+	50-100
HFCs‡	0 ppb	0.045 ppb	1430	14
Tropospheric ozone	25 ppb	34 ppb	17	hours

ppm = parts per million; **ppb** = parts per billion; ‡Hydrofluorcarbons were introduced in the last decade to replace CFCs as refrigerants; * Data from 2012-2013. ¶ Figures contrast the radiative effect of different greenhouse gases relative to CO_2 over 100 years, e.g. over 100 years, methane is 25 times more potent as a greenhouse gas than CO_2 § How long the gas persists in the atmosphere. *Data compiled from the Carbon dioxide Information Centre http://cdiac.ornl.gov/pns/current_ghg.html*

1. What is a greenhouse gas? _____

_____

_____

2. What is the greenhouse effect? _____

_____

_____

3. Calculate the increase (as a %) in the 'greenhouse gases' between the pre-industrial era and the 2012 measurements using the data from the table above. **HINT**: The calculation for carbon dioxide is: (395.4 - 280) ÷ 280 x 100 =

(a) Carbon dioxide: _____ (b) Methane: _____ (c) Nitrous oxide: _____

4. Why is the effect of water vapour often not included in greenhouse gas tables? _____

_____

_____

WEB LINK
129 130

KNOW

© 2012-2014 **BIOZONE** International
ISBN: 978-1-927173-93-0
Photocopying Prohibited

130 Global Warming

Key Idea: Global warming refers to the continuing rise in the average temperature of the Earth's surface.

Since the mid 20th century, the Earth's surface temperature has been steadily increasing. The consensus scientific view (97% of publishing climate scientists) is that this phenomenon, called **global warming**, is attributable to the increase in atmospheric levels of CO_2 and other greenhouse gases emitted as a result of human activity.

Changes in Near-Surface Temperature

This graph right shows how the mean temperature for each year from 1860-2010 (bars) compares with the average temperature between 1961 and 1990. The blue line represents the fitted curve and shows the general trend indicated by the annual data.

Most anomalies since 1977 have been above normal and warmer than the long term mean, indicating that global temperatures are tracking upwards. The decade 2001-2010 has been the warmest on record.

Source: Hadley Centre for Prediction and Research

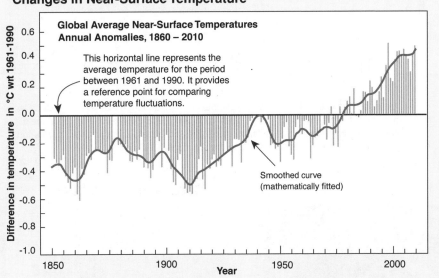

Global Average Near-Surface Temperatures Annual Anomalies, 1860 – 2010

This horizontal line represents the average temperature for the period between 1961 and 1990. It provides a reference point for comparing temperature fluctuations.

Smoothed curve (mathematically fitted)

Difference in temperature in °C wrt 1961-1990

Year

Changes in Atmospheric CO_2

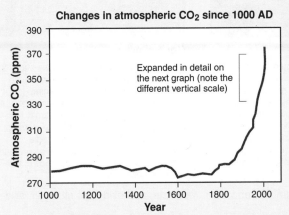

Changes in atmospheric CO_2 since 1000 AD

Expanded in detail on the next graph (note the different vertical scale)

Atmospheric CO_2 (ppm)

Year

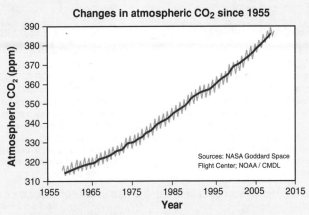

Changes in atmospheric CO_2 since 1955

Sources: NASA Goddard Space Flight Center; NOAA / CMDL

Atmospheric CO_2 (ppm)

Year

Atmospheric CO_2 has been rapidly increasing since the 1800s. In 2012, the world emitted a record (till then) 34.5 billion tonnes of CO_2 from fossil fuels. In total, humans have emitted 545 billion tonnes of CO_2. CO_2 levels fluctuate seasonally, especially in the northern hemisphere because of its much larger landmass and forests.

During the Industrial Revolution (1760-1840), coal was burned in huge quantities to power machinery. The increase in CO_2 released is attributed to an increase in average global temperatures. The combustion of fossil fuels (coal, oil, and natural gas) continues to pump CO_2 into the atmosphere and contribute to the current global warming.

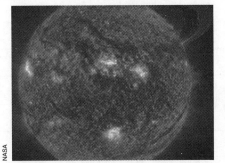

The Earth receives energy from the Sun (above) as UV, visible, and near-infrared radiation. Some is absorbed by the Earth's surface and the rest is reflected away as long-wavelength thermal radiation (heat). Much of this is trapped by the greenhouse gases and directed back to Earth, further increasing the mean surface temperature.

The oceans act as a carbon sink, absorbing the CO_2 produced from burning fossil fuels. The CO_2 reacts in the water, forming carbonic acid, lowering ocean pH, and reducing the availability of carbonate ions. This makes it harder for corals (above) to build their calcium carbonate exoskeletons and is causing significant coral reef damage.

LINK 133   LINK 132   LINK 131   WEB 130   KNOW

Climate Modelling

Predictions of 2001 models

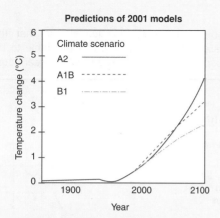

Predictions of 2012 models

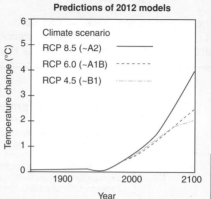

Factors affecting climate models

- Complex natural systems with large numbers of variables.
- Lack of long term accurate data.
- Uncertainty about the way human and natural activity influence climate.
- Climate models often need to be simplified in order to get them to work.

Data: IPCC assessment reports 2001/2012

Computer modelling is a tool for predicting climate change, although this is difficult because of the number of factors involved and the often conflicting data. Scientists often produce a series of climate models based on different scenarios and update them as new information becomes available. New and old models can then be compared (above).

International-mindedness: Global impact

The impact of greenhouse gas emissions is global regardless of where the emissions arise. Reducing emissions requires international cooperation.

Confusing the Debate

Media coverage

Global warming is a complex issue and most people obtain their information from the popular media. Despite the scientific consensus on the role of human activity in global warming, some media sources still provide biased or inaccurate information. In order to make an informed decision, people must read or listen to a wide range of media, or read scientific documents and make their own decision.

Lobby Groups

Lobby groups with specific interests strive constantly to influence policy makers. Reducing CO_2 emissions by restricting coal and oil use will help reduce global warming. However, fossil fuel consumption generates billions of dollars of revenue for coal and oil companies, so they lobby against legislation that penalizes fossil fuel use. If successful, lobbying could result in less effective climate change policies.

Controversy

All scientific bodies of international standing agree that human activity has contributed disproportionately to global warming. However, there are still some in the political, scientific, and commercial community who claim that global warming is not occurring. These people often command media attention and engage a poorly informed public audience, who are often suspicious of the scientific community.

1. Explain the relationship between the rise in concentrations of atmospheric CO_2, methane, and oxides of nitrogen, and global warming:

_____

_____

_____

_____

_____

2. (a) What effect did the Industrial Revolution have on atmospheric CO_2 levels? _____

 (b) Explain why this occurred: _____

 _____

3. Explain why models of climate change are constantly revised: _____

_____

_____

4. Evaluate claims by climate change sceptics that human activities are not causing climate change. Summarize your arguments and staple your summary into your workbook.

131 Global Warming and Effects on Biodiversity

Key Idea: Global warming is causing shifts in the distribution, behaviour, and even viability of plant and animal species. Global warming is changing the habitats of organisms and this may have profound effects on the biodiversity of specific regions as well as on the planet overall. As temperatures rise, organisms may be forced to move to areas better suited to their temperature tolerances. Those that cannot move or tolerate the temperature change may face extinction. Changes in precipitation as a result of climate change will also affect where organisms can live. Long term changes in climate will ultimately result in a shift in vegetation zones as some habitats contract and others expand.

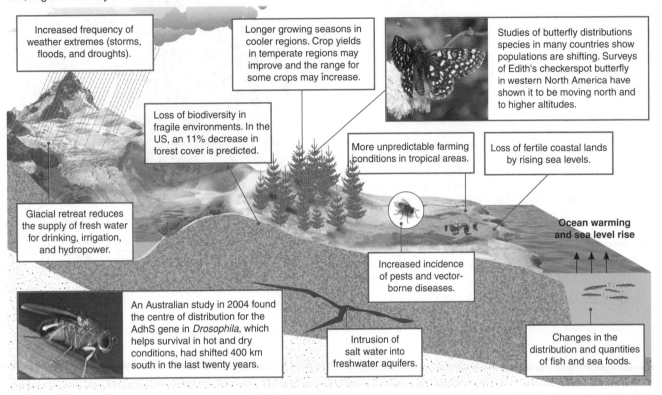

Increased frequency of weather extremes (storms, floods, and droughts).

Longer growing seasons in cooler regions. Crop yields in temperate regions may improve and the range for some crops may increase.

Studies of butterfly distributions species in many countries show populations are shifting. Surveys of Edith's checkerspot butterfly in western North America have shown it to be moving north and to higher altitudes.

Loss of biodiversity in fragile environments. In the US, an 11% decrease in forest cover is predicted.

More unpredictable farming conditions in tropical areas.

Loss of fertile coastal lands by rising sea levels.

Glacial retreat reduces the supply of fresh water for drinking, irrigation, and hydropower.

Ocean warming and sea level rise

An Australian study in 2004 found the centre of distribution for the AdhS gene in *Drosophila*, which helps survival in hot and dry conditions, had shifted 400 km south in the last twenty years.

Increased incidence of pests and vector-borne diseases.

Intrusion of salt water into freshwater aquifers.

Changes in the distribution and quantities of fish and sea foods.

Effects of Increases in Temperature on Crop Yields

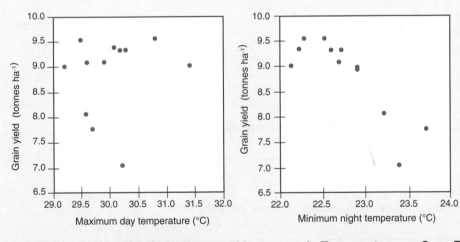

Studies on the grain production of rice have shown that maximum daytime temperatures have little effect on crop yield. However minimum night time temperatures lower crop yield by as much as 5% for every 0.5°C increase in temperature.

Source: Peng S. *et al.* PNAS 2004

Possible Effects of Increases in Temperature on Crop Damage

Source: Currano *et.al.* PNAS 2007

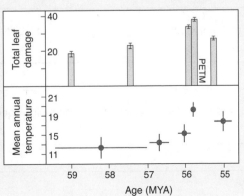

The fossil record shows that global temperatures rose sharply around 56 million years ago. Studies of fossil leaves with insect browse damage indicate that leaf damage peaked at the same time as the Paleocene Eocene Thermal Maximum (PETM). This gives some historical evidence that as temperatures increase, plant damage caused by insects also rises. This could have implications for agricultural crops.

Effects of Increases in Temperature on Animal Populations

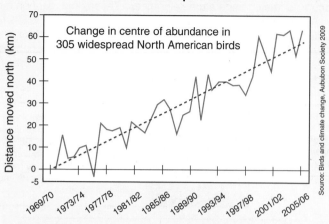

Change in centre of abundance in 305 widespread North American birds

Distance moved north (km)

Source: Birds and climate change, Aububon Society 2009

1969/70 1973/74 1977/78 1981/82 1985/86 1989/90 1993/94 1997/98 2001/02 2005/06

Animals living at altitude are also affected by warming climates and are being forced to shift their normal range. As temperatures increase, the snow line increases in altitude pushing alpine animals to higher altitudes. In some areas of North America this has resulting the local extinction of the North American pika (*Ochotona princeps*).

Seven star

A number of studies indicate that animals are beginning to be affected by increases in global temperatures. Data sets from around the world show that birds are migrating up to two weeks earlier to summer feeding grounds and are often not migrating as far south in winter.

1. Describe some of the likely effects of global warming on physical aspects of the environment: _____

_____

_____

_____

2. (a) Discuss the probable effects of global warming on plant crops: _____

_____

_____

(b) Suggest how farmers might be able to adjust to these changes: _____

_____

_____

_____

3. Discuss the evidence that insect populations are affected by global temperature: _____

_____

_____

4. (a) Describe how increases in global temperatures have affected some migratory birds: _____

_____

_____

(b) Explain how these changes in migratory patterns might affect food availability for these populations: _____

_____

_____

5. Explain how global warming could lead to the local extinction of some alpine species: _____

_____

_____

_____

132 Global Warming and Effects on the Arctic

Key Idea: Higher average temperatures melt sea-ice. Less heat is reflected back to space, warming sea temperature and promoting further melting of the ice.

The surface temperature of the Earth is in part regulated by the amount of ice on its surface, which reflects a large amount of heat into space. However, the area and thickness of the polar sea-ice has almost halved since 1980. This melting of sea-ice can trigger a cycle where less heat is reflected into space during summer, warming seawater and reducing the area and thickness of ice forming in the winter.

Arctic sea-ice summer minimum
1980: 7.8 million km^2

Arctic sea-ice summer minimum
2012: Record low, 4.1 million km^2

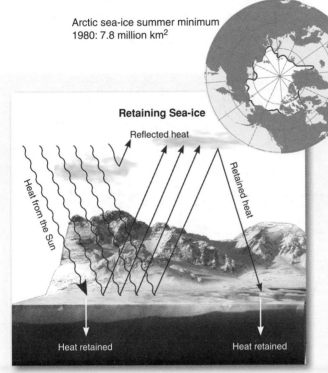

Retaining Sea-ice

Reflected heat

Heat from the Sun

Retained heat

Heat retained

Heat retained

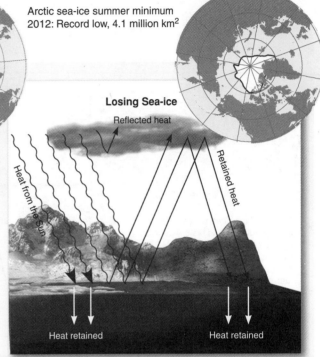

Losing Sea-ice

Reflected heat

Heat from the Sun

Retained heat

Heat retained

Heat retained

The **albedo** (reflectivity of sea-ice) helps to maintain its presence. Thin sea-ice has a lower albedo than thick sea-ice. More heat is reflected when sea-ice is thick and covers a greater area. This helps to regulate the temperature of the sea, keeping it cool.

As sea-ice retreats, more non-reflective surface is exposed. Heat is retained instead of being reflected, warming both the air and water and causing sea-ice to form later in the autumn than usual. Thinner and less reflective ice forms and perpetuates the cycle.

The temperature in the Arctic has been above average every year since 1988. Coupled with the reduction in summer sea-ice, this is having dire effects on Arctic wildlife such as polar bears, which hunt out on the ice. The reduction in sea-ice reduces their hunting range and forces them to swim longer distances to firm ice. Studies have already shown an increase in drowning deaths of polar bears.

Average* Arctic Air Temperature Fluctuations

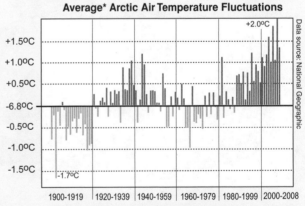

Data source: National Geographic

+2.0ºC
+1.5ºC
+1.0ºC
+0.5ºC
-6.8ºC
-0.5ºC
-1.0ºC
-1.5ºC
-1.7ºC

1900-1919 | 1920-1939 | 1940-1959 | 1960-1979 | 1980-1999 | 2000-2008

*Figure shows deviation from the average annual surface air temperature over land. Average calculated on the years 1961-2000.

1. Explain how low sea-ice albedo and volume affects the next year's sea-ice cover: _____

_____

_____

2. Discuss the effects of decreasing summer sea-ice on polar wildlife: _____

_____

_____

_____

LINK
129

WEB
132

KNOW

133 Ocean Acidification

Key Idea: Carbon dioxide reacts with water to reduce its pH. The oceans act as a carbon sink, absorbing much of the CO_2 produced from burning fossil fuels. When CO_2 reacts with water it forms carbonic acid, which decreases the pH of the oceans. This could have major effects on marine life, especially shell making organisms. Ocean acidification is relative term, referring to the oceans becoming less basic as the pH decreases. Ocean pH is still above pH 7.0

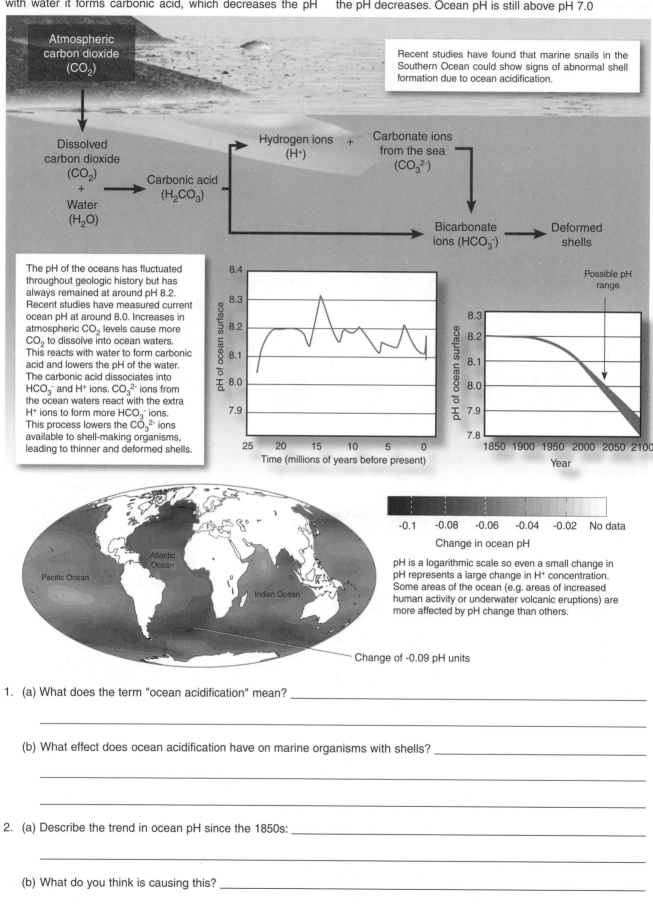

Atmospheric carbon dioxide (CO_2)

Recent studies have found that marine snails in the Southern Ocean could show signs of abnormal shell formation due to ocean acidification.

Dissolved carbon dioxide (CO_2) + Water (H_2O) → Carbonic acid (H_2CO_3) → Hydrogen ions (H^+) + Carbonate ions from the sea (CO_3^{2-}) → Bicarbonate ions (HCO_3^-) → Deformed shells

The pH of the oceans has fluctuated throughout geologic history but has always remained at around pH 8.2. Recent studies have measured current ocean pH at around 8.0. Increases in atmospheric CO_2 levels cause more CO_2 to dissolve into ocean waters. This reacts with water to form carbonic acid and lowers the pH of the water. The carbonic acid dissociates into HCO_3^- and H^+ ions. CO_3^{2-} ions from the ocean waters react with the extra H^+ ions to form more HCO_3^- ions. This process lowers the CO_3^{2-} ions available to shell-making organisms, leading to thinner and deformed shells.

pH of ocean surface — Time (millions of years before present): 25, 20, 15, 10, 5, 0

pH of ocean surface — Possible pH range — Year: 1850 1900 1950 2000 2050 2100

-0.1   -0.08   -0.06   -0.04   -0.02   No data
Change in ocean pH

Atlantic Ocean — Pacific Ocean — Indian Ocean

Change of -0.09 pH units

pH is a logarithmic scale so even a small change in pH represents a large change in H^+ concentration. Some areas of the ocean (e.g. areas of increased human activity or underwater volcanic eruptions) are more affected by pH change than others.

1. (a) What does the term "ocean acidification" mean? _____

_____

(b) What effect does ocean acidification have on marine organisms with shells? _____

_____

_____

2. (a) Describe the trend in ocean pH since the 1850s: _____

_____

(b) What do you think is causing this? _____

_____

© 2012-2014 **BIOZONE** International
ISBN: 978-1-927173-93-0
Photocopying Prohibited

134 Applying the Precautionary Principle

Key Idea: The precautionary principle is a management and policy strategy to protect against public or environmental harm. The precautionary principle requires those wanting to carry out an activity to prove that their action will not be harmful to human health or the environment. It aims to prevent harm occurring rather than having to correct a problem once it has occurred. The precautionary principle places a social responsibility on an action or change maker to protect the public or environment from harm (i.e. they must show their actions are safe before they are allowed to proceed).

When do we apply the precautionary principle?

Global warming provides a good illustrative scenario for how the precautionary principle might be applied.

- Change could occur:
  In places such as Greenland, rising global temperatures would cause ice to melt and glaciers to reduce in size. Polar ice caps would shrink. The Larson B ice shelf in Antarctica has already disintegrated.

- Persistent and irreversible harm:
  The melting of ice caps would destroy and reduce habitat for polar organisms, such as polar bears, possibly driving them to extinction. Sea level rises (caused by the thermal expansion of water and melting of the ice sheets) may flood low lying costal areas.

- Chain reactions and flow on effects:
  A single change caused by global warming may have multiple effects. For example, glacial melting changes habitat structure, causes water shortages in areas reliant on glacial melt water, and can cause localized changes in patterns of freeze and thaw.

- Difficulty in control or repair:
  Because global warming is influenced by global CO_2 production, controlling it is difficult. Unpredicted events (e.g. cyclonic events) that may be difficult (or impossible) to control could occur.

- Uncertainty:
  There is a large amount of evidence to support the fact that the climate is warming. However, there is some debate about the extent to which human activity has contributed to this. Much of this debate is created by those with vested interests in the status quo.

- Current activities can be linked to global warming:
  Human activities, such as burning fossil fuels, produce CO_2, which is a greenhouse gas, and contributes to increased global warming.

These factors indicate a need for caution when proposing activities that could enhance global warming (such as the development of a new coal-fired power station or legislation controlling greenhouse emissions). The precautionary principle acts as a simple tool to minimize impact.

The precautionary principle is an approach to decision-making in which preventative measures are justified even if there is scientific uncertainty about the outcome or risk. In natural systems, which are complex and often unpredictable, complete scientific certainty is rarely possible. This makes precaution the most reasonable response. The precautionary principle does not argue for zero risk, but for risk reduction. One difficulty in applying the precautionary principle is that solving a problem in one area may shift the risk elsewhere. For example, nuclear power generation carries significant environmental risks, but not using nuclear power may increase reliance on fossil fuels and contribute to accelerated global warming.

1. How could the precautionary principle be applied to stop exploitation of a resource in an environmentally sensitive area?

_____

_____

_____

2. Give an example, other than the one described, where using the precautionary principle may in fact cause greater harm:

_____

_____

_____

KNOW

135 Chapter Review

Summarize what you know about this topic under the headings provided. You can draw diagrams or mind maps, or write short notes to organize your thoughts. Use the images and hints and guidelines included to help you:

Modes of nutrition:
HINT: Distinguish between producers, consumers, detritivores, and saprotrophs.

Ecosystems:
HINT: Describe the biotic and abiotic factors that make up an ecosystem.

Energy flow:
HINT: Describe energy flow through an ecosystem. Remember to include the role of autotrophs and the 10% law.

Carbon cycling:
HINT: Carbon cycles through the biotic and abiotic environment.
State the importance of photosynthesis and cell respiration in the carbon cycle.

 CO_2

Climate change:
HINT: Describe global warming. Include the main contributing factors and the role of human activity.

136 KEY TERMS: Did You Get It?

1. Test your vocabulary by matching each term to its definition, as identified by its preceding letter code.

autotroph

A A sequence of steps describing how an organism derives energy from the ones before it.

carbon cycle

B The process of the Earth's surface steadily increasing in temperature. Usually attributed to the rise in gases produced by burning or use of fossil fuels and industrial processes.

consumer

C A measured and marked region used to isolate a sample area for study.

detritivore

D A complex series of interactions showing the feeding relationships between organisms in an ecosystem.

ecological pyramid

E An organism that obtains its carbon and energy from other organisms.

food chain

F Organisms that obtain their energy from other living organisms or their dead remains.

food web

G Biogeochemical cycle by which carbon is exchanged among the biotic and abiotic components of the Earth.

global warming

H An organism that obtains energy from dead material by extracellular digestion.

greenhouse gas

I A graphical representation of the numbers, energy, or biomass at each trophic level in an ecosystem. Often pyramidal in shape.

heterotroph

J An organism that obtains energy by ingesting dead material mixed with inorganic material.

quadrat

K Any of the feeding levels that energy passes through in an ecosystem.

saprotroph

L Any gas in the atmosphere that causes the retention of heat in the Earth's atmosphere. Major gases are water vapour and carbon dioxide.

trophic level

M An organism that manufactures its own food from simple inorganic substances.

2. A simple food chain for a cropland ecosystem is pictured below. Label the organisms with their trophic status (e.g. primary consumer).

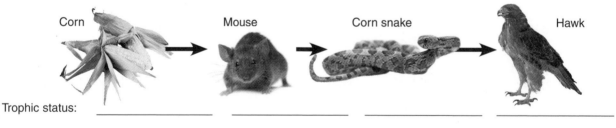

Corn   Mouse   Corn snake   Hawk

Trophic status: _____ _____ _____ _____

3. Increasing levels of atmospheric CO_2 and climate changes have been attributed to human activities such as the burning of fossil fuels (i.e. the changes are anthropogenic). Some people disagree with this viewpoint. Investigate the evidence for anthropogenic climate change, state whether you agree or disagree, and state the reasons for your opinion:

_____

_____

_____

_____

_____

_____

© 2012-2014 **BIOZONE** International
ISBN: 978-1-927173-93-0
Photocopying Prohibited

TEST

Topic 5

Topic 5

Evolution and Biodiversity

Key terms

adaptation

analogous character

Archaea (archaens)

artificial selection

Eubacteria (bacteria)

binomial nomenclature

clade

cladistics

cladogenesis

cladogram

class

dichotomous key

domain

Eukarya (eukaryotes)

evolution

family

fossil record

genus

homologous structure

kingdom

meiosis

mutation

natural selection

natural classification

order

pentadactyl limb

phyletic gradualism

phylogeny

phylum

selective breeding

sexual reproduction

shared derived characteristic

species

5.1 Evidence for evolution

Understandings, applications, skills

☐ 1 Explain what is meant by evolution and identify the processes involved in bringing about genetic change in populations. — 137

☐ 2 Use examples to show how the fossil record provides evidence for evolution. — 138

☐ 3 Explain how selective breeding of domesticated animals results in evolution. — 139

☐ 4 Describe the pentadactyl limb and explain how adaptive radiation explains differences in function in homologous structures. Compare the pentadactyl limb or mammals, birds, amphibians, and reptiles with different modes of locomotion. — 140

☐ 5 Explain how populations evolve into separate species by phyletic gradualism (anagenesis). Recognize that gradual divergence is only one model for the pace of evolutionary change in species. — 141

☐ 6 Describe phenotypic change in populations, using the example of melanistic insects in polluted areas. — 144

TOK *Experiments cannot be performed to verify past events or their causes, but there are scientific methods to establish beyond reasonable doubt the past history of species. How do these methods compare to those historians used?*

 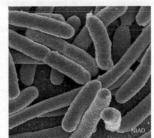 

5.2 Natural selection

Understandings, applications, skills

☐ 1 Explain the role of mutation, meiosis, and sexual reproduction in generating variation between individuals in a species. Explain natural selection in terms of variation, over-production of individuals, adaptation, and differential survival of offspring. Describe the outcome of natural selection and relate this to genetic change in a population over time (i.e. evolution). — 142

☐ 2 Explain what is meant by an adaptation and give examples. Relate adaptation to environment to survival and successful reproduction. — 143

☐ 3 Describe examples, including (but not restricted to) changes in beak size in Galápagos finches and evolution of antibiotic resistance in bacteria, to show how natural selection in the prevailing environment can lead to genetic change in populations. — 145-148

TOK *Natural selection is a theory for the mechanism by which evolution occurs. The evidence for it is overwhelming. How much evidence is needed to support a theory and what counter evidence is required to refute it?*

5.3 Classification of biodiversity

Understandings, applications, skills

Activity number

☐ 1 Identify the three domains into which species are now classified.   149

☐ 2 Describe the binomial nomenclature system for naming species.   150

> **TOK** *The adoption of binomial nomenclature is due largely to Linnaeus. He classified humans into four groups - a classification now regarded as racist. Do we need to consider social context when evaluating ethical aspects of knowledge claims?*

☐ 3 Explain how taxonomists classify species using a hierarchy of taxa. State the principal taxa for classifying eukaryotes under the five (or six) kingdom classification system. Classify one plant and one animal species from domain to species level.   150 151

☐ 4 Recognize features of some major plant phyla (divisions): Bryophyta, Filicinopyta, Coniferophyta, and Angiospermophyta.   151 152

☐ 5 Recognize features of major animal phyla: Porifera, Cnidaria, Platyhelmintha, Annelida, Mollusca, Arthropoda, and Chordata.   151 153

☐ 6 Recognize features of major vertebrate classes: birds, mammals, amphibians, reptiles, and fish.   151 153

☐ 7 Explain what is meant by a dichotomous key. Construct a dichotomous key to identify organisms to species level.   154 155

☐ 8 Explain what is meant by a natural classification. When might taxonomists reclassify a species or group of species? How do natural classifications help in identifying species and predicting characteristics shared by a taxon?   156 157

5.4 Cladistics

Understandings, applications, skills

Activity number

☐ 1 Explain what is meant by a clade and explain how species are assigned to clades on the basis of shared derived characteristics. Explain the role of molecular evidence (e.g. DNA and amino acid sequence data) in providing the evidence for assigning species to clades.   30 149

> **TOK** *The recognition by Carl Woesse, in 1977, of a separate line of descent for the Archaea raised objections from some famous scientists at the time. To what extent are conservative views in science desirable?*

☐ 2 Explain the positive correlation between the genetic differences between species and the time since their divergence from a common ancestor. Distinguish between homologous and analogous characteristics and comment on the significance of homologies in constructing phylogenies.   156 157

☐ 3 Explain what is meant by a cladogram. Analyse cladograms to deduce evolutionary relationships. Explain the basis for the recent reclassification of some taxa, e.g. primates and figworts.   156 157

137 Genes, Inheritance, and Selection

Key Idea: Selection acts on the heritable phenotypic variation that is the result of an individual's combination of alleles.

Each individual in a population is the carrier of its own particular combination of genetic material. In sexually reproducing organisms, different combinations of genes arise because of the shuffling of the chromosomes during gamete formation. New allele combinations also occur as a result of mate selection and the chance meeting of different gametes from each of the two parents. Some organisms have allele combinations well suited to the prevailing environment. Those organisms will have greater reproductive success (**fitness**) than those with less favourable allele combinations. Consequently, their genes (alleles) will be represented in a greater proportion in subsequent generations. For asexual species, offspring are essentially clones. New alleles can arise through mutation and some of these may confer a selective advantage. Of course, environments are rarely static, so new allele combinations are always being tested for success.

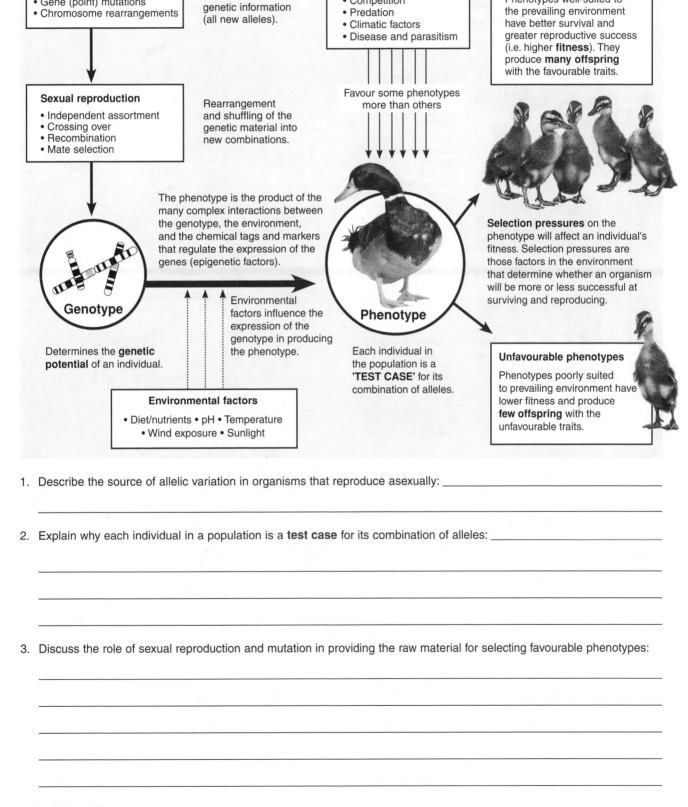

Mutations
- Gene (point) mutations
- Chromosome rearrangements

Provides the source of all new genetic information (all new alleles).

Sexual reproduction
- Independent assortment
- Crossing over
- Recombination
- Mate selection

Rearrangement and shuffling of the genetic material into new combinations.

The phenotype is the product of the many complex interactions between the genotype, the environment, and the chemical tags and markers that regulate the expression of the genes (epigenetic factors).

Genotype

Determines the **genetic potential** of an individual.

Environmental factors influence the expression of the genotype in producing the phenotype.

Environmental factors
- Diet/nutrients • pH • Temperature
- Wind exposure • Sunlight

Selection pressures
- Competition
- Predation
- Climatic factors
- Disease and parasitism

Favour some phenotypes more than others

Favourable phenotypes

Phenotypes well-suited to the prevailing environment have better survival and greater reproductive success (i.e. higher **fitness**). They produce **many offspring** with the favourable traits.

Phenotype

Each individual in the population is a **'TEST CASE'** for its combination of alleles.

Selection pressures on the phenotype will affect an individual's fitness. Selection pressures are those factors in the environment that determine whether an organism will be more or less successful at surviving and reproducing.

Unfavourable phenotypes

Phenotypes poorly suited to prevailing environment have lower fitness and produce **few offspring** with the unfavourable traits.

1. Describe the source of allelic variation in organisms that reproduce asexually: _____

_____

2. Explain why each individual in a population is a **test case** for its combination of alleles: _____

_____

_____

_____

3. Discuss the role of sexual reproduction and mutation in providing the raw material for selecting favourable phenotypes:

_____

_____

_____

_____

_____

© 2012-2014 **BIOZONE** International
ISBN: 978-1-927173-93-0
Photocopying Prohibited

KNOW

138 The Fossil Record

Key Idea: Fossils provide a record of the appearance and extinction of organisms. The fossil record can be used to establish the relative order of past events.

Fossils are the remains of long-dead organisms that have escaped decay, with their remains becoming mineralized. Fossils provide a record of the appearance and extinction of organisms, from species to whole taxonomic groups. Once this record is calibrated against a time scale (by using dating techniques), it is possible to build up a picture of the evolutionary changes that have taken place. The evolution of the horse (below right) from the ancestral *Hyracotherium* to modern *Equus* is well documented in the fossil record. The rich fossil record, which includes numerous **transitional fossils**, has enabled scientists to develop a robust model of horse **phylogeny** (evolutionary history). Horse evolution exhibits a complex tree-like lineage with many divergences. Many species coexisted for some time over the 55 million year evolutionary period. The environmental transition from forest to grasslands drove many of the changes observed in the equid fossil record.

Profile with Sedimentary Rocks Containing Fossils

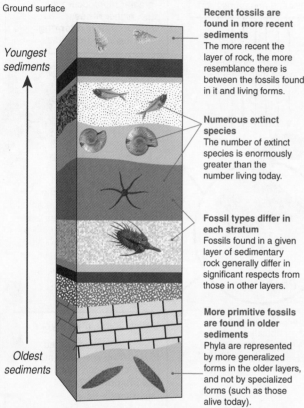

Ground surface

Youngest sediments

Oldest sediments

Recent fossils are found in more recent sediments
The more recent the layer of rock, the more resemblance there is between the fossils found in it and living forms.

Numerous extinct species
The number of extinct species is enormously greater than the number living today.

Fossil types differ in each stratum
Fossils found in a given layer of sedimentary rock generally differ in significant respects from those in other layers.

More primitive fossils are found in older sediments
Phyla are represented by more generalized forms in the older layers, and not by specialized forms (such as those alive today).

Rock Strata are Layered Through Time
Rock strata are arranged in the order that they were deposited (unless they have been disturbed by geological events). The most recent layers are near the surface and the oldest are at the bottom.

New Fossil Types Mark Changes in Environment
In the rocks marking the end of one geological period, it is common to find new fossils that become dominant in the next. Each geological period had an environment very different from those before and after. Their boundaries coincided with drastic environmental changes and the appearance of new niches. These produced new selection pressures resulting in new adaptive features in the surviving species as their populations responded to the changes.

Fossil Evidence of Horse Evolution

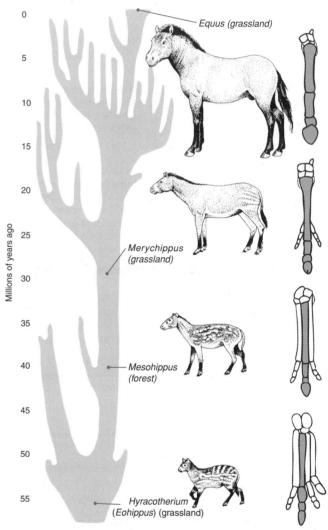

Millions of years ago

0 — *Equus* (grassland)
5
10
15
20
25 — *Merychippus* (grassland)
30
35
40 — *Mesohippus* (forest)
45
50
55 — *Hyracotherium* (*Eohippus*) (grassland)

The fossil record of the horse provides much evidence for evolution (change in body size, limb length, tooth structure, toe reduction). The genus *Equus* (the modern horse) is the only living genus of what was a large and diverse group of animals. Observation of the leg bones of various ancestors of *Equus* show a progressive loss of the outer toe bones, leaving the single middle bone and hoof supporting the animal.

1. How does the fossil record provide a record of evolutionary change over time: _____

_____

2. In which way does the equid fossil record provide a good example of the evolutionary process?_____

_____

_____

© 2012-2014 **BIOZONE** International
ISBN: 978-1-927173-93-0
Photocopying Prohibited

139 Selection and Population Change

Key Idea: Selective breeding is a method for rapidly producing change in the phenotypic characteristics of a population. Humans may create the selection pressure for evolutionary change by choosing and breeding together individuals with particular traits. The example of milk yield in Holstein cows (below) illustrates how humans have directly influenced the genetic makeup of Holstein cattle with respect to milk

production and fertility. Since the 1960s, the University of Minnesota has maintained a Holstein cattle herd that has not been subjected to any selection. They also maintain a herd that was subjected to selective breeding for increased milk production between 1965 and 1985. They compared the genetic merit of milk yield in these groups to that of the U.S.A. Holstein average.

Gain in Genetic Merit of Milk Yield

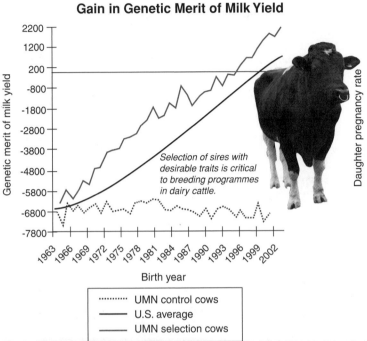

Selection of sires with desirable traits is critical to breeding programmes in dairy cattle.

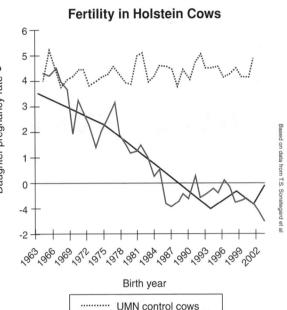

.... UMN control cows
———— U.S. average
———— UMN selection cows

Fertility in Holstein Cows

Based on data from T.S. Sonstegard et al

.... UMN control cows
———— U.S. average
———— UMN selection cows

Milk production in the University of Minnesota herd subjected to selective breeding increased in line with the U.S. average production. In real terms, milk production per cow per milking season increased by 3740 kg since 1964. The herd with no selection remained effectively constant for milk production.

Along with increased milk production there has been a distinct decrease in fertility. The fertility of the University of Minnesota herd that was not subjected to selection remained constant while the fertility of the herd selected for milk production decreased with the U.S. fertility average.

1. (a) Describe the relationship between milk yield and fertility on Holstein cows: _____

_____

_____

(b) What does this suggest about where the genes for milk production and fertility are carried? _____

_____

_____

2. What limits might this place on maximum milk yield? _____

_____

_____

3. Natural selection is the mechanism by which organisms with favourable traits become proportionally more common in the population. How does selective breeding mimic natural selection? How does the example of the Holstein cattle show that reproductive success is a compromise between many competing traits?

_____

_____

_____

_____

_____

140 Homologous Structures

Key Idea: Homologous structures (homologies) are structural similarities present as a result of common ancestry. The common structural components have been adapted to different purposes in different taxa.

The bones of the forelimb of air-breathing vertebrates are composed of similar bones arranged in a comparable pattern. This is indicative of common ancestry. The early land vertebrates were amphibians with a **pentadactyl limb** structure (a limb with five fingers or toes). All vertebrates that descended from these early amphibians have limbs with this same basic pentadactyl pattern. They also illustrate the phenomenon known as **adaptive radiation**, since the basic limb plan has been adapted to meet the requirements of different niches.

Generalized Pentadactyl Limb

The forelimbs and hind limbs have the same arrangement of bones but they have different names. In many cases bones in different parts of the limb have been highly modified to give it a specialized locomotory function.

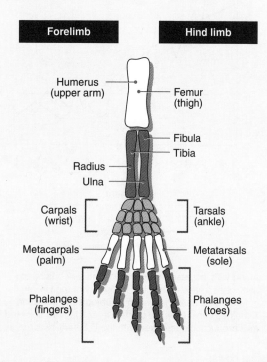

Specializations of Pentadactyl Limbs

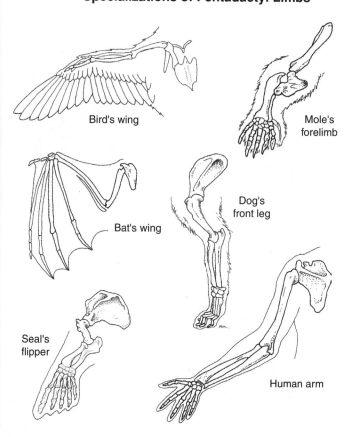

1. Briefly describe the purpose of the major anatomical change that has taken place in each of the limb examples above:

   (a) Bird wing: _Highly modified for flight. Forelimb is shaped for aerodynamic lift and feather attachment._

   (b) Human arm: _____

   (c) Seal flipper: _____

   (d) Dog leg: _____

   (e) Mole forelimb: _____

   (f) Bat wing: _____

2. Contrast the bone structure in the forelimb wing of bats and birds: _____

   _____

   _____

   _____

3. Explain how homology in the pentadactyl limb is evidence for adaptive radiation: _____

   _____

   _____

   _____

141 Divergence and Evolution

Key Idea: Populations moving into a new environment may diverge from their common ancestor and form new species. The diversification of an ancestral group into two or more species in different habitats is called **divergent evolution**. This process is shown below, where two species have diverged from a **common ancestor**. Divergence is common in evolution. When divergent evolution involves the formation of a large number of species to occupy different niches, this

is called an **adaptive radiation**. The example below (right) describes the radiation of the mammals after the extinction of the dinosaurs made new niches available. Note that the evolution of species may not necessarily involve branching. A species may accumulate genetic changes that, over time, result in the emergence of what can be recognized as a different species. This is known as **phyletic gradualism** (also called sequential evolution or anagenesis).

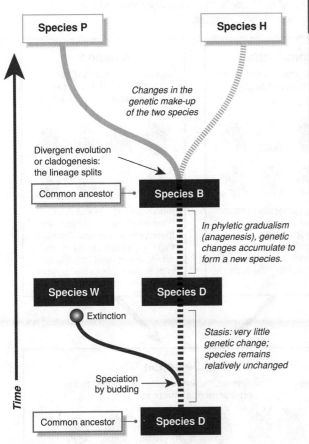

Species P **Species H**

Changes in the genetic make-up of the two species

Divergent evolution or cladogenesis: the lineage splits

Common ancestor — **Species B**

In phyletic gradualism (anagenesis), genetic changes accumulate to form a new species.

Species W **Species D**

Extinction

Stasis: very little genetic change; species remains relatively unchanged

Speciation by budding

Common ancestor — **Species D**

Time

Mammalian Adaptive Radiation

Marine predator niche
Arboreal herbivore niche
Underground herbivore niche
Terrestrial predator niche
Freshwater predator niche
Browsing/ grazing niche
Flying predator/ frugivore niche

Megazostrodon: one of the first mammals

The earliest true mammals evolved about 195 million years ago, long before they underwent their major adaptive radiation some 65-50 million years ago. These ancestors to the modern forms were very small (12 cm), many were nocturnal and fed on insects and other invertebrate prey. *Megazostrodon* (above) is a typical example. This shrew-like animal is known from fossil remains in South Africa and first appeared in the Early Jurassic period (about 195 million years ago).

It was climatic change as well as the extinction of the dinosaurs (and their related forms) that suddenly left many niches vacant for exploitation by such adaptable 'generalists'. All modern mammal orders developed very quickly and early.

1. In the hypothetical example of divergent evolution illustrated above, left:

   (a) Describe the type of evolution that produced species B from species D: _____

   (b) Describe the type of evolution that produced species P and H from species B: _____

   (c) Name all species that evolved from: **Common ancestor D**: _____ **Common ancestor B**: _____

   (d) Explain why species B, P, and H all possess a physical trait not found in species D or W: _____
   _____

2. (a) Explain the distinction between **divergence** and **adaptive radiation**: _____
   _____
   _____

   (b) Explain the difference between **sequential evolution** and **divergent evolution**: _____
   _____
   _____
   _____
   _____

142 Mechanism of Natural Selection

Key Idea: Evolution by natural selection describes how organisms that are better adapted to their environment survive to produce a greater number of offspring.

Evolution is the change in inherited characteristics in a population over generations. Evolution is the consequence of interaction between four factors: (1) The potential for populations to increase in numbers, (2) Genetic variation as a result of mutation and sexual reproduction, (3) competition for resources, and (4) proliferation of individuals with better survival and reproduction.

Natural selection is the term for the mechanism by which better adapted organisms survive to produce a greater number of viable offspring. This has the effect of increasing their proportion in the population so that they become more common. This is the basis of Darwin's theory of evolution by natural selection.

We can demonstrate the basic principles of evolution using the analogy of a 'population' of M&M's candy.

#1

In a bag of M&M's, there are many colours, which represents the variation in a population. As you and a friend eat through the bag of candy, you both leave the blue ones, which you both dislike, and return them to bag.

#2

The blue candy becomes more common...

#3

Eventually, you are left with a bag of blue M&M's. Your selective preference for the other colours changed the make-up of the M&M's population. This is the basic principle of selection that drives evolution in natural populations.

Darwin's Theory of Evolution by Natural Selection

Darwin's theory of evolution by natural selection is outlined below. It is widely accepted by the scientific community today and is one of founding principles of modern science.

Overproduction
Populations produce too many young: many must die

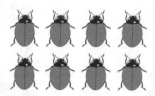

Populations generally produce more offspring than are needed to replace the parents. Natural populations normally maintain constant numbers. A certain number will die without reproducing.

Variation
Individuals show variation: some variations more favourable than others

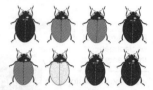

Individuals in a population have different phenotypes and therefore, genotypes. Some traits are better suited to the environment, and individuals with these have better survival and reproductive success.

Natural Selection
Natural selection favours the individuals best suited to the environment at the time

Individuals in the population compete for limited resources. Those with favourable variations will be more likely to survive. Relatively more of those without favourable variations will die.

Inherited
Variations are inherited: the best suited variants leave more offspring

The variations (both favourable and unfavourable) are passed on to offspring. Each generation will contain proportionally more descendants of individuals with favourable characters.

1. Identify the four factors that interact to bring about evolution in populations: _____

_____

_____

_____

© 2012-2014 **BIOZONE** International
ISBN: 978-1-927173-93-0
Photocopying Prohibited

Variation, Selection, and Population Change

1. Variation through mutation and sexual reproduction:
In a population of brown beetles, mutations independently produce red colouration and 2 spot marking on the wings. The individuals in the population compete for limited resources.

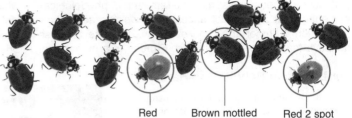

Red   Brown mottled   Red 2 spot

2. Selective predation:
Brown mottled beetles are eaten by birds but red ones are avoided.

3. Change in the genetics of the population:
Red beetles have better survival and fitness and become more numerous with each generation. Brown beetles have poor fitness and become rare.

Natural populations, like the ladybug population above, show genetic variation. This is a result of **mutation** (which creates new alleles) and sexual reproduction (which produces new combinations of alleles). Some variants are more suited to the environment of the time than others. These variants will leave more offspring, as described for the hypothetical population (right).

2. What produces the genetic variation in populations? _____

3. Define evolution: _____

_____

4. Explain how the genetic make-up of a population can change over time: _____

_____

_____

_____

_____

_____

_____

5. Complete the table below by calculating the percentage of beetles in the example above right.

Beetle population	% Brown beetles	% Red beetles	% Red beetles with spots
1			
2			
3			

143 Adaptation

Key Idea: Adaptive features enhance an individual's fitness. An **adaptation** (or adaptive feature) is any heritable trait that equips an organism to its functional position in the environment (its niche). These traits may be structural, physiological, or behavioural and reflect ancestry as well as adaptation. Adaptation is important in an evolutionary sense because adaptive features promote fitness. **Fitness** is a measure of an organism's ability to maximize the numbers of offspring surviving to reproductive age. Genetic adaptation must not be confused with physiological adjustment (acclimatization), which refers to an organism's ability to adjust during its lifetime to changing environmental conditions (e.g. a person's acclimatization to altitude). Examples of adaptive features arising through evolution are described below.

Ear Length in Rabbits and Hares

The external ears of many mammals are used as important organs to assist in thermoregulation (controlling loss and gain of body heat). The ears of rabbits and hares native to hot, dry climates, such as the jack rabbit of south-western USA and northern Mexico, are relatively very large. The Arctic hare lives in the tundra zone of Alaska, northern Canada and Greenland, and has ears that are relatively short. This reduction in the size of the extremities (ears, limbs, and noses) is typical of cold adapted species.

Arctic hare: *Lepus arcticus*

Black-tail jackrabbit: *Lepus californicus*

Body Size in Relation to Climate

Regulation of body temperature requires a large amount of energy and mammals exhibit a variety of structural and physiological adaptations to increase the effectiveness of this process. Heat production in any endotherm depends on body volume (heat generating metabolism), whereas the rate of heat loss depends on surface area. Increasing body size minimizes heat loss to the environment by reducing the surface area to volume ratio. Animals in colder regions therefore tend to be larger overall than those living in hot climates. This relationship is know as **Bergman's rule** and it is well documented in many mammalian species. Cold adapted species also tend to have more compact bodies and shorter extremities than related species in hot climates.

Fennec fox

Arctic fox

The **fennec fox** of the Sahara illustrates the adaptations typical of mammals living in hot climates: a small body size and lightweight fur, and long ears, legs, and nose. These features facilitate heat dissipation and reduce heat gain.

The **Arctic fox** shows the physical characteristics typical of cold adapted mammals: a stocky, compact body shape with small ears, short legs and nose, and dense fur. These features reduce heat loss to the environment.

Number of Horns in Rhinoceroses

Not all differences between species can be convincingly interpreted as adaptations to particular environments. Rhinoceroses charge rival males and predators, and the horn(s), when combined with the head-down posture, add effectiveness to this behaviour. Horns are obviously adaptive, but it is not clear if having one (Indian rhino) or two (black rhino) horns is related to the functionality in the environment or a reflection of evolution from a small hornless ancestor.

African black rhino

Great Indian rhino

1. Distinguish between adaptive features (genetic) and acclimatization: _____

_____

_____

2. Explain the nature of the relationship between the length of extremities (such as limbs and ears) and climate:

_____

_____

3. Explain the adaptive value of a compact body with a relatively small surface area in a colder climate: _____

_____

_____

© 2012-2014 **BIOZONE** International
ISBN: 978-1-927173-93-0

144 Melanism in Insects

Key Idea: Directional selection pressures on the peppered moth during the Industrial Revolution shifted the common phenotype from the grey form to the melanic (dark) form. Natural selection may act on the frequencies of phenotypes (and hence genotypes) in populations to shift the phenotypic mean in a particular direction. Colour change in the **peppered moth** (*Biston betularia*) during the Industrial Revolution is often used to show **directional selection** in a polymorphic population (polymorphic means having more than two forms). Intensive coal burning during this time caused trees to become dark with soot and the dark form (morph) of peppered moth became dominant.

The gene controlling colour in the peppered moth, is located on a single locus. The allele for the melanic (dark) form (**M**) is dominant over the allele for the grey (light) form (**m**).

Olaf Leillinger

Melanic form
Genotype: MM or Mm

Olaf Leillinger

Grey form
Genotype: mm

The peppered moth, *Biston betularia*, has two forms: a grey mottled form, and a dark melanic form. During the Industrial Revolution, the relative abundance of the two forms changed to favour the dark form. The change was thought to be the result of selective predation by birds. It was proposed that the grey form was more visible to birds in industrial areas where the trees were dark. As a result, birds preyed upon them more often, resulting in higher numbers of the dark form surviving.

Museum collections of the peppered moth over the last 150 years show a marked change in the frequency of the melanic form (above right). Moths collected in 1850, prior to the major onset of the Industrial Revolution in England, were mostly the grey form (above left). Fifty years later the frequency of the melanic forms had increased.

In the 1940s and 1950s, coal burning was still at intense levels around the industrial centres of Manchester and Liverpool. During this time, the melanic form of the moth was still very dominant. In the rural areas further south and west of these industrial centres, the occurrence of the grey form increased dramatically. With the decline of coal burning factories and the introduction of the Clean Air Act in cities, air quality improved between 1960 and 1980. Sulfur dioxide and smoke levels dropped to a fraction of their previous levels. This coincided with a sharp fall in the relative numbers of melanic moths (right).

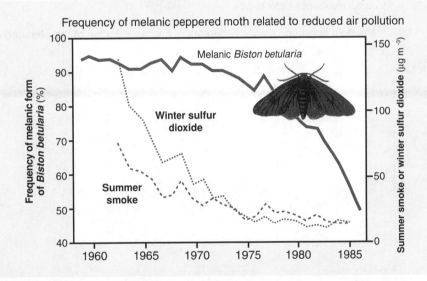

Frequency of melanic peppered moth related to reduced air pollution

1. The populations of peppered moth in England have undergone changes in the frequency of an obvious phenotypic character over the last 150 years. What is the phenotypic character?

_____

2. Describe how the selection pressure on the grey form has changed with change in environment over the last 150 years:

_____

_____

_____

3. Describe the relationship between allele frequency and phenotype frequency: _____

_____

_____

4. The level of pollution dropped around Manchester and Liverpool between 1960 and 1985. How did the frequency of the darker melanic form change during this period?

_____

145 Selection for Beak Size in Darwin's Finches

Key Idea: The effect of natural selection on a population can be verified by making quantitative measurements of phenotypic traits.

Natural selection acts on the phenotypes of a population. Individuals with phenotypes that increase their fitness produce more offspring, increasing the proportion of the genes corresponding to that phenotype in the next generation. Numerous population studies have shown natural selection can cause phenotypic changes in a population relatively quickly.

The finches on the Galápagos island (Darwin's finches) are famous in that they are commonly used as examples of how evolution produces new species. In this activity you will analyse data from the measurement of beak depths of the medium ground finch (*Geospiza fortis*) on the island of Daphne Major near the centre of the Galápagos Islands. The measurements were taken in 1976 before a major drought hit the island and in 1978 after the drought (survivors and survivors' offspring).

Beak depth (mm)	No. 1976 birds	No. 1978 survivors	Beak depth of offspring (mm)	Number of birds
7.30-7.79	1	0	7.30-7.79	2
7.80-8.29	12	1	7.80-8.29	2
8.30-8.79	30	3	8.30-8.79	5
8.80-9.29	47	3	8.80-9.29	21
9.30-9.79	45	6	9.30-9.79	34
9.80-10.29	40	9	9.80-10.29	37
10.30-10.79	25	10	10.30-10.79	19
10.80-11.29	3	1	10-80-11.29	15
11.30+	0	0	11.30+	2

1. Use the data above to draw two separate sets of histograms:

    (a) On the left hand grid draw side-by-side histograms for the number of 1976 birds per beak depth and the number of 1978 survivors per beak depth.

    (b) On the right hand grid draw a histogram of the beak depths of the offspring of the 1978 survivors.

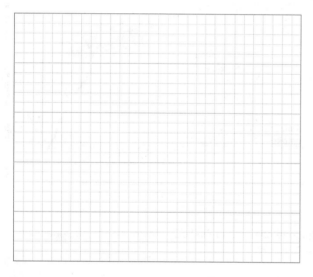

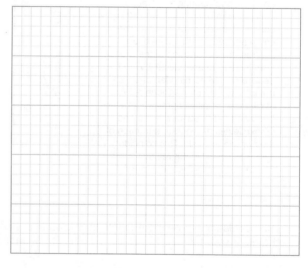

2. (a) Mark the approximate mean beak depth on the graphs of the 1976 beak depths and the 1978 offspring.

    (b) How much has the average moved from 1976 to 1978? _____

    (c) Is beak depth heritable? What does this mean for the process of natural selection in the finches?

    _____

    _____

3. The 1976 drought resulted in plants dying back and not producing seed. Based on the graphs, what can you say about competition between the birds for the remaining seeds, i.e. in what order were the seeds probably used up?

    _____

    _____

    _____

© 2012-2014 **BIOZONE** International
ISBN: 978-1-927173-93-0
Photocopying Prohibited

146 The Evolution of Antibiotic Resistance

Key Idea: Bacteria can develop resistance to antibiotics and can pass this on to the next generation and to other populations. Antibiotic resistance arises when a genetic change allows bacteria to tolerate levels of antibiotic that would normally inhibit growth. This resistance may arise spontaneously by mutation or copying error, or by transfer of genetic material between microbes. Genomic analyses from 30,000 year old permafrost sediments show that the genes for antibiotic resistance have long been present in the bacterial genome. Modern use of antibiotics has simply provided the selective environment for their proliferation. Many bacterial strains have even acquired resistance to multiple antibiotics.

The Evolution of Antibiotic Resistance in Bacteria

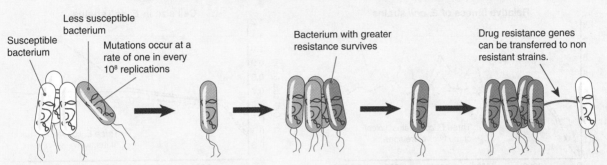

Susceptible bacterium

Less susceptible bacterium

Mutations occur at a rate of one in every 10^8 replications

Bacterium with greater resistance survives

Drug resistance genes can be transferred to non resistant strains.

Any population, including bacterial populations, includes variants with unusual traits, in this case reduced sensitivity to an antibiotic. These variants arise as a result of mutations in the bacterial chromosome. Such mutations are well documented and some are ancient.

When a person takes an antibiotic, only the most susceptible bacteria will die. The more resistant cells remain and continue dividing. Note that the antibiotic does not create the resistance; it provides the environment in which selection for resistance can take place.

If the amount of antibiotic delivered is too low, or the course of antibiotics is not completed, a population of resistant bacteria develops. Within this population too, there will be variation in susceptibility. Some will survive higher antibiotic levels.

A highly resistant population has evolved. The resistant cells can exchange genetic material with other bacteria (via horizontal gene transmission), passing on the genes for resistance. The antibiotic initially used against this bacterial strain will now be ineffective.

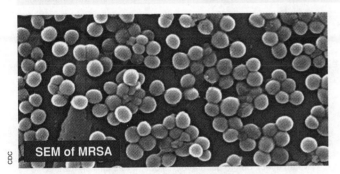

SEM of MRSA

CDC

Staphylococcus aureus is a common bacterium responsible for various minor skin infections in humans. MRSA is a variant strain that has evolved resistance to penicillin and related antibiotics. MRSA is troublesome in hospital-associated infections because patients with open wounds, invasive devices (e.g. catheters), or poor immunity are at greater risk for infection than the general public.

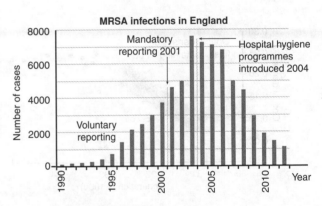

MRSA infections in England

Mandatory reporting 2001

Hospital hygiene programmes introduced 2004

Voluntary reporting

Number of cases / Year

In the UK, MRSA cases rose sharply during the early-mid 1990s, but are now declining as a result of mandatory reporting and the implementation of stringent hospital hygiene programmes.

1. Describe two ways in which antibiotic resistance can become widespread:

   (a) _____

   (b) _____

2. Genomic evidence indicates that the genes for antibiotic resistance are ancient:

   (a) How could these genes have arisen in the first place? _____

   _____

   (b) Why were they not lost from the bacterial genome? _____

   _____

   (c) Explain why these genes are proliferating now: _____

   _____

   _____

LINK
181
WEB
146
APP

147 Investigating Evolution

Key Idea: Long term experiments with *E. coli* have observed increased fitness as a result of mutation.

In 1988, Richard Lenski and his research group began the **E. coli Long Term Evolution Experiment**. They prepared 12 populations of the bacterium *E. coli* in a growth medium in which glucose was the limiting factor for growth. Each day, 1% of each population was transferred to a flask of fresh medium, and a sample of the population was frozen and stored. The experiment has now reached more than 50,000 generations. The populations were tested at intervals for mutations, but strains were not selected for any particular trait (other than being in a low glucose medium). **Fitness** (contribution to the next generation) in a low glucose medium has increased in all populations compared to the ancestral *E. coli* strain.

The *E. coli* Long Term Evolution Experiment

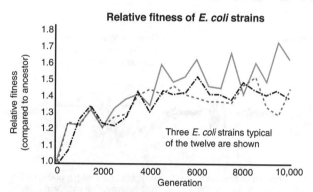

Every 500 generations, the fitness of each population was compared to the fitness of the ancestor (denoted as 1). As would be expected, the relative fitness of all the *E. coli* populations increased over time as they adapted to the low glucose environment. However, this increased fitness only applies in the low glucose environment. When placed in a different environment, fitness relative to the ancestor actually decreases. Interestingly, although the relative fitness of the populations has followed a similar course, the fitness has varied within and between populations.

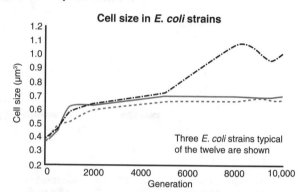

The size of the *E. coli* cells was also measured every 500 generations. Cell size increased in all populations and they became rounder. All populations increased their growth rate by about 70% on average. The population density also decreased. The increase in cell size and growth rate is probably an adaptation for acquiring the limited amount of glucose available in the solution. It has been estimated that each population generated millions of mutations over the course of the experiment, but only a few have become fixed within the populations.

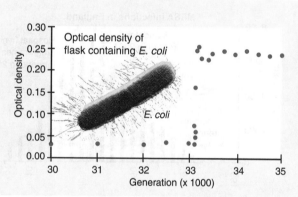

New Mutation Confers Much Greater Fitness

A particularly important mutation became evident in one of the *E. coli* populations after 31,000 generations. It became able to metabolize the citrate that was a component of the growth medium. This ability gave the strain increased fitness relative to all the other populations. The mutation for citrate metabolism was noticed when the optical density (cloudiness) of the flask containing the *E. coli* suddenly increased (left), indicating an increase in the population of bacteria.

Investigations into the citrate mutation found that many other previous mutations were required before the final mutation could have an effect. Before generation 15,000, it was unlikely for this strain to evolve the ability to metabolize citrate. After generation 15,000 it was more likely to evolve the ability to metabolize citrate.

1. (a) How did the fitness of *E. coli* change over the course of the experiment? _____

   _____

   _____

   (b) Explain why the increase in the fitness of the *E. coli* populations is only relative to a certain environment:

   _____

   _____

2. Why is an increase in growth rate an advantage to the *E. coli* populations? _____

   _____

   _____

3. Explain the significance in the mutations after the 15,000 generation mark in the development of citrate metabolism and their meaning in evolutionary development:

   _____

   _____

© 2012-2014 **BIOZONE** International
ISBN: 978-1-927173-93-0
Photocopying Prohibited

148 The Evolution of Insecticide Resistance

Key Idea: Insect resistance to insecticide is increasing as a result of ineffectual initial applications of insecticide that only kill the most susceptible insects, allowing more resistant individuals to proliferate.

Insecticides are pesticides used to control pest insects. They have been used for hundreds of years, but their use has increased since synthetic insecticides were first developed in the 1940s. When **insecticide resistance** develops, the control agent will no longer control the target species. Resistance can arise through behavioural, anatomical, biochemical, and physiological mechanisms, but the underlying process is a form of **natural selection,** in which the most resistant organisms survive to pass on their genes to their offspring. To combat increasing resistance, higher doses of more potent pesticides are sometimes used. This drives the selection process, so that increasingly higher dose rates are required to combat rising resistance. This phenomenon is made worse by the development of multiple resistance in some pest species.

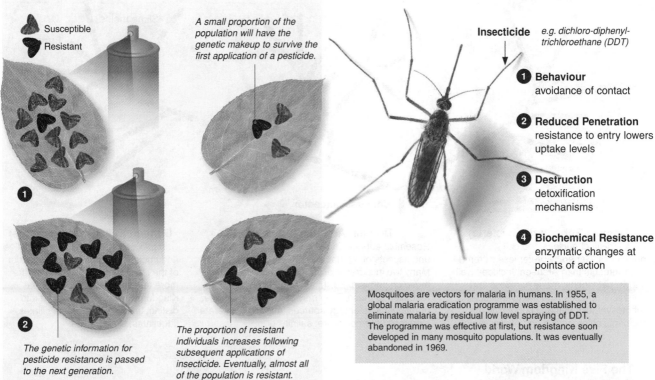

Susceptible
Resistant

A small proportion of the population will have the genetic makeup to survive the first application of a pesticide.

The genetic information for pesticide resistance is passed to the next generation.

The proportion of resistant individuals increases following subsequent applications of insecticide. Eventually, almost all of the population is resistant.

Insecticide   e.g. dichloro-diphenyl-trichloroethane (DDT)

1 Behaviour
avoidance of contact

2 Reduced Penetration
resistance to entry lowers uptake levels

3 Destruction
detoxification mechanisms

4 Biochemical Resistance
enzymatic changes at points of action

Mosquitoes are vectors for malaria in humans. In 1955, a global malaria eradication programme was established to eliminate malaria by residual low level spraying of DDT. The programme was effective at first, but resistance soon developed in many mosquito populations. It was eventually abandoned in 1969.

The Development of Resistance

The application of an insecticide can act as a potent selection pressure for resistance in pest insects. The insecticide acts as a selective agent, and only individuals with greater natural resistance survive the application to pass on their genes to the next generation. These genes (or combination of genes) may spread through all subsequent populations. Insect populations generally reproduce very quickly, so resistance can spread quickly through populations.

Mechanisms of Resistance in Insect Pests

Insecticide resistance in insects can arise through a combination of mechanisms. (1) Increased sensitivity to an insecticide will cause the pest to avoid a treated area. (2) Certain genes (e.g. the *PEN* gene) confer stronger physical barriers, decreasing the rate at which the chemical penetrates the cuticle. (3) Detoxification by enzymes within the insect's body can render the pesticide harmless, and (4) structural changes to the target enzymes make the pesticide ineffective. No single mechanism provides total immunity, but together they transform the effect from potentially lethal to insignificant.

1. Give two reasons why widespread insecticide resistance can develop very rapidly in insect populations:

   (a) _____

   (b) _____

2. Explain how repeated insecticide applications act as a selective agent for evolutionary change in insect populations:

   _____

   _____

3. Research the implications of synthetic insecticide resistance to human populations: _____

   _____

   _____

   _____

APP

149 The New Tree of Life

Key Idea: The classification of life into specific groups, or taxa, is constantly being updated in light of new information. Taxonomy is the science of classification and, like all science, constantly changing as new information is discovered. With the advent of DNA sequencing technology, scientists began to analyse the genomes of many bacteria. In 1996, the results of a scientific collaboration examining DNA evidence confirmed that life comprises three major evolutionary lineages (domains) and not two as was the convention. The recognized lineages are the Bacteria (formerly Eubacteria), the Eukarya (Eukaryotes), and the Archaea (formerly Archaebacteria). The new classification reflects the fact that there are very large differences between the archaeans and the bacteria. All three domains probably had a distant common ancestor.

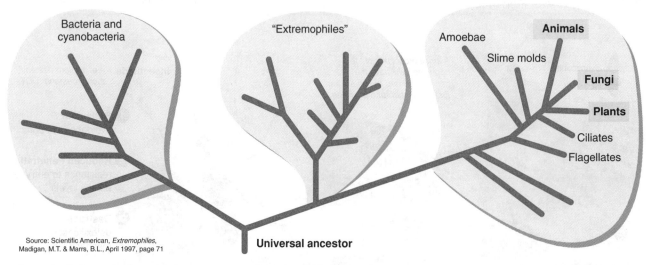

Source: Scientific American, *Extremophiles*, Madigan, M.T. & Marrs, B.L., April 1997, page 71

Domain Eubacteria (bacteria)
Lack a distinct nucleus and cell organelles. Generally prefer less extreme environments than Archaea. Includes well-known pathogens, many harmless and beneficial species, and the cyanobacteria (photosynthetic bacteria containing the pigments chlorophyll a and phycocyanin).

Domain Archaea (archeans)
Resemble eubacteria but cell wall composition and aspects of metabolism are very different. Many live in extreme environments similar to those on primeval Earth, although they are not restricted to these. They may use sulfur, methane, or halogens as energy sources, and many tolerate extremes of temperature, salinity, or pH.

Domain Eukarya (eukaryotes)
Complex cell structure with organelles and nucleus. This group contains four of the kingdoms classified under the more traditional system. Note that Kingdom Protista is separated into distinct taxa: e.g. amoebae, ciliates, flagellates, slime molds.

The Five Kingdom World
Before DNA sequencing showed that life was divided into three major domains, taxonomists divided life into five kingdoms based mainly on visible characteristics. This system is still common mainly because it is useful in separating the multicellular organisms that are familiar in our everyday experience. In this system, all prokaryotic organisms are placed in one kingdom (or sometimes two, making six kingdoms) with protists (single celled eukaryotes), fungi, plants, and animals being the other four. Clearly, it is not an accurate representation of the evolutionary relationships between organisms. In particular, the Kingdom Protista is a diverse collection of organisms that are not necessarily closely related.

In general there are eight taxa (taxonomic groups) used to classify organisms. These are: **domain, kingdom, phylum, class, order, family, genus,** and **species.** An organism's classification is not necessarily fixed, and may change as new information comes to light.

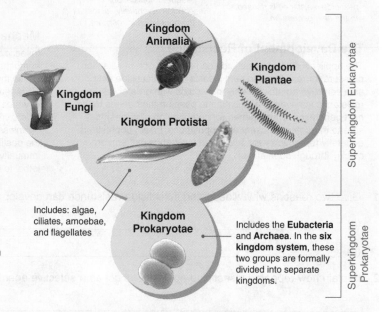

Includes: algae, ciliates, amoebae, and flagellates

Includes the **Eubacteria** and **Archaea**. In the **six kingdom system**, these two groups are formally divided into separate kingdoms.

1. Describe one feature of the three domain system that is very different from the five kingdom classification:

_____

_____

_____

_____

© 2012-2014 **BIOZONE** International
ISBN: 978-1-927173-93-0
Photocopying Prohibited

150 Classification System

Key Idea: Organisms are named and assigned to taxa based on their shared characteristics and evolutionary relationships. In biological classification, organisms are categorized into a hierarchical system of taxonomic groups (taxa) based on their shared characteristics. The fundamental unit of classification is the **species**, and each member of a species is assigned a unique two part (binomial) name that identifies it. Classification systems are nested, so that with increasing taxonomic rank, related taxa at one hierarchical level are combined into more inclusive taxa at the next higher level.

Naming an organism

Most organisms have a common name as well a scientific name. Common names may change from place to place as people from different areas name organisms differently based on both language and custom. Scientifically, every organism is given a classification that reflects its known lineage (i.e. its evolutionary history). The last two (and most specific) parts of that lineage are the **genus** and **species** names. Together these are called the scientific name and every species has its own. This two-part naming system is called **binomial nomenclature**. When typed the name is always *italicised*. If handwritten, it should be <u>underlined</u>.

The animal *Rangifer tarandus* is known as the caribou in North America, but as the reindeer in Europe. The scientific name is unambiguous.

1. The table below shows part of the classification for the English oak using the **principal taxa** of classification. For this question, use page 208 and a reference source such as the internet to complete the classification for the English oak.

   (a) Complete the list of the taxonomic groupings on the left hand side of the table below:

   (b) Complete the classification for English oak (*Quercus robur*) in the table below:

	Taxonomic Rank	English Oak Classification
1.	Domain	Eukarya
2.		
3.		
4.		
5.		
6.	Family	Fagaceae
7.		
8.		

   English oak
   Quercus robur

2. (a) What is the two part naming system for classifying organisms called? _____

   (b) What are the two parts of the name? _____

3. Give two reasons why the classification of organisms is important:

   (a) _____

   (b) _____

4. Construct an acronym or mnemonic to help you remember the principal taxonomic groupings (KPCOFGS):

   _____

5. Classification has traditionally been based on similarities in morphology, but new biochemical methods are now widely used to determine species relatedness. What contribution are these techniques making to the science of classification?

   _____

   _____

   _____

LINK 157   LINK 156   LINK 151   LINK 100   LINK 77   WEB 150

KNOW

Classification of the Ethiopian Hedgehog

The classification of the **Ethiopian hedgehog** is given below, showing the levels that can be used in classifying an organism. Not all possible subdivisions have been shown. For example, it is possible to indicate such categories as **super-class** and **sub-family**. The only natural category is the **species**, often separated into **sub-species**, which generally differ in appearance.

Kingdom: **Animalia**
Animals; one of five kingdoms

Phylum: **Chordata**
Animals with a notochord (supporting rod of cells along the upper surface)
tunicates, salps, lancelets, and vertebrates

23 other phyla

Sub-phylum: **Vertebrata**
Animals with backbones
fish, amphibians, reptiles, birds, mammals

Class: **Mammalia**
Animals that suckle their young on milk from mammary glands
placentals, marsupials, monotremes

Infra-class: **Eutheria or Placentals**
Mammals whose young develop for some time in the female's reproductive tract gaining nourishment from a placenta
placental mammals

Order: **Eulipotyphla**
The insectivore-type mammals. Once part of the now abandoned order Insectivora, the order also includes shrews, moles, desmans, and solenodons.

17 other orders

Family: **Erinaceidae**
Comprises two subfamilies: the true or spiny hedgehogs and the moonrats (gymnures). Representatives in the family include the common European hedgehog, desert hedgehog, and the moonrats. There is much debate over the true classification of the other families within Eulipotyphla. Sometimes the order is given as Erinaceomorpha with Erinaceidae as its only family.

4 other families

Genus: *Paraechinus*
One of twelve genera in this family. The genus *Paraechinus* includes three species which are distinguishable by a wide and prominent naked area on the scalp.

11 other genera

Species: *aethiopicus*
The Ethiopian hedgehog inhabits arid coastal areas. Their diet consists mainly of insects, but includes small vertebrates and the eggs of ground nesting birds.

3 other species

The advent of DNA sequencing and other molecular techniques has led to many taxonomic revisions. There is now considerable debate over the classification of species at almost all levels of classification.

The, now defunct, order Insectivora in one example. This order comprised many mammals with unspecialized features. The order was initially abandoned in 1956 but persisted (and still persists) in many textbooks. Over the next 50 years families were moved out, merged back together, split apart and reformed in other ways based on new evidence or interpretations. Do not be surprised if you see more than one classification for hedgehogs (or any other organism for that matter).

Ethiopian hedgehog
Paraechinus aethiopicus

© 2012-2014 **BIOZONE** International
ISBN: 978-1-927173-93-0

151 Features of Taxonomic Groups

Key Idea: Taxonomy is the branch of science concerned with identifying, describing, classifying, and naming organisms. Taxonomy is the science of classifying organisms. It relies on identifying and describing characteristics that clearly distinguish organisms from each other. Classification systems recognizing three domains (rather than five or six kingdoms) are now seen as better representations of the true diversity of life. However, for the purposes of describing the groups with which we are most familiar, the five kingdom system (used here) is still appropriate. The distinguishing features of some major **taxa** are provided in the following pages. The examples provided give an indication of the diversity within each taxon.

SUPERKINGDOM: PROKARYOTAE (Bacteria)

- Also known as prokaryotes. The term moneran is no longer in use.
- Two major bacterial lineages are recognized: the **Archaebacteria** (Archaea) and the more derived **Eubacteria** (Bacteria).
- All have a prokaryotic cell structure: they lack the nuclei and chromosomes of eukaryotic cells, and have smaller (70S) ribosomes.
- Have a tendency to spread genetic elements across species barriers by conjugation, viral transduction, and other processes.
- Asexual. Can reproduce rapidly by binary fission.

- Have evolved a wider variety of metabolism types than eukaryotes.
- Bacteria grow and divide or aggregate into filaments or colonies of various shapes. Colony type is often diagnostic.
- They are taxonomically identified by their appearance (form) and through biochemical differences.

Species diversity: 10,000+ Bacteria are rather difficult to classify to species level because of their relatively rampant genetic exchange, and because their reproduction is asexual.

Eubacteria

- Also known as 'true bacteria', they probably evolved from the more ancient Archaebacteria.
- Distinguished from Archaebacteria by differences in cell wall composition, nucleotide structure, and ribosome shape.
- Diverse group includes most bacteria.
- The **gram stain** is the basis for distinguishing two broad groups of bacteria. It relies on the presence of peptidoglycan in the cell wall. The stain is easily washed from the thin peptidoglycan layer of gram negative walls but is retained by the thick peptidoglycan layer of gram positive cells, staining them a dark violet colour.

Gram Positive Bacteria

The walls of gram positive bacteria consist of many layers of peptidoglycan forming a thick, single-layered structure that holds the gram stain.

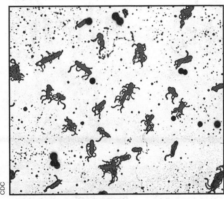

Bacillus alvei: a gram positive, flagellated bacterium. Note how the cells appear dark.

Gram Negative Bacteria

The cell walls of gram negative bacteria contain only a small proportion of peptidoglycan, so the dark violet stain is not retained by the organisms.

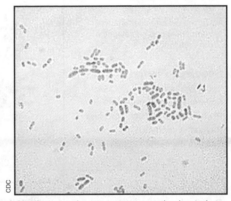

Alcaligenes odorans: a gram negative bacterium. Note how the cells appear pale.

SUPERKINGDOM: EUKARYOTAE
Kingdom: FUNGI

- Heterotrophic.
- Rigid cell wall made of chitin.
- Vary from single celled to large multicellular organisms.
- Mostly saprotrophic (ie. feeding on dead or decaying material).
- Terrestrial and immobile.

Examples:
Mushrooms/toadstools, yeasts, truffles, morels, moulds, and lichens.

Species diversity: 80,000 +

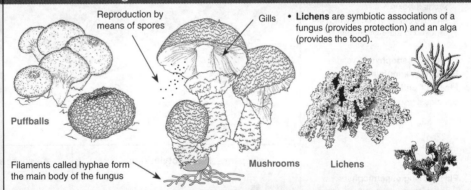

Reproduction by means of spores

Gills

Puffballs

Filaments called hyphae form the main body of the fungus

Mushrooms

Lichens

- **Lichens** are symbiotic associations of a fungus (provides protection) and an alga (provides the food).

Kingdom: PROTISTA

- A diverse group of organisms. They are polyphyletic and so better represented in the 3 domain system.
- Unicellular or simple multicellular.
- Widespread in moist or aquatic environments.

Examples of algae: green, red, and brown algae, dinoflagellates, diatoms.

Examples of protozoa: amoebas, foraminiferans, radiolarians, ciliates.

Species diversity: 55,000 +

Algae 'plant-like' protists

- Autotrophic (photosynthesis)
- Characterized by the type of chlorophyll present

Cell walls of cellulose, sometimes with silica

Diatom

Protozoa 'animal-like' protists

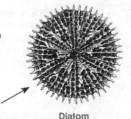

- Heterotrophic nutrition and feed via ingestion
- Most are microscopic (5 μm - 250 μm)

Move via projections called pseudopodia

Lack cell walls

Amoeba

KNOW

Kingdom: PLANTAE

- Multicellular organisms (the majority are photosynthetic and contain chlorophyll).
- Cell walls made of cellulose; food is stored as starch.
- Subdivided into two major divisions based on tissue structure: *Bryophytes* (non-vascular plants) and *Tracheophytes* (vascular plants).

Non-Vascular Plants:

- Non-vascular, lacking transport tissues (no xylem or phloem).
- Small and restricted to moist, terrestrial environments.
- Do not possess 'true' roots, stems, or leaves.

Phylum Bryophyta: Mosses, liverworts, and hornworts.

Species diversity: 18,600 +

Phylum: Bryophyta

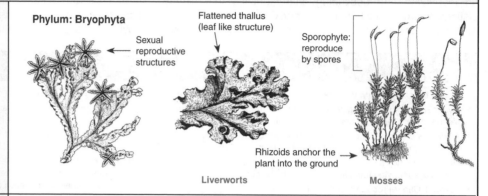

Sexual reproductive structures

Flattened thallus (leaf like structure)

Sporophyte: reproduce by spores

Rhizoids anchor the plant into the ground

Liverworts

Mosses

Vascular Plants:

- Vascular: possess transport tissues.
- Possess true roots, stems, and leaves, as well as stomata.
- Reproduce via spores, not seeds.
- Clearly defined alternation of sporophyte and gametophyte generations.

Seedless Plants:

Spore producing plants, includes:
Phylum Filicinophyta: Ferns
Phylum Sphenophyta: Horsetails
Phylum Lycophyta: Club mosses
Species diversity: 13,000 +

Phylum: Lycophyta

Leaves

Club moss

Phylum: Sphenophyta

Leaves

Horsetail

Phylum: Filicinophyta

Reproduce via spores on the underside of leaf

Large dividing leaves called fronds

Rhizome

Adventitious roots

Fern

Seed Plants:

Also called Spermatophyta. Produce seeds housing an embryo. Includes:

Gymnosperms

- Lack enclosed chambers in which seeds develop.
- Produce seeds in cones which are exposed to the environment.

Phylum Cycadophyta: Cycads
Phylum Ginkgophyta: Ginkgoes
Phylum Coniferophyta: Conifers
Species diversity: 730 +

Phylum: Cycadophyta

Palm-like leaves

Cone

Cycad

Phylum: Ginkgophyta

Flat leaves

Ginkgo

Phylum: Coniferophyta

Needle-like leaves

Male cones

Woody stems

Female cones

Conifer

Angiosperms

Phylum: Angiospermophyta

- Seeds in specialized reproductive structures called flowers.
- Female reproductive ovary develops into a fruit.
- Pollination usually via wind or animals.

Species diversity: 260,000 +

The phylum Angiospermophyta may be subdivided into two classes:

Class Monocotyledoneae (Monocots)

Class Dicotyledoneae (Dicots)

Angiosperms: Monocotyledons

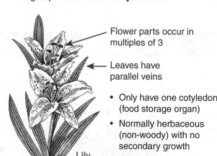

Flower parts occur in multiples of 3

Leaves have parallel veins

- Only have one cotyledon (food storage organ)
- Normally herbaceous (non-woody) with no secondary growth

Lily

Examples: cereals, lilies, daffodils, palms, grasses.

Angiosperms: Dicotyledons

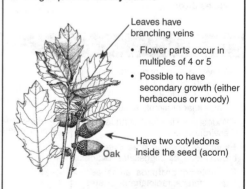

Leaves have branching veins

- Flower parts occur in multiples of 4 or 5
- Possible to have secondary growth (either herbaceous or woody)

Have two cotyledons inside the seed (acorn)

Oak

Examples: many annual plants, trees and shrubs.

Kingdom: ANIMALIA

- Over 800,000 species in 33 phyla.
- Multicellular, heterotrophic organisms.
- Animal cells lack cell walls.

- Major phyla subdivided on the basis of body symmetry, development of the coelom, and external and internal structures.

- Most animals show bilateral symmetry. Radial symmetry is a feature of cnidarians and ctenophores.

Phylum: Porifera

- Lack organs and body symmetry
- All are aquatic (mostly marine).
- Asexual reproduction by budding.
- Lack a nervous system.

Examples: sponges.

Species diversity: 8000 +

- Capable of regeneration (the replacement of lost parts)
- Possess spicules (needle-like internal structures) for support and protection

Body wall perforated by pores through which water enters

Water leaves by a larger opening - the osculum

Sponge

Tube sponge

Sessile (attach to ocean floor)

Phylum: Cnidaria

- Radial symmetry (body divisible through more than two planes.
- Diploblastic with two basic body forms:
  Medusa: umbrella shaped and free swimming by pulsating bell.
  Polyp: cylindrical, some are sedentary, others glide, or use tentacles as legs.
- Some have a life cycle that alternates between a polyp and a medusa stage.
- All are aquatic (most are marine).

Examples: Jellyfish, sea anemones, hydras, and corals.

Species diversity: 11,000 +

Some have air-filled floats

Nematocysts (stinging cells)

Single opening acts as mouth and anus

Polyps may aggregate in colonies

Polyps stick to seabed

Brain coral

Jellyfish (Portuguese man-of-war)

Sea anemone

Colonial polyps

Contraction of the bell propels the free swimming medusa

Phylum: Rotifera

- A diverse group of small, pseudocoelomates with sessile, colonial, and planktonic forms.
- Most freshwater, a few marine.
- Typically reproduce via cyclic parthenogenesis.
- Characterized by a wheel of cilia on the head used for feeding and locomotion, a large muscular pharynx (mastax) with jaw like trophi, and a foot with sticky toes.

Species diversity: 1500 +

Cilia

Head

Mastax

Foot

Toes

Bdelloid: non-planktonic, creeping rotifer

Spines for protection against predators

Lorica

Ovary

Eggs

Planktonic forms swim using their crown of cilia

Phylum: Platyhelminthes

- Unsegmented. Coelom has been lost.
- Flattened body shape.
- Mouth, but no anus.
- Many are parasitic.

Examples: Tapeworms, planarians, flukes.

Species diversity: 20,000 +

Hooks

Detail of head (scolex)

Liver fluke

Tapeworm

Planarian

Phylum: Nematoda

- Tiny, unsegmented roundworms.
- Many are plant/animal parasites

Examples: Hookworms, stomach worms, lung worms, filarial worms

Species diversity: 80,000 - 1 million

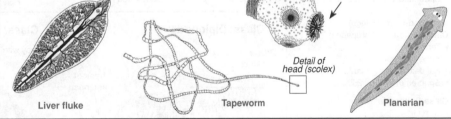

Muscular pharynx   Ovary   Anus

A roundworm parasite

Mouth   A general nematode body plan   Intestine

Phylum: Annelida

- Cylindrical, segmented body with chaetae (bristles).
- Move using hydrostatic skeleton and/or parapodia (appendages).

Examples: Earthworms, leeches, polychaetes (including tubeworms).

Species diversity: 15,000 +

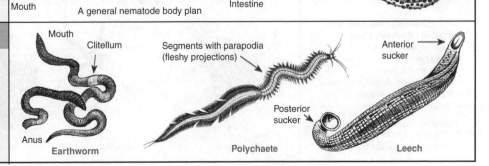

Mouth

Clitellum

Segments with parapodia (fleshy projections)

Anterior sucker

Posterior sucker

Anus

Earthworm   Polychaete   Leech

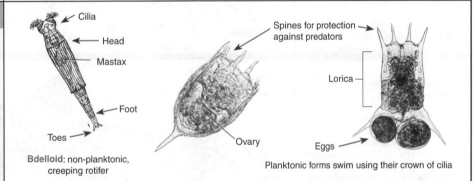
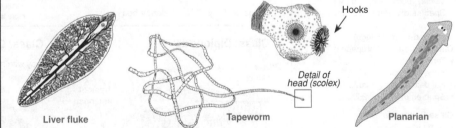

Kingdom: ANIMALIA (continued)

Phylum: Mollusca

- Soft bodied and unsegmented.
- Body comprises head, muscular foot, and visceral mass (organs).
- Most have radula (rasping tongue).
- Aquatic and terrestrial species.
- Aquatic species possess gills.

Examples: Snails, mussels, squid.
Species diversity: 110,000 +

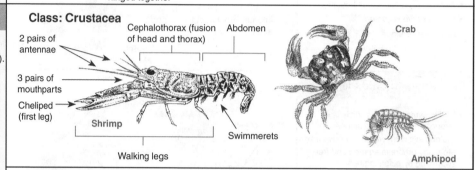

Class: Bivalvia

Radula lost in bivalves
Mantle secretes shell
Muscular foot for locomotion
Two shells hinged together
Scallop

Class: Gastropoda

Mantle secretes shell
Tentacles with eyes
Head
Muscular foot for locomotion
Land snail

Class: Cephalopoda

Well developed eyes
Squid
Foot divided into tentacles

Phylum: Arthropoda

- Exoskeleton made of chitin.
- Grow in stages after molting (ecdysis).
- Jointed appendages.
- Segmented bodies.
- Heart found on dorsal side of body.
- Open circulation system.
- Most have compound eyes.

Species diversity: 1 million +
Make up 75% of all living animals.

Arthropods are subdivided into the following classes:
Class: Crustacea (crustaceans)
- Mainly marine.
- Exoskeleton impregnated with mineral salts.
- Gills often present.
- Includes: Lobsters, crabs, barnacles, prawns, shrimps, isopods, amphipods
- **Species diversity:** 35,000 +

Class: Arachnida (chelicerates)
- Almost all are terrestrial.
- 2 body parts: cephalothorax and abdomen (except horseshoe crabs).
- Includes: spiders, scorpions, ticks, mites, horseshoe crabs.
- **Species diversity:** 57,000 +

Class: Insecta (insects)
- Mostly terrestrial.
- Most are capable of flight.
- 3 body parts: head, thorax, abdomen.
- Include: Locusts, dragonflies, cockroaches, butterflies, bees, ants, beetles, bugs, flies, and more
- **Species diversity:** 800,000 +

Myriapods (=many legs)
Class Diplopoda (millipedes)
- Terrestrial.
- Have a rounded body.
- Eat dead or living plants.
- **Species diversity:** 2000 +

Class Chilopoda (centipedes)
- Terrestrial.
- Have a flattened body.
- Poison claws for catching prey.
- Feed on insects, worms, and snails.
- **Species diversity:** 7000 +

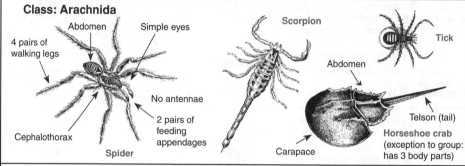

Class: Crustacea

2 pairs of antennae
Cephalothorax (fusion of head and thorax)
Abdomen
Crab
3 pairs of mouthparts
Cheliped (first leg)
Shrimp
Walking legs
Swimmerets
Amphipod

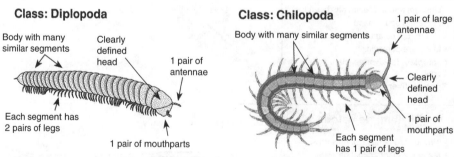

Class: Arachnida

Abdomen
Simple eyes
Scorpion
4 pairs of walking legs
No antennae
2 pairs of feeding appendages
Cephalothorax
Spider
Tick
Abdomen
Telson (tail)
Carapace
Horseshoe crab (exception to group: has 3 body parts)

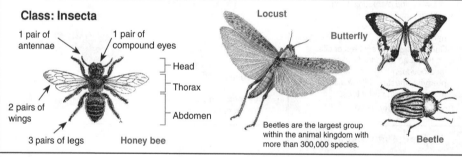

Class: Insecta

1 pair of antennae
1 pair of compound eyes
Locust
Butterfly
Head
Thorax
Abdomen
2 pairs of wings
3 pairs of legs
Honey bee
Beetles are the largest group within the animal kingdom with more than 300,000 species.
Beetle

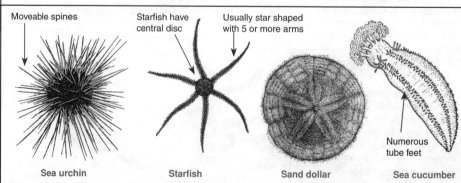

Class: Diplopoda

Body with many similar segments
Clearly defined head
1 pair of antennae
Each segment has 2 pairs of legs
1 pair of mouthparts

Class: Chilopoda

Body with many similar segments
1 pair of large antennae
Clearly defined head
1 pair of mouthparts
Each segment has 1 pair of legs

Phylum: Echinodermata

- Rigid body wall, internal skeleton made of calcareous plates.
- Many possess spines.
- Ventral mouth, dorsal anus.
- External fertilization.
- Unsegmented, marine organisms.
- Tube feet for locomotion.
- Water vascular system.

Examples: Starfish, brittlestars, feather stars, sea urchins, sea lilies.
Species diversity: 6000 +

Moveable spines
Starfish have central disc
Usually star shaped with 5 or more arms
Numerous tube feet
Sea urchin
Starfish
Sand dollar
Sea cucumber

© 2012-2014 **BIOZONE** International
ISBN: 978-1-927173-93-0
Photocopying Prohibited

Kingdom: ANIMALIA (continued)

Phylum: Chordata

- Dorsal notochord (flexible, supporting rod) present at some stage in the life history.
- Post-anal tail present at some stage in their development.
- Dorsal, tubular nerve cord.
- Pharyngeal slits present.
- Circulation system closed in most.
- Heart positioned on ventral side.

Species diversity: 48,000 +

- A very diverse group with several sub-phyla:
  - Urochordata (sea squirts, salps)
  - Cephalochordata (lancelet)
  - Craniata (vertebrates)

Sub-Phylum Vertebrata

- Internal skeleton of cartilage or bone.
- Well developed nervous system.
- Vertebral column replaces notochord.
- Two pairs of appendages (fins or limbs) attached to girdles.

Further subdivided into:

Class: Chondrichthyes (cartilaginous fish)
- Skeleton of cartilage (not bone).
- No swim bladder.
- All aquatic (mostly marine).
- Include: Sharks, rays, and skates.

Species diversity: 850 +

Class: Osteichthyes (bony fish)
- Swim bladder present.
- All aquatic (marine and fresh water).

Species diversity: 21,000 +

Class: Amphibia (amphibians)
- Lungs in adult, juveniles may have gills (retained in some adults).
- Gas exchange also through skin.
- Aquatic and terrestrial (limited to damp environments).
- Include: Frogs, toads, salamanders, and newts.

Species diversity: 3900 +

Class Reptilia (reptiles)
- Ectotherms with no larval stages.
- Teeth are all the same type.
- Eggs with soft leathery shell.
- Mostly terrestrial.
- Include: Snakes, lizards, crocodiles, turtles, and tortoises.

Species diversity: 7000 +

Class: Aves (birds)
- Terrestrial endotherms.
- Eggs with hard, calcareous shell.
- Strong, light skeleton.
- High metabolic rate.
- Gas exchange assisted by air sacs.

Species diversity: 8600 +

Class: Mammalia (mammals)
- Endotherms with hair or fur.
- Mammary glands produce milk.
- Glandular skin with hair or fur.
- External ear present.
- Teeth are of different types.
- Diaphragm between thorax/abdomen.

Species diversity: 4500 +

Subdivided into three subclasses:
Monotremes, marsupials, placentals.

Class: Chondrichthyes (cartilaginous fish)

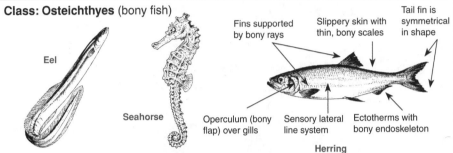

Ectotherms with endoskeleton made of cartilage

Lateral line sense organ

Asymmetrical tail fin provides lift

Skin with toothlike scales

Pelvic fin

Pectoral fin

No operculum (bony flap) over gills

Hammerhead shark

Stingray

Class: Osteichthyes (bony fish)

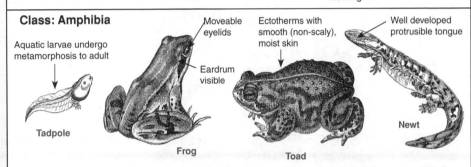

Eel

Seahorse

Fins supported by bony rays

Slippery skin with thin, bony scales

Tail fin is symmetrical in shape

Operculum (bony flap) over gills

Sensory lateral line system

Ectotherms with bony endoskeleton

Herring

Class: Amphibia

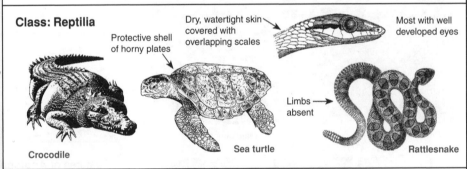

Aquatic larvae undergo metamorphosis to adult

Moveable eyelids

Ectotherms with smooth (non-scaly), moist skin

Well developed protrusible tongue

Eardrum visible

Tadpole

Frog

Toad

Newt

Class: Reptilia

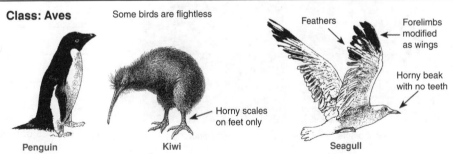

Protective shell of horny plates

Dry, watertight skin covered with overlapping scales

Most with well developed eyes

Limbs absent

Crocodile

Sea turtle

Rattlesnake

Class: Aves

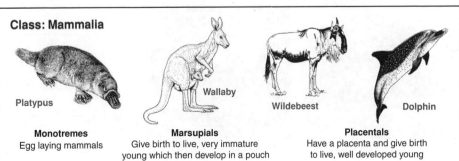

Some birds are flightless

Feathers

Forelimbs modified as wings

Horny beak with no teeth

Horny scales on feet only

Penguin

Kiwi

Seagull

Class: Mammalia

Platypus

Wallaby

Wildebeest

Dolphin

Monotremes
Egg laying mammals

Marsupials
Give birth to live, very immature young which then develop in a pouch

Placentals
Have a placenta and give birth to live, well developed young

152 Features of Plants

Key Idea: The plant kingdom is monophyletic, meaning that it is derived from a common ancestor. The variety we see in plant taxa today is a result of their enormous diversification from the first plants.

Although plants are some of the most familiar organisms in our environment, the science of plant classification has long been controversial, with a lack of agreement as to the placement of taxa. Here we recognize four major taxa. Describe the distinguishing features of each, using the photos and the activity "Features of Taxonomic Groups" to help you.

Liverwort

Moss

Fern frond

Ground fern

Pine tree with female cones

Male pine cone

Coconut palms

Corn plants

Deciduous tree

Flowering plant

1. Features of **Bryophyta** (mosses and liverworts):

_____

_____

_____

_____

_____

2. Features of **Filicinophyta** (ferns): _____

_____

_____

_____

_____

3. Features of **Coniferophyta** (conifers):

_____

_____

_____

_____

_____

_____

4. Features of **Angiospermophyta** (flowering plants):

_____

_____

_____

_____

(a) Features of monocots: _____

_____

_____

_____

(b) Features of dicots: _____

_____

_____

_____

LINK

APP 151

153 Features of Animal Taxa

Key Idea: The animal kingdom is classified into about 35 phyla of which about 10 are common.

Representatives of the more familiar animal taxa are illustrated in this activity: **poriferans** (sponges), **cnidarians** (jellyfish, sea anemones, corals), **platyhelminthes** (flatworms), **annelids** (segmented worms), **arthropods** (insects, crustaceans, spiders, scorpions, centipedes, millipedes), **molluscs** (snails, bivalves, squid, octopus), **echinoderms** (starfish, sea urchins), and **vertebrates** from the phylum chordata (fish, amphibians, reptiles, birds, mammals). Describe the **distinguishing features** of each taxon. Underline the feature you think is most important in distinguishing the taxon from others. Use the photographs, weblinks, and the activity "Features of Taxonomic Groups" to help you.

Sponge

Chimney sponge

Sea anemones

Jellyfish

Planarian

Tapeworm

Tubeworms

Earthworm

Long-horned beetle

Spider

Nautilus

Abalone

1. Features of phylum **Porifera**: _____

2. Features of phylum **Cnidaria**: _____

3. Features of phylum **Platyhelminthes**: _____

4. Features of phylum **Annelida**: _____

5. Features of phylum **Arthropoda**: _____

6. Features of phylum **Mollusca**: _____

LINK 151   WEB 153   APP

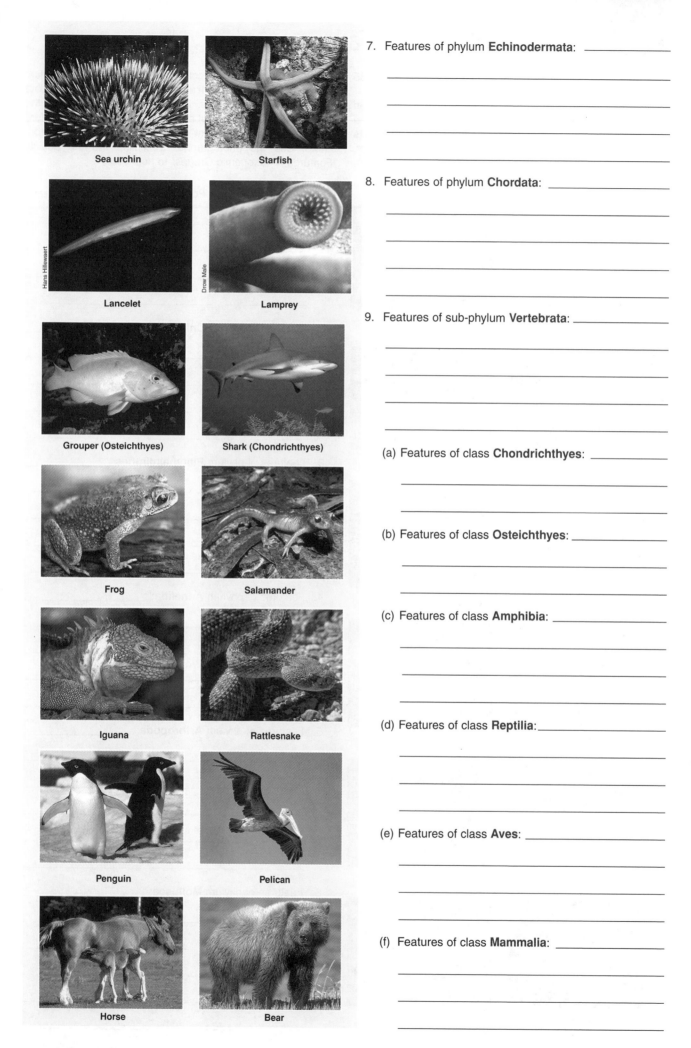

Sea urchin

Starfish

Lancelet

Lamprey

Grouper (Osteichthyes)

Shark (Chondrichthyes)

Frog

Salamander

Iguana

Rattlesnake

Penguin

Pelican

Horse

Bear

7. Features of phylum **Echinodermata**: _____

_____

_____

_____

_____

8. Features of phylum **Chordata**: _____

_____

_____

_____

_____

9. Features of sub-phylum **Vertebrata**: _____

_____

_____

_____

_____

(a) Features of class **Chondrichthyes**: _____

_____

_____

(b) Features of class **Osteichthyes**: _____

_____

_____

(c) Features of class **Amphibia**: _____

_____

_____

_____

(d) Features of class **Reptilia**: _____

_____

_____

_____

(e) Features of class **Aves**: _____

_____

_____

_____

(f) Features of class **Mammalia**: _____

_____

_____

_____

154 Classification Keys

Key Idea: Classification keys are used to identify an organism based on its distinguishing features and assign it to a species. An organism's classification should include a clear, unambiguous **description**, an accurate **diagram**, and its unique name, denoted by the **genus** and **species**. Classification keys are used to identify an organism and assign it to the correct species (assuming that the organism has already been formally classified and is included in the key). Typically, keys are dichotomous and involve a series of linked steps. At each step, a choice is made between two features. Each alternative leads to another question until an identification is made. If the organism cannot be identified, it may be a new species or the key may need revision. This activity describes two examples of **dichotomous keys**. The first describes features for identifying the larvae of genera within the order Trichoptera (caddisflies). The second is a key to the identification of aquatic insect orders.

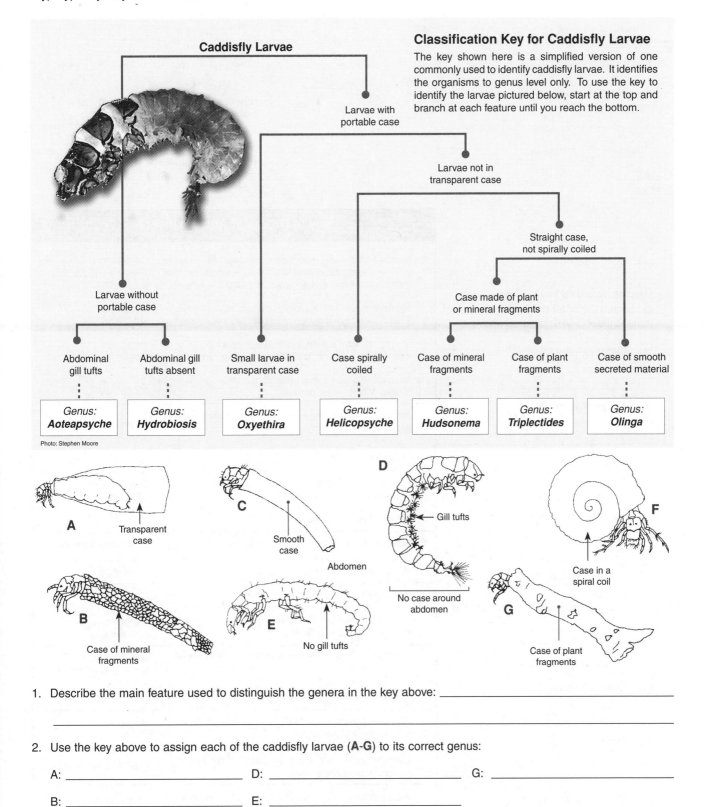

Classification Key for Caddisfly Larvae

The key shown here is a simplified version of one commonly used to identify caddisfly larvae. It identifies the organisms to genus level only. To use the key to identify the larvae pictured below, start at the top and branch at each feature until you reach the bottom.

Caddisfly Larvae

- Larvae with portable case
  - Larvae not in transparent case
    - Straight case, not spirally coiled
      - Case made of plant or mineral fragments
- Larvae without portable case

Larvae without portable case:
- Abdominal gill tufts → **Genus: Aoteapsyche**
- Abdominal gill tufts absent → **Genus: Hydrobiosis**

- Small larvae in transparent case → **Genus: Oxyethira**
- Case spirally coiled → **Genus: Helicopsyche**
- Case of mineral fragments → **Genus: Hudsonema**
- Case of plant fragments → **Genus: Triplectides**
- Case of smooth secreted material → **Genus: Olinga**

Photo: Stephen Moore

A — Transparent case
B — Case of mineral fragments
C — Smooth case, Abdomen
D — Gill tufts, No case around abdomen
E — No gill tufts
F — Case in a spiral coil
G — Case of plant fragments

1. Describe the main feature used to distinguish the genera in the key above: _____

_____

2. Use the key above to assign each of the caddisfly larvae (**A-G**) to its correct genus:

A: _____   D: _____   G: _____

B: _____   E: _____

C: _____   F: _____

© 2012-2014 **BIOZONE** International
ISBN: 978-1-927173-93-0
Photocopying Prohibited

SKILL

216

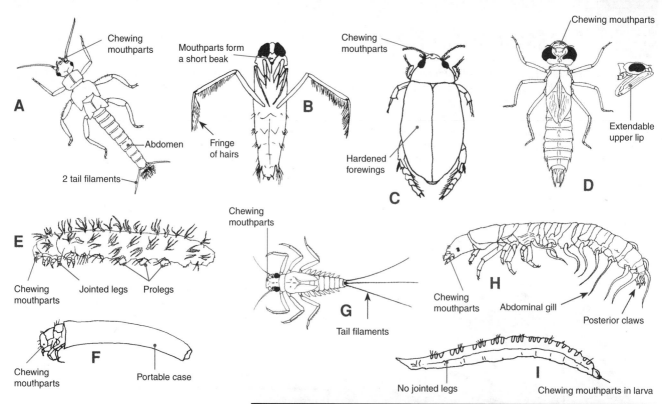

A — Chewing mouthparts — Abdomen — 2 tail filaments

B — Mouthparts form a short beak — Fringe of hairs

C — Chewing mouthparts — Hardened forewings

D — Chewing mouthparts — Extendable upper lip

E — Chewing mouthparts — Jointed legs — Prolegs

F — Chewing mouthparts — Portable case

G — Chewing mouthparts — Tail filaments

H — Chewing mouthparts — Abdominal gill — Posterior claws

I — No jointed legs — Chewing mouthparts in larva

Key to Orders of Aquatic Insects

1	Insects with chewing mouthparts; forewings are hardened and meet along the midline of the body when at rest (they may cover the entire abdomen or be reduced in length).	**Coleoptera** (beetles)
	Mouthparts piercing or sucking and form a pointed cone	*Go to 2*
	With chewing mouthparts, but without hardened forewings	*Go to 3*
2	Mouthparts form a short, pointed beak; legs fringed for swimming or long and spaced for suspension on water.	**Hemiptera** (bugs)
	Mouthparts do not form a beak; legs (if present) not fringed or long, or spaced apart.	*Go to 3*
3	Prominent upper lip (labium) extendable, forming a food capturing structure longer than the head.	**Odonata** (dragonflies & damselflies)
	Without a prominent, extendable labium	*Go to 4*
4	Abdomen terminating in three tail filaments which may be long and thin, or with fringes of hairs.	**Ephemeroptera** (mayflies)
	Without three tail filaments	*Go to 5*
5	Abdomen terminating in two tail filaments	**Plecoptera** (stoneflies)
	Without long tail filaments	*Go to 6*
6	With three pairs of jointed legs on thorax	*Go to 7*
	Without jointed, thoracic legs (although non-segmented prolegs or false legs may be present).	**Diptera** (true flies)
7	Abdomen with pairs of non-segmented prolegs bearing rows of fine hooks.	**Lepidoptera** (moths and butterflies)
	Without pairs of abdominal prolegs	*Go to 8*
8	With eight pairs of finger-like abdominal gills; abdomen with two pairs of posterior claws.	**Megaloptera** (dobsonflies)
	Either, without paired, abdominal gills, or, if such gills are present, without posterior claws.	*Go to 9*
9	Abdomen with a pair of posterior prolegs bearing claws with subsidiary hooks; sometimes a portable case.	**Trichoptera** (caddisflies)

3. Use the simplified key to identify each of the orders (by order or common name) of aquatic insects (**A-I**) pictured above:

(a) Order of insect A:

_____

(b) Order of insect B:

_____

(c) Order of insect C:

_____

(d) Order of insect D:

_____

(e) Order of insect E:

_____

(f) Order of insect F:

_____

(g) Order of insect G:

_____

(h) Order of insect H:

_____

(i) Order of insect I:

_____

© 2012-2014 **BIOZONE** International
ISBN: 978-1-927173-93-0
Photocopying Prohibited

155 Keying Out Plant Species

Key Idea: Being able to create a dichotomous key is an important skill and helps to identify the important features of a group of organisms.

Dichotomous keys are a useful tool in biology and can enable identification to species level provided the characteristics chosen are appropriate for separating species. The following simple activity requires you to devise a key that will identify the five species of the genus *Acer* by the illustrations of the leaves. It provides valuable practice in identifying characteristic features to identify plants to species level.

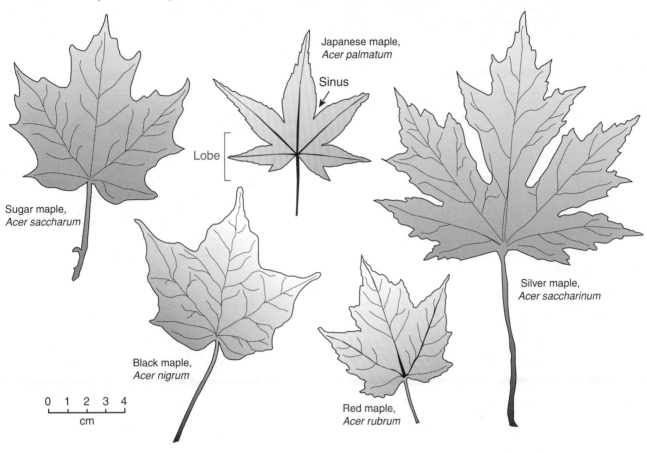

Japanese maple, *Acer palmatum*

Sinus

Lobe

Sugar maple, *Acer saccharum*

Silver maple, *Acer saccharinum*

Black maple, *Acer nigrum*

Red maple, *Acer rubrum*

0 1 2 3 4
cm

1. Use the diagrams of the leaves of the common species of *Acer* to create a dichotomous key in the space below that will identify all five species shown. Your key should be able to be used by someone else to correctly identify all species. You could create a branched key (e.g. page 215) or a written key (e.g. page 216).

156 Cladograms and Phylogenetic Trees

Key Idea: There are many ways to construct phylogenetic trees (evolutionary histories). Cladistics is a method based on shared derived characteristics.

Phylogenetic systematics is a science in which the fields of taxonomy (the study of naming organisms) and phylogenetics (the study of evolutionary history) overlap. Traditional methods for establishing **phylogenetic trees** have emphasized the physical (morphological) similarities between organisms in order to group species into genera and other higher level taxa. In contrast, **cladistics** is a method that relies on **shared derived characteristics** (synapomorphies) and ignores features that are not the result of shared ancestry. A cladogram is a phylogenetic tree constructed using cladistics. Although cladistics has traditionally relied on morphological data, molecular data (e.g. DNA sequences) are increasingly being used to construct cladograms.

Constructing a Simple Cladogram

A table listing the features for comparison allows us to identify where we should make branches in the **cladogram**. An outgroup (one which is known to have no or little relationship to the other organisms) is used as a basis for comparison.

Taxa

Comparative features	Jawless fish (outgroup)	Bony fish	Amphibians	Lizards	Birds	Mammals
Vertebral column	✔	✔	✔	✔	✔	✔
Jaws	✗	✔	✔	✔	✔	✔
Four supporting limbs	✗	✗	✔	✔	✔	✔
Amniotic egg	✗	✗	✗	✔	✔	✔
Diapsid skull	✗	✗	✗	✔	✔	✗
Feathers	✗	✗	✗	✗	✔	✗
Hair	✗	✗	✗	✗	✗	✔

The table above lists features shared by selected taxa. The outgroup (jawless fish) shares just one feature (vertebral column), so it gives a reference for comparison and the first branch of the cladogram (tree).

As the number of taxa in the table increases, the number of possible trees that could be drawn increases exponentially. To determine the most likely relationships, the rule of **parsimony** is used. This assumes that the tree with the least number of evolutionary events is most likely to show the correct evolutionary relationship.

Three possible cladograms are shown on the right. The top cladogram requires six events while the other two require seven events. Applying the rule of parsimony, the top cladogram must be taken as correct.

Parsimony can lead to some confusion. Some evolutionary events have occurred multiple times. An example is the evolution of the four chambered heart, which occurred separately in both birds and mammals. The use of fossil evidence and DNA analysis can help to solve problems like this.

Possible Cladograms

Using DNA Data

DNA analysis has allowed scientists to confirm many phylogenies and refute or redraw others. In a similar way to morphological differences, DNA sequences can be tabulated and analysed. The ancestry of whales has been in debate since Darwin. The radically different morphologies of whales and other mammals makes it difficult work out the correct phylogenetic tree. However recently discovered fossil ankle bones, as well as DNA studies, show whales are more closely related to hippopotami than to any other mammal. Coupled with molecular clocks, DNA data can also give the time between each split in the lineage.

The DNA sequences on the right show part of the nucleotide subset 141-200 and some of the matching nucleotides used to draw the cladogram. Although whales were once thought most closely related to pigs, based on the DNA analysis the most parsimonious tree disputes this.

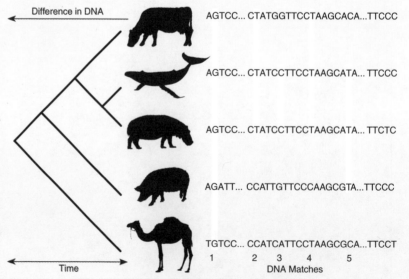

WEB 156 LINK 157

KNOW

© 2012-2014 **BIOZONE** International
ISBN: 978-1-927173-93-0

	CHARACTER												
Taxon	1	2	3	4	5	6	7	8	9	10	11	12	13
Zebra-perch sea chub	0	0	0	0	0	0	0	0	0	0	0	0	0
Barred surfperch	1	0	0	0	0	0	0	0	0	1	1	0	0
Walleye surfperch	1	0	0	0	0	1	0	1	0	1	1	0	0
Black perch	1	1	1	0	0	0	0	0	0	0	0	1	0
Rainbow seaperch	1	1	1	0	0	0	0	0	0	0	0	1	0
Rubberlip surfperch	1	1	1	1	1	0	0	0	0	0	0	0	1
Pile surfperch	1	1	1	1	1	0	0	0	0	0	0	0	1
White seaperch	1	1	1	1	1	0	0	0	0	0	0	0	0
Shiner perch	1	1	1	1	1	1	0	0	0	0	0	0	0
Pink seaperch	1	1	1	1	1	1	1	1	0	0	0	0	0
Kelp perch	1	1	1	1	1	1	1	1	1	0	0	0	0
Reef perch	1	1	1	1	1	1	1	1	1	0	0	0	0

Juvenile surfperch

Steve Lonhart (SIMoN / MBNMS) PD NOAA

Surfperches are viviparous (live bearing) and the females give birth to relatively well developed young. Some of the characters (below, left) relate to adaptations of the male for internal fertilization. Others relate to deterring or detecting predators. In the matrix, characters are assigned a 0 or 1 depending on whether they represent the ancestral (0) or derived (1) state. This coding is common in cladistics because it allows the data to be analysed by computer.

Data after Cailliel et al, 1986

Selected characters for cladogram assembly

1.	Viviparity (live bearing)	0 No	1 Yes
2.	Males with flask organ	0 No	1 Yes
3.	Orbit without bony front wall	0 Yes	1 No
4.	Tail length	0 Short	1 Long
5.	Body depth	0 Deep	1 Narrow
6.	Body size	0 Large	1 Small
7.	Length of dorsal fin base	0 Long	1 Short
8.	Eye diameter	0 Moderate	1 Large
9.	Males with anal crescent	0 No	1 Yes
10	Pectoral bone with process	0 No	1 Yes
11.	Length of dorsal sheath	0 Long	1 Short
12.	Body mostly darkish	0 No	1 Yes
13.	Flanks with large black bars	0 No	1 Yes

1. This activity provides the taxa and character matrix for 11 genera of marine fishes in the family of surfperches. The outgroup given is a representative of a sister family of rudderfishes (zebra-perch sea chub), which are not live-bearing. Your task is to create the most parsimonious cladogram from the matrix of character states provided. To help you, we have organized the matrix with genera having the smallest blocks of derived character states (1) at the top following the outgroup representative. Use a separate sheet of graph paper, working from left to right to assemble your cladogram. Identify the origin of derived character states with horizontal bars, as shown in the examples earlier in this activity. CLUE: You should end up with 15 steps. Two derived character states arise twice independently. Staple your cladogram to this page.

2. In the cladogram you have constructed for the surfperches, two characters have evolved twice independently:

   (a) Identify these two characters: _____

   (b) What selection pressures do you think might have been important in the evolution of these two derived states?

   _____

   _____

3. What assumption is made when applying the rule of parsimony in constructing a cladogram? _____

   _____

   _____

4. Explain the difference between a shared characteristic and a shared derived characteristic: _____

   _____

   _____

5. In the DNA data for the whale cladogram (previous page) identify the DNA match that shows a mutation event must have happened twice in evolutionary history.

   _____

   _____

6. A phylogenetic tree is a hypothesis for an evolutionary history. How could you test it? _____

   _____

   _____

   _____

   _____

157 Cladistics

Key Idea: The phylogeny and classification of many organisms have been reinterpreted as a result of cladistic analysis. Cladistic analysis is one method by which we can construct an evolutionary history for a taxonomic group. Phylogenies constructed using traditional and cladistic methods do not necessarily conflict, but cladistics' emphasis on molecular data has led to reclassifications of a number of taxa (including primates and many plants). In **natural classifications**, all members of the group have descended from a common ancestor (they are **monophyletic**). Molecular evidence has shown that many traditional groups (e.g. reptiles and figworts (described below)) are not descended from a common ancestor, but are **paraphyletic** or even **polyphyletic** (see below). Popular classifications will probably continue to reflect similarities and differences in appearance, rather than a strict evolutionary history. In this respect, they are a compromise between phylogeny and the need for a convenient filing system for species diversity.

Determining Phylogenetic Relationships

Increasingly, analyses to determine evolutionary relationships rely on cladistic analyses of character states. Cladism groups species according to their most recent common ancestor on the basis of shared derived characteristics. A phylogeny constructed using cladistics thus includes only **monophyletic groups**, i.e. the common ancestor and all of its descendants. It excludes both paraphyletic and polyphyletic groups (right). It is important to understand these terms when constructing cladograms. The cladist restriction to using only shared derived characteristics creates an unambiguous branching tree. One problem with this approach is that a strictly cladistic classification could theoretically have an impractically large number of taxonomic ranks and may be incompatible with a Linnaean (traditional classification) system.

Cladistic schemes have traditionally used morphological characteristics, with gain (or loss) of a character indicating a derived state. Increasingly, molecular comparisons are being used, particularly for highly conserved genes such as those coding for ribosomal RNA. For prokaryotes, molecular phylogeny studies have been the most important tool in revealing evolutionary relationships and revolutionizing traditional classification schemes.

Taxon 2 is **polyphyletic** as it includes organisms with different ancestors. The group "warm-blooded (endothermic) animals" is polyphyletic as it includes birds and mammals.

Taxon 3 is **paraphyletic**. It includes species A without including all of A's descendants. The traditional grouping of reptiles is paraphyletic because it does not include birds.

Taxon 1

Taxon 2

Species F   Species G   Species H   Species I

Taxon 3

Species J   Species K

Species C   Species D

Species E

Species B

Taxon 1 is **monophyletic** as all the organisms are related to species B (the common ancestor). All the descendants of the first reptiles form a monophyletic group.

Species A

Cladistics and the Reclassification of Figworts

The angiosperm order Lamiales (which includes lavender, lilac, olive, jasmine, and snapdragons) has around 24,000 members. It contains several families, one of the largest once being Scrophulariaceae (**figworts**). This family once included snapdragons, foxgloves, veronica, and monkeyflowers. While the other families in Lamiales have relatively well defined characteristics, the figworts were mostly assigned to their family based on their lack of characteristic traits. This suggested that the plants in the figwort family were not monophyletic.

Investigations using three genes (*rbcL*, *ndhF*, and *rps2*) from chloroplast DNA revealed that there were at least five distinct monophyletic groups (and probably more) within the figworts, each worthy of family status.

Common figwort, *Scrophularia nodosa*

Luc Viatour cc 3.0

Green figwort, *Scrophularia umbrosa*

Roger Griffith

1. (a) What is meant by a natural classification?_____

_____

_____

(b) What is the basis for a natural classification (how is the taxonomic hierarchy established): _____

_____

_____

© 2012-2014 **BIOZONE** International
ISBN: 978-1-927173-93-0
Photocopying Prohibited

Classification of the "Great Apes"

A Classical Taxonomic View	A Cladistic View

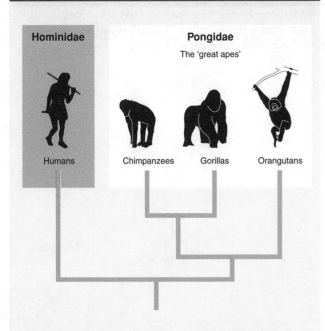

Hominidae

Humans

Pongidae

The 'great apes'

Chimpanzees Gorillas Orangutans

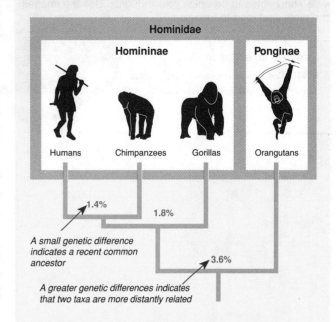

Hominidae

Homininae

Humans Chimpanzees Gorillas

Ponginae

Orangutans

1.4%

A small genetic difference indicates a recent common ancestor

1.8%

3.6%

A greater genetic differences indicates that two taxa are more distantly related

On the basis of overall anatomical similarity (e.g. bones and limb length, teeth, musculature), apes are grouped into a family (Pongidae) that is separate from humans and their immediate ancestors (Hominidae). The family Pongidae (the great apes) is not monophyletic (of one phylogeny), because it stems from an ancestor that also gave rise to a species in another family (i.e. humans). This traditional classification scheme is now at odds with schemes derived after considering genetic evidence.

Based on the evidence of genetic differences (% values above), chimpanzees and gorillas are more closely related to humans than to orangutans, and chimpanzees are more closely related to humans than they are to gorillas. Under this scheme there is no true family of great apes. The family Hominidae includes two subfamilies: Ponginae and Homininae (humans, chimpanzees, and gorillas). This classification is monophyletic: the Hominidae includes all the species that arise from a common ancestor.

2. Explain why cladistics provides a more likely evolutionary tree for any particular group of organisms than some traditional classification methods:

_____

_____

_____

_____

_____

3. (a) Describe the contribution of biochemical evidence to clarifying some evolutionary relationships:

_____

_____

_____

(b) When might it not be advisable to rely on biochemical evidence alone?

_____

_____

_____

4. Based on the diagram above, state the family to which the chimpanzees belong under:

(a) A traditional scheme: _____

(b) A cladistic scheme: _____

158 Chapter Review

Summarize what you know about this topic under the headings provided. You can draw diagrams or mind maps, or write short notes to organize your thoughts. Use the images and hints, included to help you:

Evidence for evolution

HINT: The fossil record, homologous structures, and heritable traits.

Natural selection

HINT: Describe the four factors that bring about evolution in a population. Illustrate using the examples of finch evolution and antibiotic resistance.

REVISE

Classification of biodiversity

HINT: Binomial nomenclature, the principal eukaryote taxa, and features of the major plant and animal phyla.

Cladistics

HINT: Define mono-, para-, and polyphyletic. Use of cladistics in taxonomy.

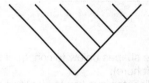

159 KEY TERMS: Did You Get It?

1. Test your vocabulary by matching each term to its definition, as identified by its preceding letter code.

adaptation

cladogram

dichotomous key

Eukaryotes

fossil record

genus

homologous structure

kingdom

natural selection

order

phylogeny

shared derived characteristic

A Key that gives two options at each step of the identification process.

B A phylogenetic tree based on shared derived characters.

C Historically, the highest rank in biological taxonomy.

D Unit of classification used to group together related families.

E A unit of classification. A group of related species.

F Group of organisms with cells containing organelles. DNA is present as chromosomes.

G Any heritable trait that equips an organism to its functional position in the environment.

H A specialized character, shared by two or more modern groups, which originated in their last common ancestor. The modern groups are not necessarily closely related.

I The sum total of current paleontological knowledge. It is all the fossils that have existed throughout life's history, whether they have been found or not.

J Structures of organisms that are the result of common ancestry are this.

K The term for the mechanism by which better adapted organisms survive to produce a greater number of viable offspring.

L The evolutionary history or genealogy of a group of organisms. Often represented as a 'tree' showing descent of new species from the ancestral one.

2. Use the shapes below to construct the following in the space below (Use a separate piece of paper if you need to and staple it here):
   (a) A dichotomous key to identify each shape:
   (b) A cladogram that shows their phylogenetic relationship (hint: A is the outgroup):

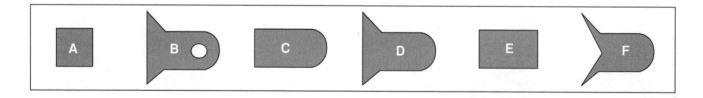

© 2012-2014 **BIOZONE** International
ISBN: 978-1-927173-93-0
Photocopying Prohibited

TEST

Topic 6

Human Physiology

Key terms

absorption

action potential

alveolus (*pl.* alveoli)

antibiotic

antibody

artery

atrium (pl. atria)

blood clot

breathing

capillary

cardiac cycle

coronary occlusion

depolarization

diabetes

digestive tract

gas exchange

heart

homeostasis

hormone

intestinal villi

liver

lung

lymphocyte

menstrual cycle

negative feedback

neuron

neurotransmitter

pancreas

pathogen

peristalsis

phagocyte

platelets

respiratory gas

resting potential

sinoatrial node

small intestine

spirometry

stomach

synapse

threshold potential

vein

ventilation

ventricle

6.1 Digestion and absorption

Understandings, applications, skills

Activity number

☐ 1 Draw an annotated diagram to describe the structure of the human digestive tract. — 160

☐ 2 Explain how food is moved through the gut by peristalsis. Identify tissue layers of the small intestine in transverse section in micrographs or with a light microscope. — 161

☐ 3 Describe digestion in the small intestine with reference to the role of enzyme secretions from the pancreas, breakdown of macromolecules into monomers, and absorption and transport of the products of digestion. Include reference to the role of intestinal villi and liver. — 162-167

☐ 4 Use dialysis tubing to model absorption of digested food in the small intestine. — 166

6.2 The blood system

Understandings, applications, skills

Activity number

☐ 1 Describe the human circulation system, with separate pulmonary and systemic circuits. Describe William Harvey's discovery of how the blood is circulated. — 168

☐ 2 Name the major categories of blood vessels in humans, state their function, and recognize them by their structure. Describe the structure of each type of blood vessel and relate its structure to its function in the circulatory system. — 169 170

☐ 3 Describe the basic structure of the heart. Recognize the chambers and valves of the heart in diagrams and dissections. — 171 174 175

☐ 4 Describe the intrinsic control of heart beat by the sinoatrial node. Explain how the beat is initiated and how the signal is propagated through the heart. Explain how the heart's basic rhythm is influenced through nervous and hormonal input. — 172

☐ 5 Describe the events in the cardiac cycle. Describe the pressure changes in the left atrium, left ventricle, and aorta during the cardiac cycle. — 173 174

TOK *We know now that emotions are brain-based, not the result of heart activity. Is science-based knowledge more valid than belief based on intuition?*

☐ 6 Describe the causes and consequences of coronary occlusions. — 176

6.3 Defence against infectious disease

Understandings, applications, skills

Activity number

☐ 1 Describe the role of the skin and mucous membranes as a first line of defence against pathogens. — 177

☐ 2 Describe the mechanism and role of blood clotting, including the role of platelets, clotting factors, thrombin, and fibrinogen. Describe the causes and consequences of blood clot formation in coronary arteries. — 176 178

☐ 3 Describe non-specific defences against pathogens by phagocytic white blood cells. Describe the role of inflammation in enhancing the activity of phagocytes. — 179 180

☐ 4 Describe the basis of specific immunity against pathogens. Recognize that some lymphocytes act as memory cells and state their role. — 177

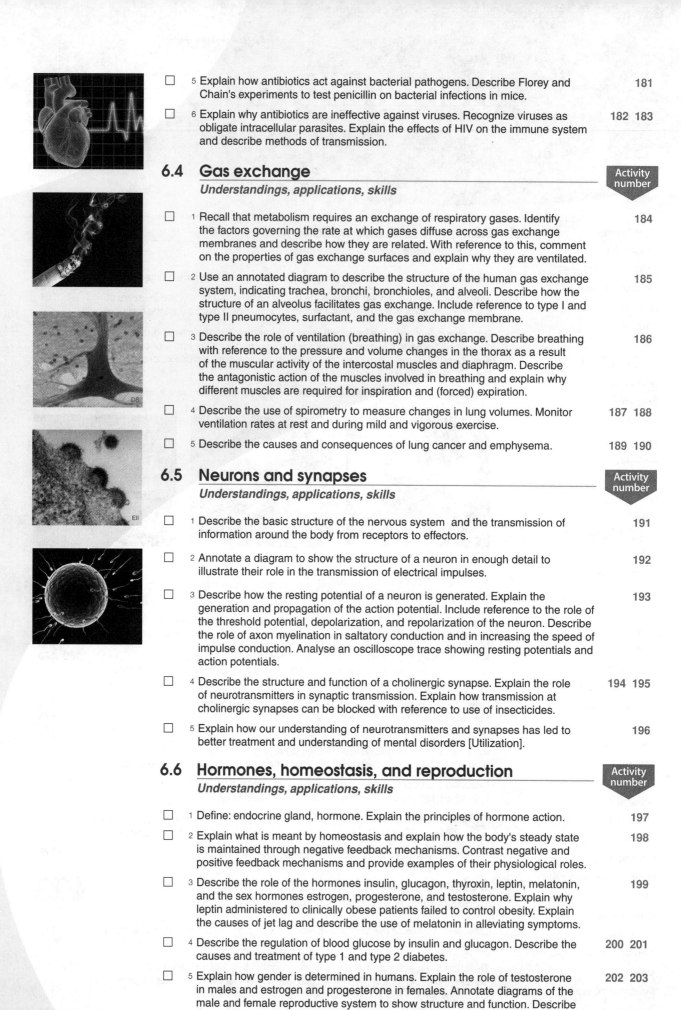

□ 5 Explain how antibiotics act against bacterial pathogens. Describe Florey and Chain's experiments to test penicillin on bacterial infections in mice. — 181

□ 6 Explain why antibiotics are ineffective against viruses. Recognize viruses as obligate intracellular parasites. Explain the effects of HIV on the immune system and describe methods of transmission. — 182 183

6.4 Gas exchange

Understandings, applications, skills

Activity number

□ 1 Recall that metabolism requires an exchange of respiratory gases. Identify the factors governing the rate at which gases diffuse across gas exchange membranes and describe how they are related. With reference to this, comment on the properties of gas exchange surfaces and explain why they are ventilated. — 184

□ 2 Use an annotated diagram to describe the structure of the human gas exchange system, indicating trachea, bronchi, bronchioles, and alveoli. Describe how the structure of an alveolus facilitates gas exchange. Include reference to type I and type II pneumocytes, surfactant, and the gas exchange membrane. — 185

□ 3 Describe the role of ventilation (breathing) in gas exchange. Describe breathing with reference to the pressure and volume changes in the thorax as a result of the muscular activity of the intercostal muscles and diaphragm. Describe the antagonistic action of the muscles involved in breathing and explain why different muscles are required for inspiration and (forced) expiration. — 186

□ 4 Describe the use of spirometry to measure changes in lung volumes. Monitor ventilation rates at rest and during mild and vigorous exercise. — 187 188

□ 5 Describe the causes and consequences of lung cancer and emphysema. — 189 190

6.5 Neurons and synapses

Understandings, applications, skills

Activity number

□ 1 Describe the basic structure of the nervous system and the transmission of information around the body from receptors to effectors. — 191

□ 2 Annotate a diagram to show the structure of a neuron in enough detail to illustrate their role in the transmission of electrical impulses. — 192

□ 3 Describe how the resting potential of a neuron is generated. Explain the generation and propagation of the action potential. Include reference to the role of the threshold potential, depolarization, and repolarization of the neuron. Describe the role of axon myelination in saltatory conduction and in increasing the speed of impulse conduction. Analyse an oscilloscope trace showing resting potentials and action potentials. — 193

□ 4 Describe the structure and function of a cholinergic synapse. Explain the role of neurotransmitters in synaptic transmission. Explain how transmission at cholinergic synapses can be blocked with reference to use of insecticides. — 194 195

□ 5 Explain how our understanding of neurotransmitters and synapses has led to better treatment and understanding of mental disorders [Utilization]. — 196

6.6 Hormones, homeostasis, and reproduction

Understandings, applications, skills

Activity number

□ 1 Define: endocrine gland, hormone. Explain the principles of hormone action. — 197

□ 2 Explain what is meant by homeostasis and explain how the body's steady state is maintained through negative feedback mechanisms. Contrast negative and positive feedback mechanisms and provide examples of their physiological roles. — 198

□ 3 Describe the role of the hormones insulin, glucagon, thyroxin, leptin, melatonin, and the sex hormones estrogen, progesterone, and testosterone. Explain why leptin administered to clinically obese patients failed to control obesity. Explain the causes of jet lag and describe the use of melatonin in alleviating symptoms. — 199

□ 4 Describe the regulation of blood glucose by insulin and glucagon. Describe the causes and treatment of type 1 and type 2 diabetes. — 200 201

□ 5 Explain how gender is determined in humans. Explain the role of testosterone in males and estrogen and progesterone in females. Annotate diagrams of the male and female reproductive system to show structure and function. Describe the significance of William Harvey's investigations of sexual reproduction. — 202 203

□ 6 Describe the menstrual cycle and its control by ovarian and pituitary hormones. — 204

□ 7 Describe how hormonal cycles are manipulated and artificial doses of hormones are used to establish a pregnancy in IVF. — 205

160 The Role of the Digestive System

Key Idea: The digestive tract is specialized to maximize the physical and chemical breakdown of food (digestion), absorption of nutrients, and elimination of undigested material. Nutrients are substances required by the body for energy, metabolism, and tissue growth and repair. Nutrients occur in food, which must be broken down by mechanical and chemical processes before the nutrients can be absorbed into the bloodstream and assimilated by the body. Digestion in humans is extracellular (occurs outside the cells). The breakdown products are then absorbed by the cells.

Processes in the Digestive System

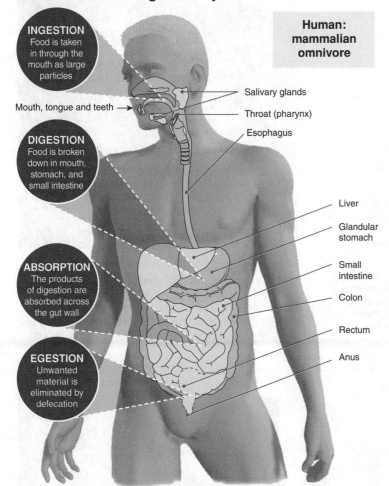

INGESTION
Food is taken in through the mouth as large particles

Mouth, tongue and teeth →

DIGESTION
Food is broken down in mouth, stomach, and small intestine

ABSORPTION
The products of digestion are absorbed across the gut wall

EGESTION
Unwanted material is eliminated by defecation

Human: mammalian omnivore

Salivary glands
Throat (pharynx)
Esophagus
Liver
Glandular stomach
Small intestine
Colon
Rectum
Anus

Before the body can incorporate the food we eat into its own tissues (a process called assimilation), it must be broken down into smaller components that can be absorbed across the intestinal wall. The breakdown of proteins, fats, and carbohydrates is achieved by enzymes and mechanical processes (such as chewing).

Some foods are specially designed to be quickly absorbed (e.g. sports gels and drinks). Energy foods (such as the one shown above) contain a mixture of simple monomers (e.g. monosaccharides) for quick absorption and larger polymers (e.g. polysaccharides) for longer lasting energy release.

Enzymes in the Digestive System

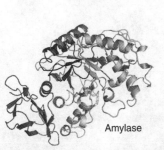
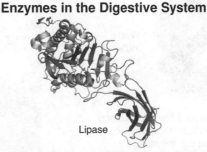
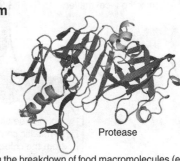

Amylase   Lipase   Protease

Enzymes play a key role in the digestive of food. They increase the speed of digestion by catalyzing the breakdown of food macromolecules (e.g. protein) into smaller monomers (e.g. amino acids) that can be absorbed by the intestinal villi of the small intestine. There are three main types of digestive enzymes; **amylases** (hydrolyse carbohydrates), **proteases** (hydrolyse protein or peptides), and **lipases** (hydrolyse lipids).

1. (a) How is food broken down during the digestive process? _____

   _____

   (b) Why must large food molecules be broken down into smaller molecules? _____

   _____

2. Explain the role of enzymes in the digestive system: _____

   _____

LINK 167 LINK 166 WEB 160

KNOW

161 Moving Food Through the Gut

Key Idea: Solid food is chewed into a small mass called a bolus and swallowed. Further digestion produces chyme. Food is moved through the gut by waves of muscular contraction called peristalsis.

Ingested food is chewed and mixed with saliva to form a small mass called a bolus. Wave-like muscular contractions called **peristalsis** moves the food, first as a bolus and then as semi-fluid chyme, through the digestive tract as described below.

Peristalsis

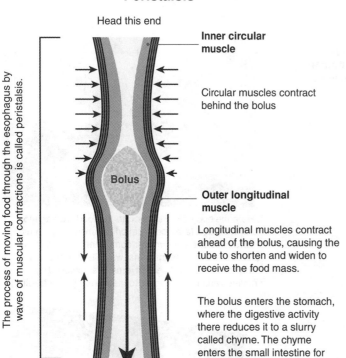

The process of moving food through the esophagus by waves of muscular contractions is called peristalsis.

Head this end

Inner circular muscle

Circular muscles contract behind the bolus

Bolus

Outer longitudinal muscle

Longitudinal muscles contract ahead of the bolus, causing the tube to shorten and widen to receive the food mass.

The bolus enters the stomach, where the digestive activity there reduces it to a slurry called chyme. The chyme enters the small intestine for further digestion.

Bolus movement

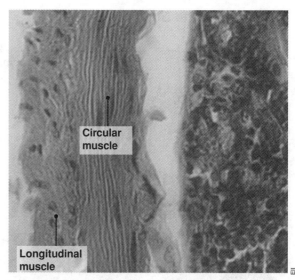

Circular muscle

Longitudinal muscle

Cross Section Through the Small Intestine

A cross section through the small intestine shows the outer longitudinal and inner circular muscles involved in peristalsis. In a cross sectional view, the longitudinal muscles appear circular because they are viewed end on, while the circular muscle appears in longitudinal section.

Peristaltic Movement in the Colon

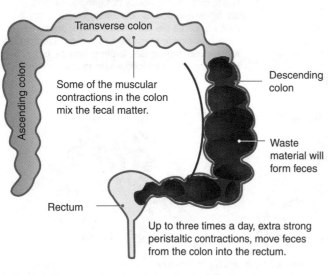

Transverse colon

Ascending colon

Some of the muscular contractions in the colon mix the fecal matter.

Descending colon

Waste material will form feces

Rectum

Up to three times a day, extra strong peristaltic contractions, move feces from the colon into the rectum.

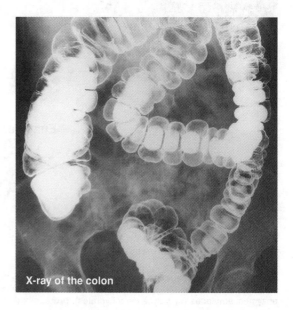

X-ray of the colon

1. Describe how **peristalsis** moves food through the gut: _____

_____

_____

2. What are the two main functions of peristalsis? _____

_____

3. Draw arrows on the X-ray of the colon (above right) to show the direction of movement of the fecal matter. Circle the areas of fecal matter.

© 2012-2014 **BIOZONE** International
ISBN: 978-1-927173-93-0
Photocopying Prohibited

162 The Stomach

Key Idea: The stomach produces acid and a protein-digesting enzyme, which break food down into a slurry, called chyme.

The **stomach** is a hollow, muscular organ between the esophagus and small intestine. In the stomach, food is mixed in an acidic environment to produce a semi-fluid mixture called chyme. The low pH of the stomach destroys microbes, denatures proteins, and activates a protein-digesting enzyme precursor. There is very little absorption in the stomach, although small molecules (glucose, alcohol) are absorbed across the stomach wall into the surrounding blood vessels.

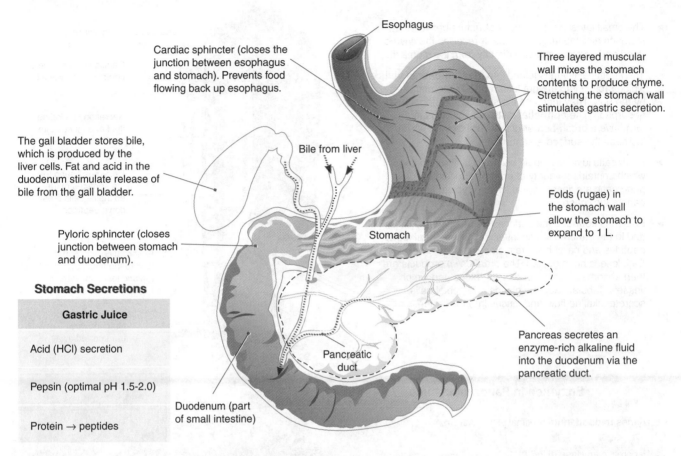

Esophagus

Cardiac sphincter (closes the junction between esophagus and stomach). Prevents food flowing back up esophagus.

Three layered muscular wall mixes the stomach contents to produce chyme. Stretching the stomach wall stimulates gastric secretion.

The gall bladder stores bile, which is produced by the liver cells. Fat and acid in the duodenum stimulate release of bile from the gall bladder.

Bile from liver

Folds (rugae) in the stomach wall allow the stomach to expand to 1 L.

Pyloric sphincter (closes junction between stomach and duodenum).

Stomach

Stomach Secretions

Gastric Juice
Acid (HCl) secretion
Pepsin (optimal pH 1.5-2.0)
Protein → peptides

Pancreas secretes an enzyme-rich alkaline fluid into the duodenum via the pancreatic duct.

Pancreatic duct

Duodenum (part of small intestine)

Detail of a Gastric Gland (Stomach Wall)

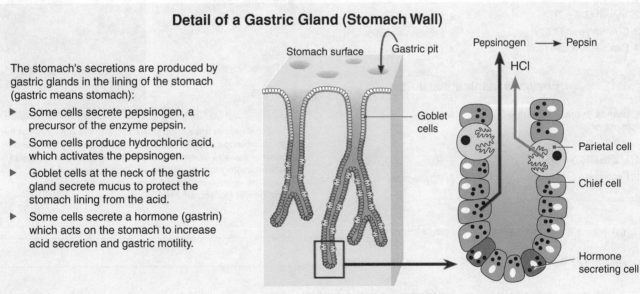

Stomach surface   Gastric pit

Pepsinogen → Pepsin

HCl

The stomach's secretions are produced by gastric glands in the lining of the stomach (gastric means stomach):

▶ Some cells secrete pepsinogen, a precursor of the enzyme pepsin.

▶ Some cells produce hydrochloric acid, which activates the pepsinogen.

▶ Goblet cells at the neck of the gastric gland secrete mucus to protect the stomach lining from the acid.

▶ Some cells secrete a hormone (gastrin) which acts on the stomach to increase acid secretion and gastric motility.

Goblet cells

Parietal cell

Chief cell

Hormone secreting cell

1. What is the role of the stomach? _____

2. What is the purpose of the hydrochloric acid produced by the parietal cells of the stomach? _____

_____

3. How does the stomach achieve the mixing of acid and enzymes with food? _____

_____

LINK 167   LINK 53   LINK 51   WEB 162

KNOW

163 The Small Intestine

Key Idea: The small intestine is the site of further enzymic break down of food and absorption of nutrients. Intestinal villi and microvilli increase surface area for nutrient absorption. The small intestine is divided into three sections: the **duodenum**, where most chemical digestion occurs, and the **jejunum** and the **ileum**, where most absorption occurs. The small intestine's role is to complete the chemical digestion of food and absorb nutrient molecules into the blood. The presence of intestinal villi in the small intestine greatly increase the surface area for absorption.

▶ The small intestine is a tubular structure between the stomach and the large intestine. It receives the chyme (food and enzyme mixture) directly from the stomach.

▶ The intestinal lining is folded into many **intestinal villi**, which project into the gut lumen (the space enclosed by the gut). They increase the surface area for nutrient absorption. The **epithelial cells** of each villus in turn have a brush-border of many **microvilli**, which increase the surface area further.

▶ Pancreatic juice is a liquid secreted by the pancreas, which contains a variety of enzymes, including pancreatic amylase, trypsin, chymotrypsin, and pancreatic lipase.

▶ Enzymes bound to the surfaces of the epithelial cells, and in the pancreatic and intestinal juices, break down peptides and carbohydrate molecules (tables below). Cellulose is not digested. The breakdown products are then absorbed into the underlying blood and lymph vessels. Tubular exocrine glands and goblet cells secrete alkaline fluid and mucus into the lumen.

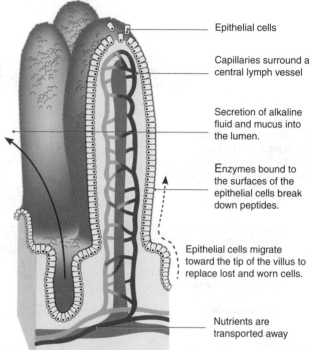

Epithelial cells

Capillaries surround a central lymph vessel

Secretion of alkaline fluid and mucus into the lumen.

Enzymes bound to the surfaces of the epithelial cells break down peptides.

Epithelial cells migrate toward the tip of the villus to replace lost and worn cells.

Nutrients are transported away

Enzymes in Pancreatic Juice

Enzymes in duodenum (optimal pH)	Action
1. Pancreatic amylase (6.7-7.0)	1. Starch → maltose
2. Trypsin (7.8-8.7)	2. Protein → peptides
3. Chymotrypsin (7.8)	3. Protein → peptides
4. Pancreatic lipase (8.0)	4. Fats → fatty acids & glycerol

Enzymes in Intestinal Juice

Enzymes in small intestine (optimal pH)	Action
1. Maltase (6.0-6.5)	1. Maltose → glucose
2. Peptidases (~ 8.0)	2. Polypeptides → amino acids

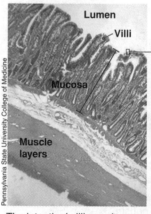

Lumen

Villi

Mucosa

Muscle layers

Pennsylvania State University College of Medicine

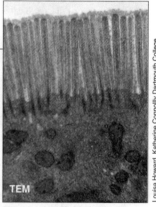

TEM

Louisa Howard, Katherine Connolly Dartmouth College

The intestinal villi are shown projecting into the gut lumen in the light microscope image (left). The microvilli forming the brush border of a single intestinal cell are shown in the transmission electron micrograph (right). The mucosa consists of three layers: the epithelium, underlying connective tissue, and thin muscle layer (muscularis mucosae).

1. (a) Name the three regions of the small intestine: _____

   (b) Identify a functional difference between these regions: _____

   _____

2. What is the purpose of the intestinal villi? _____

   _____

3. Where are enzymes found in the small intestine? _____

   _____

4. In general do the pancreatic enzymes act in acidic or alkaline conditions? _____

© 2012-2014 **BIOZONE** International
ISBN: 978-1-927173-93-0
Photocopying Prohibited

164 The Large Intestine, Rectum, and Anus

Key Idea: The large intestine absorbs water and solidifies the indigestible material before passing it to the rectum. Undigested waste are egested as feces from the anus.
After most of the nutrients have been absorbed in the small intestine, the remaining semi-fluid contents pass into the large intestine (appendix, cecum, colon, and rectum). The large intestine's main role is to reabsorb water and electrolytes, and to consolidate the waste material into feces, which are collected in the rectum before being expelled from the anus (a process called egestion).

▶ After most of the nutrients have been absorbed in the small intestine, the remaining semi-fluid contents pass into the large intestine (appendix, cecum, colon, and rectum). This mixture includes undigested or indigestible food, (such as **cellulose**), bacteria, dead cells, mucus, bile, ions, and water. In humans and other omnivores, the large intestine's main role is to reabsorb water and electrolytes and to consolidate the undigested material into feces. This consolidated material is then eliminated via the anus.

▶ The rectum stores the waste fecal material before it is discharged out the anus. Fullness in the rectum produces the urge to defecate. If too little water is absorbed, the feces will be watery as in diarrhea. If too much water is absorbed the feces will become compacted and difficult to pass.

▶ Defecation is controlled by the anal sphincters, whose usual state is to be contracted (closing the orifice). Defecation is under nervous control.

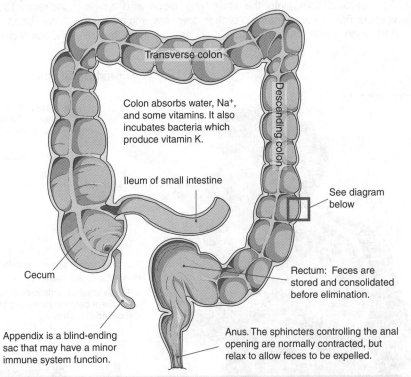

Transverse colon

Descending colon

Colon absorbs water, Na+, and some vitamins. It also incubates bacteria which produce vitamin K.

Ileum of small intestine

See diagram below

Cecum

Rectum: Feces are stored and consolidated before elimination.

Appendix is a blind-ending sac that may have a minor immune system function.

Anus. The sphincters controlling the anal opening are normally contracted, but relax to allow feces to be expelled.

Lining of the Large Intestine

The lining of the large intestine has a simple epithelium containing tubular glands (crypts) with many mucus-secreting cells. The mucus lubricates the colon wall and helps to form and move the feces. In the photograph, some of the crypts are in XS and some are in LS.

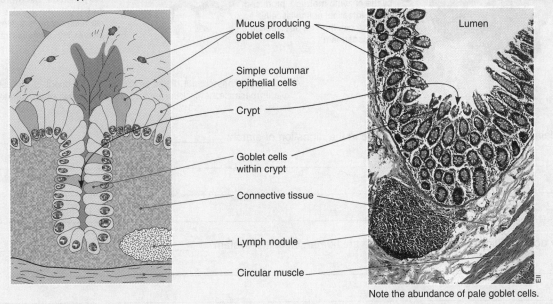

Mucus producing goblet cells

Lumen

Simple columnar epithelial cells

Crypt

Goblet cells within crypt

Connective tissue

Lymph nodule

Circular muscle

Note the abundance of pale goblet cells.

1. What is the main purpose of the large intestine? _____

_____

2. What are the effects of absorbing too little and too much water in the large intestine? _____

_____

165 Digestion, Absorption, and Transport

Key Idea: Food must be digested into smaller components that can be absorbed by the body's cells and assimilated. Nutrient absorption involves both active and passive transport. Digestion breaks down food molecules into forms (simple sugars, amino acids, and fatty acids) that can pass through the intestinal lining into the underlying blood and lymph vessels. For example, starch is broken down first into maltose and short chain carbohydrates such as dextrose, before being hydrolysed to glucose (below). Breakdown products of other foodstuffs include amino acids (from proteins), and fatty acids, glycerol, and acylglycerols (from fats). The passage of these molecules from the gut into the blood or lymph is called absorption. Nutrients are then transported directly or indirectly to the liver for storage or processing. After they have been **absorbed** nutrients can be **assimilated**, i.e incorporated into the substance of the body itself.

Digestion of Starch

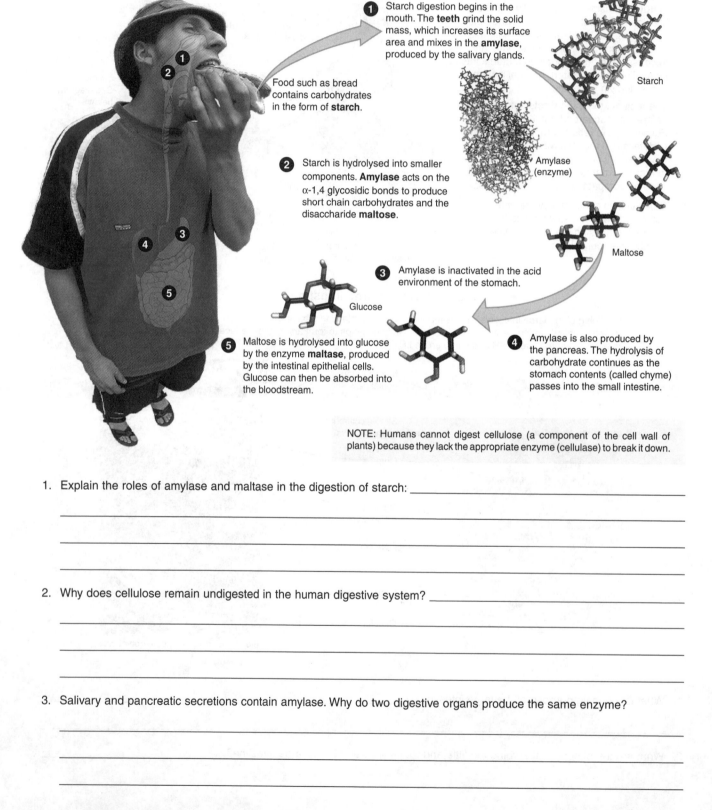

1 Starch digestion begins in the mouth. The **teeth** grind the solid mass, which increases its surface area and mixes in the **amylase**, produced by the salivary glands.

Food such as bread contains carbohydrates in the form of **starch**.

Starch

Amylase (enzyme)

2 Starch is hydrolysed into smaller components. **Amylase** acts on the α-1,4 glycosidic bonds to produce short chain carbohydrates and the disaccharide **maltose**.

Maltose

3 Amylase is inactivated in the acid environment of the stomach.

Glucose

5 Maltose is hydrolysed into glucose by the enzyme **maltase**, produced by the intestinal epithelial cells. Glucose can then be absorbed into the bloodstream.

4 Amylase is also produced by the pancreas. The hydrolysis of carbohydrate continues as the stomach contents (called chyme) passes into the small intestine.

NOTE: Humans cannot digest cellulose (a component of the cell wall of plants) because they lack the appropriate enzyme (cellulase) to break it down.

1. Explain the roles of amylase and maltase in the digestion of starch: _____

_____

_____

_____

2. Why does cellulose remain undigested in the human digestive system? _____

_____

_____

_____

3. Salivary and pancreatic secretions contain amylase. Why do two digestive organs produce the same enzyme?

_____

_____

_____

4. Describe how each of the following nutrients are absorbed by the intestinal villi:

(a) Glucose and galactose:

_____

(b) Fructose:

_____

(c) Amino acids:

_____

(d) Dipeptides:

_____

(e) Tripeptides:

_____

(f) Short chain fatty acids:

_____

(g) Monoglycerides:

_____

(h) Fat soluble vitamins:

_____

Nutrient Absorption by Intestinal Villi

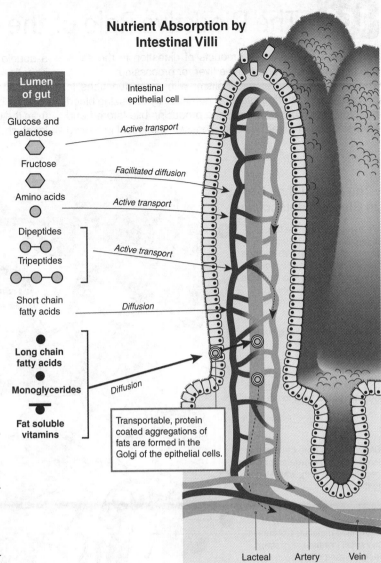

Cross section through a villus, showing how the products of digestion are absorbed across the intestinal epithelium into the capillaries or into the lacteals of the lymphatic system. The nutrients are delivered to the liver.

5. How are concentration gradients maintained for the absorption of nutrients by diffusion:

_____

_____

_____

_____

6. The experiment on the right demonstrates why food must be digested before it can be absorbed.

(a) Why was there no starch in the distilled water?

_____

_____

_____

_____

(b) Use this model to explain why food must be digested before it can be absorbed.

_____

_____

_____

_____

_____

Modelling Nutrient Absorption

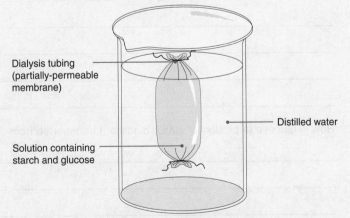

To model absorption of nutrients in the small intestine, a dialysis tubing was filled with a solution of starch and glucose. The outside of the tubing was washed with distilled water to remove any starch or glucose that spilled on to the outer surface during filling. The tubing was placed into a beaker of distilled water. After one hour, the distilled water was tested for the presence of starch or glucose. Only glucose was present in the distilled water.

NOTE: Dialysis tubing comes in many different pore sizes. It only allows molecules smaller than the size of the pore to pass through.

166 The Digestive Role of the Liver

Key Idea: The absorbed products of digestion in the small intestine are transported to the liver for processing.

The liver is central to metabolism, with many functions in processing and storing essential nutrients, synthesizing blood proteins, metabolizing toxins, and producing bile (stored and released from the gall bladder). The liver is unusual in having a double blood supply. It receives oxygen-rich blood via the hepatic arteries, but also receives venous blood from the gastrointestinal tract via the hepatic portal system. The portal venous blood contains the products of digestion absorbed from the gut, which are processed in the liver before the blood is returned (via the hepatic veins) into the general circulation.

The Role of the Liver in Glucose Metabolism

Excess glucose in the blood is removed by the liver and stored as glycogen. When blood glucose levels become too low it is released back into the blood as glucose (see *Glucose and Glycogen in the Liver* right).

The liver monitors the glucose content of the blood.

Digested nutrient molecules (e.g. glucose) travel through the **hepatic portal vein** to the liver.

Blood vessels from the small and large intestine merge to form the hepatic portal vein, which leads to the liver.

Digestion and absorption of many food molecules occurs in the small intestine, e.g. starch is hydrolysed into glucose and absorbed across the intestinal epithelium.

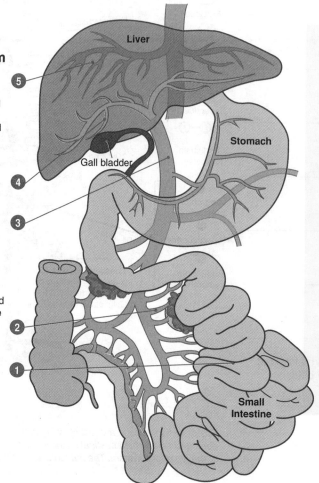

Glucose and Glycogen in the Liver

▶ **Glycogenesis**
Excess glucose in the blood is converted to **glycogen** by glycogenesis in response to high blood glucose. Glycogen is stored in the liver and muscle tissue.

▶ **Glycogenolysis**
Conversion of stored glycogen to glucose (glycogen breakdown). The free glucose is released into the blood in response to low blood glucose.

▶ **Gluconeogenesis**
Production of glucose from non-carbohydrate sources (e.g. glycerol, pyruvate, lactate, and amino acids). Occurs in response to fasting, starvation, or prolonged periods of exercise when glycogen stores are exhausted. It is also part of the general adaptation syndrome in response to stress.

1. What role does the liver play in digestion? _____

_____

_____

_____

_____

2. How is glucose (a product of starch digestion) transported from the small intestine to the liver? _____

_____

_____

_____

3. What role does the liver play in maintaining glucose balance in the blood? _____

_____

_____

_____

_____

© 2012-2014 **BIOZONE** International
ISBN: 978-1-927173-93-0
Photocopying Prohibited

167 Summary of the Human Digestive Tract

Key Idea: The human digestive tract has specialized regions, which each perform a certain task.

This activity will help to consolidate your knowledge of the structure and function of the human digestive system.

1. In the spaces provided on the diagram below, identify the parts labeled **A-L** (choose from the word list provided). Match each of the **functions** described (a)-(g) with the letter representing the corresponding structure on the diagram.

Feeding and satiety centers in the hypothalamus regulate eating

Structures of the Human Gut

Word list: *Liver, small intestine, gall bladder, stomach, salivary glands, colon (large intestine), esophagus, pancreas, mouth and teeth, anus, rectum, appendix.*

A	G
B	H
C	I
D	J
E	K
F	L

The Functions of Gut Structures

In the boxes provided, write the letter (A-L) that represents the part of the gut responsible for each of the functions summarized below:

(a) Main region for enzymatic digestion & nutrient absorption

(b) Consolidation of the feces before elimination

(c) Main function (humans) is water and mineral absorption

(d) Secretes acid and pepsin, stores and mixes food

(e) A gland which produces an alkaline, enzyme-rich fluid

(f) Produces bile and has many homeostatic functions

(g) Produces saliva which contains the enzyme amylase

2. The micrograph on the right is a section through the small intestine (the duodenum). Label the regions (A-D) on the diagram using the following word list to help you: *Epithelium, circular muscle, longitudinal muscle, mucosa.*

(a) _____

(b) _____

(c) _____

(d) _____

3. Label the lumen on the photograph.

4. Why do the two layers of smooth muscle have different appearance in the cross section?

_____

_____

_____

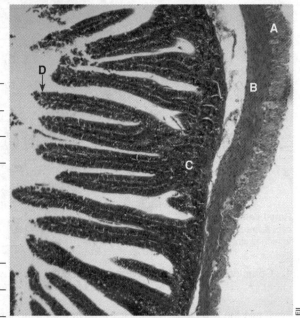

LINK 164 LINK 163 LINK 162

SKILL

168 The Circulatory System

Key Idea: The human circulatory system is an efficient, double circuit system comprising a pulmonary circuit (heart to lungs) and a systemic circuit (heart to body).

The blood vessels of the circulatory system form a network that carries blood away from the heart, transports it to the body's tissues, and then returns it to the heart. The vessels are organized into specific routes to circulate the blood throughout the body. Humans have a double circulatory system. The **pulmonary circuit** carries blood between the heart and lungs. The **systemic circuit** carries blood between the heart and the rest of the body. **William Harvey**, an English physician, dissected animals to study their circulatory systems. He was the first (in 1628) to accurately describe how the heart pumped blood around the body.

Schematic Overview of the Human Circulatory System

Deoxygenated blood (coloured blue below) travels to the right side of the heart via the vena cavae. The heart pumps the deoxygenated blood to the lungs where it releases carbon dioxide and receives oxygen. The oxygenated blood (coloured white below) travels via the pulmonary vein back to the heart from where it is pumped to all parts of the body. The **venous system** (figure, left) returns blood from the capillaries to the heart. The **arterial system** (figure right) carries blood from the heart to the capillaries. **Portal systems** carry blood between two capillary beds.

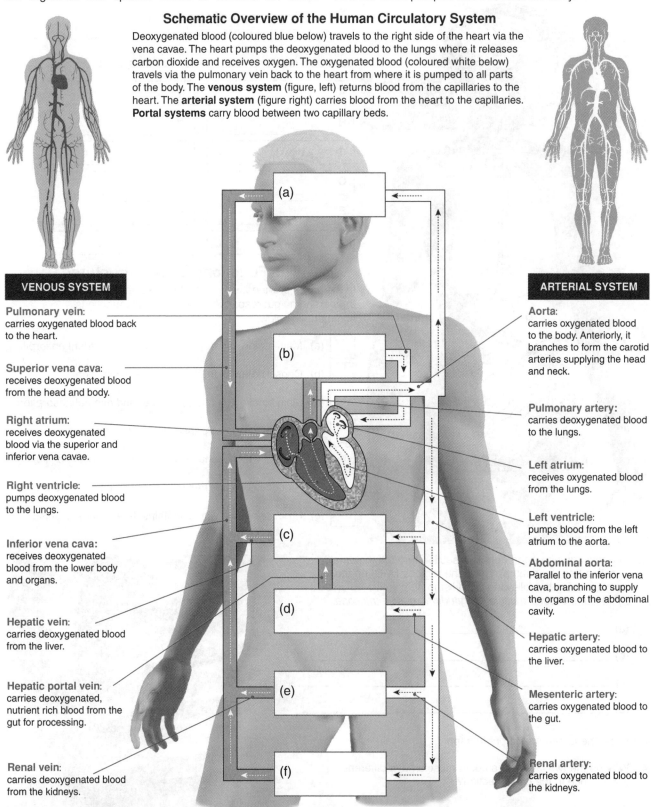

VENOUS SYSTEM

Pulmonary vein: carries oxygenated blood back to the heart.

Superior vena cava: receives deoxygenated blood from the head and body.

Right atrium: receives deoxygenated blood via the superior and inferior vena cavae.

Right ventricle: pumps deoxygenated blood to the lungs.

Inferior vena cava: receives deoxygenated blood from the lower body and organs.

Hepatic vein: carries deoxygenated blood from the liver.

Hepatic portal vein: carries deoxygenated, nutrient rich blood from the gut for processing.

Renal vein: carries deoxygenated blood from the kidneys.

ARTERIAL SYSTEM

Aorta: carries oxygenated blood to the body. Anteriorly, it branches to form the carotid arteries supplying the head and neck.

Pulmonary artery: carries deoxygenated blood to the lungs.

Left atrium: receives oxygenated blood from the lungs.

Left ventricle: pumps blood from the left atrium to the aorta.

Abdominal aorta: Parallel to the inferior vena cava, branching to supply the organs of the abdominal cavity.

Hepatic artery: carries oxygenated blood to the liver.

Mesenteric artery: carries oxygenated blood to the gut.

Renal artery: carries oxygenated blood to the kidneys.

1. Complete the diagram above by labelling the boxes with the correct organs:
   lungs, liver, head, intestines, genitals/lower body, kidneys.

2. Circle the two blood vessels involved in the pulmonary circuit.

© 2012-2014 **BIOZONE** International
ISBN: 978-1-927173-93-0
Photocopying Prohibited

169 Blood Vessels

Key Idea: Arteries transport blood away from the heart and veins transport blood back to the heart. Capillaries allow the exchange of material between the blood and the tissues. The blood vessels of the circulatory system connect the body's cells to the organs that exchange gases, absorb nutrients, and dispose of wastes. The structural differences between blood vessels are related to their functional roles.

Capillaries, whose functional role is exchange, consist only of a thin endothelium. They are small blood vessels with a diameter of 4-10 μm. Red blood cells (7-8 μm) can only just squeeze through. Blood flow is very slow through the capillaries (less than 1 mm per second) allowing the exchange of nutrients and wastes between the blood and tissues. Capillaries form large networks, especially in tissues and organs with high metabolic rates.

Thin endothelium (one cell thick)

CO₂ and wastes move out of the cells into the blood

Red blood cell

O₂ and nutrients move into the cells of the tissues

Capillaries

Blood flow

Relatively thin outer layer of elastic and connective tissue anchors artery to surrounding structures.

A thick layer of elastic tissue and smooth muscle allows the artery to stretch and contract so that blood flow is smoother and can be regulated.

Artery

Thin endothelium

Lumen (space inside vessel)

Thin endothelium

Blood flow

Vein

One-way valves prevent blood flowing in the wrong direction.

Relatively thick outer layer of connective tissue

Thin central layer of elastic and muscle tissue

Arteries carry blood away from the heart to the capillaries within the tissues. Arteries have a structure that enables them to withstand and maintain the high pressure of the blood coming from the heart.

Veins collect the blood from the capillaries and return it to the heart. Veins are similar in structure to arteries, but have less elastic and muscle tissue. Although veins are less elastic than arteries, they can still expand enough to adapt to changes in the pressure and volume of the blood passing through them.

1. State the function of the following:

    (a) The arteries: _____

    (b) The capillaries: _____

    (c) The veins: _____

2. Describe the role of the valves in assisting the veins to return blood back to the heart: _____
    _____
    _____

3. Identify the blood vessels labelled **A** and **B** on the photo (right). Give reasons for your answer:

    A: _____

    _____

    B: _____

    _____

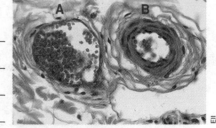

LINK 171   LINK 170   WEB 169   **SKILL**

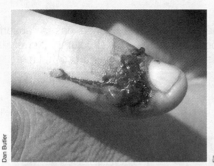

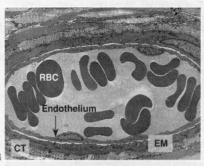

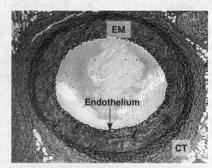

If a vein is cut, as in the severe finger wound shown above, the blood oozes out slowly in an even flow, and usually clots quickly as it leaves. In contrast, if a cut is made into an artery, the arterial blood spurts rapidly and requires pressure to staunch the flow.

This TEM shows the structure of a typical vein. Note the red blood cells (RBC) in the lumen of the vessel, the inner layer of of epithelial cells (the endothelium), the central layer of elastic and muscle tissue (EM), and the outer connective tissue (CT) layer.

Arteries have a thick central layer of elastic and smooth muscle tissue (EM). Near the heart, arteries have more elastic tissue. This enables them to withstand high blood pressure. Arteries further from the heart have more smooth muscle; this helps them to maintain blood pressure.

4. Describe the contrasting structure of veins and arteries for each of the following properties:

   (a) Thickness of muscle and elastic tissue: _____

   _____

   _____

   (b) Size of the lumen (inside of the vessel): _____

   _____

5. Explain the reasons for the differences you have described above: _____

   _____

   _____

   _____

6. (a) Describe the structure of capillaries, explaining how it differs from that of veins and arteries: _____

   _____

   _____

   _____

   (b) What are the reasons for these differences? _____

   _____

   _____

   _____

7. Explain the structural and functional reasons for the slow blood flow through capillaries: _____

   _____

   _____

8. Explain why blood oozes from a venous wound, rather than spurting as it does from an arterial wound: _____

   _____

   _____

9. Why do capillaries form dense networks in tissues with a high metabolic rate?_____

   _____

   _____

© 2012-2014 **BIOZONE** International
ISBN: 978-1-927173-93-0

170 Capillary Networks

Key Idea: Capillaries form branching networks where exchanges between the blood and tissues take place.

The flow of blood through a capillary bed is called microcirculation. In most parts of the body, there are two types of vessels in a capillary bed: the true capillaries, where exchanges take place, and a vessel called a vascular shunt, which connects the arteriole and venule at either end of the bed. The shunt diverts blood past the true capillaries when the metabolic demands of the tissue are low. When tissue activity increases, the entire network fills with blood.

1. Describe the structure of a capillary network:

_____

_____

_____

_____

_____

_____

2. Explain the role of the smooth muscle sphincters and the vascular shunt in a capillary network:

_____

_____

_____

_____

_____

_____

3. (a) Describe a situation where the capillary bed would be in the condition labelled A:

_____

_____

_____

_____

(b) Describe a situation where the capillary bed would be in the condition labelled B:

_____

_____

_____

_____

4. How does a portal venous system differ from other capillary systems?

_____

_____

_____

_____

_____

_____

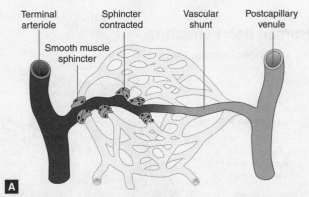

A

When the sphincters contract (close), blood is diverted via the vascular shunt to the postcapillary venule, bypassing the exchange capillaries.

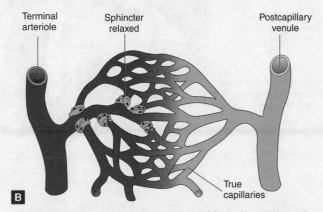

B

When the sphincters are relaxed (open), blood flows through the entire capillary bed allowing exchanges with the cells of the surrounding tissue.

Connecting Capillary Beds
The role of portal venous systems

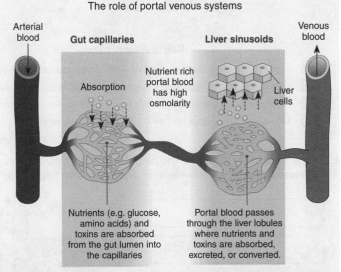

A portal venous system occurs when a capillary bed drains into another capillary bed through veins, without first going through the heart. Portal systems are relatively uncommon. Most capillary beds drain into veins which then drain into the heart, not into another capillary bed. The diagram above depicts the hepatic portal system, which includes both capillary beds and the blood vessels connecting them.

171 The Heart

Key Idea: Humans have a four chambered heart divided into left and right halves. It acts as a double pump.

The heart is the centre of the human cardiovascular system. It is a hollow, muscular organ made up of four chambers (two **atria** and two **ventricles**) that alternately fill and empty of blood, acting as a double pump. The left side (systemic circuit) pumps blood to the body tissues and the right side (pulmonary circuit) pumps blood to the lungs. The heart lies between the lungs, to the left of the midline, and is surrounded by a double layered pericardium of connective tissue, which prevents over distension of the heart and anchors it within the central compartment of the thoracic cavity.

Human heart structure

(sectioned, anterior view)

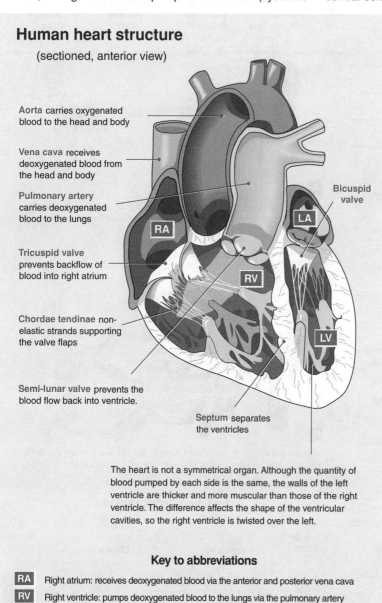

Aorta carries oxygenated blood to the head and body

Vena cava receives deoxygenated blood from the head and body

Pulmonary artery carries deoxygenated blood to the lungs

Tricuspid valve prevents backflow of blood into right atrium

Chordae tendinae non-elastic strands supporting the valve flaps

Semi-lunar valve prevents the blood flow back into ventricle.

Bicuspid valve

RA

LA

RV

LV

Septum separates the ventricles

The heart is not a symmetrical organ. Although the quantity of blood pumped by each side is the same, the walls of the left ventricle are thicker and more muscular than those of the right ventricle. The difference affects the shape of the ventricular cavities, so the right ventricle is twisted over the left.

Key to abbreviations

RA Right atrium: receives deoxygenated blood via the anterior and posterior vena cava

RV Right ventricle: pumps deoxygenated blood to the lungs via the pulmonary artery

LA Left atrium: receives blood returning to the heart from the lungs via the pulmonary veins

LV Left ventricle: pumps oxygenated blood to the head and body via the aorta

Top view of a heart in section, showing valves

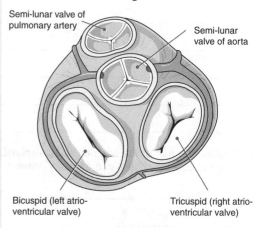

Semi-lunar valve of pulmonary artery

Semi-lunar valve of aorta

Bicuspid (left atrio-ventricular valve)

Tricuspid (right atrio-ventricular valve)

Posterior view of heart

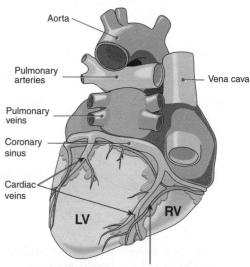

Aorta

Pulmonary arteries

Vena cava

Pulmonary veins

Coronary sinus

Cardiac veins

LV

RV

Coronary arteries: The high oxygen demands of the heart muscle are met by a dense capillary network. Coronary arteries arise from the aorta and spread over the surface of the heart supplying the cardiac muscle with oxygenated blood. Deoxygenated blood is collected by cardiac veins and returned to the right atrium via a large coronary sinus.

1. In the schematic diagram of the heart, below, label the four chambers and the main vessels entering and leaving them. The arrows indicate the direction of blood flow. Use large coloured circles to mark the position of each of the four valves.

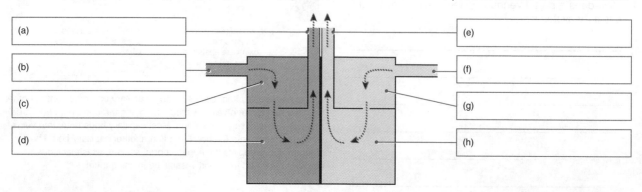

(a)

(b)

(c)

(d)

(e)

(f)

(g)

(h)

© 2012-2014 **BIOZONE** International
ISBN: 978-1-927173-93-0
Photocopying Prohibited

Pressure Changes and the Asymmetry of the Heart

The heart is not a symmetrical organ. The left ventricle and its associated arteries are thicker and more muscular than the corresponding structures on the right side. This asymmetry is related to the necessary pressure differences between the pulmonary (lung) and systemic (body) circulations (not to the distance over which the blood is pumped *per se*). The graph below shows changes in blood pressure in each of the major blood vessel types in the systemic and pulmonary circuits (the horizontal distance not to scale). The pulmonary circuit must operate at a much lower pressure than the systemic circuit to prevent fluid from accumulating in the alveoli of the lungs. The left side of the heart must develop enough "spare" pressure to enable increased blood flow to the muscles of the body and maintain kidney filtration rates without decreasing the blood supply to the brain.

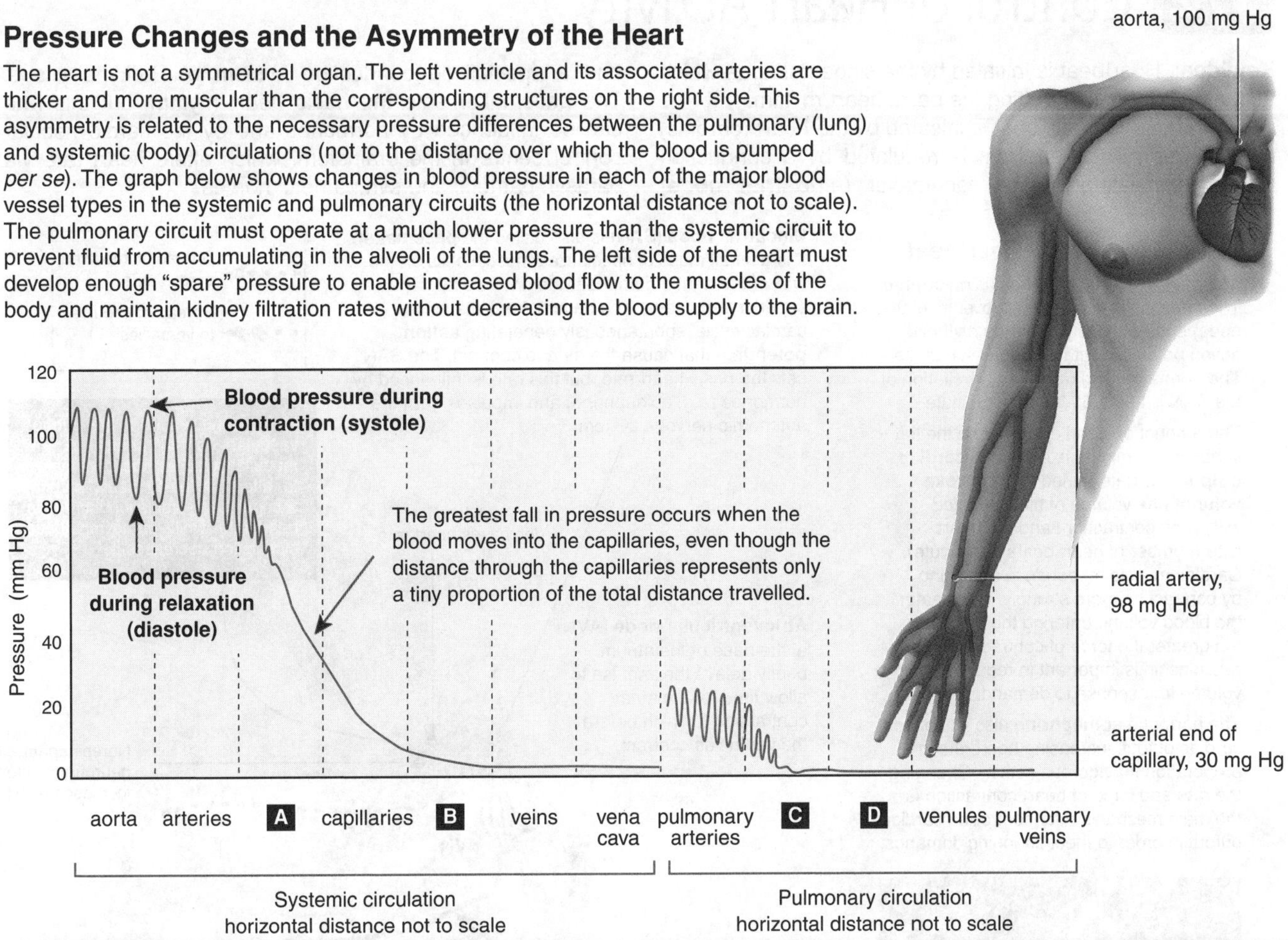
aorta, 100 mg Hg
radial artery, 98 mg Hg
arterial end of capillary, 30 mg Hg
120
100
80
60
40
20
0
Pressure (mm Hg)
Blood pressure during contraction (systole)
Blood pressure during relaxation (diastole)
The greatest fall in pressure occurs when the blood moves into the capillaries, even though the distance through the capillaries represents only a tiny proportion of the total distance travelled.
aorta
arteries
A
capillaries
B
veins
vena cava
pulmonary arteries
C
D
venules
pulmonary veins
Systemic circulation
horizontal distance not to scale
Pulmonary circulation
horizontal distance not to scale

2. What is the purpose of the valves in the heart? ______________________

____

3. The heart is full of blood, yet it requires its own blood supply. Suggest two reasons why this is the case:

(a) ______________________

(b) ______________________

4. Predict the effect on the heart if blood flow through a coronary artery is restricted or blocked: ______________________

____

____

5. Identify the vessels corresponding to the letters **A-D** on the graph above:

A: ____________ B: ____________ C: ____________ D: ____________

6. (a) Why must the pulmonary circuit operate at a lower pressure than the systemic system? ______________________

____

____

(b) Relate this to differences in the thickness of the wall of the left and right ventricles of the heart: ______________________

____

____

7. What are you recording when you take a pulse? ______________________

____

ISBN: 978-1-927173-93-0

172 Control of Heart Activity

Key Idea: Heartbeat is initiated by the sinoatrial node which acts as a pacemaker setting the basic heart rhythm.

The origin of the heart-beat is initiated by the heart (cardiac) muscle itself. The heartbeat is regulated by a conduction system consisting of the pacemaker (**sinoatrial node**) and a specialized conduction system of Purkyne tissue. The pacemaker sets the basic heart rhythm, but this rate can be influenced by hormones and by the cardiovascular control centre in the brainstem, which alters heart rate via parasympathetic and sympathetic nerves.

Generation of the Heartbeat

The basic rhythmic heartbeat is **myogenic**. The nodal cells (SAN and atrioventricular node) spontaneously generate rhythmic action potentials without neural stimulation. The normal resting rate of self-excitation of the SAN is about 50 beats per minute.

The amount of blood ejected from the left ventricle per minute is called the **cardiac output**. It is determined by the **stroke volume** (the volume of blood ejected with each contraction) and the **heart rate** (number of heart beats per minute). Cardiac muscle responds to stretching by contracting more strongly. The greater the blood volume entering the ventricle, the greater the force of contraction. This relationship is important in regulating stroke volume in response to demand.

The hormone **epinephrine** also influences cardiac output, increasing heart rate in preparation for vigorous activity. Changing the rate and force of heart contraction is the main mechanism for controlling cardiac output in order to meet changing demands.

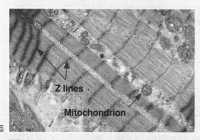

TEM of cardiac muscle showing striations in a fibre (muscle cell). The Z lines that delineate the contractile units of the rod-like units of the fibre. The fibres are joined by specialized electrical junctions called Intercalated discs, which allow impulses to spread rapidly through the heart muscle.

Sinoatrial node (SAN) is also called the **pacemaker**. It is a small mass of specialized muscle cells on the wall of the right atrium, near the entry point of the superior vena cava. The pacemaker initiates the cardiac cycle, spontaneously generating **action potentials** that cause the atria to contract. The SAN sets the basic heart rate, but this rate is influenced by hormones (e.g. epinephrine) and impulses from the autonomic nervous system.

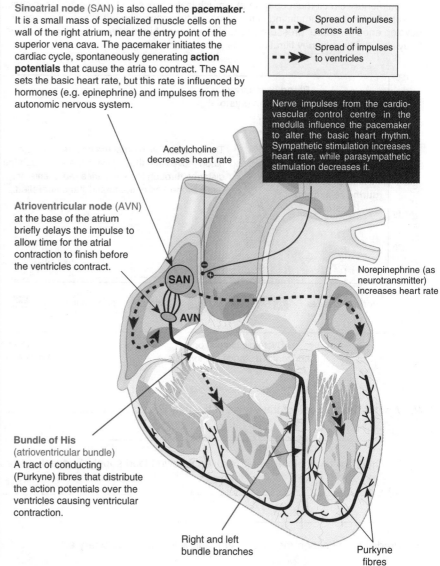

- - - ➤ Spread of impulses across atria

- - - ➤ Spread of impulses to ventricles

Nerve impulses from the cardio-vascular control centre in the medulla influence the pacemaker to alter the basic heart rhythm. Sympathetic stimulation increases heart rate, while parasympathetic stimulation decreases it.

Acetylcholine decreases heart rate

Atrioventricular node (AVN) at the base of the atrium briefly delays the impulse to allow time for the atrial contraction to finish before the ventricles contract.

Norepinephrine (as neurotransmitter) increases heart rate

SAN

AVN

Bundle of His (atrioventricular bundle) A tract of conducting (Purkyne) fibres that distribute the action potentials over the ventricles causing ventricular contraction.

Right and left bundle branches

Purkyne fibres

1. Describe the role of each of the following in heart activity:

   (a) The sinoatrial node: _____

   (b) The atrioventricular node: _____

   (c) The bundle of His: _____

   (d) Intercalated discs: _____

2. What is the significance of delaying the impulse at the AVN? _____

   _____

3. The heart-beat is intrinsic. Why is it important to be able to influence this basic rhythm via nerves and hormones?

   _____

   _____

4. (a) What is the effect of the hormone epinephrine on heart rate? _____

   (b) What sympathetic neurotransmitter has the same effect? _____

© 2012-2014 **BIOZONE** International
ISBN: 978-1-927173-93-0
Photocopying Prohibited

173 The Cardiac Cycle

Key Idea: The cardiac (heart) cycle refers to the sequence of events of a heartbeat and involves three main stages: atrial systole, ventricular systole, and complete cardiac diastole. The heart pumps with alternate contractions (**systole**) and relaxations (**diastole**). Heartbeat occurs in a cycle involving three stages: atrial systole, ventricular systole, and complete cardiac diastole. Pressure changes within the heart's chambers generated by the cycle of contraction and relaxation are responsible for blood movement and cause the heart valves to open and close, preventing the backflow of blood. The heartbeat occurs in response to electrical impulses, which can be recorded as a trace called an electrocardiogram or ECG (bottom).

The Cardiac Cycle

The **pulse** results from the rhythmic expansion of the arteries as the blood spurts from the left ventricle. Pulse rate therefore corresponds to heart rate.

Stage 1: Atrial contraction and ventricular filling
The ventricles relax and blood flows into them from the atria. Note that 70% of the blood from the atria flows passively into the ventricles. It is during the last third of ventricular filling that the atria contract.

Stage 2: Ventricular contraction
The atria relax, the ventricles contract, and blood is pumped from the ventricles into the aorta and the pulmonary artery. The start of ventricular contraction coincides with the first heart sound.

Stage 3: (not shown) There is a short period of atrial and ventricular relaxation. Semilunar valves (**SLV**) close to prevent backflow into the ventricles (see diagram, left). The cycle begins again. For a heart beating at 75 beats per minute, one cardiac cycle lasts about 0.8 seconds.

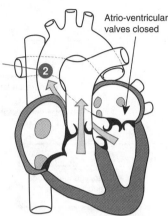

Atrio-ventricular valves closed

Heart during ventricular filling

Heart during ventricular contraction

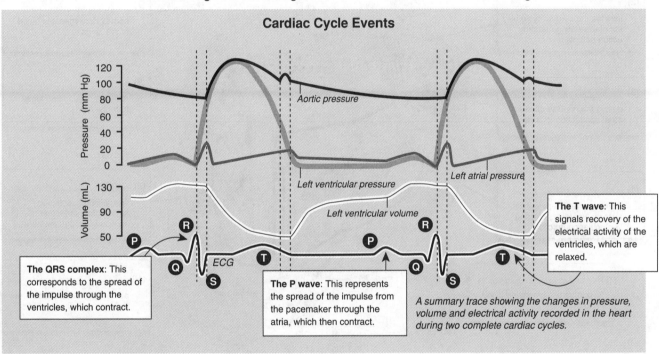

Cardiac Cycle Events

Aortic pressure

Left ventricular pressure

Left atrial pressure

Left ventricular volume

The T wave: This signals recovery of the electrical activity of the ventricles, which are relaxed.

The QRS complex: This corresponds to the spread of the impulse through the ventricles, which contract.

The P wave: This represents the spread of the impulse from the pacemaker through the atria, which then contract.

A summary trace showing the changes in pressure, volume and electrical activity recorded in the heart during two complete cardiac cycles.

1. On the trace above:

   (a) When is the aortic pressure highest? _____

   (b) Which electrical event immediately precedes the increase in ventricular pressure?_____

   (c) What is happening when the pressure of the left ventricle is lowest? _____

2. Suggest the physiological reason for the period of electrical recovery experienced each cycle (the T wave):

   _____

3. Using the letters indicated, mark the points on trace above corresponding to each of the following:

   (a) E: Ejection of blood from the ventricle   (c) FV: Filling of the ventricle

   (b) BVC: Closing of the bicuspid valve   (d) BVO: Opening of the bicuspid valve

LINK
171
WEB
173
APP

174 Review of the Heart

Key Idea: The human heart comprises four chambers, which act as a double pump. Its contraction is myogenic, but can be influenced by other factors.

This activity summarizes key features of the structure and function of the human heart. The necessary information to answer these can be found in earlier activities in this topic.

1. On the diagram below, label the identified components of heart structure and intrinsic control (**1-10**), and some of the components of extrinsic control of heart rate (**A-B**).

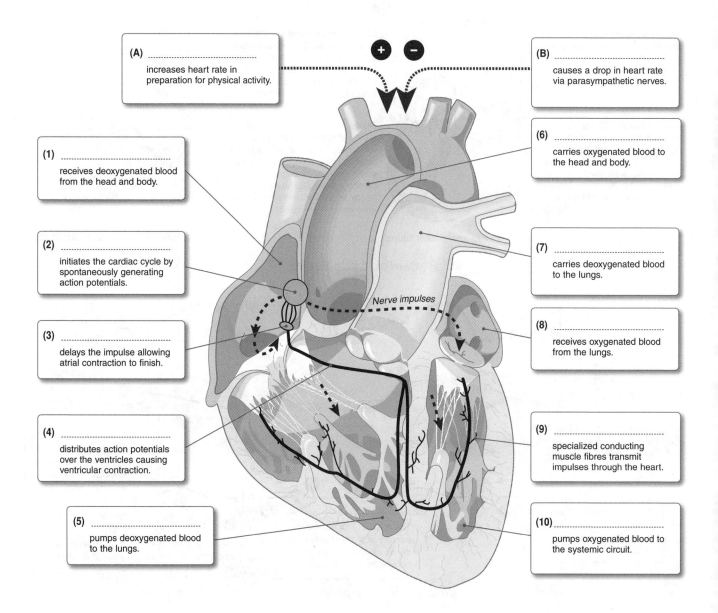

(A) ....
increases heart rate in preparation for physical activity.

(B) ....
causes a drop in heart rate via parasympathetic nerves.

(1) ....
receives deoxygenated blood from the head and body.

(6) ....
carries oxygenated blood to the head and body.

(2) ....
initiates the cardiac cycle by spontaneously generating action potentials.

(7) ....
carries deoxygenated blood to the lungs.

(3) ....
delays the impulse allowing atrial contraction to finish.

(8) ....
receives oxygenated blood from the lungs.

(4) ....
distributes action potentials over the ventricles causing ventricular contraction.

(9) ....
specialized conducting muscle fibres transmit impulses through the heart.

(5) ....
pumps deoxygenated blood to the lungs.

(10) ....
pumps oxygenated blood to the systemic circuit.

Nerve impulses

2. An **ECG** is the result of different impulses produced at each phase of the **cardiac cycle** (the sequence of events in a heartbeat). For each electrical event indicated in the ECG below, describe the corresponding event in the cardiac cycle:

A --
The spread of the impulse from the pacemaker (sinoatrial node) through the atria.

B --
The spread of the impulse through the ventricles.

C --
Recovery of the electrical activity of the ventricles.

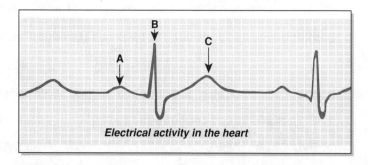

Electrical activity in the heart

© 2012-2014 **BIOZONE** International
ISBN: 978-1-927173-93-0
Photocopying Prohibited

175 Dissecting a Mammalian Heart

Key Idea: Dissecting a sheep's heart allows hands-on exploration of a mammalian heart.

The dissection of a sheep's heart is a common practical activity and allows hands-on exploration of the appearance and structure of a mammalian heart. A diagram of a heart is an idealized representation of an organ that may look quite different in reality. You must learn to transfer what you know from a diagram to the interpretation of the real organ.

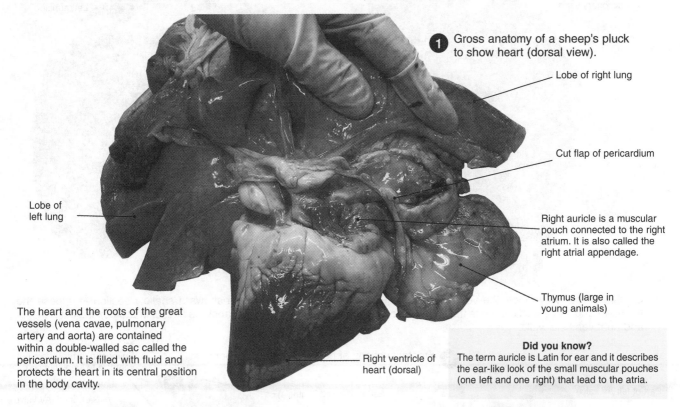

1 Gross anatomy of a sheep's pluck to show heart (dorsal view).

Lobe of right lung

Cut flap of pericardium

Right auricle is a muscular pouch connected to the right atrium. It is also called the right atrial appendage.

Thymus (large in young animals)

Lobe of left lung

The heart and the roots of the great vessels (vena cavae, pulmonary artery and aorta) are contained within a double-walled sac called the pericardium. It is filled with fluid and protects the heart in its central position in the body cavity.

Right ventricle of heart (dorsal)

Did you know?
The term auricle is Latin for ear and it describes the ear-like look of the small muscular pouches (one left and one right) that lead to the atria.

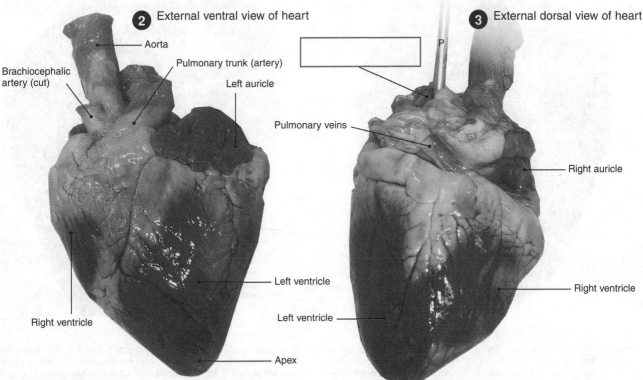

2 External ventral view of heart

Aorta

Brachiocephalic artery (cut)

Pulmonary trunk (artery)

Left auricle

Left ventricle

Right ventricle

Left ventricle

Apex

3 External dorsal view of heart

P

Pulmonary veins

Right auricle

Right ventricle

Note the main surface features of an isolated heart. The narrow pointed end forms the **apex** of the heart, while the wider end, where the blood vessels enter is the **base**. The ventral surface of the heart (above) is identified by a groove, the **interventricular sulcus**, which marks the division between the left and right ventricles.

1. Use coloured lines to indicate the interventricular sulcus and the base of the heart. Label the coronary arteries.

On the dorsal surface of the heart, above, locate the large thin-walled **vena cavae** and **pulmonary veins**. You may be able to distinguish between the anterior and posterior vessels. On the right side of the dorsal surface (as you look at the heart) at the base of the heart is the **right atrium**, with the **right ventricle** below it.

2. On this photograph, label the vessel indicated by the probe (P).

LINK
171

SKILL

4 Dorsal view of heart

5 Shallow section, ventral view of heart

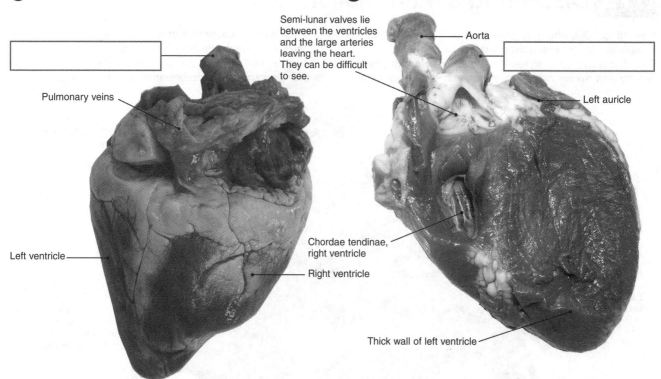

Semi-lunar valves lie between the ventricles and the large arteries leaving the heart. They can be difficult to see.

Pulmonary veins

Aorta

Left auricle

Left ventricle

Chordae tendinae, right ventricle

Right ventricle

Thick wall of left ventricle

3. On this **dorsal view**, label the vessel indicated. Palpate the heart and feel the difference in the thickness of the left and right ventricle walls.

4. This photograph shows a shallow section to expose the right ventricle. Label the vessel in the box indicated.

6 Frontal sections of heart to show chambers

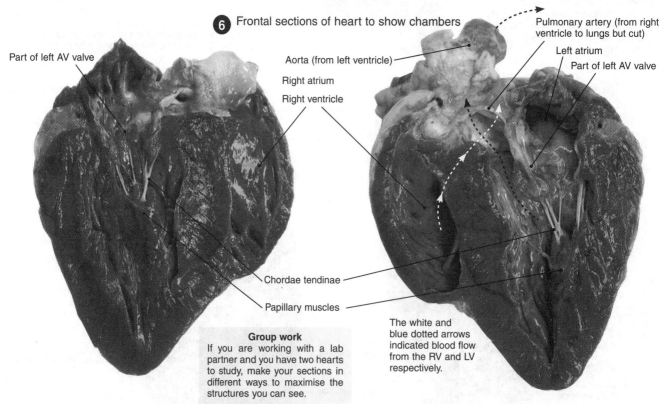

Part of left AV valve

Aorta (from left ventricle)

Right atrium

Right ventricle

Pulmonary artery (from right ventricle to lungs but cut)

Left atrium

Part of left AV valve

Chordae tendinae

Papillary muscles

The white and blue dotted arrows indicated blood flow from the RV and LV respectively.

Group work

If you are working with a lab partner and you have two hearts to study, make your sections in different ways to maximise the structures you can see.

If the heart is sectioned and the two halves opened, the valves of the heart can be seen. Each side of the heart has a one-way valve between the atrium and the ventricle known as the **atrioventricular valve**. They close during ventricular contraction to prevent back flow of the blood into the lower pressure atria.

The atrioventricular (AV) valves of the two sides of the heart are similar in structure except that the right AV valve has three cusps (tricuspid) while the left atrioventricular valve has two cusps (bicuspid or mitral valve). Connective tissue (**chordae tendineae**) run from the cusps to **papillary muscles** on the ventricular wall.

5. Judging by their position and structure, what do you suppose is the function of the chordae tendinae?

_____

_____

6. What feature shown here most clearly distinguishes the left and right ventricles?.

_____

_____

176 Coronary Occlusions

Key Idea: A coronary occlusion is the partial or complete obstruction of blood flow in a coronary artery.

Coronary occlusions are blockages of the coronary arteries (arteries that carry blood to the heart). They may be partial or complete, and may occur suddenly or develop over time.

Occlusions restrict the blood flow to the heart and, without adequate oxygen and nutrients, the heart tissue is damaged and may even die. Atherosclerosis, a condition where the arteries become narrow and hardened as a result of plaque build up in the artery wall, is a common cause of occlusion.

Arteries are lined with endothelium, a thin layer of cells which makes the artery smooth and allows the blood to flow through it easily. If fatty deposits (plaques) form on the artery, the blood flow through the artery is disrupted. The flow of blood to the heart can become limited as the plaque increases in size. An inadequate supply of blood to the heart may result in angina. People suffering angina often feel breathless and have chest pain, as the heart beats harder and faster to meet its oxygen demands.

Risk factors for coronary artery disease

► High blood pressure damages arterial walls.

► High levels of LDL cholesterol contribute to plaque formation because less cholesterol is transported to the liver and more is deposited in the artery walls.

► Smoking damages blood vessels, raises blood pressure, and reduces oxygen availability.

► High blood sugar levels damages blood vessels.

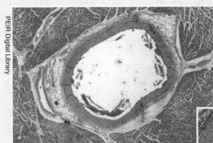

PEIR Digital Library

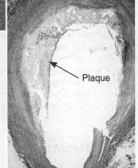

Plaque

Normal unobstructed coronary artery (left), and a coronary artery with moderately severe blockage (below). Note the formation of the plaque in the arterial wall.

Plaques may break off causing a blood clot to form on the plaque's surface. If it is large enough, the blood clot can completely block the artery and cause a **heart attack**. If blood flow is not restored to the heart within 20-40 minutes, the heart muscle begins to die. Heart attacks can be fatal.

Plaque Development and Coronary Occlusions

Increasing size of plaque buildup

Decreasing blood flow

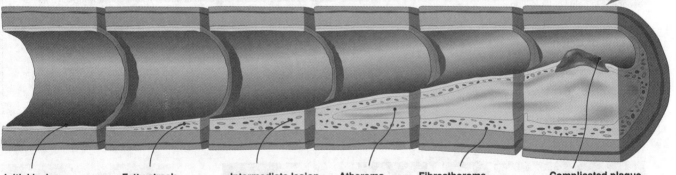

Initial lesion
Blood chemicals or long term high blood pressure damages the artery wall.

Fatty streak
Low density lipoproteins (LDLs) accumulate on the damaged artery wall.

Intermediate lesion
Greasy yellow lesions called atherosclerotic plaques form.

Atheroma
The plaque grows as more lipids become deposited.

Fibroatheroma
Smooth muscle increases and dies. Fibres deteriorate and are replaced with scar tissue.

Complicated plaque
Plaque hardens and may break away, causing blood cells to clot at the damaged site.

1. (a) What is a coronary occlusion? _____

_____

(b) How can a coronary occlusion lead to heart damage and failure? _____

_____

_____

_____

2. List some factors that increase the chances of a coronary occlusion: _____

_____

_____

LINK
46

WEB
176

APP

177 The Body's Defences

Key Idea: The human body has a tiered system of defences against disease-causing organisms.

The body has several lines of defence against disease causing organisms (**pathogens**). The first line of defence consists of external barrier to stop pathogens entering the body. If this fails, a second line of defence targets any foreign bodies (including pathogens) that enter. Lastly, the immune system provides specific or targeted defence against the pathogen. The ability to ward off disease through the various

defence mechanisms is called **resistance**. **Non-specific resistance** protects against a broad range of pathogens and is provided by the first and second lines of defence. **Specific resistance** (the immune response) is the third tier of defence and is specific to a particular pathogen. Part of the immune response involves the production of **antibodies** (proteins that identify and neutralize foreign material). Antibodies recognize and respond to **antigens**, foreign or harmful substances that cause an immune response.

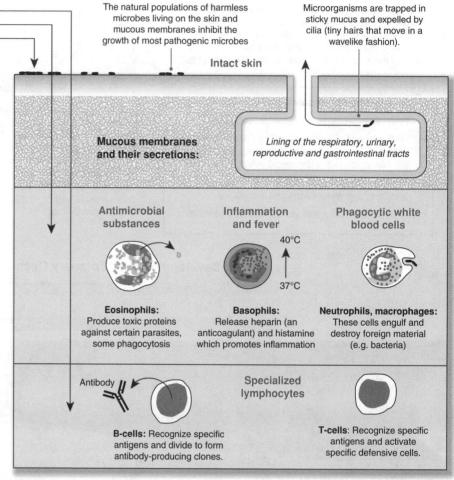

Most microorganisms find it difficult to get inside the body. If they succeed, they face a range of other defences.

The natural populations of harmless microbes living on the skin and mucous membranes inhibit the growth of most pathogenic microbes

Microorganisms are trapped in sticky mucus and expelled by cilia (tiny hairs that move in a wavelike fashion).

Intact skin

Mucous membranes and their secretions:

Lining of the respiratory, urinary, reproductive and gastrointestinal tracts

1st Line of Defence

The **skin** provides a physical barrier to the entry of pathogens. Healthy skin is rarely penetrated by microorganisms. Its low pH is unfavorable to the growth of many bacteria and its chemical secretions (e.g. sebum, antimicrobial peptides) inhibit growth of bacteria and fungi. Tears, mucus, and saliva also help to wash bacteria away.

2nd Line of Defence

A range of defence mechanisms operate inside the body to inhibit or destroy pathogens. These responses react to the presence of any pathogen, regardless of which species it is. White blood cells are involved in most of these responses.

It includes the **complement system** whereby plasma proteins work together to bind pathogens and induce an inflammatory responses to help fight infection.

3rd Line of Defence

Once the pathogen has been identified by the immune system, **lymphocytes** (a type of white blood cell) launch a range of specific responses to the pathogen, including the production of **antibodies**. Each type of antibody is produced by a B-cell clone and is specific against a particular **antigen**.

Antimicrobial substances

Eosinophils: Produce toxic proteins against certain parasites, some phagocytosis

Inflammation and fever

40°C

37°C

Basophils: Release heparin (an anticoagulant) and histamine which promotes inflammation

Phagocytic white blood cells

Neutrophils, macrophages: These cells engulf and destroy foreign material (e.g. bacteria)

Antibody

Specialized lymphocytes

B-cells: Recognize specific antigens and divide to form antibody-producing clones.

T-cells: Recognize specific antigens and activate specific defensive cells.

Tears contain antimicrobial substances as well as washing contaminants from the eyes.

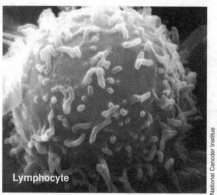

Lymphocyte

Some B-cells (a type of lymphocyte) have **immunological memory**, and quickly react to an antigen the next time it is re-encountered.

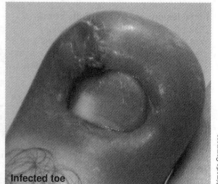

Infected toe

Inflammation is a localized response to infection characterized by swelling, pain, and redness.

1. How does the skin act as a barrier to prevent pathogens entering the body? _____

_____

_____

© 2012-2014 **BIOZONE** International
ISBN: 978-1-927173-93-0
Photocopying Prohibited

The Importance of the First Line of Defence

The skin forms an important physical barrier against the entry of pathogens into the body. A natural population of harmless microbes live on the skin, but most other microbes find the skin inhospitable. The continual shedding of old skin cells (arrow, right) physically removes bacteria from the surface of the skin. Sebaceous glands in the skin (labelled right) produce sebum, which has antimicrobial properties, and the slightly acidic secretions of sweat inhibit microbial growth.

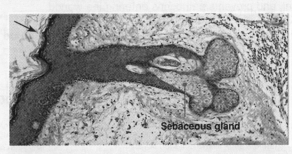

Cilia line the epithelium of the nasal passage (below right). Their wave-like movement sweeps foreign material out and keeps the passage free of microorganisms, preventing them from colonizing the body.

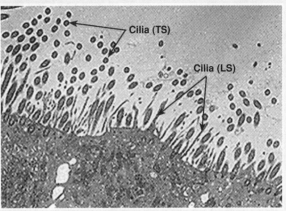

Antimicrobial chemicals are present in many bodily secretions. Tears, saliva, nasal secretions, and human breast milk all contain lysozymes and phospholipases. Lysozymes kill bacterial cells by catalysing the hydrolysis of cell wall linkages, whereas phospholipases hydrolyse the phospholipids in bacterial cell membranes, causing bacterial death. Low pH gastric secretions also inhibit microbial growth, and reduce the number of pathogens establishing colonies in the gastrointestinal tract.

2. Describe the role of each of the following in non-specific defence:

   (a) Phospholipases: _____

   _____

   (b) Cilia: _____

   _____

   (c) Sebum: _____

   _____

3. Distinguish between an **antibody** and an **antigen**: _____

   _____

   _____

4. Describe the functional role of each of the following defence mechanisms:

   (a) Phagocytosis by white blood cells: _____

   _____

   (b) Antimicrobial substances: _____

   _____

   (c) Antibody production: _____

   _____

5. Explain the value of a three tiered system of defence against microbial invasion: _____

   _____

   _____

   _____

© 2012-2014 **BIOZONE** International
ISBN: 978-1-927173-93-0

178 Blood Clotting and Defence

Key Idea: Blood clotting restricts blood loss from a torn blood vessel, and prevents pathogens entering the wound.

Blood has a role in the body's defence against infection. Tearing or puncturing of a blood vessel initiates **blood clotting** through a cascade effect involving platelets, clotting factors, and plasma proteins (below). Clotting quickly seals off the tear, preventing blood loss and the invasion of bacteria into the site. Clot formation is triggered by the release of clotting factors from the damaged cells at the site of the damage. A hardened clot forms a scab, which acts to prevent further blood loss and acts as a mechanical barrier to the entry of pathogens.

Blood clotting

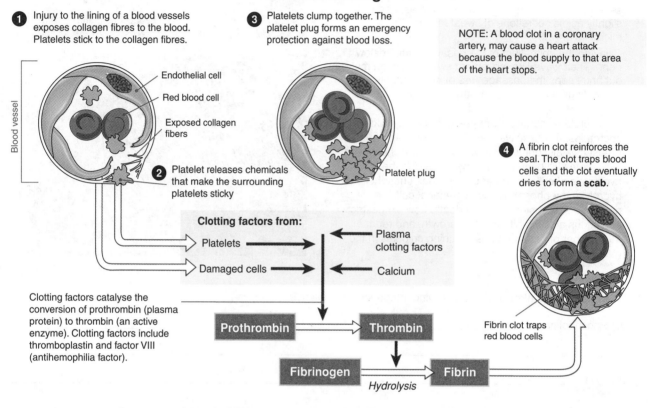

❶ Injury to the lining of a blood vessels exposes collagen fibres to the blood. Platelets stick to the collagen fibres.

❸ Platelets clump together. The platelet plug forms an emergency protection against blood loss.

NOTE: A blood clot in a coronary artery, may cause a heart attack because the blood supply to that area of the heart stops.

Endothelial cell

Red blood cell

Exposed collagen fibers

❷ Platelet releases chemicals that make the surrounding platelets sticky

Platelet plug

❹ A fibrin clot reinforces the seal. The clot traps blood cells and the clot eventually dries to form a **scab**.

Blood vessel

Clotting factors from:

Platelets

Damaged cells

Plasma clotting factors

Calcium

Clotting factors catalyse the conversion of prothrombin (plasma protein) to thrombin (an active enzyme). Clotting factors include thromboplastin and factor VIII (antihemophilia factor).

Fibrin clot traps red blood cells

Prothrombin → Thrombin

Fibrinogen → Fibrin

Hydrolysis

1. What role does blood clotting have in internal defence? _____

_____

2. Explain the role of each of the following in the sequence of events leading to a blood clot:

(a) Injury: _____

_____

(b) Release of chemicals from platelets: _____

_____

(c) Clumping of platelets at the wound site: _____

_____

(d) Formation of a fibrin clot: _____

_____

3. (a) What is the role of clotting factors in the blood in formation of the clot? _____

_____

(b) Why are these clotting factors not normally present in the plasma? _____

_____

© 2012-2014 **BIOZONE** International
ISBN: 978-1-927173-93-0
Photocopying Prohibited

179 The Action of Phagocytes

Key Idea: Phagocytes are types of white blood cells that ingest microbes and digest them by phagocytosis.
Phagocytes are types of white blood cells that provide non-specific resistance by ingesting (phagocytosing) harmful

microbes. During infections, especially bacterial infections, the total number of white blood cells increases up to four times the normal number. The ratio of different types of white blood cell types changes during the course of an infection.

How a Phagocyte Destroys Microbes

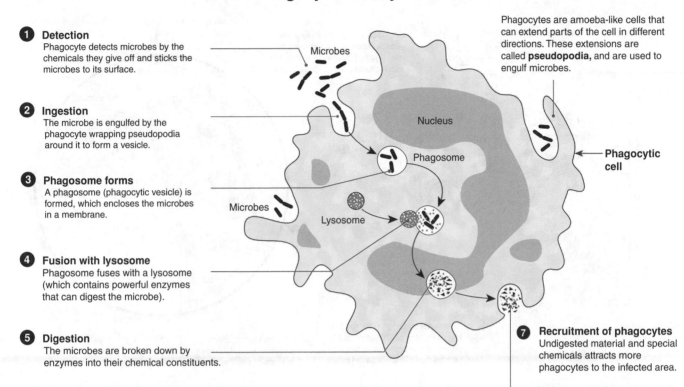

1 Detection
Phagocyte detects microbes by the chemicals they give off and sticks the microbes to its surface.

2 Ingestion
The microbe is engulfed by the phagocyte wrapping pseudopodia around it to form a vesicle.

3 Phagosome forms
A phagosome (phagocytic vesicle) is formed, which encloses the microbes in a membrane.

4 Fusion with lysosome
Phagosome fuses with a lysosome (which contains powerful enzymes that can digest the microbe).

5 Digestion
The microbes are broken down by enzymes into their chemical constituents.

6 Discharge
Indigestible material is discharged from the phagocyte cell.

Phagocytes are amoeba-like cells that can extend parts of the cell in different directions. These extensions are called **pseudopodia**, and are used to engulf microbes.

Microbes

Nucleus

Phagosome

Microbes

Lysosome

Phagocytic cell

7 Recruitment of phagocytes
Undigested material and special chemicals attracts more phagocytes to the infected area.

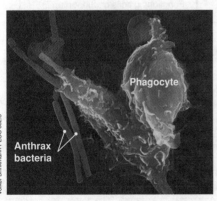

During phagocytosis, foreign bodies (e.g bacteria) are engulfed and destroyed by phagocytes. In the example on the left, anthrax bacteria are being engulfed. The anthrax is bound to receptors on the surface of the phagocyte. The phagocyte stretches around the bacterium, engulfing and killing it.

Some microbes (e.g. the malaria parasite, right) evade the immune system by entering the host's cells. When this occurs in a phagocyte, the microbe prevents fusion of the lysosome with the phagosome. The pathogen multiplies inside the phagocyte, almost filling it.

Volker Brinkmann PLOS cc2.5

Phagocyte

Anthrax bacteria

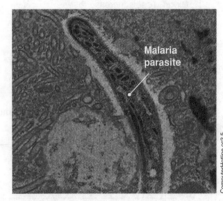

Malaria parasite

ComputerHotline cc2.5

1. Briefly outline the process of phagocytosis: _____

_____

_____

2. Why is phagocytosis classed as a non-specific type of resistance? _____

_____

3. How can the number of white blood cells in a blood sample be an indication of a microbial infection?

_____

_____

LINK 282 LINK 177 WEB 179

KNOW

180 Inflammation

Key Idea: Inflammation is a type of non-specific resistance in response to harmful stimuli, such as pathogens.

Damage to the body's tissues (e.g. by sharp objects, heat, or microbial infection) triggers a defensive response called **inflammation**. It is usually characterized by four symptoms: pain, redness, heat, and swelling. The inflammatory response is beneficial and has the following functions: (1) to destroy the cause of the infection and remove it and its products from the body; (2) if this fails, to limit the effects on the body by confining the infection to a small area; (3) replacing or repairing tissue damaged by the infection. Inflammation can be divided into three distinct phases, shown below.

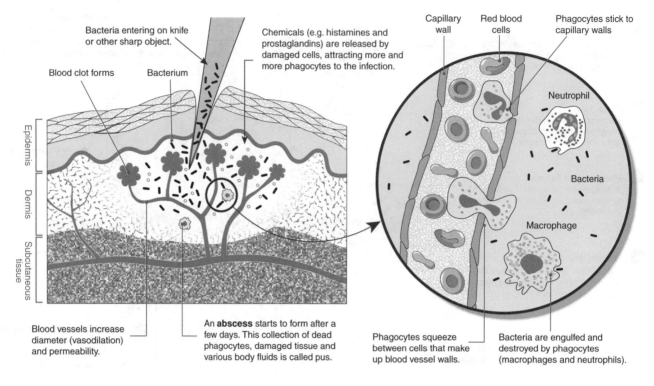

Bacteria entering on knife or other sharp object.

Chemicals (e.g. histamines and prostaglandins) are released by damaged cells, attracting more and more phagocytes to the infection.

Blood clot forms

Bacterium

Epidermis

Dermis

Subcutaneous tissue

Blood vessels increase diameter (vasodilation) and permeability.

An **abscess** starts to form after a few days. This collection of dead phagocytes, damaged tissue and various body fluids is called pus.

Capillary wall

Red blood cells

Phagocytes stick to capillary walls

Neutrophil

Bacteria

Macrophage

Phagocytes squeeze between cells that make up blood vessel walls.

Bacteria are engulfed and destroyed by phagocytes (macrophages and neutrophils).

Stages in inflammation

Increased Diameter and Permeability of Blood Vessels	**Phagocyte Migration and Phagocytosis**	**Tissue Repair**
Blood vessels increase their diameter and permeability in the area of damage. This increases blood flow to the area and allows defensive substances to leak into tissue spaces.	Within one hour of injury, phagocytes appear at the site of the injury. They squeeze between cells of blood vessel walls to reach the damaged area where they destroy invading microbes.	Functioning cells or supporting connective cells create new tissue to replace dead or damaged cells. Some tissue regenerates easily (skin) while others do not at all (cardiac muscle).

1. Outline the three stages of inflammation and identify the beneficial role of each stage:

(a) _____

_____

(b) _____

_____

(c) _____

_____

2. Identify two features of phagocytes important in the response to microbial invasion: _____

_____

3. State the role of histamines and prostaglandins in inflammation: _____

4. Why is inflammation classed as non-specific resistance ? _____

_____

WEB 180   LINK 177

© 2012-2014 **BIOZONE** International
ISBN: 978-1-927173-93-0
Photocopying Prohibited

181 Antibiotics

Key Idea: Antibiotics are chemicals that kill bacteria or inhibit their growth. They are ineffective against viruses.

Antibiotics are chemical substances act against bacterial infections by either killing the bacteria (**bactericidal**) or preventing them from growing (**bacteriostatic**). The cells of eukaryotes are not affected by antibiotics because they have different structures and metabolic pathways to bacterial (prokaryotic) cells. Antibiotics are ineffective against viruses. Antibiotics are produced naturally by bacteria and fungi, but some are produced synthetically (manufactured).

How Antimicrobial Drugs Work

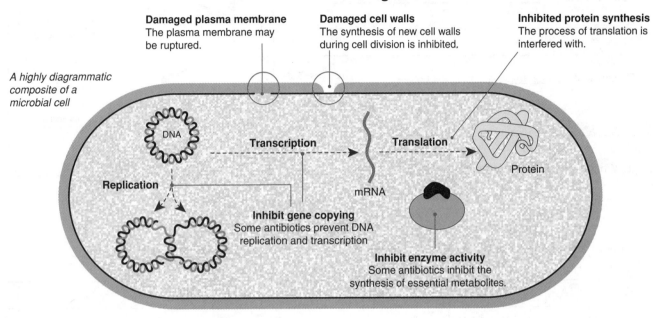

Damaged plasma membrane
The plasma membrane may be ruptured.

Damaged cell walls
The synthesis of new cell walls during cell division is inhibited.

Inhibited protein synthesis
The process of translation is interfered with.

A highly diagrammatic composite of a microbial cell

DNA

Transcription

Translation

Protein

mRNA

Replication

Inhibit gene copying
Some antibiotics prevent DNA replication and transcription

Inhibit enzyme activity
Some antibiotics inhibit the synthesis of essential metabolites.

Bacteria can Become Resistant to Antibiotics

Antibiotic resistance arises when a genetic change allows bacteria to tolerate levels of an antibiotic that would normally adversely affect it. Drug resistance makes it difficult to treat and control some diseases and patients with infections caused by drug-resistant bacteria are more likely to suffer medical complications and death. Some bacteria have resistance to multiple antibiotics. Such bacteria are called superbugs. For example, methicillin resistant strains of *Staphylococcus aureus* (MRSA) have acquired genes for resistance to all penicillins. Superbugs are now widespread. The infections they cause are very difficult to treat, and are more easily spread through the population.

The bacterium *Mycobacterium tuberculosis* causes tuberculosis (TB), and has developed resistance to several drugs. Today, one in seven new TB cases is resistant to the two drugs most commonly used as treatments, and 5% of these patients die.

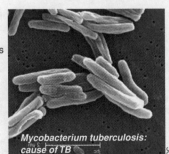

Mycobacterium tuberculosis: cause of TB

Resistant strains of *M. tuberculosis* usually develop because patients have failed to complete their long course of antibiotics (left), or the dose of antibiotics prescribed has been too low.

Application: Florey and Chain's Penicillin Experiments

The antibiotic properties of penicillins, a group of antibiotics produced by the *Penicillium* fungi, were discovered by Alexander Fleming in 1928. However, it wasn't until the 1940s that it was, grown, purified, and tested in significant quantities by a team of scientists including **Howard Florey** and **Ernst Chain**. Florey and Chain experimented on mice (below) to treat streptococcal infections with penicillin. Its success quickly lead to human trials, and penicillin was eventually used with great success to treat World War 2 soldiers suffering from infected wounds. It saved millions of lives and is documented at the first successful antibiotic treatment.

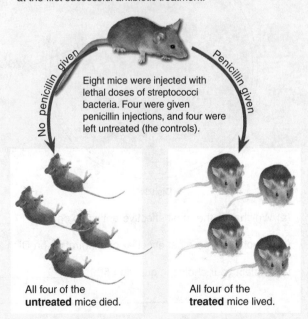

No penicillin given

Penicillin given

Eight mice were injected with lethal doses of streptococci bacteria. Four were given penicillin injections, and four were left untreated (the controls).

All four of the **untreated** mice died.

All four of the **treated** mice lived.

1. Why are eukaryotic cells (such as human cells) not affected by antibiotics? _____

_____

_____

2. The graph right shows the effects of two antibiotics. Identify the antibiotic with a bacteriostatic action and the antibiotic with a bactericidal action. Explain your choice:

Bacteriostatic: _____

_____

_____

Bactericidal: _____

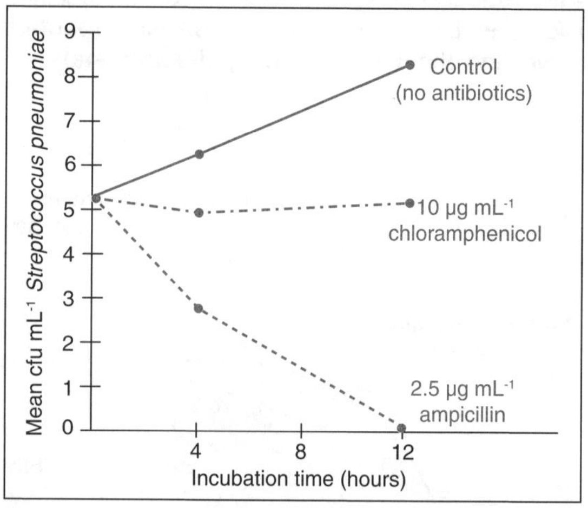

_____

_____

3. (a) What is **antibiotic resistance**? _____

_____

_____

(b) What are the implications to humans of bacteria acquiring resistance to several different antibiotics? _____

_____

_____

_____

4. Why were Florey and Chain's penicillin experiments such an important medical breakthrough? _____

_____

_____

5. Two students carried out an experiment to determine the effect of antibiotics on bacteria. They placed discs saturated with antibiotic on petri dishes evenly coated with bacterial colonies. Dish 1 contained four different antibiotics labelled A to D and a control labelled CL. Dish 2 contained four different concentrations of a single antibiotic and a control labelled CL.

Dish 1

Dish 2

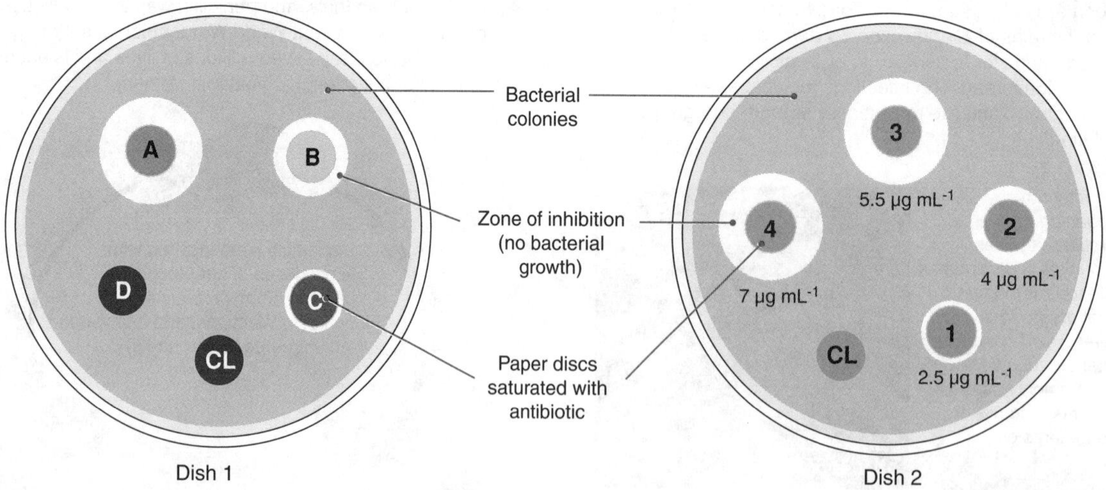

(a) Which was the most effective antibiotic on Dish 1? _____

(b) Which was the most effective concentration on Dish 2? _____

(c) Explain your choice in question 5(b): _____

_____

_____

182 Viral Diseases

Key Idea: Viruses are infectious agents and are responsible for a large number of diseases. They can be difficult to treat because antibiotics are ineffective against them.

A **virus** is a highly infectious pathogen that replicates only inside the living cells of other organisms. Viral infections cause a wide range of commonly occurring diseases, e.g. the common cold, influenza, cold sores, and life-threatening diseases such as HIV/AIDS and ebola. Viral infections can spread rapidly and, once contracted, are difficult to treat and must usually are left to run their course. Recovery from infection is usually associated with the host's immune system combating the infection. Developing effective antiviral drugs is difficult because they must target the virus but not affect the host cell. Antibiotics are ineffective against viruses because they only target specific aspects of bacterial metabolism. Currently, immunization to induce an immune response provides the best protection against viral disease. However, viruses mutate rapidly and this immunity may not be lifelong.

Types of Viruses Affecting Humans

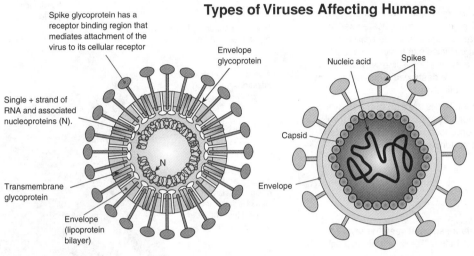

Spike glycoprotein has a receptor binding region that mediates attachment of the virus to its cellular receptor

Envelope glycoprotein

Single + strand of RNA and associated nucleoproteins (N).

Transmembrane glycoprotein

Envelope (lipoprotein bilayer)

Structure of a coronavirus

Nucleic acid

Spikes

Capsid

Envelope

General structure of an enveloped animal virus, e.g. Herpesvirus

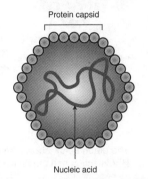

Viruses are not organisms!

Viruses are metabolically inert until they are inside the host cell and hijacking its metabolic machinery to make new viral particles.

Protein capsid

Nucleic acid

General structure of a non-enveloped animal virus, e.g. Papillomavirus (wart virus)

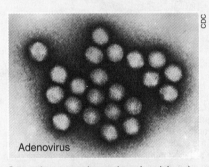

Adenovirus

Some viruses, such as adenovirus (above) can survive outside the body for prolonged periods of time. This can make it very difficult to stop them from infecting a new host and spreading through the population. The adenovirus causes respiratory illness.

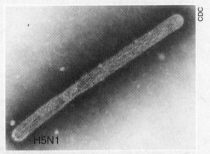

H5N1

Most viruses are very specific to the organism and type of tissue they infect, but sometimes viruses can spread between species. The avian influenza H5N1 virus (above) causes serious illness in birds, but is also capable of infecting and killing humans.

International-mindedness: Containment of Disease

Bird flu (avian influenza) is a highly infectious viral disease which has infected the human population several times since it was first reported in 1997. International collaboration between a number of agencies (e.g. the WHO and UN) and key nations ensures that the disease is closely monitored, and any new outbreaks quickly reported. Heightened surveillance, accurate reporting, and fast communication between agencies is designed to minimize the risk of further human infections. Failing that, containment strategies are in place to reduce spread of the disease.

1. What is a virus? _____

_____

2. Why are antibiotics ineffective against a viral disease? _____

_____

_____

3. What role does international collaboration play in helping to prevent a disease spreading? _____

_____

_____

_____

183 HIV and AIDS

Key Idea: The human immunodeficiency virus infects lymphocyte cells, eventually causing AIDS, a fatal disease, which acts by impairing the immune system.

HIV (human immunodeficiency virus) is a retrovirus, a single-stranded RNA virus which infects lymphocytes called helper T-cells. Over time, a disease called **AIDS** (acquired immunodeficiency syndrome) develops and the immune system loses its ability to fight off infections as more helper T-cells are destroyed. There is no cure or vaccine for HIV, but some drugs have been developed which can slow the progress of the disease.

HIV Infects Lymphocytes

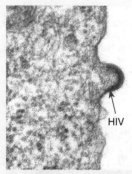

CDC

HIV budding from a lymphocyte

HIV infects helper T-cell lymphocytes. It uses the cells to replicate itself in great numbers, then the newly formed viral particles exit the cell to infect more helper T-cells. Many helper T-cells are destroyed in the process of HIV replication.

Helper T-cells are part of the body's immune system, so when their levels become too low, the immune system can no longer fight off infections.

The graph below shows the relationship between the level of HIV infection and the number of helper T-cells in an individual.

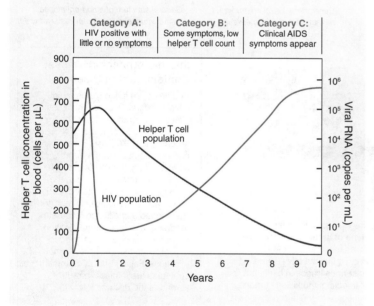

Category A: HIV positive with little or no symptoms	Category B: Some symptoms, low helper T cell count	Category C: Clinical AIDS symptoms appear

AIDS: The End Stage of an HIV Infection

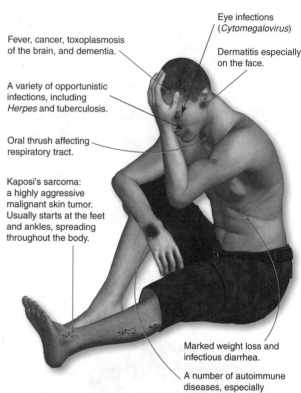

Fever, cancer, toxoplasmosis of the brain, and dementia.

A variety of opportunistic infections, including *Herpes* and tuberculosis.

Oral thrush affecting respiratory tract.

Kaposi's sarcoma: a highly aggressive malignant skin tumor. Usually starts at the feet and ankles, spreading throughout the body.

Eye infections (*Cytomegalovirus*)

Dermatitis especially on the face.

Marked weight loss and infectious diarrhea.

A number of autoimmune diseases, especially destruction of platelets.

The range of symptoms resulting from HIV infection is huge, but are not the result of the HIV infection directly. The symptoms arise from secondary infections that gain a foothold in the body due to the weakened immune system (due to the reduced number of helper T-cells). People with healthy immune systems can be exposed to pathogens and not suffer serious effects because their immune system fights them off. However, people with HIV are susceptible to all pathogens because their immune system is too weak to fight them off. As the immune system become progressively weaker, the infected person becomes sicker.

1. (a) What type of cells does HIV infect? _____

    (b) What effect does HIV have on the body's immune system? _____

    _____

    _____

    _____

2. Study the graph above showing how HIV affects the number of helper T-cells.

    (a) Describe how the virus population changes with the progression of the disease: _____

    _____

    _____

© 2012-2014 **BIOZONE** International
ISBN: 978-1-927173-93-0
Photocopying Prohibited

Transmission, Diagnosis, Treatment, and Prevention of HIV

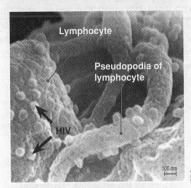

A SEM shows spherical HIV-1 virions on the surface of a human lymphocyte.

HIV is easily transmitted between intravenous drug users who share needles.

Modes of Transmission

1. HIV is transmitted in blood, vaginal secretions, semen, breast milk, and across the placenta.

2. In developed countries, blood transfusions are no longer a likely source of infection because blood is tested for HIV antibodies prior to use.

3. Historically, transmission of HIV in developed countries has been primarily through intravenous drug use and homosexual activity, but heterosexual transmission is increasing.

4. Transmission via heterosexual activity has been particularly important to the spread of HIV in Asia and southern Africa, partly because of the high prevalence of risky sexual behaviour in these regions.

Treatment

There is currently no cure or vaccine available for HIV, but several drugs help slow down the effects and spread of an HIV infection.

Drugs that work against HIV are called antiretroviral therapy (ART). There are currently five different "classes" ARTs, and each class of drug attacks the virus at different points of its life cycle. HIV patients are given a mixture (cocktail) of different ARTs. This provides the most effective way of controlling the virus and reducing the rate it spreads, and also helps to prevent drug resistance. Drug resistance arises due to HIV's high mutation rate and short generation times, and also through the misuse of antiviral drugs.

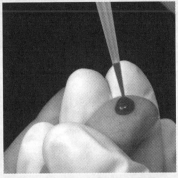

Diagnosis of HIV is possible using a simple antibody-based test on a blood sample.

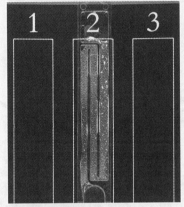

A positive HIV rapid test result shows clumping (aggregation) where HIV antibodies have reacted with HIV protein-coated latex beads.

(b) How do the helper T-cells respond to the HIV infection? _____

_____

_____

_____

3. (a) Describe three common ways in which HIV can be transmitted from one person to another: _____

_____

_____

(b) Why have the rates of HIV infections from blood products dropped in developed countries? _____

_____

_____

4. Why are a cocktail of different antiretroviral therapy drugs used to treat HIV? _____

_____

_____

5. Drug resistance in HIV is becoming more common. How can drug resistance develop, and why is it a problem? _____

_____

_____

_____

184 Introduction to Gas Exchange

Key Idea: Gas exchange is the process by which oxygen and carbon dioxide are exchanged between the cells of an organism and the environment.

Living cells require energy for the activities of life. Energy is released in cells by the breakdown of sugars and other substances in cellular respiration. As a consequence of this process, gases (carbon dioxide and oxygen) need to be exchanged by diffusion. Gas exchange occurs across a gas exchange surface (membrane) between the lungs and the external environment. Diffusion gradients are maintained by transport of gases away from the gas exchange surface. Gas exchange membranes must be in close proximity to the blood for this to occur effectively.

The Need for Gas Exchange

Gas exchange is the process by which oxygen is acquired and carbon dioxide is removed. Cell respiration creates a constant demand for oxygen (O_2) and a need to eliminate carbon dioxide gas (CO_2). The inputs and outputs for cell respiration are shown below for a generalized animal cell.

Gas exchange surfaces provide a means for gases to enter and leave the body. In humans, the gas exchange surfaces are within the **lungs**, which are housed within the chest cavity. The lung tissue provides a large surface area for the exchange of gases by diffusion. The lungs are protected from drying out because they are internal.

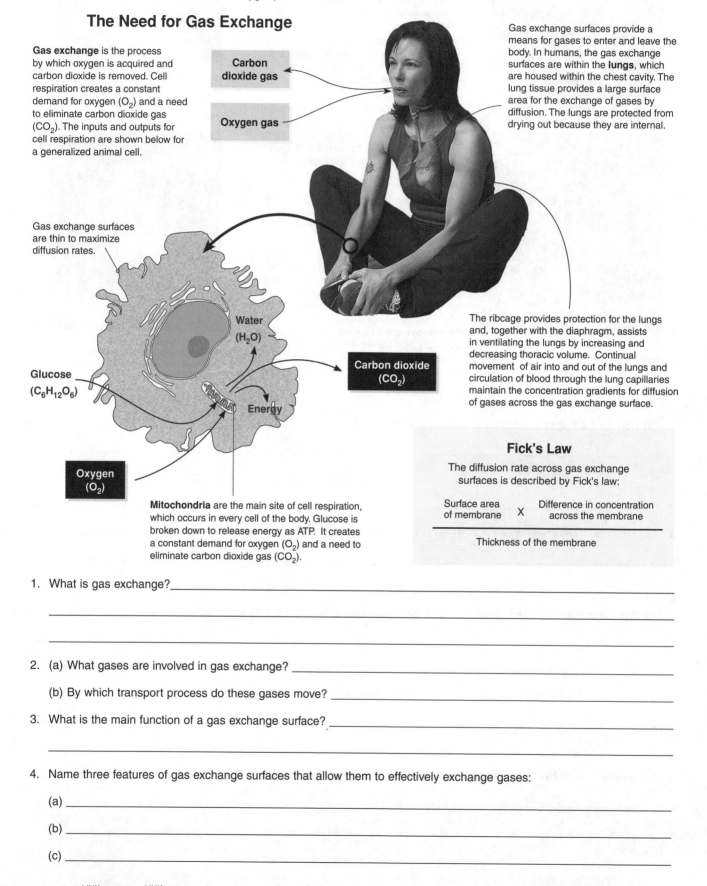

Carbon dioxide gas

Oxygen gas

Gas exchange surfaces are thin to maximize diffusion rates.

Water (H_2O)

Glucose ($C_6H_{12}O_6$)

Carbon dioxide (CO_2)

Energy

Oxygen (O_2)

The ribcage provides protection for the lungs and, together with the diaphragm, assists in ventilating the lungs by increasing and decreasing thoracic volume. Continual movement of air into and out of the lungs and circulation of blood through the lung capillaries maintain the concentration gradients for diffusion of gases across the gas exchange surface.

Mitochondria are the main site of cell respiration, which occurs in every cell of the body. Glucose is broken down to release energy as ATP. It creates a constant demand for oxygen (O_2) and a need to eliminate carbon dioxide gas (CO_2).

Fick's Law

The diffusion rate across gas exchange surfaces is described by Fick's law:

$$\frac{\text{Surface area of membrane} \times \text{Difference in concentration across the membrane}}{\text{Thickness of the membrane}}$$

1. What is gas exchange? _____

_____

_____

2. (a) What gases are involved in gas exchange? _____

   (b) By which transport process do these gases move? _____

3. What is the main function of a gas exchange surface? _____

_____

4. Name three features of gas exchange surfaces that allow them to effectively exchange gases:

   (a) _____

   (b) _____

   (c) _____

© 2012-2014 **BIOZONE** International
ISBN: 978-1-927173-93-0
Photocopying Prohibited

185 The Gas Exchange System

Key Idea: Lungs are internal sac-like organs connected to the outside by a system of airways. The smallest airways end in thin-walled alveoli, where gas exchange occurs.

The gas exchange (or respiratory system) includes all the structures associated with exchanging respiratory gases with the environment. In humans, this system consists of paired lungs connected to the outside air by way of a system of tubular passageways: the trachea, bronchi, and bronchioles. The details of exchanges across the gas exchange membrane are described on the next page.

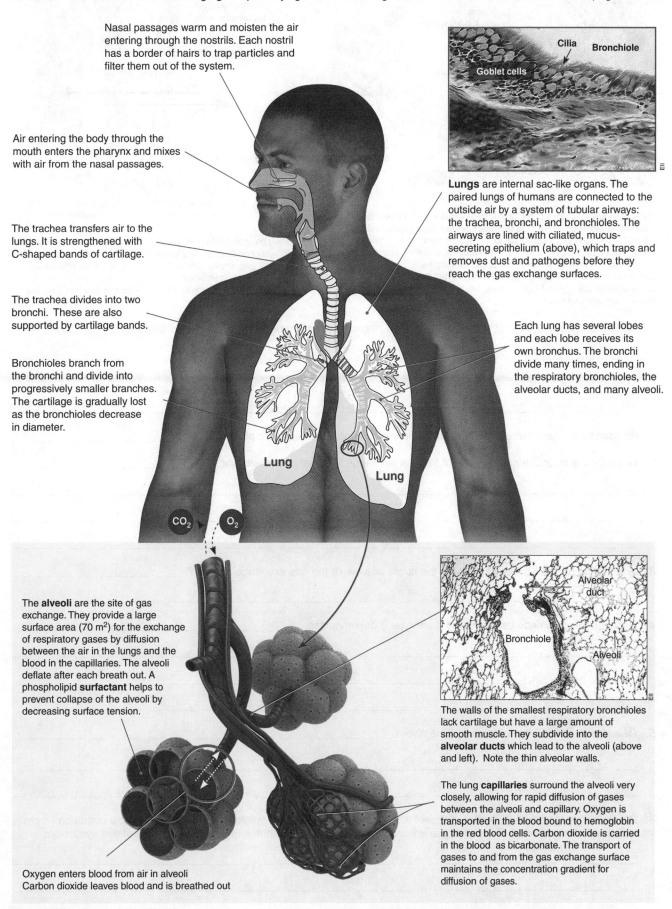

Nasal passages warm and moisten the air entering through the nostrils. Each nostril has a border of hairs to trap particles and filter them out of the system.

Air entering the body through the mouth enters the pharynx and mixes with air from the nasal passages.

The trachea transfers air to the lungs. It is strengthened with C-shaped bands of cartilage.

The trachea divides into two bronchi. These are also supported by cartilage bands.

Bronchioles branch from the bronchi and divide into progressively smaller branches. The cartilage is gradually lost as the bronchioles decrease in diameter.

Cilia / Bronchiole / Goblet cells

Lungs are internal sac-like organs. The paired lungs of humans are connected to the outside air by a system of tubular airways: the trachea, bronchi, and bronchioles. The airways are lined with ciliated, mucus-secreting epithelium (above), which traps and removes dust and pathogens before they reach the gas exchange surfaces.

Each lung has several lobes and each lobe receives its own bronchus. The bronchi divide many times, ending in the respiratory bronchioles, the alveolar ducts, and many alveoli.

Lung / Lung

CO_2 / O_2

The **alveoli** are the site of gas exchange. They provide a large surface area (70 m^2) for the exchange of respiratory gases by diffusion between the air in the lungs and the blood in the capillaries. The alveoli deflate after each breath out. A phospholipid **surfactant** helps to prevent collapse of the alveoli by decreasing surface tension.

Alveolar duct / Bronchiole / Alveoli

The walls of the smallest respiratory bronchioles lack cartilage but have a large amount of smooth muscle. They subdivide into the **alveolar ducts** which lead to the alveoli (above and left). Note the thin alveolar walls.

The lung **capillaries** surround the alveoli very closely, allowing for rapid diffusion of gases between the alveoli and capillary. Oxygen is transported in the blood bound to hemoglobin in the red blood cells. Carbon dioxide is carried in the blood as bicarbonate. The transport of gases to and from the gas exchange surface maintains the concentration gradient for diffusion of gases.

Oxygen enters blood from air in alveoli
Carbon dioxide leaves blood and is breathed out

LINK 190   LINK 189   LINK 187   WEB 185   KNOW

Cross Section Through an Alveolus

Alveolar macrophage (defensive role)

Alveolus

Connective tissue cell

Monocyte (defensive role)

Surfactant secreted by type II pneumocytes

Nucleus of type I pneumocyte

Gas exchange membrane

Alveolus

Red blood cell in capillary

Connective tissue containing elastic fibres

Capillary

The alveoli are very close to the blood-filled capillaries. The alveolus is lined with alveolar epithelial cells or **pneumocytes**. Type I pneumocytes (90-95% of aveolar cells) contribute to the gas exchange membrane (right). Type II pneumocytes secrete a **surfactant**, which decreases surface tension within the alveoli and prevents them from collapsing and sticking to each other. Macrophages and monocytes defend the lung tissue against pathogens. Elastic connective tissue gives the alveoli their ability to expand and recoil.

The Gas Exchange Membrane

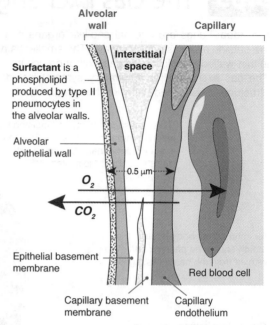

Alveolar wall

Capillary

Interstitial space

Surfactant is a phospholipid produced by type II pneumocytes in the alveolar walls.

Alveolar epithelial wall

O_2

CO_2

0.5 μm

Epithelial basement membrane

Red blood cell

Capillary basement membrane

Capillary endothelium

The **gas exchange membrane** is the term for the layered junction between the alveolar epithelial cells (pneumocytes), the endothelial cells of the capillary, and their associated basement membranes (thin, collagenous layers that underlie the epithelial tissues). Gases move freely across this membrane.

1. (a) Explain how the basic structure of the human respiratory system provides such a large area for gas exchange:

    _____

    _____

    _____

    (b) Identify the general region of the lung where exchange of gases takes place: _____

2. Describe the structure and purpose of the gas exchange (respiratory) membrane: _____

    _____

    _____

    _____

3. What is the purpose of the cartilage in the larger airways of the gas exchange system? _____

    _____

4. Describe the difference between type I and type II pneumocytes: _____

    _____

    _____

    _____

5. Describe the role of the surfactant in the alveoli: _____

    _____

    _____

6. Babies born prematurely are often deficient in surfactant. This causes respiratory distress syndrome; a condition where breathing is very difficult. From what you know about the role of surfactant, explain the symptoms of this syndrome:

    _____

    _____

© 2012-2014 **BIOZONE** International
ISBN: 978-1-927173-93-0
Photocopying Prohibited

186 Breathing

Key Idea: Breathing provides a continual supply of air to the lungs to maintain the concentration gradients for gas exchange. Different muscles are used in inspiration and expiration to force air in and out of the lungs.

Breathing (ventilation) provides a continual supply of oxygen-rich air to the lungs and expels air high in carbon dioxide. Together with the cardiovascular system, which transports respiratory gases between the alveolar and the cells of the body, breathing maintains concentration gradients for gas exchange. Breathing is achieved by the action of muscles.

1. Explain the purpose of breathing: _____
   _____
   _____

2. In general terms, how is breathing achieved?
   _____
   _____
   _____
   _____

3. (a) Describe the sequence of events involved in quiet breathing:
   _____
   _____
   _____
   _____
   _____

   (b) What is the essential difference between this and the situation during forced breathing:
   _____
   _____
   _____

4. During inspiration, which muscles are:

   (a) Contracting: _____
   _____

   (b) Relaxed: _____
   _____

5. During forced expiration, which muscles are:

   (a) Contracting: _____
   _____

   (b) Relaxed: _____
   _____

6. Explain the role of antagonistic muscles in breathing:
   _____
   _____
   _____

Breathing and Muscle Action

Muscles can only do work by contracting, so they can only perform movement in one direction. To achieve motion in two directions, muscles work as antagonistic pairs. Antagonistic pairs of muscles have opposing actions and create movement when one contracts and the other relaxes. Breathing in humans involves two sets of antagonistic muscles. The external and internal intercostal muscles of the ribcage, and the diaphragm and abdominal muscles.

Inspiration (Inhalation or Breathing In)

During quiet breathing, inspiration is achieved by increasing the thoracic volume (therefore decreasing the pressure inside the lungs). Air then flows into the lungs in response to the decreased pressure inside the lung. Inspiration is always an active process involving muscle contraction.

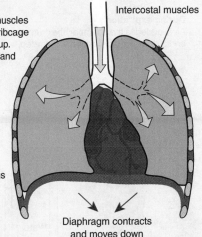

1 External intercostal muscles contract causing the ribcage to expand and move up. Diaphragm contracts and moves down.

2 Thoracic volume increases, lungs expand, and the pressure inside the lungs decreases.

3 Air flows into the lungs in response to the pressure gradient.

Intercostal muscles

Diaphragm contracts and moves down

Expiration (Exhalation or Breathing Out)

In quiet breathing, expiration is a passive process, achieved when the external intercostals and diaphragm relax and thoracic volume decreases. Air flows passively out of the lungs to equalize with the air pressure. In active breathing, muscle contraction is involved in bringing about both inspiration and expiration.

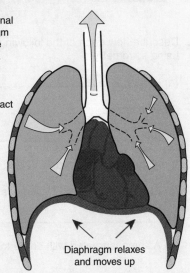

1 In **quiet breathing**, external intercostals and diaphragm relax. The elasticity of the lung tissue causes recoil.

In **forced breathing**, the internal intercostals and abdominal muscles contract to compress the thoracic cavity and increase the force of the expiration.

2 Thoracic volume decreases and the pressure inside the lungs increases.

3 Air flows passively out of the lungs in response to the pressure gradient.

Diaphragm relaxes and moves up

LINK 188 LINK 187 WEB 186 APP

187 Measuring Lung Function

Key Idea: A lung function test, called spirometry, measures changes in lung volume and can be used diagnostically.

The volume of gases exchanged during breathing varies according to the physiological demands placed on the body (e.g. by exercise) and an individual's lung function. **Spirometry** measures changes in lung volume by measuring how much air a person can breathe in and out and how fast the air can be expelled. Spirometry can measure changes in ventilation rates during exercise and can be used to assess impairments in lung function, as might occur as a result of disease. In humans, the total adult lung capacity varies between 4 and 6 litres (L or dm^3) and is greater in males.

Determining Changes in Lung Volume Using Spirometry

The apparatus used to measure the amount of air exchanged during breathing and the rate of breathing is a **spirometer** (also called a respirometer). A simple spirometer consists of a weighted drum, containing oxygen or air, inverted over a chamber of water. A tube connects the air-filled chamber with the subject's mouth, and soda lime in the system absorbs the carbon dioxide breathed out. Breathing results in a trace called a spirogram, from which lung volumes can be measured directly.

During inspiration
Air is removed from the chamber, the drum sinks, and an upward deflection is recorded on the paper on the rotating drum.

During expiration
Air is added to the chamber, the drum rises, and a downward deflection is recorded.

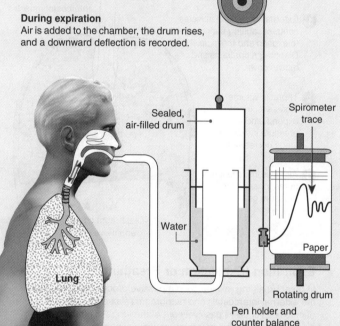

Pulley

Sealed, air-filled drum

Spirometer trace

Water

Paper

Lung

Rotating drum

Pen holder and counter balance

Lung Volumes and Capacities

The air in the lungs can be divided into volumes. Lung capacities are combinations of volumes.

DESCRIPTION OF VOLUME	Vol (L)
Tidal volume (TV) Volume of air breathed in and out in a single breath	0.5
Inspiratory reserve volume (IRV) Volume breathed in by a maximum inspiration at the end of a normal inspiration	3.3
Expiratory reserve volume (ERV) Volume breathed out by a maximum effort at the end of a normal expiration	1.0
Residual volume (RV) Volume of air remaining in the lungs at the end of a maximum expiration	1.2

DESCRIPTION OF CAPACITY	
Inspiratory capacity (IC) = TV + IRV Volume breathed in by a maximum inspiration at the end of a normal expiration	3.8
Vital capacity (VC) = IRV + TV + ERV Volume that can be exhaled after a maximum inspiration.	4.8
Total lung capacity (TLC) = VC + RV The total volume of the lungs. Only a fraction of TLC is used in normal breathing	6.0

PRIMARY INDICATORS OF LUNG FUNCTION

Forced expiratory volume in 1 second (FEV_1)
The volume of air that is maximally exhaled in the first second of exhalation.

Forced vital capacity (FVC)
The total volume of air that can be forcibly exhaled after a maximum inspiration.

1. Describe how each of the following might be expected to influence values for lung volumes and capacities obtained using spirometry:

   (a) Height: _____

   (b) Gender: _____

   (c) Age: _____

2. A percentage decline in FEV_1 and FVC (to <80% of normal) are indicators of impaired lung function, e.g in asthma:

   (a) Explain why a forced volume is a more useful indicator of lung function than tidal volume:

   _____

   _____

   (b) Asthma is treated with drugs to relax the airways. Suggest how spirometry could be used during asthma treatment:

   _____

© 2012-2014 **BIOZONE** International
 ISBN: 978-1-927173-93-0
 Photocopying Prohibited

Respiratory gas	Approximate percentages of O_2 and CO_2		
	Inhaled air	Air in lungs	Exhaled air
O_2	21.0	13.8	16.4
CO_2	0.04	5.5	3.6

Above: The percentages of respiratory gases in air (by volume) during normal breathing. The percentage volume of oxygen in the alveolar air (in the lung) is lower than that in the exhaled air because of the influence of the **dead air volume** (the air in the spaces of the nose, throat, larynx, trachea and bronchi). This air (about 30% of the air inhaled) is unavailable for gas exchange.

Left: During exercise, the breathing rate, tidal volume, and **pulmonary ventilation** rate or **PV** (the amount of air exchanged with the environment per minute) increase up to a maximum (as indicated below).

Spirogram for a male during quiet and forced breathing, and during exercise

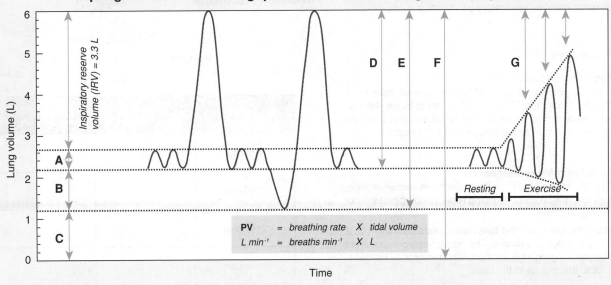

PV = breathing rate X tidal volume
$L\ min^{-1}$ = breaths min^{-1} X L

3. Using the definitions given on the previous page, identify the volumes and capacities indicated by the letters **A-F** on the spirogram above. For each, indicate the volume (vol) in liters (L). The inspiratory reserve volume has been identified:

(a) **A**: _____ Vol: _____ (d) **D**: _____ Vol: _____

(b) **B**: _____ Vol: _____ (e) **E**: _____ Vol: _____

(c) **C**: _____ Vol: _____ (f) **F**: _____ Vol: _____

4. Explain what is happening in the sequence indicated by the letter **G**: _____

_____

5. Calculate PV when breathing rate is 15 breaths per minute and tidal volume is 0.4 L: _____

6. (a) Describe what would happen to PV during strenuous exercise: _____

(b) Explain how this is achieved: _____

7. The table above gives approximate percentages for respiratory gases during breathing. Study the data and then:

(a) Calculate the difference in CO_2 between inhaled and exhaled air: _____

(b) Explain where this 'extra' CO_2 comes from: _____

(c) Explain why the dead air volume raises the oxygen content of exhaled air above that in the lungs: _____

_____

_____

188 Exercise and Breathing

Key Idea: Breathing rate and heart rate both increase during exercise to meet the body's increased metabolic demands.

During exercise, the body's metabolic rate increases and the demand for oxygen increases. Oxygen is required for cellular respiration and ATP production. Increasing the rate of breathing delivers more oxygen to working tissues and

enables them to make the ATP they need to keep working. An increased breathing rate also increases the rate at which carbon dioxide is expelled from the body. Heart rate also increases so blood can be moved around the body more quickly. This allows for faster delivery of oxygen and removal of carbon dioxide.

In this practical, you will work in groups of three to see how exercise affects breathing and heart rate. Choose one person to carry out the exercise and one person each to record heart rate and breathing rate.

Heart rate (beats per minute) is obtained by measuring the pulse (right) for 15 seconds and multiplying by four.

Breathing rate (breaths per minute) is measured by counting the number of breaths taken in 15 seconds and multiplying it by four.

CAUTION: The person exercising should have no known pre-existing heart or respiratory conditions.

Measuring the carotid pulse

Gently press your index and middle fingers, not your thumb, against the carotid artery in the neck (just under the jaw) or the radial artery (on the wrist just under the thumb) until you feel a pulse.

Measuring the radial pulse

Procedure

Resting Measurements
Have the person carrying out the exercise sit down on a chair for 5 minutes. They should try not to move. After 5 minutes of sitting, measure their heart rate and breathing rate. Record the resting data on the table (right).

Exercising Measurements
Choose an exercise to perform. Some examples include step ups onto a chair, skipping rope, jumping jacks, and running in place.

Begin the exercise, and take measurements after 1, 2, 3, and 4 minutes of exercise. The person exercising should stop just long enough for the measurements to be taken. Record the results in the table.

Post Exercise Measurements
After the exercise period has finished, have the exerciser sit down in a chair. Take their measurements 1 and 5 minutes after finishing the exercise. Record the results on the table.

	Heart rate (beats minute^{-1})	Breathing rate (breaths minute^{-1})
Resting		
1 minute		
2 minutes		
3 minutes		
4 minutes		
1 minute after		
5 minutes after		

1. (a) Graph your results on separate piece of paper. You will need to use one axis for heart rate and another for breathing rate. When you have finished answering the questions below, attach it to this page.

   (b) Analyse your graph and describe what happened to heart rate and breathing rate **during exercise**: _____

   _____

   _____

   _____

   _____

2. (a) Describe what happened to heart rate and breathing rate **after exercise**: _____

   _____

   _____

   (b) Why did this change occur? _____

   _____

SKILL

189 Smoking and Lung Cancer

Key Idea: Cigarette smoke contains harmful substances which cause lung cancer. Lung cancer is usually fatal.

Tobacco smoke contains many carcinogens, harmful substances which cause cancer. **Lung cancer** is the most widely known harmful effect of smoking and is the leading cause of cancer deaths worldwide. The carcinogens damage the DNA in cells of the lung, resulting in the formation of a tumour (a tissue mass). Lung cancer is usually fatal because the cancer spreads from the lung to other parts of the body before it is detected. Cigarette smoking is the principal risk factor for development of lung cancer, but also causes other diseases such as CVD, chronic bronchitis, and emphysema.

Causes of Lung Cancer

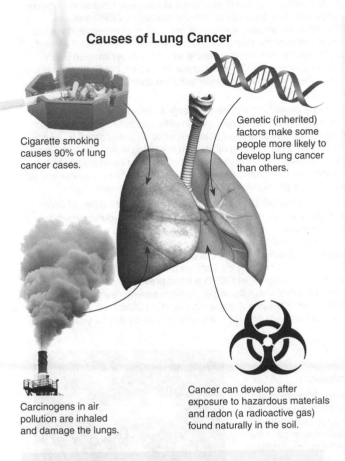

Cigarette smoking causes 90% of lung cancer cases.

Genetic (inherited) factors make some people more likely to develop lung cancer than others.

Carcinogens in air pollution are inhaled and damage the lungs.

Cancer can develop after exposure to hazardous materials and radon (a radioactive gas) found naturally in the soil.

Economic Costs of Lung Cancer

Lung cancer has large economic costs to society. These may be direct costs relating to health care (e.g. hospital stays, treatment, and drugs). Direct costs are estimated to cost over $10 billion a year in the US alone. Lung cancer also has indirect costs such as loss of production due to absences from work because of illness or early death.

How Smoking Damages the Lungs

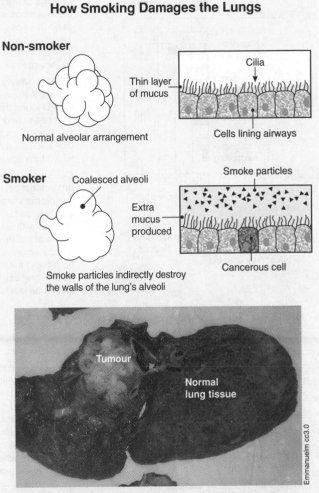

Non-smoker

Thin layer of mucus

Cilia

Cells lining airways

Normal alveolar arrangement

Smoker

Coalesced alveoli

Extra mucus produced

Smoke particles

Cancerous cell

Smoke particles indirectly destroy the walls of the lung's alveoli

Tumour

Normal lung tissue

Emmanuelm cc3.0

Lung cancer occurs when the cells of the lungs grow quickly and uncontrollably, and eventually forms a tumour (the large white mass in the photo above). As the tumour grows, it destroys healthy lung tissue, and may block off an airway (bronchi) and prevent the affected section of the lung from functioning normally.

---

1. Analyse the graph (right) comparing deaths from lung cancer in smokers and non-smokers. What conclusion can you draw about the effect of smoking on lung cancer?

_____

_____

_____

_____

2. Smoking causes the lung tissue to lose its elasticity and tar from the tobacco smoke clogs the airways and damages the alveoli. Use the diagram (top right) to explain why smoking reduces the gas exchange capacity of the lung tissue:

_____

_____

_____

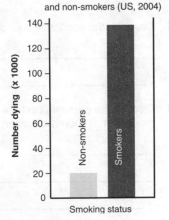

Deaths from lung cancer in smokers and non-smokers (US, 2004)

Number dying (x 1000)

140
120
100
80
60
40
20
0

Non-smokers

Smokers

Smoking status

LINK **74**   LINK **35**   WEB **189**   **APP**

190 Emphysema

Key Idea: Emphysema is a lung disease that results in the breakdown of lung tissue and damage to the alveoli. Emphysema is one of a group of lung diseases collectively called chronic obstructive pulmonary disease (COPD). They have serious consequences on a person's health including shortness of breath and a higher risk of heart disease.

Activity Limitation in People With and Without COPD

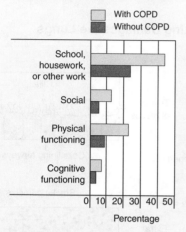

- With COPD
- Without COPD

- School, housework, or other work
- Social
- Physical functioning
- Cognitive functioning

Percentage: 0 10 20 30 40 50

Cognitive, physical, social and activity-related limitations are more common among people with chronic obstructive pulmonary disease.

The Impact of COPD in the US

Chronic obstructive pulmonary disease (COPD) is a serious lung disease which makes it difficult for people to breathe because their airways are partially blocked or because the alveoli of the lungs lose their elasticity or become damaged. COPD includes **chronic bronchitis** and **emphysema**, and affects about 12 million people in the US. Most of those affected are over the age of 40 and smoking is the cause in the vast majority of cases. This relationship is clear; people who have never smoked rarely develop COPD. The symptoms of COPD and asthma are similar, but COPD causes permanent damage to the airways, and so symptoms are chronic (persistent) and treatment is limited.

COPD severely limits the capacity of sufferers to carry out even a normal daily level of activity. A survey by the American Lung Association of hundreds of people living with COPD found that nearly half became short of breath while washing, dressing, or doing light housework (left). Over 25% reported difficulty in breathing while sitting or lying still. Lack of oxygen also places those with COPD at high risk of heart failure. As the disease becomes more severe, sufferers usually require long-term oxygen therapy, in which they are more or less permanently attached to an oxygen supply.

COPD is estimated to cost the US $32 billion dollars each year, $14 billion of which are indirect costs, such as lost working days. A 'flare-up' of COPD, during which the symptoms worsen, is one of the commonest reasons for admission to hospital and the disease places a substantial burden on health services. COPD is the only major disease with an increasing death rate in the US, rising 16%, and is the third leading cause of death. At least 120,000 people die each year from the end stages of COPD, but the actual number may be higher as COPD is often present in patients who die from heart failure and stroke. Many of these people have several years of ill health before they die. Being able to breathe is something we don't often think about. What must it be like to struggle for each breath, every minute of every day for years?

A Personal Story

Deborah Ripley's message from her mother Jenny (used with permission)

"Fear, anxiety, depression, and carbon monoxide are ruining whatever life my mother has left. I posted this portrait of my Mum on the photo website Flickr because she wants to send a warning to anyone who's still smoking. I've just returned from visiting her in a nursing home where she's virtually shackled to the bed. Getting up to go to the bathroom practically kills her. She was admitted to hospital after a bout of pneumonia, which required intensive antibiotic therapy and left her hardly able to breathe. She has moderate dementia caused by a series of mini-strokes, which is aggravated by the pneumonia. She has no recollection of who has visited her or when, so consequently thinks she's alone most of the time, which is upsetting and disturbing for her.

This is all caused by damage to her brain and lungs as a result of 65 years of smoking. In those moments when she is lucid, she asks me who she can warn that this could happen to them. She said 'if people could see me lying here like this it would put them off...' None of her other known blood relatives suffered this sort of decline in their old age and, as far as I know, none of them smoked".

Thankfully, Jenny's pneumonia has since subsided and her COPD is being well managed. However, constant vigilance is important because flare-ups are common with COPD and recovery from lung infections is difficult when breathing is already compromised.

1. Describe the economic impact of smoking-related COPD: _____

_____

_____

2. Discuss the personal costs of a smoking-related disease and comment on the value of personal testimonials such as those from Deborah's mother:

_____

_____

_____

_____

© 2012-2014 **BIOZONE** International
ISBN: 978-1-927173-93-0
Photocopying Prohibited

191 Nervous Regulatory Systems

Key Idea: The nervous and endocrine systems work together to maintain homeostasis. Neurons of the nervous system transmit information as nerve impulses to the central nervous system, which coordinates appropriate responses to stimuli. In humans, the nervous and endocrine (hormonal) systems work together to regulate the internal environment and maintain homeostasis in a fluctuating environment. The

nervous system contains cells called **neurons** (nerve cells) which are specialized to transmit information in the form of electrochemical impulses (action potentials). The nervous system is a signalling network with branches carrying information directly to and from specific target tissues. Impulses can be transmitted over considerable distances and the response is very precise and rapid.

Coordination by the Nervous System

The vertebrate nervous system consists of the **central nervous system** (brain and spinal cord), and the nerves and receptors outside it (**peripheral nervous system**). Sensory input to receptors comes via stimuli. Information about the effect of a response is provided by feedback mechanisms so that the system can be readjusted. The basic organization of the nervous system can be simplified into a few key components: the sensory receptors, a central nervous system processing point, and the effectors which bring about the response (below).

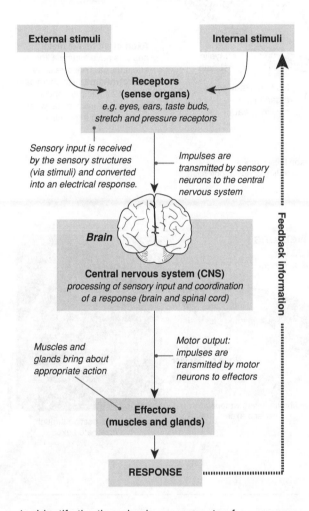

External stimuli → Internal stimuli

Receptors (sense organs)
e.g. eyes, ears, taste buds, stretch and pressure receptors

Sensory input is received by the sensory structures (via stimuli) and converted into an electrical response.

Impulses are transmitted by sensory neurons to the central nervous system

Brain

Central nervous system (CNS)
processing of sensory input and coordination of a response (brain and spinal cord)

Muscles and glands bring about appropriate action

Motor output: impulses are transmitted by motor neurons to effectors

Effectors (muscles and glands)

RESPONSE

Feedback information

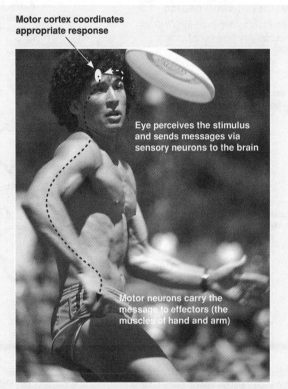

Motor cortex coordinates appropriate response

Eye perceives the stimulus and sends messages via sensory neurons to the brain

Motor neurons carry the message to effectors (the muscles of hand and arm)

In the example above, the frisbee's approach is perceived by the eye. The motor cortex of the brain integrates the sensory message. Coordination of hand and body orientation is brought about through motor neurons to the muscles.

Comparison of Nervous and Hormonal Control

	Nervous Control	Hormonal Control
Communication	Impulses across synapses	Hormones in the blood
Speed	Very rapid (within a few milliseconds)	Relatively slow (over minutes, hours, or longer)
Duration	Short term and reversible	Longer lasting effects
Target pathway	Specific (through nerves) to specific cells	Hormones broadcast to target cells everywhere
Action	Causes glands to secrete or muscles to contract	Causes changes in metabolic activity

1. Identify the three basic components of a nervous system and describe their role:

    (a) _____

    (b) _____

    (c) _____

2. Comment on the significance of the differences between the speed and duration of nervous and hormonal controls:

    _____

    _____

    _____

© 2012-2014 **BIOZONE** International
ISBN: 978-1-927173-93-0
Photocopying Prohibited

LINK 197   WEB 191   KNOW

192 Neuron Structure and Function

Key Idea: Neurons are electrically excitable cells that are specialized to process and transmit information via electrical and chemical signals. Increased axon diameter and myelination both increase conduction speed along a neuron. **Neurons** transmit information in the form of electrochemical signals from receptors (in the central nervous system) to effectors. Neurons consist of a cell body (soma) and long processes (dendrites and axons). Conduction speed increases with axon diameter and with myelination. Faster conduction speeds enable more rapid responses to stimuli.

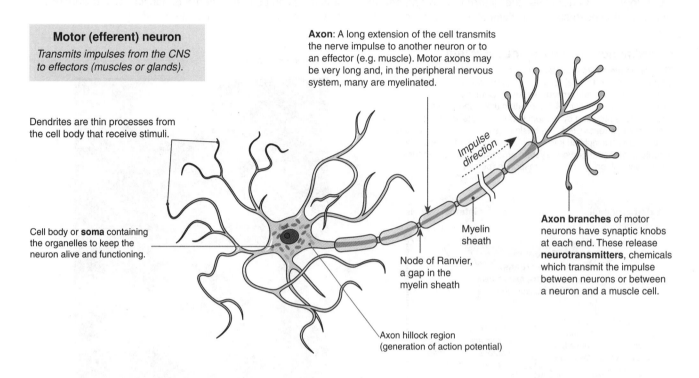

Motor (efferent) neuron

Transmits impulses from the CNS to effectors (muscles or glands).

Axon: A long extension of the cell transmits the nerve impulse to another neuron or to an effector (e.g. muscle). Motor axons may be very long and, in the peripheral nervous system, many are myelinated.

Dendrites are thin processes from the cell body that receive stimuli.

Impulse direction

Cell body or **soma** containing the organelles to keep the neuron alive and functioning.

Myelin sheath

Node of Ranvier, a gap in the myelin sheath

Axon branches of motor neurons have synaptic knobs at each end. These release **neurotransmitters**, chemicals which transmit the impulse between neurons or between a neuron and a muscle cell.

Axon hillock region (generation of action potential)

Where conduction speed is important, the axons of neurons are sheathed within a lipid and protein rich substance called **myelin**. Myelin is produced by oligodendrocytes in the central nervous system (CNS) and by Schwann cells in the peripheral nervous system (PNS). At intervals along the axons of myelinated neurons, there are gaps between neighbouring Schwann cells and their sheaths. These are called **nodes of Ranvier**. Myelin acts as an insulator, increasing the speed at which nerve impulses travel because it prevents ion flow across the neuron membrane and forces the current to "jump" along the axon from node to node.

Myelinated Neurons
Diameter: 1-25 μm
Conduction speed: 6-120 ms⁻¹

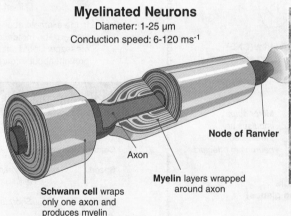

Node of Ranvier

Axon

Schwann cell wraps only one axon and produces myelin

Myelin layers wrapped around axon

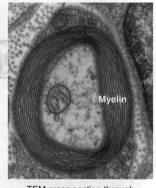

Myelin

Roadnottaken

TEM cross section through a myelinated axon

---

1. What is the function of a neuron? _____

_____

2. What factors increase the speed of conduction along a neuron? _____

_____

3. How does myelination increase the speed of nerve impulse conduction? _____

_____

_____

4. What is the advantage of faster conduction of nerve impulses? _____

_____

© 2012-2014 **BIOZONE** International
ISBN: 978-1-927173-93-0
Photocopying Prohibited

193 The Nerve Impulse

Key Idea: A nerve impulse involves the movement of an action potential along a neuron as a series of electrical depolarization events in response to a stimulus.

The plasma membranes of cells, including neurons, contain **sodium-potassium ion pumps** which actively pump sodium ions (Na^+) out of the cell and potassium ions (K^+) into the cell. The action of these ion pumps in neurons creates a separation of charge (a potential difference or voltage) either side of the membrane and makes the cells **electrically excitable**. It

is this property that enables neurons to transmit electrical impulses. The **resting state** of a neuron, with a net negative charge inside, is maintained by the sodium-potassium pumps, which actively move two K^+ into the neuron for every three Na^+ moved out (below left). When a nerve is stimulated, a brief increase in membrane permeability to Na^+ temporarily reverses the membrane polarity (a **depolarization**). After the nerve impulse passes, the sodium-potassium pump restores the resting potential.

The Resting Neuron

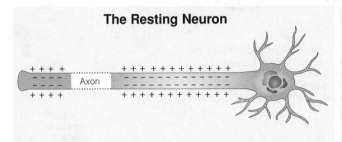

When a neuron is not transmitting an impulse, the inside of the cell is negatively charged relative to the outside and the cell is said to be electrically polarized. The potential difference (voltage) across the membrane is called the **resting potential**. For most nerve cells this is about -70 mV. Nerve transmission is possible because this membrane potential exists.

The Nerve Impulse

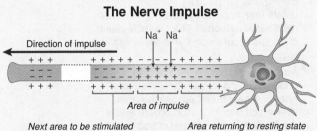

When a neuron is stimulated, the distribution of charges on each side of the membrane briefly reverses. This process of **depolarization** causes a burst of electrical activity to pass along the axon of the neuron as an **action potential**. As the charge reversal reaches one region, local currents depolarize the next region and the impulse spreads along the axon.

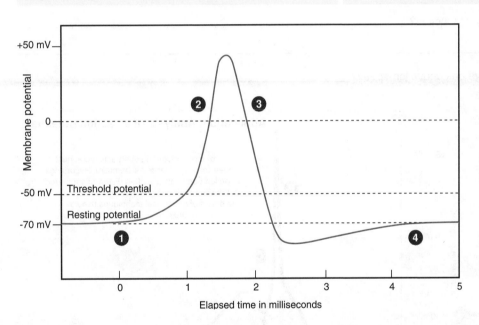

The depolarization in an axon can be shown as a change in membrane potential (in millivolts). A stimulus must be strong enough to reach the **threshold potential** before an action potential is generated. This is the voltage at which the depolarization of the membrane becomes unstoppable.

The action potential is **all or nothing** in its generation and because of this, impulses (once generated) always reach threshold and move along the axon without attenuation. The resting potential is restored by the movement of potassium ions (K^+) out of the cell. During this **refractory period**, the nerve cannot respond, so nerve impulses are discrete.

Voltage-Gated Ion Channels and the Course of an Action Potential

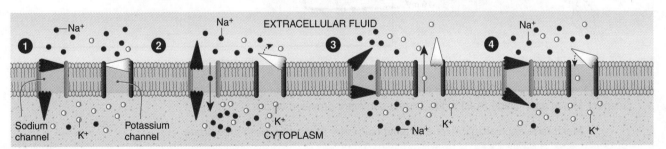

Resting state:
Voltage activated Na^+ and K^+ channels are closed. Negative interior is maintained by the Na^+/K^+ pump.

Depolarization:
Voltage activated Na^+ channels open and there is a rapid influx of Na^+ ions. The interior of the neuron becomes positive relative to the outside.

Repolarization:
Voltage activated Na^+ channels close and the K^+ channels open; K^+ moves out of the cell, restoring the negative charge to the cell interior.

Returning to resting state:
Voltage activated Na^+ and K^+ channels close and the Na^+/K^+ pump restores the original balance of ions, returning the neuron to its resting state.

© 2012-2014 **BIOZONE** International
ISBN: 978-1-927173-93-0
Photocopying Prohibited

LINK **195**   LINK **194**   WEB **193**   **KNOW**

270

Axon myelination is a feature of vertebrate nervous systems and it enables them to achieve very rapid speeds of nerve conduction. Myelinated neurons conduct impulses by **saltatory conduction**, a term that describes how the impulse jumps along the fibre. In a myelinated neuron, action potentials are generated only at the nodes, which is where the voltage gated channels occur. The axon is insulated so the action potential at one node is sufficient to trigger an action potential in the next node and the impulse jumps along the fibre. This differs from impulse transmission in a non-myelinated neuron in which voltage-gated channels occur along the entire length of the axon.

As well as increasing the speed of conduction, the myelin sheath reduces energy expenditure because the area over which depolarization occurs is less (and therefore the number of sodium and potassium ions that need to be pumped to restore the resting potential is fewer).

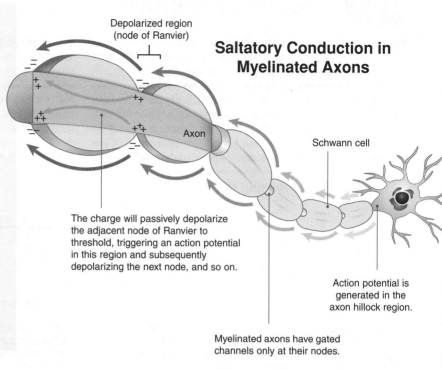

Saltatory Conduction in Myelinated Axons

Depolarized region (node of Ranvier)

Axon

Schwann cell

The charge will passively depolarize the adjacent node of Ranvier to threshold, triggering an action potential in this region and subsequently depolarizing the next node, and so on.

Action potential is generated in the axon hillock region.

Myelinated axons have gated channels only at their nodes.

1. What is an action potential? _____

_____

2. (a) What occurs during saltatory conduction? _____

_____

(b) What influence does this have on conduction speed? _____

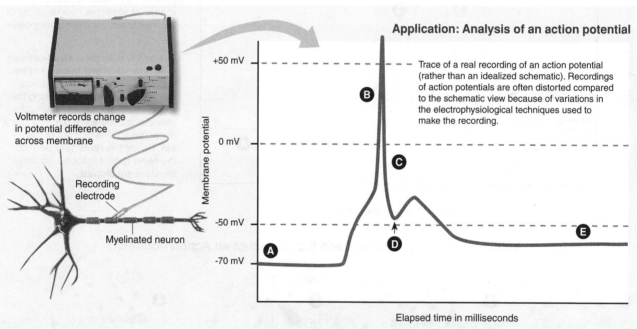

Voltmeter records change in potential difference across membrane

Recording electrode

Myelinated neuron

Application: Analysis of an action potential

+50 mV

0 mV

-50 mV

-70 mV

Membrane potential

Elapsed time in milliseconds

Trace of a real recording of an action potential (rather than an idealized schematic). Recordings of action potentials are often distorted compared to the schematic view because of variations in the electrophysiological techniques used to make the recording.

3. The graph above shows a recording of the changes in membrane potential in an axon during transmission of an action potential. Match each stage (A-E) to the correct summary provided below.

[ ] Membrane depolarization (due to rapid Na+ entry across the axon membrane.

[ ] Hyperpolarization (an overshoot caused by the delay in closing of the K+ channels.

[ ] Return to resting potential after the stimulus has passed.

[ ] Repolarization as the Na+ channels close and slower K+ channels begin to open.

[ ] The membrane's resting potential.

© 2012-2014 **BIOZONE** International
ISBN: 978-1-927173-93-0
Photocopying Prohibited

194 Neurotransmitters

Key Idea: Neurotransmitters are chemicals that allow the transmission of signals between neurons.

Neurotransmitters are chemicals that transmit signals between neurons. They are found in the axon endings of neurons and are released into the space between one neuron and the next (the synaptic cleft) after depolarization or hyperpolarization of the nerve ending. Different neurotransmitters may produce different responses depending on their location in the body. They can be excitatory (likely to cause an action potential in the receiving neuron) or inhibitory (causing hyperpolarization) depending on the receptor they activate.

Neurotransmitters Carry Signals Between Neurons

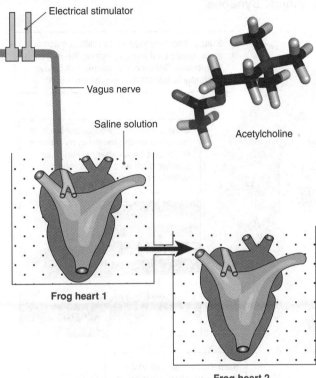

Electrical stimulator

Vagus nerve

Saline solution

Acetylcholine

Frog heart 1

Frog heart 2

Chemical signalling between neurons was first demonstrated in 1921 by Otto Loewi. In his experiment, the still beating hearts of two frogs were placed in connected flasks filled with saline solution. The vagus nerve of the first heart was still attached and was stimulated by electricity to reduce the heart's rate of beating. After a delay, the rate of beating in the second heart also slowed. Increasing the beating rate in the first heart caused an increase in the beating rate in the second heart. Electrical stimulus of the first heart caused it to release a chemical into the saline that then affected the heartbeat of the second heart. The chemical was found to be **acetylcholine,** now known to be a common neurotransmitter.

The Effect of Insecticides on Neurotransmitters

The discovery of neurotransmitters and how they work has allowed scientists to exploit their properties to develop useful applications, including insecticides. Insecticides are chemical substances used to control pest insect numbers. Many insecticides work by affecting the signalling of nerve cells by either blocking uptake of signalling molecules or facilitating the uptake of far greater amounts than normal.

Neonicotoid insecticides are a group of insectides which mimic the action of acetylcholine in synapses. They bind irreversibly to the postsynaptic nicotinic acetylcholine receptors causing the over-stimulation of the neuron, which results in death of the insect. The effects are cumulative, meaning the build up over time, so even at low doses they are fatal.

Pea aphid sucking sap from a pea plant.

Shipher Wu cc2.5

Neonicotoid insecticides are particularly effective against sucking insects, such as aphids (above), which cause large scale damage to many commercial crops.

1. What is the purpose of a neurotransmitter? _____

_____

2. (a) Name the neurotransmitter discovered by Loewi in his frog heart experiment: _____

   (b) Why was there a delay before the second heart in the experiment reduced its beating rate? _____

_____

_____

_____

3. How do **neonicotoid insecticides** interact with chemical synapses? _____

_____

_____

LINK
195

APP

195 Chemical Synapses

Key Idea: Chemical synapses are junctions between neurons, or between neurons and receptor or effector cells.

Action potentials are transmitted across junctions called **synapses**. Synapses can occur between two neurons, or between a neuron and an effector cell (e.g. muscle or gland). The axon terminal is a swollen knob, and a small gap (synaptic cleft) separates it from the receiving neuron. The synaptic knobs are filled with tiny packets of chemicals

called **neurotransmitters**. The neurotransmitter diffuses across the gap, where it interacts with the receiving (post-synaptic) membrane and causes an electrical response. In the example below, the neurotransmitter causes a membrane depolarization and the generation of an action potential. Some neurotransmitters have the opposite effect and cause inhibition (e.g. slowing heart rate). Chemical synapses are the most widespread type of synapse in nervous systems.

The Structure of a Chemical Synapse

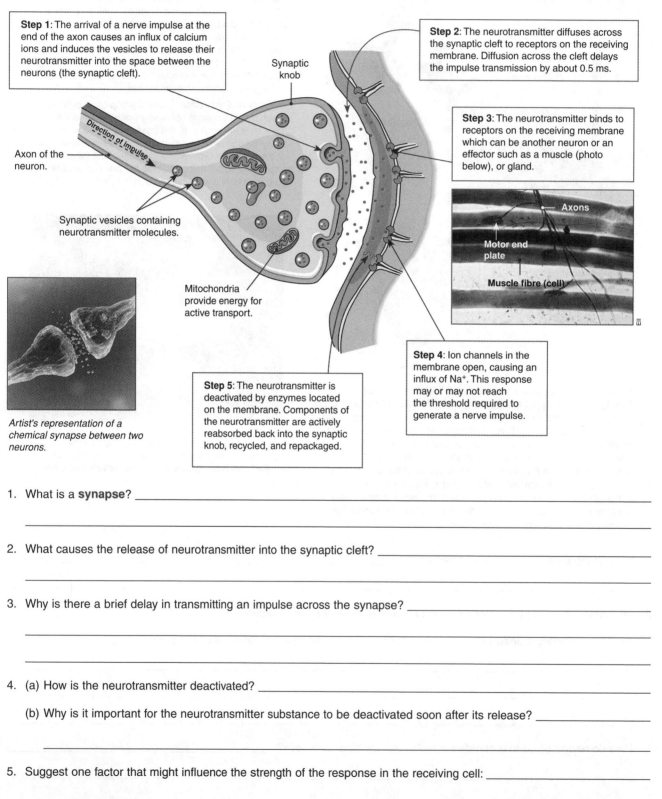

Step 1: The arrival of a nerve impulse at the end of the axon causes an influx of calcium ions and induces the vesicles to release their neurotransmitter into the space between the neurons (the synaptic cleft).

Step 2: The neurotransmitter diffuses across the synaptic cleft to receptors on the receiving membrane. Diffusion across the cleft delays the impulse transmission by about 0.5 ms.

Step 3: The neurotransmitter binds to receptors on the receiving membrane which can be another neuron or an effector such as a muscle (photo below), or gland.

Step 4: Ion channels in the membrane open, causing an influx of Na$^+$. This response may or may not reach the threshold required to generate a nerve impulse.

Step 5: The neurotransmitter is deactivated by enzymes located on the membrane. Components of the neurotransmitter are actively reabsorbed back into the synaptic knob, recycled, and repackaged.

Synaptic knob

Direction of impulse

Axon of the neuron.

Synaptic vesicles containing neurotransmitter molecules.

Mitochondria provide energy for active transport.

Axons

Motor end plate

Muscle fibre (cell)

Artist's representation of a chemical synapse between two neurons.

1. What is a **synapse**? _____

_____

2. What causes the release of neurotransmitter into the synaptic cleft? _____

_____

3. Why is there a brief delay in transmitting an impulse across the synapse? _____

_____

_____

4. (a) How is the neurotransmitter deactivated? _____

(b) Why is it important for the neurotransmitter substance to be deactivated soon after its release? _____

_____

5. Suggest one factor that might influence the strength of the response in the receiving cell: _____

_____

© 2012-2014 **BIOZONE** International
ISBN: 978-1-927173-93-0
Photocopying Prohibited

196 Chemical Imbalances in the Brain

Key Idea: Some mental disorders can be treated with drugs that act on the synapses in the brain.

Many types of mental illness result from disturbances to natural levels of specific neurotransmitters, and can lead to

the failure of specific neural pathways. Scientists have used their knowledge of neurotransmitters and chemical synapses to develop drugs that either replace or boost levels of specific neurotransmitters and help treat specific brain disorders.

Parkinson's Disease

Patients with **Parkinson's disease** show decreased stimulation in the motor cortex of the brain. This results from reduced dopamine production in the substantia nigra region (right) where dopamine is produced. This is usually the result of the death of nerve cells. Symptoms, slow physical movement and spasmodic tremors, often don't begin to appear until a person has lost 70% of their dopamine-producing cells.

Treating Parkinson's Disease

Parkinson's disease is caused by reduced dopamine production and low dopamine levels in the brain pathways involved with movement. Treatments for Parkinson's have focused on increasing the body's dopamine levels. Dopamine is unable to cross the blood-brain barrier, so cannot be administered as a treatment. However, **L-dopa** is a dopamine precursor that can cross the blood-brain barrier and enter the brain. Once in the brain, it is converted to dopamine. L-dopa has been shown to reduce some of the symptoms of Parkinson's disease.

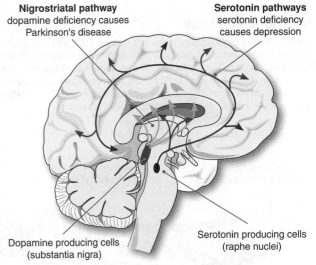

Nigrostriatal pathway
dopamine deficiency causes Parkinson's disease

Serotonin pathways
serotonin deficiency causes depression

Dopamine producing cells (substantia nigra)

Serotonin producing cells (raphe nuclei)

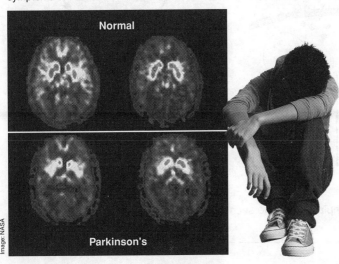

Normal

Parkinson's

Image: NASA

Positron emission tomography (PET) measures the activity of dopamine neurons in the substantia nigra area of the brain. Parkinson's patients (lower panel) show reduced activity in the dopamine neurons compared with normal patients.

Depression

A person with **depression** (left) experiences prolonged periods of extremely low mood, including low self esteem, regret, guilt, and feelings of hopelessness. Depression may be caused by a mixture of environmental factors (e.g. stress) and biological factors (e.g. low **serotonin** production by the raphe nuclei in the brain, above).

Treating Depression

Recognition of the link between **serotonin** and **depression** has resulted in the development of **antidepressant drugs** that alter serotonin levels. Monoamine oxidase inhibitors (MAOI) are commonly used antidepressants that increase serotonin levels by preventing its breakdown in the brain. Newer drugs, called Selective Serotonin Re-uptake Inhibitors (SSRIs), stop serotonin re-uptake by presynaptic cells. This increases the levels of extracellular serotonin, making more available to bind to the postsynaptic cells, and stabilizing serotonin levels in the brain. SSRIs have fewer side effects than other antidepressants because they specifically target serotonin and no other neurotransmitters.

1. What role do neurotransmitters have in mental illness? _____

_____

_____

2. Describe the cause of the following diseases and describe how they can be treated using pharmaceuticals:

(a) Parkinson's disease: _____

_____

_____

(b) Depression: _____

_____

_____

LINK
194

WEB
196

APP

197 Hormonal Regulatory Systems

Key Idea: The endocrine system regulates physiological processes by releasing blood borne chemical messengers (called hormones) which interact with target cells.

The endocrine system is made up of endocrine cells (organized into endocrine glands) and the hormones they produce. Hormones are potent chemical regulators. They are produced in very small quantities but can exert a very large effect on metabolism. Endocrine glands secrete hormones directly into the bloodstream rather than through a duct or tube. The basis of hormonal control and the role of negative feedback mechanisms in regulating hormone levels are described below.

How Hormones Work

Endocrine cells produce hormones and secrete them into the bloodstream where they are distributed throughout the body. Although hormones are sent throughout the body, they affect only specific target cells. These target cells have receptors on the plasma membrane which recognize and bind the hormone (see inset, below). The binding of hormone and receptor triggers the response in the target cell. Cells are unresponsive to a hormone if they do not have the appropriate receptors.

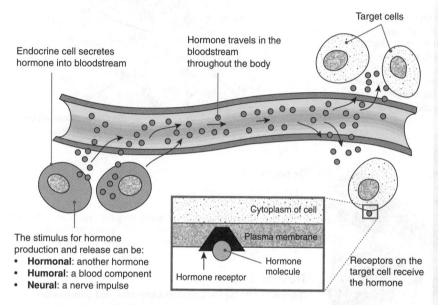

Target cells

Hormone travels in the bloodstream throughout the body

Endocrine cell secretes hormone into bloodstream

The stimulus for hormone production and release can be:
- **Hormonal**: another hormone
- **Humoral**: a blood component
- **Neural**: a nerve impulse

Cytoplasm of cell

Plasma membrane

Hormone molecule

Hormone receptor

Receptors on the target cell receive the hormone

Antagonistic Hormones

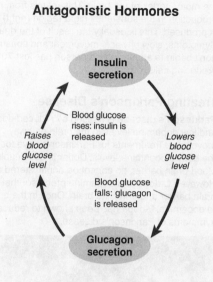

Insulin secretion

Blood glucose rises: insulin is released

Raises blood glucose level

Lowers blood glucose level

Blood glucose falls: glucagon is released

Glucagon secretion

The effects of one hormone are often counteracted by an opposing hormone. Feedback mechanisms adjust the balance of the two hormones to maintain a physiological function. Example: insulin acts to decrease blood glucose and glucagon acts to raise it.

1. (a) What are **antagonistic hormones**? Describe an example of how two such hormones operate:

_____

_____

_____

_____

(b) Describe the role of feedback mechanisms in adjusting hormone levels (explain using an example if this is helpful):

_____

_____

2. How can a hormone influence only the target cells even though all cells may receive the hormone?

_____

_____

3. Explain why hormonal control differs from nervous system control with respect to the following:

(a) The speed of hormonal responses is slower: _____

_____

(b) Hormonal responses are generally longer lasting: _____

_____

© 2012-2014 **BIOZONE** International
ISBN: 978-1-927173-93-0
Photocopying Prohibited

198 Principles of Homeostasis

Key Idea: Homeostasis is the relatively constant state of the internal environment, which is maintained by regulatory mechanisms despite changes in the external environment. Maintaining **homeostasis** (a constant internal environment) occurs through interactions of the body systems. Homeostatic control systems have three functional components: a receptor to detect change, a control centre, and an effector to direct an appropriate response. **Negative feedback** is a control system which maintains the body's internal environment at a steady state. Negative feedback has a stabilizing effect and acts to discourage variations from a set point. It works by returning internal conditions back to a steady state when variations are detected. Most body systems achieve homeostasis through negative feedback.

Organ systems maintain a constant internal environment that provides for the needs of all the body's cells, making it possible for animals to move through different and often highly variable external environments. This representation shows how organ systems permit exchanges with the environment. The exchange surfaces are usually internal, but may be connected to the environment via openings on the body surface.

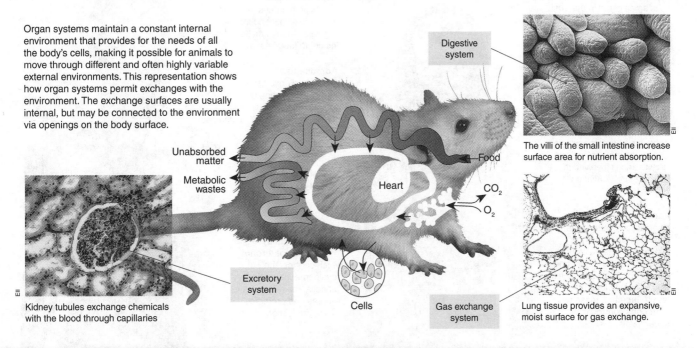

Digestive system

Unabsorbed matter

Metabolic wastes

Heart

Food

CO_2

O_2

Excretory system

Cells

Gas exchange system

The villi of the small intestine increase surface area for nutrient absorption.

Kidney tubules exchange chemicals with the blood through capillaries

Lung tissue provides an expansive, moist surface for gas exchange.

Negative Feedback and Control Systems

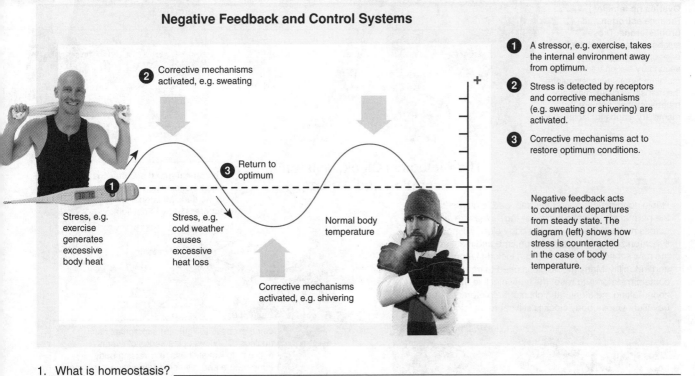

2 Corrective mechanisms activated, e.g. sweating

3 Return to optimum

1 Stress, e.g. exercise generates excessive body heat

Stress, e.g. cold weather causes excessive heat loss

Normal body temperature

Corrective mechanisms activated, e.g. shivering

1 A stressor, e.g. exercise, takes the internal environment away from optimum.

2 Stress is detected by receptors and corrective mechanisms (e.g. sweating or shivering) are activated.

3 Corrective mechanisms act to restore optimum conditions.

Negative feedback acts to counteract departures from steady state. The diagram (left) shows how stress is counteracted in the case of body temperature.

1. What is homeostasis? _____

_____

2. How do negative feedback mechanisms maintain homeostasis in a variable environment? _____

_____

_____

_____

LINK
200

WEB
198

KNOW

199 The Endocrine System

Key Idea: The endocrine system regulates specific aspects of metabolism through hormones, which are secreted from endocrine glands.

The **endocrine glands** are ductless glands distributed throughout the body. They secrete **hormones** (chemical messengers) which are carried in the blood to exert a specific effect on target cells. They are then broken down and excreted. The hypothalamus is part of the brain and not strictly an endocrine gland, but it links the nervous and endocrine systems and helps to regulate the body's activities.

Pineal
This small gland in the brain secretes **melatonin**, which regulates the sleep-wake cycle. Melatonin secretion follows a circadian (24 hour) rhythm and coordinates reproductive hormones too.

Thyroid gland
Secretes **thyroxin**, an iodine containing hormone which stimulates metabolism and growth and is involved in temperature regulation.

Pancreas
Specialized α endocrine cells secrete **glucagon** which acts to raise blood glucose levels. β endocrine cells produce **insulin** which acts to lower blood glucose levels. Together, these two hormones control blood sugar levels.

Ovaries (in females)
Produce **estrogen** and **progesterone**. These sex hormones control the development of primary and secondary sex characteristics, maintain femaleness, stimulate the menstrual cycle, maintain pregnancy, and prepare the mammary glands for lactation.

Hypothalamus
The hypothalamus monitors hormone levels and indirectly regulates many functions, including food intake.

Adipose tissue
Cells in adipose tissue secrete **leptin**, a hormone which acts on the hypothalamus to inhibit appetite. It acts on other sites to increase energy expenditure.

Testes (in males)
Produce **testosterone**, which controls the development of male genitalia, maintains male features and promotes sperm production.

The Biological Clock, Melatonin Secretion and Jet Lag

Rapid, long distance air travel can lead to disruption of the normal sleep-wake cycle (**jet lag**). When travelling across multiple time zones, the body clock will not be synchronized with the destination time and must adjust to the new schedule. Symptoms can include fatigue, insomnia, and irritability. Many medications used to treat jet lag contain melatonin to reset the body clock to the new time zone. Taking melatonin at night can help promote sleep in travellers whose body clock is still set to the old time zone.

The severity of jet lag is linked to the west–east distance traveled, rather than the length of flight.

The **pineal gland** secretes the sleep-inducing hormone **melatonin** in the dark. Melatonin production is suppressed by bright light.

Melatonin

Eye

Once exposed to light, the suprachiasmatic nucleus (SCN) sets off a series of events to promote wakefulness (e.g. raising body temperature and releasing stimulating hormones). It also communicates with the pineal gland, suppressing melatonin production until it is dark.

The **biological clock** is responsible for regulating the natural sleep-wake cycle, which involves being awake and active during the day and sleeping at night when it is dark. The clock is made up of a collection of cells in the hypothalamus, called the suprachiasmatic nucleus (**SCN**), just behind the eyes. Light from the eyes helps to regulate SCN activity. To keep the cycle synchronized with the 24 hour-day cycle, the clock needs to be reset each day.

© 2012-2014 **BIOZONE** International
ISBN: 978-1-927173-93-0

The Role of Leptin

Increase in fat mass

↑ Leptin
↑ Body temperature
↑ Energy expenditure
↓ Appetite
Result: weight loss

Decrease in fat mass

↓ Leptin
↓ Body temperature
↓ Energy expenditure
↑ Appetite
Result: weight gain

Leptin is an appetite-suppressing hormone secreted by fat cells. Leptin acts on receptors in the hypothalamus, regulating appetite, food intake, and metabolism. It is important in maintaining normal body weight (above). Mice unable to produce leptin become morbidly obese (right).

US Govt

Leptin and Obesity

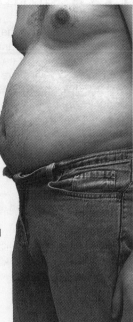

When the effects of leptin were first identified, it was hoped that leptin supplements could be used to treat obesity. However, when researchers gave obese people leptin they didn't lose weight as expected.

Further research showed that obese people had pre-existing elevated leptin levels and had developed leptin resistance. Over time, the brain's ability to respond to leptin had become reduced, so that leptin no longer worked to suppress appetite.

1. Use the information opposite to complete the following table.

Hormone(s)	Secreted by	Role
Insulin		
	Adipose tissue	
	Ovaries	
Melatonin		
Testosterone		
		Increases blood glucose levels
Thyroxin		

2. (a) How does long distance travel disrupt an individual's normal sleep-wake cycle? _____

_____

_____

(b) How could taking **melatonin** supplements help reduce the effects of jet lag? _____

_____

_____

_____

3. Explain the role of **leptin** in maintaining body weight: _____

_____

_____

4. Leptin levels in obese people are very high, yet it does not decrease their appetite. Explain why this occurs: _____

_____

_____

_____

200 Control of Blood Glucose

Key Idea: The endocrine portion of the pancreas produces the hormones insulin and glucagon, which maintain blood glucose at a steady state through negative feedback.

Blood glucose levels are controlled by **negative feedback** involving two hormones, insulin and glucagon. These hormones are produced by the islet cells of the pancreas, and act in opposition to control blood glucose levels. **Insulin** is secreted from β cells, and lowers blood glucose by promoting the uptake of glucose from the blood into the body's cells, and conversion of glucose into the storage molecule glycogen in the liver. **Glucagon** is secreted from α cells and increases blood glucose by stimulating the breakdown of stored glycogen and the synthesis of glucose from amino acids. Negative feedback stops hormone secretion when normal blood glucose levels are restored.

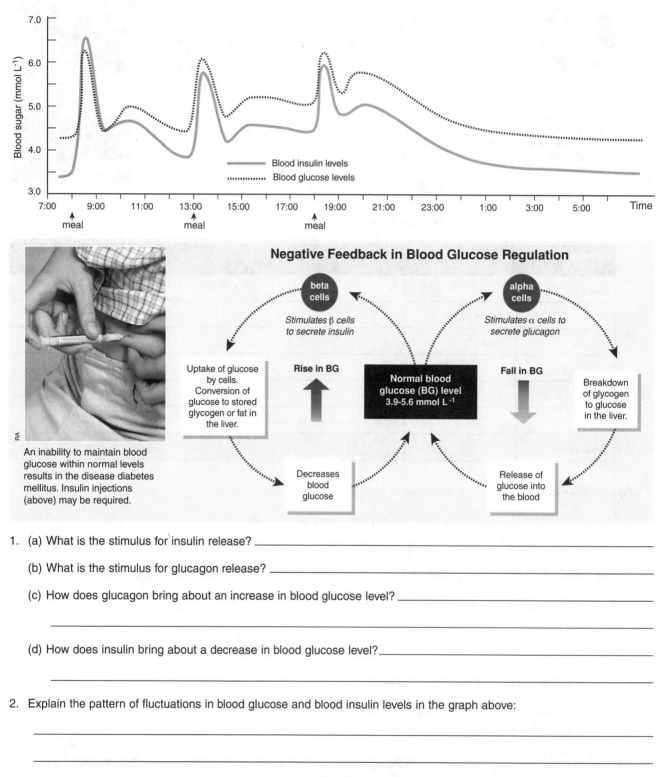

An inability to maintain blood glucose within normal levels results in the disease diabetes mellitus. Insulin injections (above) may be required.

Negative Feedback in Blood Glucose Regulation

1. (a) What is the stimulus for insulin release? _____

   (b) What is the stimulus for glucagon release? _____

   (c) How does glucagon bring about an increase in blood glucose level? _____

   _____

   (d) How does insulin bring about a decrease in blood glucose level? _____

   _____

2. Explain the pattern of fluctuations in blood glucose and blood insulin levels in the graph above:

   _____

   _____

   _____

   _____

3. Identify the mechanism regulating insulin and glucagon secretion (humoral, hormonal, neural): _____

© 2012-2014 **BIOZONE** International
ISBN: 978-1-927173-93-0
Photocopying Prohibited

201 Diabetes Mellitus

Key Idea: Diabetes mellitus is a condition in which blood glucose levels are elevated, either because of a lack of insulin (type 1) or because of resistance to insulin's effects (type 2). **Diabetes mellitus** (diabetes) is a condition in which blood glucose is too high because the body's cells cannot take up glucose in the normal way. It is usually detected by glucose appearing in the urine (glucose is normally reabsorbed and does not enter the urine). The two types of diabetes, type 1 and type 2, have different causes and treatments, but both are life threatening conditions if untreated (below).

Type 1 Diabetes

No insulin is produced because the insulin-producing cells of the pancreas are damaged.

Type 2 Diabetes

Insulin is produced. However, either not enough insulin is made, or the body's cells do not react to it.

The **beta cells** of the pancreatic islets (outlined above) produce insulin.

Type 1 Diabetes Mellitus (insulin dependent)

Age at onset: Early in life; often in childhood. Often called juvenile onset diabetes

Cause: Absolute deficiency of insulin due to lack of insulin production (pancreatic β cells are destroyed in an autoimmune reaction). There is a genetic component, but usually a childhood viral infection (e.g. mumps or rubella) triggers the development of type 1 diabetes.

Treatment: Blood glucose is monitored regularly and insulin injections combined with dietary management are used to keep blood sugar levels stable.

Therapies involving pancreatic transplants, or transplants of insulin-producing islet cells or stem cells are currently being investigated.

Type 2 Diabetes Mellitus (insulin resistant)

Age at onset: Usually in adults over the age of 40, but its incidence is increasing in younger adults and obese children.

Cause: Type 2 diabetes occurs most commonly as a result of lifestyle factors. Obesity (BMI > 27), a sedentary lifestyle, hypertension, high blood lipids, and a poor diet all contribute to make a person more susceptible to developing type 2 diabetes. Ethnicity and genetic factors may also be a contributing factor.

Treatment: Increasing physical activity, losing weight (especially abdominal fat), and improving diet may be sufficient to control type 2 diabetes.

The use of prescribed anti-diabetic drugs and insulin therapy (injections) may be required if lifestyle changes are insufficient on their own.

1. Why does diabetes mellitus result in high blood sugar levels? _____

_____

_____

2. Discuss the differences between type 1 and type 2 diabetes, including causes and treatments: _____

_____

_____

_____

_____

_____

_____

© 2012-2014 **BIOZONE** International
ISBN: 978-1-927173-93-0
Photocopying Prohibited

LINK **305**   LINK **200**   WEB **201**   **APP**

202 The Male Reproductive System

Key Idea: The reproductive role of the male is to produce the sperm and deliver them to the female.

When a sperm combines with an egg, it contributes half the genetic material of the offspring and, in humans and other mammals, determines its sex. The reproductive structures of the human male is shown below.

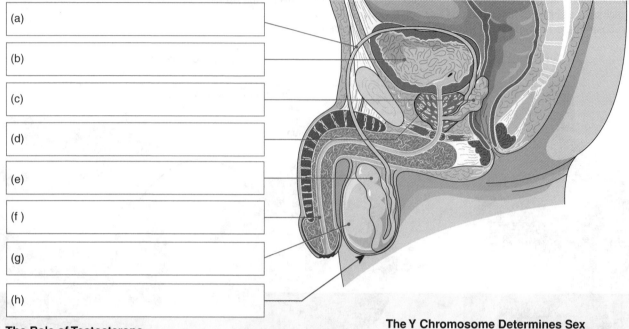

(a)

(b)

(c)

(d)

(e)

(f)

(g)

(h)

The Role of Testosterone

Testosterone plays three major roles in male development.

It causes the development of the male primary sexual characteristics (male genitalia) in the embryo.

At puberty, testosterone causes the development of the male secondary sexual characteristics (development of sperm, growth of body hair, development of muscles, deepening of voice).

In adult males testosterone plays a role in maintaining the sex drive and production of sperm.

Y chromosome

The Y Chromosome Determines Sex

In humans, males have two sex chromosomes, one X and one Y chromosome. The determination of sex is based on the presence or absence of the Y chromosome; without the Y chromosome, an individual will develop into a female.

One of the genes located on the Y chromosome is the SRY gene (sex determining region Y). This gene produces a protein that causes the gonads to develop into testes and produce testosterone.

1. The male human reproductive system and associated structures are shown above. Using the following word list, and the weblinks provide below, identify the labeled parts (write your answers in the spaces provided on the diagram).
   Word list: *bladder, scrotal sac, sperm duct (vas deferens), epididymis, seminal vesicle, testis, urethra, prostate gland*

2. In a short sentence, state the function of each of the structures labelled (a)-(h) in the diagram above:

   (a) _____

   (b) _____

   (c) _____

   (d) _____

   (e) _____

   (f) _____

   (g) _____

   (h) _____

3. What roles does testosterone play in male development and the male reproductive system? _____

   _____

   _____

4. How is sex determined in humans? _____

   _____

© 2012-2014 **BIOZONE** International
ISBN: 978-1-927173-93-0
Photocopying Prohibited

203 The Female Reproductive System

Key Idea: The female reproductive system produces eggs, receives the penis and sperm during sexual intercourse, protects and houses the developing fetus, and produces milk to nourish the young after birth.

The female reproductive system consists of the ovaries, Fallopian tubes, uterus, the vagina and external genitalia, and the breasts. The reproductive structures of the human female is shown below.

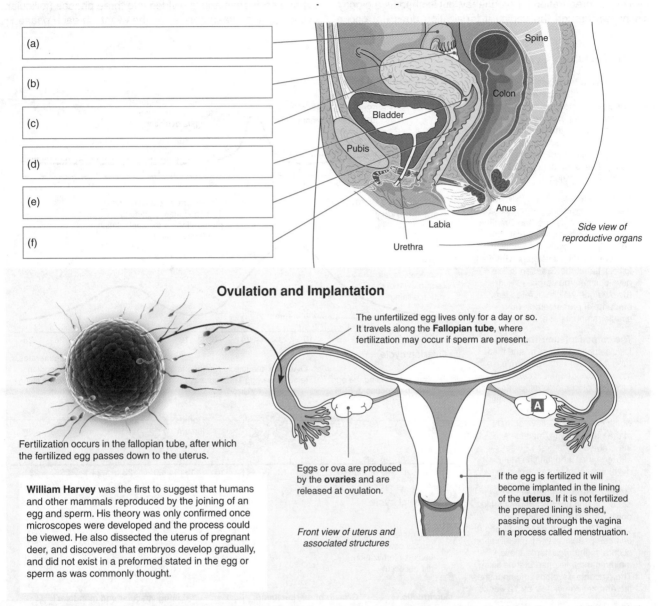

(a)

(b)

(c)

(d)

(e)

(f)

Spine

Colon

Bladder

Pubis

Labia

Anus

Urethra

Side view of reproductive organs

Ovulation and Implantation

The unfertilized egg lives only for a day or so. It travels along the **Fallopian tube**, where fertilization may occur if sperm are present.

Fertilization occurs in the fallopian tube, after which the fertilized egg passes down to the uterus.

Eggs or ova are produced by the **ovaries** and are released at ovulation.

If the egg is fertilized it will become implanted in the lining of the **uterus**. If it is not fertilized the prepared lining is shed, passing out through the vagina in a process called menstruation.

Front view of uterus and associated structures

William Harvey was the first to suggest that humans and other mammals reproduced by the joining of an egg and sperm. His theory was only confirmed once microscopes were developed and the process could be viewed. He also dissected the uterus of pregnant deer, and discovered that embryos develop gradually, and did not exist in a preformed stated in the egg or sperm as was commonly thought.

1. The female human reproductive system and associated structures are illustrated above. Using the word list, and the weblinks below to identify the labeled parts. **Word list**: *ovary, uterus (womb), vagina, fallopian tube (oviduct), cervix, clitoris.*

2. In a few words or a short sentence, state the function of each of the structures labeled (a) - (e) in the above diagram:

   (a) _____

   (b) _____

   (c) _____

   (d) _____

   (e) _____

3. (a) Name the organ labelled (**A**) in the diagram: _____

   (b) Name the event associated with this organ that occurs every month: _____

   (c) Name the process by which mature ova are produced: _____

4. Where does fertilization occur? _____

LINK **311**   LINK **257**   LINK **199**   WEB **203** SKILL

204 The Menstrual Cycle

Key Idea: The menstrual cycle involves cyclical changes in the ovaries and uterus to prepare for fertilization of an egg. In humans, fertilization of the ovum (egg) is most likely to occur around the time of ovulation. The uterine lining (endometrium) thickens in preparation for pregnancy, but is shed as a bloody discharge through the vagina if fertilization does not occur.

This event, called **menstruation**, characterizes the human reproductive or **menstrual cycle**. The menstrual cycle starts from the first day of bleeding and lasts for about 28 days. It involves predictable changes in response to pituitary and ovarian hormones and is divided into three phases (follicular, ovulatory, and luteal) defined by the events in each phase.

The Menstrual Cycle

Luteinizing hormone (LH) and follicle stimulating hormone (FSH): These hormones from the anterior pituitary have numerous effects. FSH stimulates the development of the ovarian follicles resulting in the release of estrogen. Estrogen levels peak, stimulating a surge in LH and triggering ovulation.

Hormone levels: Of the follicles that begin developing in response to FSH, usually only one (the Graafian follicle) becomes dominant. In the first half of the cycle, estrogen is secreted by this developing Graafian follicle. Later, the Graafian follicle develops into the corpus luteum (below right) which secretes large amounts of progesterone (and smaller amounts of estrogen).

The corpus luteum: The Graafian follicle continues to grow and then (about day 14) ruptures to release the egg (ovulation). LH causes the ruptured follicle to develop into a corpus luteum (yellow body). The corpus luteum secretes progesterone which promotes full development of the uterine lining, maintains the embryo in the first 12 weeks of pregnancy, and inhibits the development of more follicles.

Menstruation: If fertilization does not occur, the corpus luteum breaks down. Progesterone secretion declines, causing the uterine lining to be shed (menstruation). If fertilization occurs, high progesterone levels maintain the thickened uterine lining. The placenta develops and nourishes the embryo completely by 12 weeks.

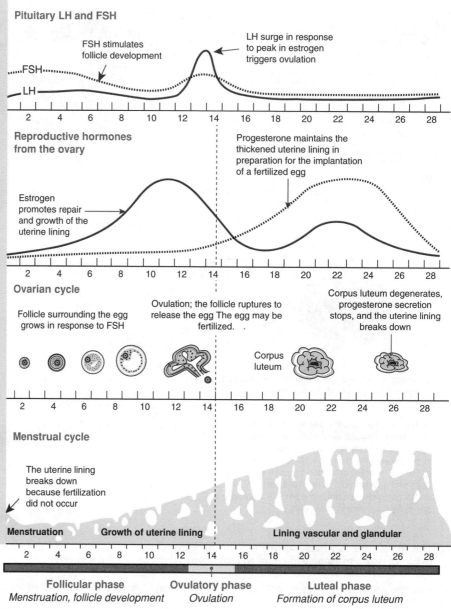

Pituitary LH and FSH

FSH stimulates follicle development

LH surge in response to peak in estrogen triggers ovulation

FSH

LH

Reproductive hormones from the ovary

Estrogen promotes repair and growth of the uterine lining

Progesterone maintains the thickened uterine lining in preparation for the implantation of a fertilized egg

Ovarian cycle

Follicle surrounding the egg grows in response to FSH

Ovulation; the follicle ruptures to release the egg The egg may be fertilized.

Corpus luteum degenerates, progesterone secretion stops, and the uterine lining breaks down

Corpus luteum

Menstrual cycle

The uterine lining breaks down because fertilization did not occur

Menstruation   Growth of uterine lining   Lining vascular and glandular

Day of the cycle: 2   4   6   8   10   12   14   16   18   20   22   24   26   28

Follicular phase
Menstruation, follicle development

Ovulatory phase
Ovulation

Luteal phase
Formation of corpus luteum

1. Name the hormone responsible for:

   (a) Follicle growth: _____   (b) Ovulation: _____

2. Each month, several ovarian follicles begin development, but only one (the Graafian follicle) develops fully:

   (a) Name the hormone secreted by the developing follicle: _____

   (b) State the role of this hormone during the follicular phase: _____

   (c) Suggest what happens to the follicles that do not continue developing: _____

3. (a) Identify the principal hormone secreted by the corpus luteum: _____

   (b) State the purpose of this hormone: _____

4. State the hormonal trigger for menstruation: _____

© 2012-2014 **BIOZONE** International
ISBN: 978-1-927173-93-0
Photocopying Prohibited

205 Using Hormones to Treat Infertility

Key Idea: *In vitro* fertilization involves fertilizing eggs with sperm in the laboratory and placing them in into the woman's uterus to develop. The process uses fertility hormones.
***In vitro* fertilization** (IVF) can be used to treat some infertility problems. During IVF, eggs are removed from the ovaries and fertilized with sperm in a laboratory. The fertilized egg is implanted into the mother's uterus for development. Most treatments for female infertility involve the use of synthetic female hormones, which stimulate ovulation, boost egg production, and induce egg release.

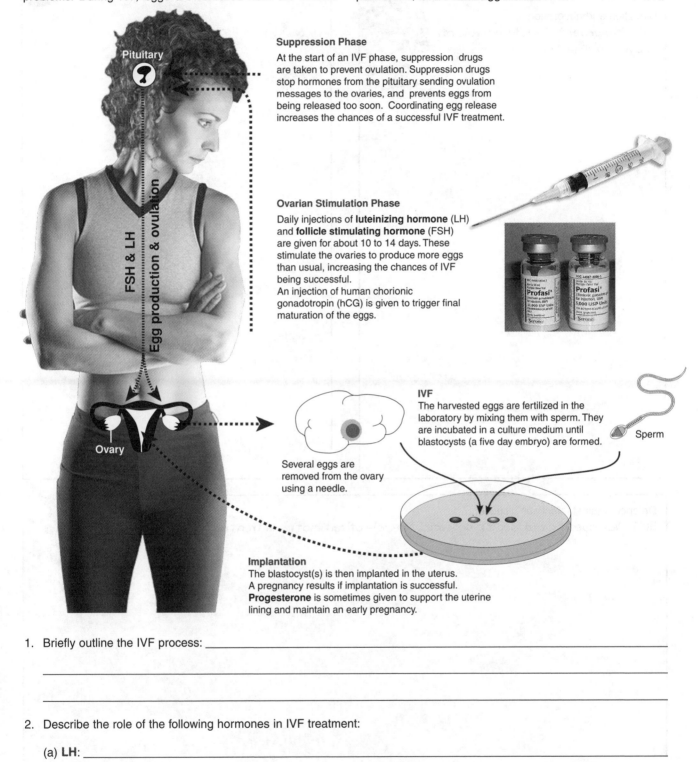

Suppression Phase

At the start of an IVF phase, suppression drugs are taken to prevent ovulation. Suppression drugs stop hormones from the pituitary sending ovulation messages to the ovaries, and prevents eggs from being released too soon. Coordinating egg release increases the chances of a successful IVF treatment.

Ovarian Stimulation Phase

Daily injections of **luteinizing hormone** (LH) and **follicle stimulating hormone** (FSH) are given for about 10 to 14 days. These stimulate the ovaries to produce more eggs than usual, increasing the chances of IVF being successful.
An injection of human chorionic gonadotropin (hCG) is given to trigger final maturation of the eggs.

IVF
The harvested eggs are fertilized in the laboratory by mixing them with sperm. They are incubated in a culture medium until blastocysts (a five day embryo) are formed.

Sperm

Several eggs are removed from the ovary using a needle.

Ovary

Pituitary

FSH & LH

Egg production & ovulation

Implantation
The blastocyst(s) is then implanted in the uterus. A pregnancy results if implantation is successful.
Progesterone is sometimes given to support the uterine lining and maintain an early pregnancy.

1. Briefly outline the IVF process: _____

_____

_____

2. Describe the role of the following hormones in IVF treatment:

   (a) **LH**: _____

   (b) **FSH**: _____

   (c) **hCG**: _____

   (d) **Progesterone**: _____

3. Why is it important that ovulation is suppressed at the beginning of the IVF process? _____

_____

LINK
312

LINK
199

WEB
205

KNOW

206 Chapter Review

Summarize what you know about this topic under the headings provided. You can draw diagrams or mind maps, or write short notes to organize your thoughts. Use the images and hints and guidelines included to help you:

The blood system:
HINT: Describe the structure of the heart and the blood vessels.

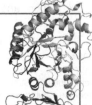

Digestion and absorption:
HINT: Remember to include the role of enzymes in digestion.

Defence against infectious disease:
HINT: Non-specific and specific defences. The role of antibiotics in treating disease.

© 2012-2014 **BIOZONE** International
ISBN: 978-1-927173-93-0
Photocopying Prohibited

REVISE

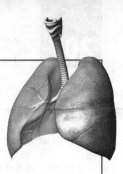

Gas exchange:

HINT: Mechanisms and role of breathing. How does lung disease affect gas exchange?

Neurons and synapses:

HINT: How are electrical impulses transmitted in and between neurons?

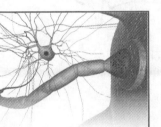

Hormones, homeostasis, and reproduction:

HINT: The role of hormones in homeostasis and reproduction.

207 KEY TERMS: Did You Get It?

1. (a) What process moves food through the gut? _____

   (b) In what region of the digestive system does most absorption occur? _____

   (c) What is the function of the villi in the small intestine? _____

   (d) What organ secretes amylase into the small intestine? _____

2. (a) What type of blood vessel transports blood away from the heart? _____

   (b) What type of blood vessel transports blood to the heart? _____

   (c) What type of blood vessel enables exchanges between the blood and tissues? _____

3. (a) What component of blood is involved in blood clotting? _____

   (b) What cell types in blood are involved in defence against pathogens? _____

   (c) What group of organisms are antibiotics effective against? _____

4. The image below shows an electrocardiogram (ECG). Mark on the diagram one full cardiac cycle.

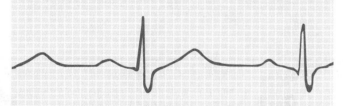

5. (a) Label the components of this neuron (right) using the following word list: *cell body, axon, dendrites, node of Ranvier.*

   (b) Is this neuron myelinated or unmyelinated?(delete one)

   (c) Explain your answer: _____

   _____

   _____

   (d) In what form do electrical signals travel in this cell?

   _____

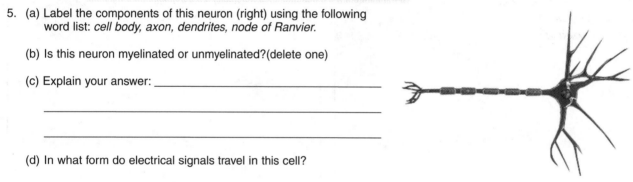

6. Test your vocabulary by matching each term to its definition, as identified by its preceding letter code.

alveoli	**A** A pulmonary function test measuring how much air a person can breathe in and out and how fast the air can be expelled.
breathing	**B** A mechanism in which the output of a system acts to oppose changes to the input of the system; the net effect is to stabilize the system against change.
diabetes	**C** The exchange of oxygen and carbon dioxide across the gas exchange membrane.
gas exchange	**D** The act of inhaling air into and exhaling air from the lungs.
lungs	**E** Internal gas exchange structures found in most vertebrates.
negative feedback	**F** Any gas that takes part in the respiratory process (usually oxygen or carbon dioxide).
pancreas	**G** A condition in which the blood glucose level is elevated above normal levels.
respiratory gas	**H** An organ that produces the hormones insulin and glucagon.
spirometry	**I** Microscopic structures in vertebrate lungs, which form the terminus of the bronchioles. The site of gas exchange.

TEST

Topic 7

Nucleic Acids

Key terms

DNA methylation
DNA polymerase I
DNA polymerase III
DNA primase
DNA replication
elongation
exon
gyrase
helicase
initiation
intron
nucleosome
polysome
primary structure
promoter
quaternary structure
secondary structure
termination
tertiary structure
transcription
transcription factors
translation
tRNA

7.1 DNA structure and replication

Understandings, applications, skills

Activity number

☐ 1 Explain how nuclear DNA is packaged with reference to the role of nucleosomes. Use molecular visualization software to analyse nucleosome structure. — 209

☐ 2 Recall the structure of DNA and explain how X-ray diffraction provided crucial evidence to support Watson and Crick's model. In what way did the structure of DNA provide a mechanism for its self-replication? — 209

☐ 3 Distinguish between exons and introns in DNA. Introns contain repeating sequences. Explain how these tandem repeats are used in profiling. Explain why intronic DNA is no longer regarded as 'junk' DNA, i.e. it has a function. — 100 209

☐ **TOK** *Highly repetitive sequences were once classified as junk DNA. To what extent do the labels we use affect the knowledge we obtain?*

☐ 4 Analyse the results of the Hersey and Chase experiment providing evidence that DNA was the genetic information. — 210

☐ 5 Describe DNA replication with reference to the directional activity of DNA polymerases, the leading and lagging strands, and the role of enzymes. Explain the use of nucleotides containing dideoxyribonucleic acid to stop DNA replication in preparing samples for base sequencing (Sanger method). — 211

7.2 Transcription and gene expression

Understandings, applications, skills

Activity number

☐ 1 Describe DNA transcription, including the role of the promoter, the direction of transcription and the role of RNA polymerase. — 212

☐ 2 Describe the post-transcriptional modification of mRNA, explaining how alternative splicing of mRNA increases the number of proteins an organism can make. — 213

☐ 3 Recall the role of nucleosomes in DNA packaging and explain how they regulate which regions of DNA are transcribed. — 208

☐ 4 Explain how gene expression is regulated in prokaryotes and eukaryotes. — 214

☐ 5 Describe the role of the environment of the cell or organism on gene expression. Explain how DNA methylation regulates gene expression and analyse DNA methylation patterns in genes with high and low rates of expression. — 215 216

☐ **TOK** *The nature (genes) vs nurture (environment) debate is still active. Is it important for science to attempt to answer this question?*

7.3 Translation

Understandings, applications, skills

Activity number

☐ 1 Describe translation, including initiation, elongation, and termination. Distinguish between the destination of proteins synthesized on free ribosomes and those synthesized on bound ribosomes. Contrast the location and speed of translation in prokaryotes and eukaryotes. Identify polysomes in electron micrographs. — 217 218

☐ 2 Describe tRNA activation as an example of enzyme-substrate specificity and the role of phosphorylation. — 217

☐ 3 Describe the levels of structural organization in proteins: primary, secondary, and tertiary. Recognize that proteins consisting of more than one polypeptide chain have a quaternary structure and describe this for one example. — 219

208 Packaging DNA in the Nucleus

Key Idea: A chromosome consists of DNA complexed with proteins to form a highly organized, tightly coiled structure. The DNA in eukaryotes is complexed with histone proteins to form chromatin. The histones assist in packaging the DNA efficiently so that it can fit into the nucleus. Prior to cell division, the chromatin is at its most compact, forming the condensed metaphase chromosomes that can be seen with a light microscope.

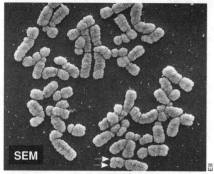

SEM

A cluster of human chromosomes seen during metaphase of cell division. Individual chromatids (arrowed) are difficult to discern on these double chromatid chromosomes.

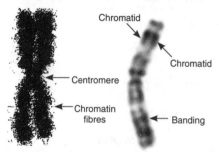

Chromatid

Chromatid

Centromere

Chromatin fibres

Banding

Chromosome TEM   Human chromosome 3

A human chromosome from a dividing white blood cell (above left). Note the compact organization of the chromatin in the two chromatids. The LM photograph (above right) shows the banding visible on human chromosome 3.

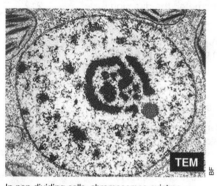

TEM

In non-dividing cells, chromosomes exist as single-armed structures. They are not visible as coiled structures, but are 'unwound' to make the genes accessible for transcription (above).

Looped domains

The evidence for the existence of looped domains comes from the study of giant lampbrush chromosomes in amphibian oocytes (above). Under electron microscopy, the lateral loops of the DNA-protein complex appear brushlike.

The Packaging of Chromatin

Chromatin structure is based on successive levels of DNA packing. **Histone proteins** are responsible for packing the DNA into a compact form. Without them, the DNA could not fit into the nucleus. Five types of histone proteins form a complex with DNA, in a way that resembles "beads on a string". These beads, or **nucleosomes**, form the basic unit of DNA packing.

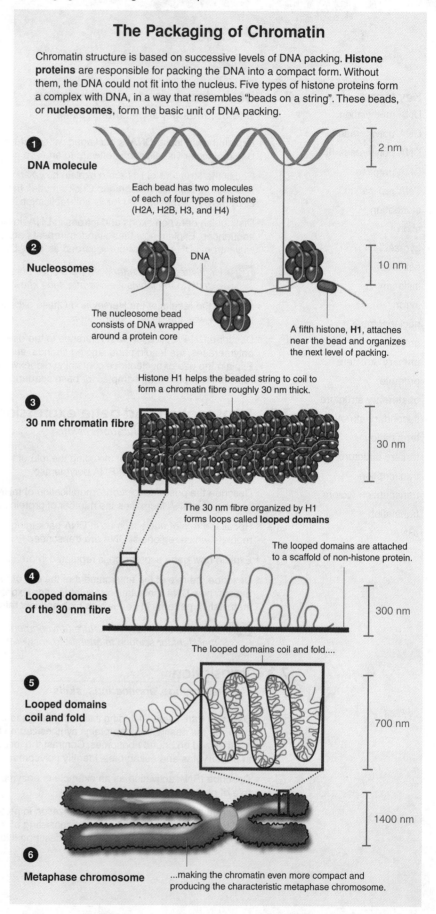

1 DNA molecule

2 nm

Each bead has two molecules of each of four types of histone (H2A, H2B, H3, and H4)

2 Nucleosomes

DNA

10 nm

The nucleosome bead consists of DNA wrapped around a protein core

A fifth histone, **H1**, attaches near the bead and organizes the next level of packing.

Histone H1 helps the beaded string to coil to form a chromatin fibre roughly 30 nm thick.

3 30 nm chromatin fibre

30 nm

The 30 nm fibre organized by H1 forms loops called **looped domains**

The looped domains are attached to a scaffold of non-histone protein.

4 Looped domains of the 30 nm fibre

300 nm

The looped domains coil and fold....

5 Looped domains coil and fold

700 nm

6 Metaphase chromosome

1400 nm

...making the chromatin even more compact and producing the characteristic metaphase chromosome.

© 2012-2014 **BIOZONE** International
ISBN: 978-1-927173-93-0
Photocopying Prohibited

Modifying DNA Packaging

The packaging of DNA affects when (and if) genes are expressed either by making the nucleosomes in the chromatin pack together tightly (**heterochromatin**) or more loosely (**euchromatin**). This affects whether or not RNA polymerase can attach to the DNA and transcribe the DNA to mRNA. Packaging of DNA is affected by histone modification and DNA methylation.

Histone Modification

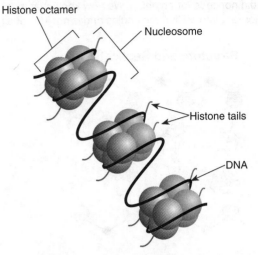

Histones may be modified by a number of processes, including methylation of the histone tail. Depending on the type of modification, the chromatin may pack together more tightly or more loosely, affecting the cell's ability to transcribe genes.

DNA Methylation

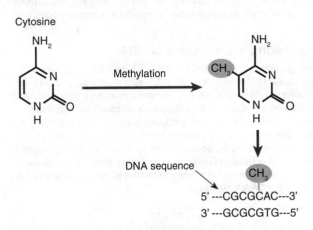

Cytosine methylation is an important process in DNA packaging and gene expression. Cytosine methylation affects gene expression in two ways: it may physically impede the binding of transcription factors or it may cause the chromatin to bind tightly together so that genes cannot be transcribed.

1. Explain the significance of the following terms used to describe the structure of chromosomes:

   (a) DNA: _____

   _____

   (b) Chromatin: _____

   _____

   (c) Histone: _____

   _____

   (d) Nucleosome: _____

   _____

2. (a) Describe the effect of histone modification and DNA methylation on DNA packaging: _____

   _____

   _____

   (b) How does this affect transcription of the DNA? _____

   _____

   _____

3. Each human cell has about 1 metre of DNA in its nucleus. Explain how this DNA is packaged into the nucleus:

   _____

   _____

   _____

   _____

   _____

209 DNA Molecules

Key Idea: Once the structure of DNA was known, it immediately suggested a mechanism for its replication.

It took the work of many scientists working in different areas many years to determine the structure of DNA. The final pieces of evidence came from a photographic technique called X-ray crystallography in which X-rays are shone through crystallized molecules to produce a pattern on a film.

The pattern can be used to understand the structure of the molecule. The focus of much subsequent research on DNA has been on protein-coding DNA sequences, yet protein-coding DNA accounts for less than 2% of the DNA in human chromosomes. The rest of the DNA was once called 'junk', meaning it did not code for anything. We now know that much of it codes for regulatory RNA molecules and is not junk at all.

Discovering the Structure of DNA

Although James Watson and Francis Crick are often credited with the discovery of the structure of DNA, credit must also go to at least two other scientists who were instrumental in acquiring the images that Watson and Crick were to base their discovery on. Maurice Wilkins and Rosalind Franklin produced X-ray diffraction patterns of the DNA molecule. The patterns provided measurements of different parts of the molecule and the position of different groups of atoms. Wilkins showed Franklin's X-ray image (photo 51) to Watson and Crick who then correctly interpreted the image and produced a model of the DNA molecule.

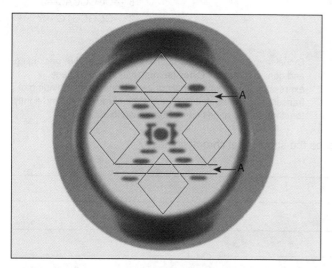

Diagram representing the image produced by Rosalind Franklin

Numerous distinct parts of the X-ray image indicate specific qualities of the DNA. The distinct X pattern indicates a helix structure. However, Watson and Crick realized that the apparent gaps in the X (labelled **A**) were due to the repeating pattern of a *double* helix. The diamond shapes (shown in blue) indicate the helix is continuous and of constant dimensions and that the sugar-phosphate backbone is on the outside of the helix. The distance between the dark horizontal bands allows the calculation of the length of one full turn of the helix.

Structure and Replication

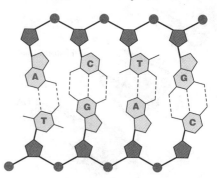

The realization that DNA was a double helix consisting of antiparallel strands made of bases that followed a strict base pairing rule suggested a mechanism for its replication. Watson and Crick proposed that each strand served as a template and that DNA replication was semi-conservative, producing two daughter strands consisting of half new and half parent material. The proposal was confirmed by Meselson and Stahl.

Exons and Introns

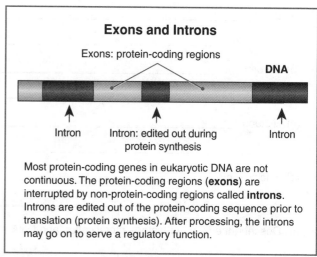

Most protein-coding genes in eukaryotic DNA are not continuous. The protein-coding regions (**exons**) are interrupted by non-protein-coding regions called **introns**. Introns are edited out of the protein-coding sequence prior to translation (protein synthesis). After processing, the introns may go on to serve a regulatory function.

1. What made Watson and Crick realize that DNA was a double helix? _____

_____

_____

2. What proportion of DNA in a eukaryotic cell is used to code for proteins? _____

3. (a) Describe the organization of protein-coding regions in eukaryotic DNA: _____

_____

_____

   (b) What might be the purpose of the introns? _____

_____

_____

© 2012-2014 **BIOZONE** International
ISBN: 978-1-927173-93-0
Photocopying Prohibited

210 DNA Carries the Code

Key Idea: A series of experiments in the 1940s and 1950s confirmed that it was DNA that carried the genetic information.

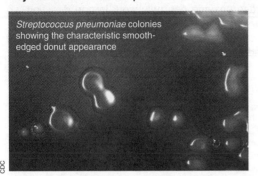

Streptococcus pneumoniae colonies showing the characteristic smooth-edged donut appearance

Scientists had known about DNA since the end of the 19th century, but its role in storing information remained unknown until the 1940s, and its structure remained a mystery for another decade after that. In 1928, experiments by British scientist Frederick Griffith gave the first indications that DNA was responsible for passing on information. Griffith had been working with two strains of the bacteria *Streptococcus pneumoniae*. Only one strain (the pathogenic strain) caused pneumonia and it was easily identified because it formed colonies with smooth edges. The other, benign strain formed colonies with rough edges. When mice were injected with the pathogenic strain they developed pneumonia and died.

The mice injected with the benign strain did not. Mice injected with the heat-killed pathogenic strain did not develop pneumonia either. This showed that the disease was not caused by a chemical associated with the bacteria, or a response by the body to the bacteria, it was the bacterial cells themselves. In a second experiment, Griffith mixed the benign strain with the heat-killed pathogenic strain and injected it into healthy mice. To his surprise, the mice developed pneumonia. When bacteria from the mice were recovered and cultured they produced colonies identical to the pathogenic strain. Somehow the harmless bacteria had acquired information from the dead pathogenic strain. Griffith called this process **transformation**.

In 1944, American scientists, led by Oswald Avery, continued with Griffith's experiments. They made an extract from the heat-killed pathogenic strain and treated it with chemicals to destroy any lipids, carbohydrates, or proteins. This was mixed with the benign strain and transformation still occurred. This established that no proteins, lipids, or carbohydrates were responsible for the transformation. When another identical extract was treated with chemicals that break down DNA, the transformation did not take place - the benign strain failed to acquire the information required to cause pneumonia. From this it was deduced that DNA was the unit that was carrying the information from one bacteria to another.

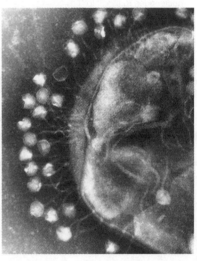

Another experiment in 1952 by Alfred Hershey and Martha Chase, confirmed what the other two experiments had shown. Hershey and Chase worked with viruses, which were known to have DNA and to transfer information to their host. However, there was debate over whether the information was transferred by the DNA or by the protein coat of the virus. Hershey and Chase used radioactive sulfur and radioactive phosphorus to mark different parts of the virus. The sulfur was incorporated into the protein coat while the phosphorus was incorporated into the viral DNA. The viruses were then mixed with bacteria and the infected bacteria analysed. The bacteria were found to contain radioactive phosphorus but not radioactive sulfur, showing that the virus had indeed passed information to its host by injecting its own DNA.

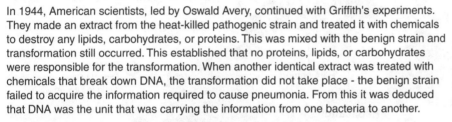

1. How did Griffith confirm that it was the bacteria causing the pneumonia and not something else?

_____

_____

_____

_____

2. Why were sulfur and phosphorus used in Hershey's experiment?_____

_____

_____

_____

3. Why is it important to conduct two different experiments (e.g. Avery's and Hershey's) when investigating a hypothesis?

_____

_____

_____

211 Enzyme Control of DNA Replication

Key Idea: The process of DNA replication is controlled by many different enzymes.

DNA replication involves many enzyme-controlled steps. They are shown below as separate, but many of the enzymes are clustered together as enzyme complexes. As the DNA is replicated, enzymes 'proof-read' it and correct mistakes. The polymerase enzyme can only work in one direction, so that one new strand is constructed as a continuous length (the leading strand) while the other new strand (the lagging strand) is made in short segments to be later joined together.

DNA replication occurs during interphase of the cell cycle at an astounding rate. As many as 4000 nucleotides per second are replicated. This explains how bacterial cells, with as many as 4 million nucleotides, can complete a cell cycle in about 20 minutes. Note that the nucleotides are present as deoxynucleoside triphosphates. When hydrolysed, these provide the energy for incorporating the nucleotide into the strand.

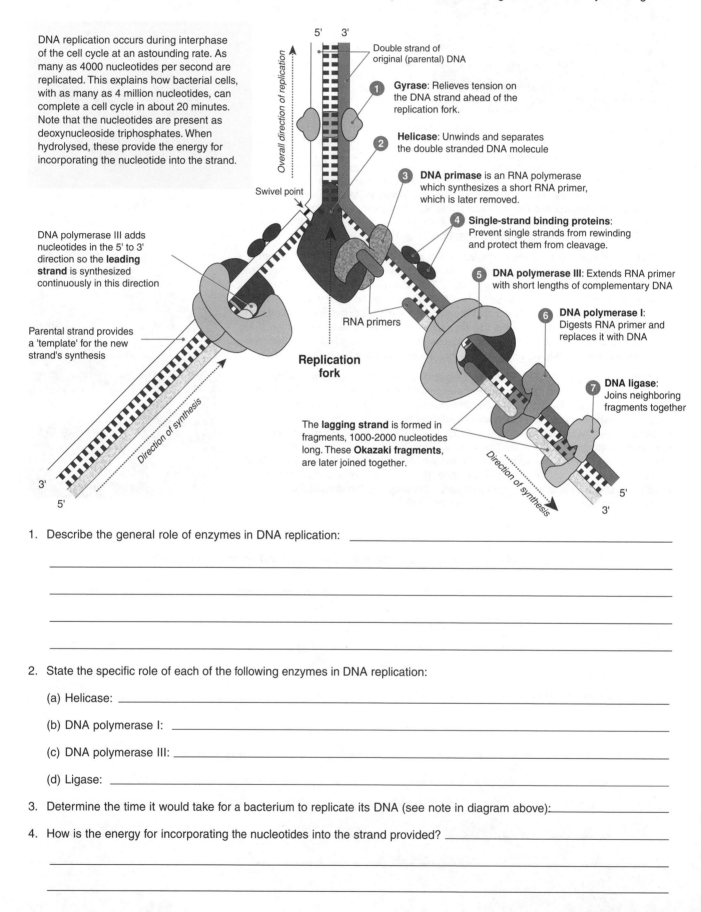

Double strand of original (parental) DNA

1 **Gyrase**: Relieves tension on the DNA strand ahead of the replication fork.

2 **Helicase**: Unwinds and separates the double stranded DNA molecule

3 **DNA primase** is an RNA polymerase which synthesizes a short RNA primer, which is later removed.

4 **Single-strand binding proteins**: Prevent single strands from rewinding and protect them from cleavage.

5 **DNA polymerase III**: Extends RNA primer with short lengths of complementary DNA

6 **DNA polymerase I**: Digests RNA primer and replaces it with DNA

7 **DNA ligase**: Joins neighboring fragments together

Swivel point

Overall direction of replication

DNA polymerase III adds nucleotides in the 5' to 3' direction so the **leading strand** is synthesized continuously in this direction

Parental strand provides a 'template' for the new strand's synthesis

Direction of synthesis

RNA primers

Replication fork

The **lagging strand** is formed in fragments, 1000-2000 nucleotides long. These **Okazaki fragments**, are later joined together.

Direction of synthesis

1. Describe the general role of enzymes in DNA replication: _____

_____

_____

_____

_____

2. State the specific role of each of the following enzymes in DNA replication:

(a) Helicase: _____

(b) DNA polymerase I: _____

(c) DNA polymerase III: _____

(d) Ligase: _____

3. Determine the time it would take for a bacterium to replicate its DNA (see note in diagram above):_____

4. How is the energy for incorporating the nucleotides into the strand provided? _____

_____

_____

LINK 56   LINK 85

© 2012-2014 **BIOZONE** International
ISBN: 978-1-927173-93-0
Photocopying Prohibited

DNA Sequencing

The way DNA replicated suggested a method for DNA sequencing. If replication could be stopped at each base on the DNA then different lengths of DNA would be produced, each ending at a specific base (A,T, C, or G). The lengths of DNA could be put in order to reveal the original DNA sequence. The method (called the Sanger method after its inventor) is illustrated below. Four separate reactions are run, each containing a modified nucleotide mixed with its normal counterpart, as well as the three other normal nucleotides. When a modified nucleotide is added to the growing complementary DNA, synthesis stops. The fragments of DNA produced from the four reactions are separated by electrophoresis and analysed by autoradiography to determine the DNA sequence.

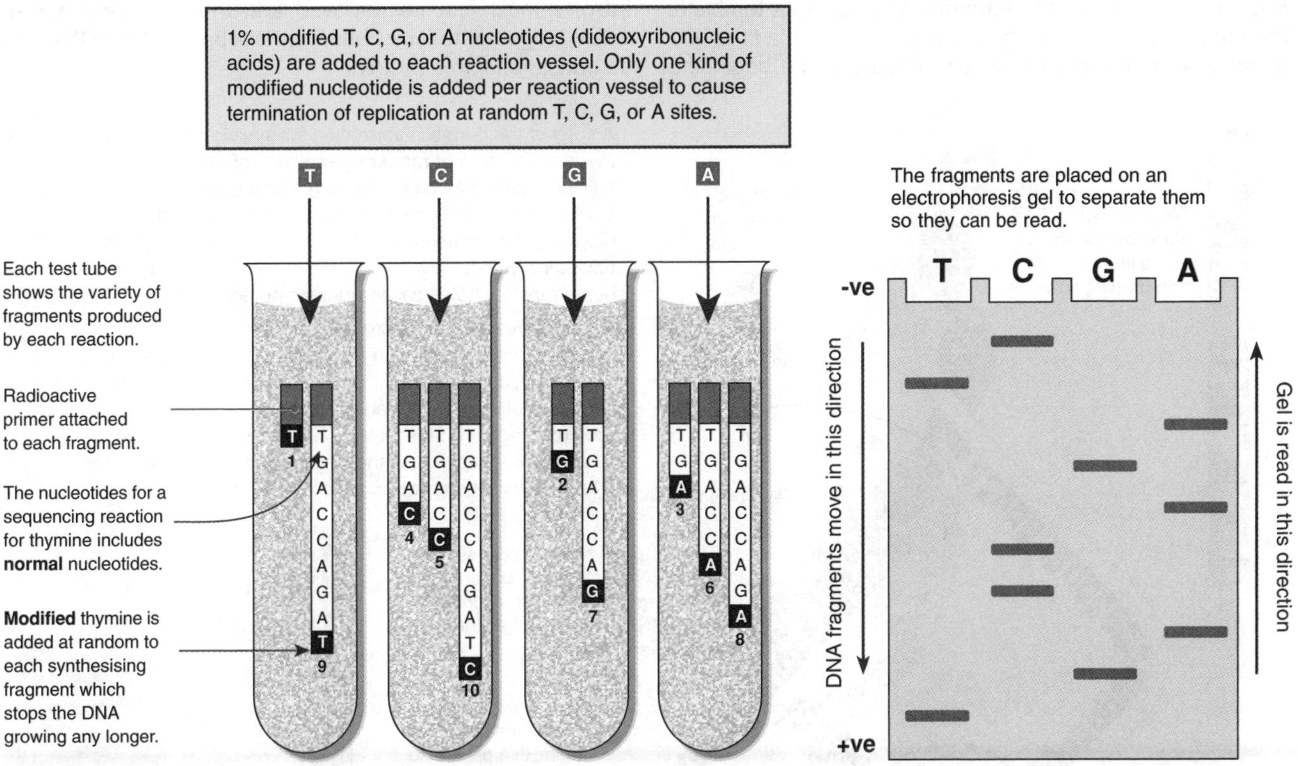
1% modified T, C, G, or A nucleotides (dideoxyribonucleic acids) are added to each reaction vessel. Only one kind of modified nucleotide is added per reaction vessel to cause termination of replication at random T, C, G, or A sites.
T
C
G
A
Each test tube shows the variety of fragments produced by each reaction.
Radioactive primer attached to each fragment.
The nucleotides for a sequencing reaction for thymine includes **normal** nucleotides.
Modified thymine is added at random to each synthesising fragment which stops the DNA growing any longer.
1
2
3
4
5
6
7
8
9
10
The fragments are placed on an electrophoresis gel to separate them so they can be read.
-ve
+ve
T C G A
DNA fragments move in this direction
Gel is read in this direction

5. Explain why DNA on the lagging strand must be replicated in short sections: ______________________

____

____

____

6. (a) In the Sanger method, what type of molecule is added to stop replication? ______________________

(b) On the gel diagram above right label the bands 1 -10 in the order they would appear in the DNA:

(c) On the diagram circle the shortest fragment:

(d) Write the sequence of the copied DNA: ______________________

(e) Write the sequence of the original DNA: ______________________

7. Why must this method of DNA sequencing use four separate reaction vessels? ______________________

____

____

____

8. Why is only 1% of the reaction mix modified DNA? ______________________

____

____

____

____

ISBN: 978-1-927173-93-0

212 Transcription

Key Idea: Transcription is the first step of gene expression. A segment of DNA is transcribed (rewritten) into mRNA. In eukaryotes, transcription takes place in the nucleus.

The enzyme that directly controls transcription is RNA polymerase, which makes a strand of mRNA using the single strand of DNA (the **template strand**) as a template. The enzyme transcribes a gene length of DNA at a time and recognizes start and stop signals (codes) at the beginning and end of the gene. Only RNA polymerase is involved in mRNA synthesis as it unwinds the DNA as well. It is common to find several RNA polymerase enzyme molecules on the same gene at any one time, allowing a high rate of mRNA synthesis to occur. In eukaryotes, non-coding sections called **introns** must first be removed and the remaining **exons** spliced together to form mature mRNA before the mRNA can be translated into a protein.

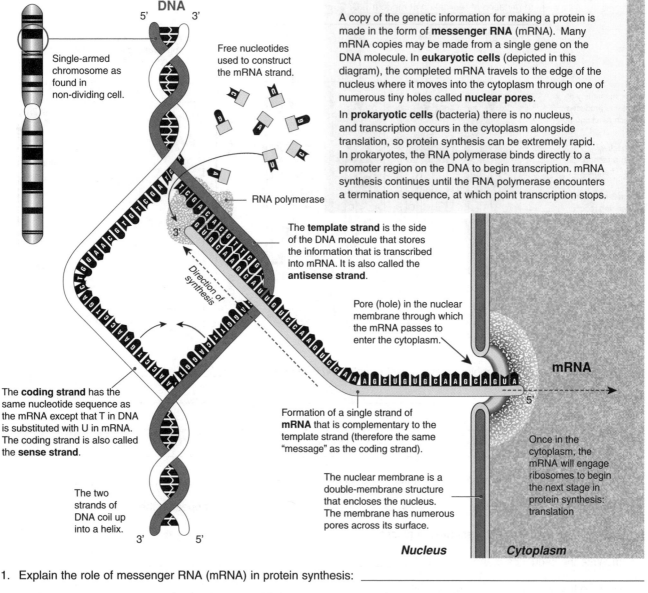

A copy of the genetic information for making a protein is made in the form of **messenger RNA** (mRNA). Many mRNA copies may be made from a single gene on the DNA molecule. In **eukaryotic cells** (depicted in this diagram), the completed mRNA travels to the edge of the nucleus where it moves into the cytoplasm through one of numerous tiny holes called **nuclear pores**.

In **prokaryotic cells** (bacteria) there is no nucleus, and transcription occurs in the cytoplasm alongside translation, so protein synthesis can be extremely rapid. In prokaryotes, the RNA polymerase binds directly to a promoter region on the DNA to begin transcription. mRNA synthesis continues until the RNA polymerase encounters a termination sequence, at which point transcription stops.

DNA

Single-armed chromosome as found in non-dividing cell.

Free nucleotides used to construct the mRNA strand.

RNA polymerase

The **template strand** is the side of the DNA molecule that stores the information that is transcribed into mRNA. It is also called the **antisense strand**.

Direction of synthesis

Pore (hole) in the nuclear membrane through which the mRNA passes to enter the cytoplasm.

mRNA

The **coding strand** has the same nucleotide sequence as the mRNA except that T in DNA is substituted with U in mRNA. The coding strand is also called the **sense strand**.

Formation of a single strand of **mRNA** that is complementary to the template strand (therefore the same "message" as the coding strand).

Once in the cytoplasm, the mRNA will engage ribosomes to begin the next stage in protein synthesis: translation

The two strands of DNA coil up into a helix.

The nuclear membrane is a double-membrane structure that encloses the nucleus. The membrane has numerous pores across its surface.

Nucleus   *Cytoplasm*

1. Explain the role of messenger RNA (mRNA) in protein synthesis: _____

_____

_____

2. The genetic code contains punctuation codons to mark the starting and finishing points of the code for synthesis of polypeptide chains and proteins. Consult the mRNA–amino acid table earlier in this workbook and state the codes for:

   (a) Start codon: _____ (b) Stop (termination) codons: _____

3. For the following triplets on the DNA, determine the **codon** sequence for the mRNA that would be synthesized:

   (a) Triplets on the DNA: T A C   T A G   C C G   C G A   T T T

   Codons on the mRNA: _____

   (b) Triplets on the DNA: T A C   A A G   C C T   A T A   A A A

   Codons on the mRNA: _____

© 2012-2014 **BIOZONE** International
ISBN: 978-1-927173-93-0
Photocopying Prohibited

213 Post Transcriptional Modification

Key Idea: Primary mRNA molecules are modified before they are translated into proteins.

Human DNA contains only 25,000 genes, but produces possibly up to 1 million different proteins. Each gene must therefore produce more than one protein. This is achieved by both **post transcriptional** and **post translational** **modification**. Primary mRNA contains exons and introns. Usually introns are removed after transcription and the exons are spliced together. However, the number of exons and the way they are spliced varies. This creates variations in the translated polypeptide chain. These mechanisms allow for the production of the diverse range of proteins.

Post Transcriptional Modification

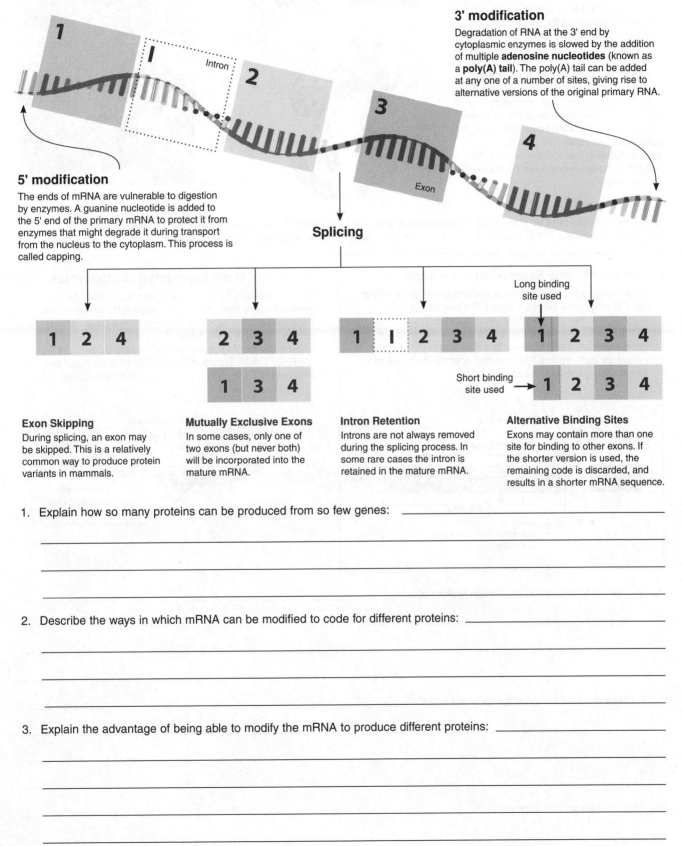

3' modification

Degradation of RNA at the 3' end by cytoplasmic enzymes is slowed by the addition of multiple **adenosine nucleotides** (known as a **poly(A) tail**). The poly(A) tail can be added at any one of a number of sites, giving rise to alternative versions of the original primary RNA.

5' modification

The ends of mRNA are vulnerable to digestion by enzymes. A guanine nucleotide is added to the 5' end of the primary mRNA to protect it from enzymes that might degrade it during transport from the nucleus to the cytoplasm. This process is called capping.

Splicing

Exon Skipping

During splicing, an exon may be skipped. This is a relatively common way to produce protein variants in mammals.

Mutually Exclusive Exons

In some cases, only one of two exons (but never both) will be incorporated into the mature mRNA.

Intron Retention

Introns are not always removed during the splicing process. In some rare cases the intron is retained in the mature mRNA.

Alternative Binding Sites

Exons may contain more than one site for binding to other exons. If the shorter version is used, the remaining code is discarded, and results in a shorter mRNA sequence.

1. Explain how so many proteins can be produced from so few genes: _____

_____

_____

2. Describe the ways in which mRNA can be modified to code for different proteins: _____

_____

_____

_____

3. Explain the advantage of being able to modify the mRNA to produce different proteins: _____

_____

_____

_____

LINK **209** WEB **213** **KNOW**

214 Controlling Gene Expression

Key Idea: Gene expression is tightly regulated. It begins when RNA polymerase attaches to the promoter region of a gene. All the cells in your body contain identical copies of your genetic instructions. Yet these cells appear very different (e.g. muscle, nerve, and epithelial cells have little in common). These morphological differences reflect profound differences in the expression of genes during the cell's development. For example, muscle cells express the genes for the proteins that make up the contractile elements of the muscle fibre. This diversity of cell structure and function reflects the precise control over the time, location, and extent of expression of a huge variety of genes.

The physical state of the DNA in or near a gene is important in helping to control whether the gene is even available for transcription. When the **heterochromatin** is condensed, the transcription proteins cannot reach the DNA and the gene is not expressed. To be transcribed, a gene must first be unpacked from its condensed state. Once unpacked, control of gene expression involves the interaction of **transcription factors** (shown attached to the DNA molecule right) with DNA sequences that control the specific gene. Initiation of transcription is the most important and universally used control point in gene expression.

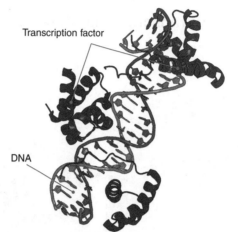

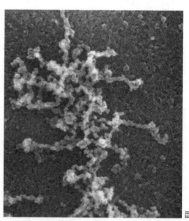

Unravelled mRNA after transcription.

Before transcription can take place, RNA polymerase must attached to the **promoter** region of the DNA. The promoter is upstream of the DNA to be transcribed. The process by which RNA polymerase binds to the promoter is different in prokaryotes and eukaryotes (below).

Control of Gene Expression in Prokaryotes

In **prokaryotes**, the RNA polymerase and associated proteins bind directly to the promoter region. The promoter is normally very close to the DNA region to be transcribed. Transcription of the structural gene is controlled by a **regulator** gene, which produces a repressor molecule that may bind to the operator and block transcription. The promoter, operator, and DNA to be transcribed are called an **operon**.

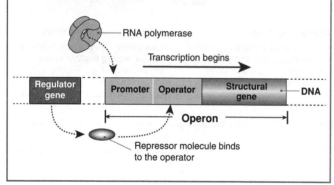

Control of Gene Expression in Eukaryotes

Eukaryote genes do not exist as operons. In **eukaryotes**, several **transcription factors** are required to bind the RNA polymerase to the promoter (the RNA polymerase cannot bind directly). Some of these transcription factors may also bind to the **enhancer** region which may be quite distant from the promoter. The transcription factors cause the promoter and enhancer to come together, which initiates transcription.

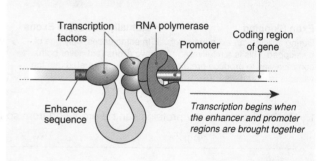

1. (a) How is transcription initiated in prokaryotes? _____

_____

_____

_____

(b) How is transcription initiated in eukaryotes? _____

_____

_____

_____

2. What is the role of the promoter? _____

_____

© 2012-2014 **BIOZONE** International
ISBN: 978-1-927173-93-0

215 Gene-Environment Interactions

Key Idea: An organism's phenotype is influenced by the environment in which it develops, even though the genotype remains unaffected.

External environmental factors can modify the phenotype encoded by genes. This can occur both during development and later in life. Even identical twins have minor differences in their appearance due to environmental factors such as diet and intrauterine environment before birth. Environmental factors that affect the phenotype include nutrients or diet, temperature, and the presence of other organisms.

Sources of Variation in Organisms

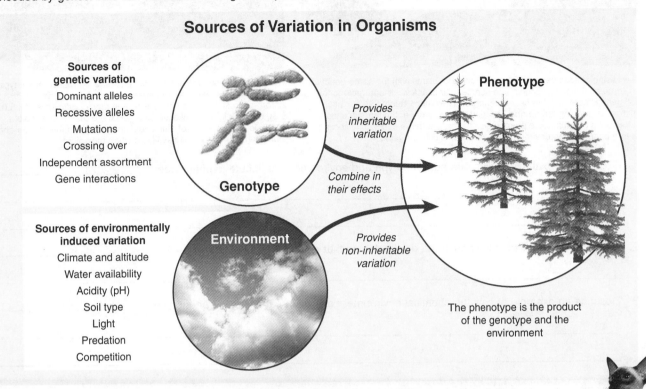

Sources of genetic variation

Dominant alleles
Recessive alleles
Mutations
Crossing over
Independent assortment
Gene interactions

Genotype

Provides inheritable variation

Combine in their effects

Phenotype

Sources of environmentally induced variation

Climate and altitude
Water availability
Acidity (pH)
Soil type
Light
Predation
Competition

Environment

Provides non-inheritable variation

The phenotype is the product of the genotype and the environment

The Effect of Temperature

The sex of some animals is determined by the incubation temperature during their embryonic development. Examples include turtles, crocodiles, and the American alligator. In some species, high incubation temperatures produce males and low temperatures produce females. In other species, the opposite is true. Temperature regulated sex determination may be advantageous by preventing inbreeding (since all siblings will tend to be of the same sex).

Colour-pointing in breeds of cats and rabbits (e.g. Siamese, Himalyan) is a result of a temperature sensitive mutation in one of the enzymes in the metabolic pathway from tyrosine to melanin. The dark pigment is only produced in the cooler areas of the body (face, ears, feet, and tail), while the rest of the body is a paler version of the same colour, or white.

The Effect of Other Organisms

Female

Male

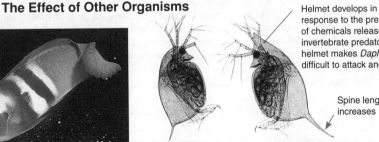

Helmet develops in response to the presence of chemicals released by invertebrate predators. The helmet makes *Daphnia* more difficult to attack and handle.

Spine length increases

Non-helmeted form

Helmeted form with long tail spine

The presence of other individuals of the same species may control sex determination for some animals. Some fish species, including some in the wrasse family (e.g. *Coris sandageri*, above), show this phenomenon. The fish live in groups consisting of a single male with attendant females and juveniles. In the presence of a male, all juvenile fish of this species grow into females. When the male dies, the dominant female will undergo physiological changes to become a male. The male has distinctive bands, whereas the female is pale in colour and has very faint markings.

Some organisms respond to the presence of other, potentially harmful, organisms by changing their morphology or body shape. Invertebrates such as *Daphnia* will grow a large helmet when a predatory midge larva is present. Such responses are usually mediated through chemicals produced by the predator (or competitor), and are common in plants as well as animals.

298

The Effect of Altitude

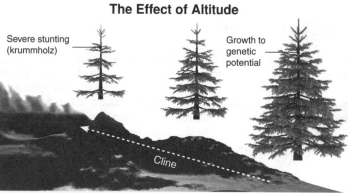

Severe stunting (krummholz)

Growth to genetic potential

Cline

The Effect of Chemical Environment

Increasing altitude can stunt the phenotype of plants with the same genotype. In some conifers, e.g. Engelmann spruce, plants at low altitude grow to their full genetic potential, but become progressively more stunted as elevation increases, forming krummholz (gnarled bushy forms) at the highest sites. A continuous gradation in a phenotypic character within a species, associated with a change in an environmental variable, is called a **cline**.

The chemical environment can influence the expressed phenotype in plants and animals. In hydrangeas, flower colour varies according to soil pH. Flowers are blue in more acidic soils (pH 5.0-5.5), but pink in more alkaline soils (pH 6.0-6.5). The blue colour is due to the presence of aluminium compounds in the flowers and aluminium is more readily available when the soil pH is low.

1. Describe an example to illustrate how genotype and environment contribute to phenotype: _____

    _____

    _____

2. What are the physical factors associated with altitude that could affect plant phenotype? _____

    _____

3. Describe an example of how the chemical environment of a plant can influence phenotype: _____

    _____

    _____

    _____

4. Why are the darker patches of fur in colour-pointed cats and rabbits found only on the face, paws and tail:

    _____

    _____

5. There has been much amusement over the size of record-breaking vegetables, such as enormous pumpkins, produced for competitions. How could you improve the chance that a vegetable would reach its maximum genetic potential?

    _____

    _____

6. (a) What is a **cline**? _____

    _____

    (b) On a windswept portion of a coast, two different species of plant (species A and species B) were found growing together. Both had a low growing (prostrate) phenotype. One of each plant type was transferred to a greenhouse where "ideal" conditions were provided to allow maximum growth. In this controlled environment, species B continued to grow in its original prostrate form, but species A changed its growing pattern and became erect in form. Identify the **cause** of the prostrate phenotype in each of the coastal grown plant species and explain your answer:

    Plant species A: _____

    _____

    _____

    Plant species B: _____

    _____

    _____

    (c) Which of these species (A or B) would be most likely to exhibit clinal variation? _____

© 2012-2014 **BIOZONE** International
ISBN: 978-1-927173-93-0

216 DNA Methylation

Key Idea: Methylation of the DNA can alter gene expression and therefore phenotype.

Methylation of DNA is an important way of controlling gene expression. Methylated DNA is usually silenced, meaning genes are not transcribed to mRNA. Methylation of cytosine turns off gene expression by changing the state of the chromatin so that transcribing proteins are not able to bind to the DNA. Methylation is also important in X-inactivation.

Methylation and Gene Expression

Gene expression changes as an organism develops. During the development of the embryo there are many genes that are switched on and off as development of tissues and organs proceeds. Some of this control is achieved by DNA methylation. Enzymes can add or remove methyl groups from cytosine bases and so activate or silence genes. Often methylation is affected by changes in the environment and this provides the developing organism with a rapid response mechanism.

Methylation in Mammalian Development

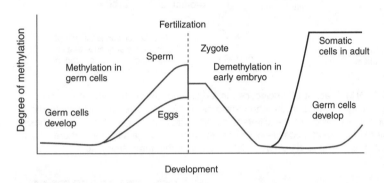

Methylation vs Gene Expression

A comparison of methylation and gene expression finds that highly expressed genes have very little methylation while genes that are not expressed have very high levels of methylation.

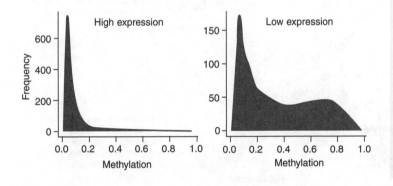

Methylation and Imprinted Genes

Genomic imprinting is a phenomenon in which the pattern of gene expression is different depending on whether the gene comes from the mother or the father. Imprinted genes are silenced by methylation and histone modification. A gene inherited from the father may be silenced while the gene inherited from the mother may be active or vice versa. Evidence of this is seen in two human genetic disorders, Angelman syndrome and Prader-Willi syndrome. Both are caused by the same mutation; a specific deletion on chromosome 15. Which syndrome is expressed depends on whether the mutation occurs on the maternal or paternal chromosome.

A liger

The effect of genomic imprinting can be seen in other mammals. Ligers (a cross between a male lion and a female tiger) although not occurring in the wild, are the biggest of the big cats. However a tigon (a cross between a female lion and a male tiger) is no bigger than a normal lion. It is thought this difference in phenotype is due to the male lion carrying imprinted genes that result in larger offspring which are normally counteracted by genes from the female. Similarly the differences between a mule (male donkey + female horse) and a hinny (female donkey + male horse) may be to do with genomic imprinting.

1. (a) What is DNA methylation? _____

_____

(b) How does DNA methylation affect gene expression?: _____

_____

_____

2. Prader-Willi syndrome is caused when a mutated gene on chromosome 15 is inherited from the father. How does this tell us that the mother must therefore have donated the imprinted gene?

_____

_____

_____

LINK
208
WEB
216
APP

217 Translation

Key Idea: Translation is the second step of gene expression. It occurs in the cytoplasm, where ribosomes read the mRNA code and decode it to synthesize protein.

In eukaryotes, translation occurs in the cytoplasm associated with free ribosomes or ribosomes on the rough endoplasmic reticulum. The diagram below shows how a mRNA molecule can be 'serviced' by many ribosomes at the same time. The role of the tRNA molecules is to bring in the individual amino acids. The anticodon of each tRNA must make a perfect complementary match with the mRNA codon before the amino acid is released. Once released, the amino acid is added to the growing polypeptide chain by enzymes.

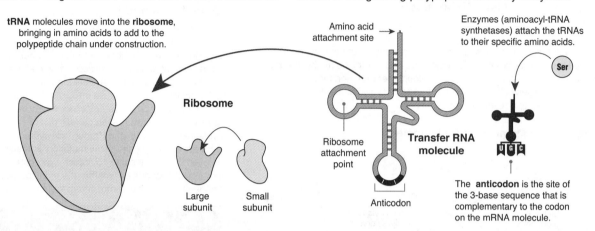

tRNA molecules move into the **ribosome**, bringing in amino acids to add to the polypeptide chain under construction.

Ribosome

Large subunit   Small subunit

Amino acid attachment site

Enzymes (aminoacyl-tRNA synthetases) attach the tRNAs to their specific amino acids.

Ser

Ribosome attachment point

Transfer RNA molecule

Anticodon

The **anticodon** is the site of the 3-base sequence that is complementary to the codon on the mRNA molecule.

Ribosomes are made up of a complex of ribosomal RNA (rRNA) and proteins. They exist as two separate sub-units (above) until they are attracted to a binding site on the mRNA molecule, when they join together. Ribosomes have binding sites that attract transfer RNA (**tRNA**) molecules loaded with amino acids. The tRNA molecules are about 80 nucleotides in length and are made under the direction of genes in the chromosomes. There is a different tRNA molecule for each of the different possible anticodons (see the diagram below) and, because of the degeneracy of the genetic code, there may be up to six different tRNAs carrying the same amino acid.

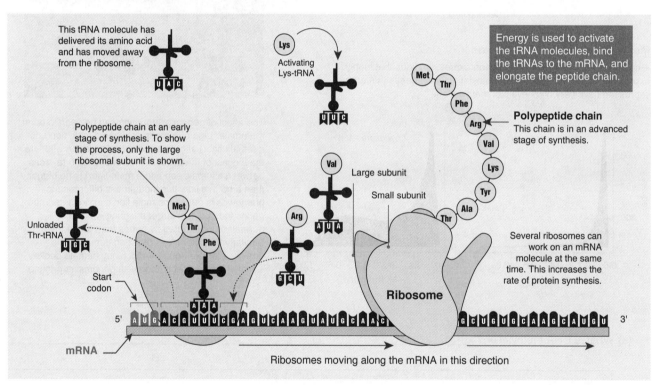

This tRNA molecule has delivered its amino acid and has moved away from the ribosome.

Lys

Activating Lys-tRNA

Energy is used to activate the tRNA molecules, bind the tRNAs to the mRNA, and elongate the peptide chain.

Met
Thr
Phe
Arg

Polypeptide chain
This chain is in an advanced stage of synthesis.

Val
Lys
Tyr
Ala
Thr

Polypeptide chain at an early stage of synthesis. To show the process, only the large ribosomal subunit is shown.

Val

Large subunit

Small subunit

Arg

Met
Thr
Phe

Unloaded Thr-tRNA

Start codon

Several ribosomes can work on an mRNA molecule at the same time. This increases the rate of protein synthesis.

Ribosome

5' AUGACGUUUCGAGUCAAGUAUGCAAC ... GCUGUGCAAGCAUGU 3'

mRNA

Ribosomes moving along the mRNA in this direction

1. For the following codons on the mRNA, determine the **anticodons** for each tRNA that would deliver the amino acids:

   Codons on the mRNA: U A C   U A G   C C G   C G A   U U U

   Anticodons on the tRNAs: _____

2. There are many different types of tRNA molecules, each with a different anticodon (HINT: see the mRNA table).

   (a) How many different tRNA types are there, each with a unique anticodon? _____

   (b) Explain your answer: _____

   _____

© 2012-2014 **BIOZONE** International
ISBN: 978-1-927173-93-0
Photocopying Prohibited

218 Protein Synthesis Summary

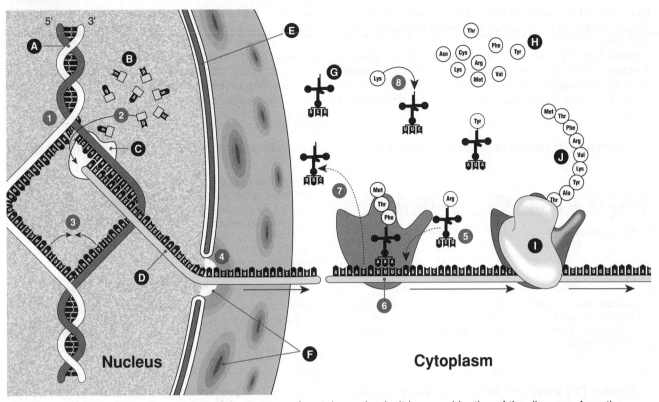

The diagram above shows an overview of the process of protein synthesis. It is a combination of the diagrams from the previous two pages. Each of the major steps in the process are numbered, while structures are labelled with letters.

1. Briefly describe each of the numbered processes in the diagram above:

   (a) Process 1: _____

   (b) Process 2: _____

   (c) Process 3: _____

   (d) Process 4: _____

   (e) Process 5: _____

   (f) Process 6: _____

   (g) Process 7: _____

   (h) Process 8: _____

2. Identify each of the structures marked with a letter and write their names below in the spaces provided:

   (a) Structure A: _____   (f) Structure F: _____

   (b) Structure B: _____   (g) Structure G: _____

   (c) Structure C: _____   (h) Structure H: _____

   (d) Structure D: _____   (i) Structure I: _____

   (e) Structure E: _____   (j) Structure J: _____

3. Describe two factors that would determine whether or not a particular protein is produced in the cell:

   (a) _____

   _____

   (b) _____

   _____

© 2012-2014 **BIOZONE** International
ISBN: 978-1-927173-93-0

TEST

219 Proteins

Key Idea: The sequence and type of animo acids in a protein determines the protein's three-dimensional shape and function.

Proteins are large, complex **macromolecules**, built up from a linear sequence of repeating units called **amino acids**. Proteins are molecules of central importance in the chemistry of life. They account for more than 50% of the dry weight of most cells, and they are important in virtually every cellular process. The folding of a protein into its functional form creates a three dimensional arrangement of the active 'R' groups. It is this **tertiary structure** that gives a protein its unique chemical properties. If a protein loses this precise structure (through **denaturation**), it is usually unable to carry out its biological function.

Primary (1°) structure (amino acid sequence)

| Phe | Glu | Tyr | Ser | Iso | Met | Ala | Ala | Ser |

Peptide bond   Amino acid

Hundreds of amino acids are linked together by peptide bonds to form polypeptide chains. The attractive and repulsive charges on the amino acids determines how the protein is organized, and its biological function.

Secondary (2°) structure (α-helix or β pleated sheet)

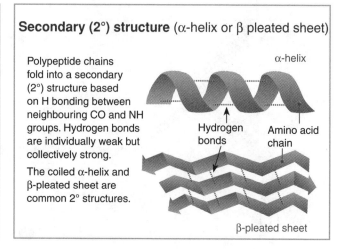

Polypeptide chains fold into a secondary (2°) structure based on H bonding between neighbouring CO and NH groups. Hydrogen bonds are individually weak but collectively strong.

The coiled α-helix and β-pleated sheet are common 2° structures.

α-helix

Hydrogen bonds   Amino acid chain

β-pleated sheet

Tertiary (3°) structure (folding of the 2° structure)

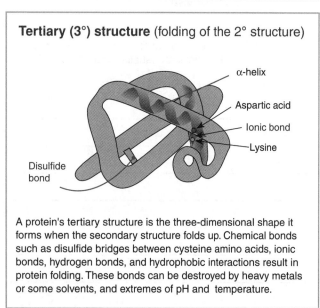

α-helix
Aspartic acid
Ionic bond
Lysine
Disulfide bond

A protein's tertiary structure is the three-dimensional shape it forms when the secondary structure folds up. Chemical bonds such as disulfide bridges between cysteine amino acids, ionic bonds, hydrogen bonds, and hydrophobic interactions result in protein folding. These bonds can be destroyed by heavy metals or some solvents, and extremes of pH and temperature.

Quaternary (4°) structure

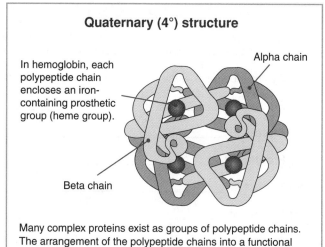

In hemoglobin, each polypeptide chain encloses an iron-containing prosthetic group (heme group).

Alpha chain
Beta chain

Many complex proteins exist as groups of polypeptide chains. The arrangement of the polypeptide chains into a functional protein is termed the quaternary structure. The example (above) shows hemoglobin.

---

1. Describe the main features that aid the formation of each part of a protein's structure:

   (a) Primary structure: _____

   (b) Secondary structure: _____

   _____

   (c) Tertiary structure: _____

   _____

   (d) Quaternary structure: _____

   _____

2. How are proteins built up into a functional structure? _____

   _____

   _____

© 2012-2014 **BIOZONE** International
ISBN: 978-1-927173-93-0
Photocopying Prohibited

220 Chapter Review

Summarize what you know about this topic under the headings provided. You can draw diagrams or mind maps, or write short notes to organize your thoughts. Use the hints and guidelines included to help you:

DNA structure and replication

HINT: Review DNA packaging and replication.

Transcription and gene expression

HINT: Translation, post-translational modification, gene expression, and imprinting.

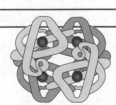

Translation

HINT: The process of translation, importance of ribosomes, and protein structure.

REVISE

221 KEY TERMS: Did You Get It?

1. Test your vocabulary by matching each term to its definition, as identified by its preceding letter code.

Term		Definition
DNA replication	A	The folding of a polypeptide chain due to bonds between the amino acids to form alpha helices and beta sheets.
exon	B	The arrangement of polypeptide chains into a functional protein.
helicase	C	Protein coding segment of DNA that alternates with non-protein coding segments called introns.
intron	D	The process by which a new copy of a DNA molecule is made.
nucleosome	E	A unit of DNA packaging consisting of a length of DNA wound around eight histone proteins.
polysome	F	The stage of gene expression in which mRNA is decoded to produce a specific polypeptide.
quaternary structure	G	A non-coding segment of DNA that is removed before a gene is translated into a protein.
secondary structure	H	An enzyme which can unwind the DNA double helix.
transcription	I	A cluster of ribosomes that are able to translate a mRNA molecule simultaneously and so produce many polypeptide chains at once.
translation	J	The stage of gene expression in which a strand of DNA is rewritten into mRNA.

2. Match the statements in the three columns below to form complete sentences, then use them to construct a coherent paragraph. The first column is in order, the centre and right columns are not. The centre column provides appropriate joining words to link the first and second parts of the sentence.

DNA is a large molecule...	with	... units called genes.
It stores the genetic code...	to	... form multi-unit proteins with a quaternary structure.
Each gene is composed...	in	... forms a double helix.
A codon is a sequence...	or	... have a regulatory or catalytic role.
During transcription, the gene...	of	... their corresponding amino acids, binding them together to form polypeptides.
The mRNA is read by ribosomes, which match up the codons...	is	... a number of codons.
Polypeptides may associate together ...	which	... copied into mRNA.
Proteins may be used structurally...	of	... three adjacent nucleotides.

Write your completed paragraph here:_____

_____

_____

_____

_____

3. Explain how gene expression can be affected by the environment: _____

_____

_____

_____

_____

_____

_____

© 2012-2014 **BIOZONE** International
ISBN: 978-1-927173-93-0
Photocopying Prohibited

TEST

Topic 8

Cellular Metabolism

Key terms

activation energy

Calvin cycle

carboxylation

cell respiration

chemiosmosis

chloroplast

competitive inhibition

cristae

decarboxylation

electron carrier

electron transport chain

end product inhibition

enzyme

enzyme inhibition

glycolysis

induced fit model

Krebs cycle

light dependent reactions

light independent reactions

matrix

metabolic pathway

mitochondrion

non-competitive inhibition

oxidative phosphorylation

photolysis

photosystem I

photosystem II

proton gradient

redox reaction

ribulose bisphosphate

RuBisCo

stroma

thylakoid discs

8.1 Metabolism

Understandings, applications, skills

Activity number

☐ 1 Explain what is meant by a metabolic pathway and describe examples. — 222

☐ 2 Explain how enzymes catalyse chemical reactions. Calculate and plot rates of enzyme-catalysed reactions from raw data. — 223

☐ 3 Describe enzyme inhibition. Distinguish competitive and non-competitive inhibition in graphs at specified substrate concentrations. Explain how an understanding of metabolic pathways can be used to develop new drugs (e.g. anti-malarial drugs). — 224

☐ 4 Explain the control of metabolic pathways by end-product inhibition. Describe end-product inhibition of the pathway that converts threonine to isoleucine. — 225

TOK *Experimental work has led to many metabolic pathways being described. How can investigating component parts give us knowledge of the whole?*

8.2 Cell respiration

Understandings, applications, skills

Activity number

☐ 1 Describe cell respiration, summarizing the events in glycolysis, the link reaction, Krebs cycle, and the electron transport chain (ETC). Include reference to the small net ATP gain in glycolysis, the formation of pyruvate, the oxidation of acetyl groups and reduction of hydrogen carriers in the Krebs cycle, and electron transfer and ATP generation in the ETC. Explain the role of water as the final electron acceptor in the ETC. — 226

☐ 2 Identify the location of each of the steps in cell respiration. Annotate a diagram of a mitochondrion to relate its structure to function. Explain how electron tomography is used to produce images of active mitochondria. — 226

☐ 3 Analyse a diagram of pathways of aerobic respiration to determine where decarboxylation and oxidation reactions occur. — 226

☐ 4 Describe how transfer of electrons between carriers in the ETC is coupled to proton pumping. Explain how the proton gradient is used to generate ATP (chemiosmosis). Recognize that chemiosmosis is a process common to both cell respiration and photosynthesis. — 227

TOK *Chemiosmotic theory encountered years of opposition before acceptance. Why are old theories not always rejected immediately after falsification?*

8.3 Photosynthesis

Understandings, applications, skills

Activity number

☐ 1 Recall the process of photosynthesis, recognizing the light dependent and the light independent phases and their locations. Annotate a diagram to describe how the structure of a chloroplast is adapted to its function. — 66 228

☐ 2 Describe the light dependent reactions of photosynthesis, including reference to the absorption of light by the photosystems, the transfer of excited electrons between carriers in the thylakoid membranes, the generation of ATP and NADPH, and the photolysis of water to generate replacement electrons. — 229

☐ 3 Describe the light independent reactions (Calvin cycle), including the role of the catalysing enzyme RuBisCo, the carboxylation of ribulose bisphosphate (RuBP), and the production of triose phosphate using reduced NADPH and ATP. Describe the fate of triose phosphates generated in the Calvin cycle and explain how the RuBP is regenerated. — 230

☐ 4 Describe Calvin's experiment to elucidate the carboxylation of RuBP. — 231

TOK *Calvin's experiment was very creative. To what extent are such elegant protocols similar to the creation of a work of art?*

222 Metabolic Pathways

Key Idea: A metabolic pathway is a series of linked biochemical reactions.

Metabolic pathways are linked biochemical reactions that occur within living organisms to maintain life. **Enzymes** activate (catalyse) each step of a metabolic pathway and, in turn, each enzyme is encoded by specific genes. Metabolic pathways are controlled by regulating the amount of enzyme present (by switching the genes encoding that enzyme on or off) or by controlling enzyme activity. Each step in a metabolic pathway is part of a sequence, where the product from one step becomes the substrate for the next.

A Simplified Metabolic Pathway

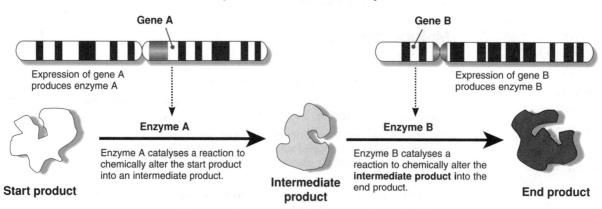

Expression of gene A produces enzyme A

Expression of gene B produces enzyme B

Enzyme A

Enzyme A catalyses a reaction to chemically alter the start product into an intermediate product.

Enzyme B

Enzyme B catalyses a reaction to chemically alter the **intermediate product** into the end product.

Start product

Intermediate product

End product

Cyclic Pathways

Some metabolic pathways flow in a cycle. Each component of the cycle is the substrate for the next reaction in the cycle and the final product is the substrate for the original first reaction (e.g. the Krebs cycle and urea cycle below). In addition, each step of the cycle is catalysed by an enzyme or a group of enzymes (sometimes called an enzyme complex).

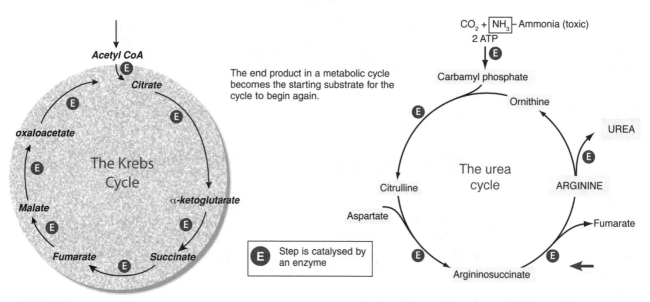

The end product in a metabolic cycle becomes the starting substrate for the cycle to begin again.

E Step is catalysed by an enzyme

1. Define the term **metabolic pathway**: _____

_____

_____

2. What is the role of enzymes in metabolic pathways? _____

_____

_____

3. Which products would not be made if the enzyme marked with the blue arrow in the urea cycle was not functional?

_____

_____

© 2012-2014 **BIOZONE** International
ISBN: 978-1-927173-93-0
Photocopying Prohibited

223 How Enzymes Work

Key Idea: Enzymes are biological catalysts. They speed up biological reactions by lowering a reaction's activation energy. Chemical reactions in cells are accompanied by energy changes. The amount of energy released or taken up is directly related to the tendency of a reaction to run to completion (for all the reactants to form products). Any reaction needs to raise the energy of the substrate to an unstable transition state before the reaction will proceed (below left). The amount of energy needed to do this is the **activation energy** (Ea). Enzymes lower the Ea by destabilizing bonds in the substrate so that it is more reactive. The current 'induced-fit' model of enzyme function is supported by studies of enzyme inhibitors, which show that enzymes are flexible and change shape when interacting with the substrate.

Enzymes Lower the Activation Energy

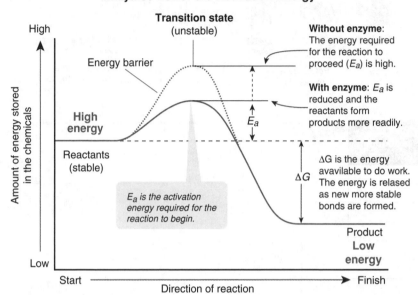

High

Transition state
(unstable)

Energy barrier

Amount of energy stored in the chemicals

High energy

Reactants
(stable)

E_a is the activation energy required for the reaction to begin.

Without enzyme:
The energy required for the reaction to proceed (E_a) is high.

With enzyme: E_a is reduced and the reactants form products more readily.

E_a

ΔG

ΔG is the energy available to do work. The energy is released as new more stable bonds are formed.

Low

Start

Product
Low energy

Finish

Direction of reaction

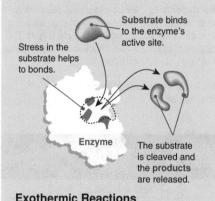

Substrate binds to the enzyme's active site.

Stress in the substrate helps to bonds.

Enzyme

The substrate is cleaved and the **products** are released.

Exothermic Reactions

Some enzymes can cause a single substrate molecule to be drawn into the active site. Chemical bonds are broken, causing the substrate molecule to break apart to become two separate molecules. Catabolic reactions break down complex molecules into simpler ones and involve a net release of energy, so they are called exothermic.
Examples: *hydrolysis, cell respiration.*

The Current Model: Induced Fit

An enzyme's interaction with its substrate is best regarded as an induced fit (below). The shape of the enzyme changes when the substrate fits into the cleft. The reactants become bound to the enzyme by weak chemical bonds. This binding can weaken bonds within the reactants themselves, allowing the reaction to proceed more readily. The current induced-fit model of enzyme function is supported by studies of enzyme inhibitors, which show that enzymes are flexible and change shape when interacting with the substrate.

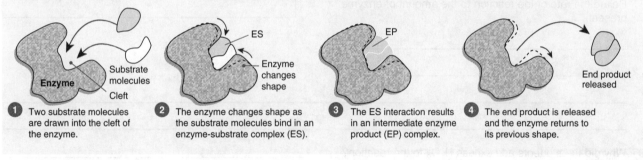

Enzyme
Substrate molecules
Cleft
ES
Enzyme changes shape
EP
End product released

1. Two substrate molecules are drawn into the cleft of the enzyme.

2. The enzyme changes shape as the substrate molecules bind in an enzyme-substrate complex (ES).

3. The ES interaction results in an intermediate enzyme product (EP) complex.

4. The end product is released and the enzyme returns to its previous shape.

1. Explain how enzymes act as **biological catalysts**: _____

_____

_____

_____

_____

2. Describe the '**induced fit**' model of enzyme action: _____

_____

_____

_____

_____

LINK **321**　LINK **52**　LINK **51**　LINK **50**　WEB **223**　**KNOW**

Skill: Calculating and Plotting
Enzyme Reaction Rates

A group of students decided to use cubes of potato, which naturally contain the enzyme catalase, placed in hydrogen peroxide to test the effect of enzyme concentration on reaction rate. The reaction rate could be measured by the volume of oxygen produced as the hydrogen peroxide was decomposed into oxygen and water.

The students cut raw potato into cubes with a mass of one gram. These were placed a conical flask with excess hydrogen peroxide (right). The reaction was left for five minutes and the volume of oxygen produced recorded.

The students recorded the results in the table below:

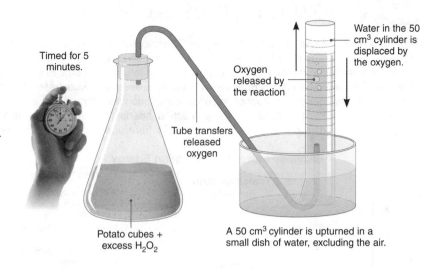

Timed for 5 minutes.

Oxygen released by the reaction

Water in the 50 cm³ cylinder is displaced by the oxygen.

Tube transfers released oxygen

Potato cubes + excess H_2O_2

A 50 cm³ cylinder is upturned in a small dish of water, excluding the air.

Mass of potato (g)	Volume oxygen (cm³) (5 minutes)			Mean	Rate of O_2 production (cm³ min⁻¹)
	Test 1	Test 2	Test 3		
1	6	5	6		
2	10	9	9		
3	14	15	15		
4	21	20	20		
5	24	23	25		

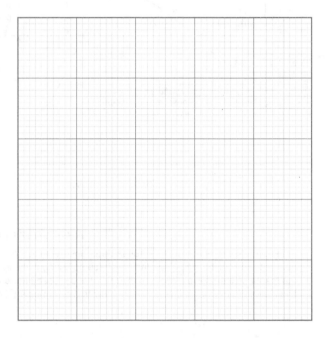

3. Complete the table by filling in the mean volume of oxygen produced and the rate of oxygen production.

4. Plot the mass of the potato vs the rate of production on the grid right:

5. Relate the rate of the reaction to the amount of enzyme present.

_____

_____

_____

_____

6. Why did the students add excess H_2O_2 to the reaction? _____

_____

7. State one extra reaction that could (or perhaps should) have been carried out by the students: _____

_____

8. The students decide to cook some potato and carry out the test again with two grams of potato. Predict the result:

_____

_____

9. Explain this result: _____

_____

_____

© 2012-2014 **BIOZONE** International
ISBN: 978-1-927173-93-0
Photocopying Prohibited

224 Enzyme Inhibition

Key Idea: Enzyme activity can be regulated by chemicals that compete for an enzyme's active site or bind in some way to the enzyme.

Enzymes may be deactivated, temporarily or permanently, by chemicals called enzyme inhibitors. Competitive inhibitors

compete directly with the substrate for the active site, and their effect can be overcome by increasing the concentration of available substrate. A non-competitive inhibitor does not occupy the active site, but distorts it so that the substrate and enzyme can no longer interact

Competitive Inhibition

Competitive inhibitors compete with the normal substrate for the enzyme's active site.

A competitive inhibitor occupies the active site only temporarily and so the inhibition is reversible. The metal **mercury** (Hg) is a competitive inhibitor.

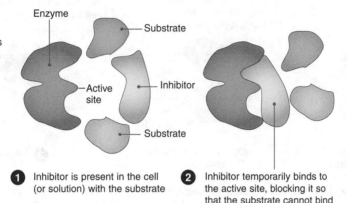

1 Inhibitor is present in the cell (or solution) with the substrate

2 Inhibitor temporarily binds to the active site, blocking it so that the substrate cannot bind

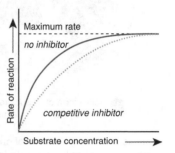

Fig.1 Effect of competitive inhibition on enzyme reaction rate at different substrate concentration

Non-competitive Inhibition

Non-competitive inhibitors bind with the enzyme at a site other than the active site. They inactivate the enzyme by altering its shape. The amino acid alanine is a non-competitive inhibitor of the enzyme pyruvate kinase.

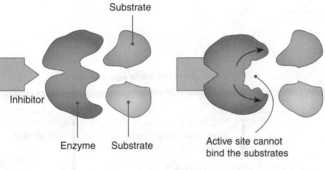

1 Without the inhibitor bound, the enzyme can bind the substrate

2 When the inhibitor binds, the enzyme changes shape.

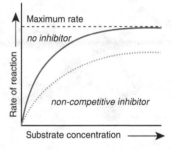

Fig.2 Effect of non-competitive inhibition on enzyme reaction rate at different substrate concentration

Developments in scientific research and the use of worldwide enzyme and genetic databases have helped research into enzymes and the metabolic pathways they control. The **MEP** metabolic pathway is an important pathway in the malaria causing parasite *Plasmodium falciparum,* but is not found in humans. The enzymes in the pathway, most notably the enzyme DXR, have therefore been targeted for the development of antimalarial drugs that inhibit their action. Research shows that the antibiotic fosmidomycin acts as both a competitive and non-competitive inhibitor of DXR.

The malaria causing parasite *Plasmodium falciparum* is transferred to humans by the *Anopheles* mosquito.

1. Distinguish between **competitive** and **non-competitive** inhibition: _____

_____

_____

2. How could you distinguish between competitive and non-competitive inhibition in an isolated system?

_____

_____

3. Why is the MEP pathway of particular interest to companies researching antimalarial drugs? _____

_____

_____

225 Control of Metabolic Pathways

Key Idea: The end product of a metabolic pathway can regulate the pathway itself.

Metabolism refers to all the chemical activities (metabolic reactions) of life. They form a tremendously complex network of reactions that is necessary in order to 'maintain' the organism. Often the products of a **metabolic pathway** regulate the pathway itself. This might be achieved by the end product of the pathway inhibiting the reactions in the pathway so that no more product is produced. This can be achieved by **allosteric enzyme regulation** (below).

Enzyme Regulation

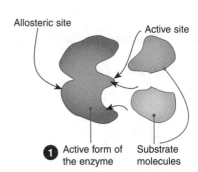

Allosteric site

Active site

1 Active form of the enzyme

Substrate molecules

Enzyme catalyses the reaction between the substrates producing a new molecule.

2 Enzyme-substrate complex

The new molecule (e.g. end product) attaches to the allosteric site of the enzyme, inhibiting the enzyme's activity.

3 Inactive form of the enzyme

A metabolic pathway is a series of enzyme-catalysed chemical reactions in a cell. Metabolic pathways can be regulated by their end products through **allosteric regulation** in a process called feedback inhibition (above). When concentrations of the end product are high it will bind to the **allosteric site** of the first enzyme in the pathway, inhibiting the enzyme and shutting down the pathway. When the concentration of the end product is reduced, it is released from the allosteric site and the pathway is activated again.

Isoleucine Synthesis from Threonine

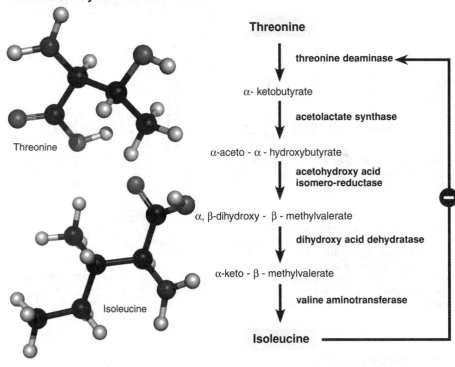

Threonine

Isoleucine

Threonine

↓ **threonine deaminase** ←

α- ketobutyrate

↓ **acetolactate synthase**

α-aceto - α - hydroxybutyrate

↓ **acetohydroxy acid isomero-reductase**

α, β-dihydroxy - β - methylvalerate

↓ **dihydroxy acid dehydratase**

α-keto - β - methylvalerate

↓ **valine aminotransferase**

Isoleucine ──

Isoleucine is an essential amino acid. It can only be synthesized by bacteria and plants. Animals must obtain it in their diet.

The pathway for the biosynthesis of isoleucine from the amino acid threonine is controlled by **end product inhibition (negative feedback)**. Threonine is converted to the intermediate molecule α-ketobutyrate by **threonine deaminase**. Threonine deaminase is inhibited by isoleucine. When there is a high concentration of isoleucine the pathway is inhibited, but as the concentration of isoleucine decreases the threonine deaminase is no longer inhibited and the pathway begins again.

1. With reference to the threonine-isoleucine pathway, explain how **end product inhibition** works: _____

_____

_____

_____

2. Explain the role of an allosteric regulator in end product inhibition: _____

_____

_____

_____

© 2012-2014 **BIOZONE** International
ISBN: 978-1-927173-93-0
Photocopying Prohibited

226 The Biochemistry of Respiration

Key Idea: During cell respiration, the energy in glucose is transferred to ATP in a series of enzyme controlled steps.

The oxidation of glucose is a catabolic, energy yielding pathway. The breakdown of glucose and other organic fuels (such as fats and proteins) to simpler molecules releases energy for ATP synthesis. Glycolysis and the Krebs cycle supply electrons to the electron transport chain, which drives **oxidative phosphorylation**. Glycolysis nets two ATP. The conversion of pyruvate (the end product of glycolysis) to

acetyl CoA links glycolysis to the Krebs cycle. One "turn" of the cycle releases carbon dioxide, forms one ATP, and passes electrons to three NAD^+ and one FAD. Most of the ATP generated in cellular respiration is produced by oxidative phosphorylation when $NADH + H^+$ and $FADH_2$ donate electrons to the series of electron carriers in the electron transport chain. At the end of the chain, electrons are passed to molecular oxygen, reducing it to water. Electron transport is coupled to ATP synthesis.

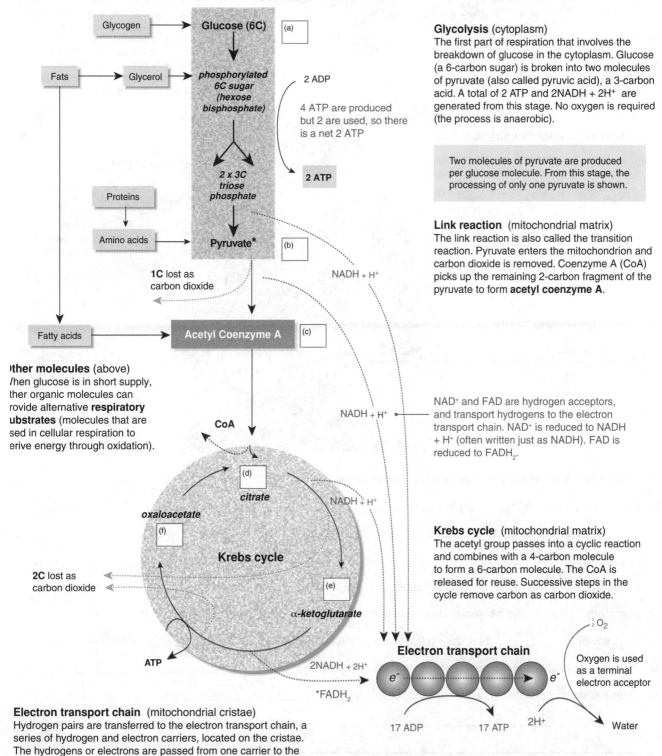

Glycolysis (cytoplasm)
The first part of respiration that involves the breakdown of glucose in the cytoplasm. Glucose (a 6-carbon sugar) is broken into two molecules of pyruvate (also called pyruvic acid), a 3-carbon acid. A total of 2 ATP and $2NADH + 2H^+$ are generated from this stage. No oxygen is required (the process is anaerobic).

Two molecules of pyruvate are produced per glucose molecule. From this stage, the processing of only one pyruvate is shown.

Link reaction (mitochondrial matrix)
The link reaction is also called the transition reaction. Pyruvate enters the mitochondrion and carbon dioxide is removed. Coenzyme A (CoA) picks up the remaining 2-carbon fragment of the pyruvate to form **acetyl coenzyme A**.

NAD^+ and FAD are hydrogen acceptors, and transport hydrogens to the electron transport chain. NAD^+ is reduced to $NADH + H^+$ (often written just as NADH). FAD is reduced to $FADH_2$.

Krebs cycle (mitochondrial matrix)
The acetyl group passes into a cyclic reaction and combines with a 4-carbon molecule to form a 6-carbon molecule. The CoA is released for reuse. Successive steps in the cycle remove carbon as carbon dioxide.

Other molecules (above)
When glucose is in short supply, other organic molecules can provide alternative **respiratory substrates** (molecules that are used in cellular respiration to derive energy through oxidation).

Oxygen is used as a terminal electron acceptor

Electron transport chain (mitochondrial cristae)
Hydrogen pairs are transferred to the electron transport chain, a series of hydrogen and electron carriers, located on the cristae. The hydrogens or electrons are passed from one carrier to the next, in a series of redox reactions, losing energy as they go. The energy released in this stepwise process is used to **phosphorylate** ADP to form ATP. Oxygen is the final electron acceptor and is reduced to water (hence the term **oxidative phosphorylation**). **Note** FAD enters the electron transport chain at a lower energy level than NAD, and only 2ATP are generated per $FADH_2$.

Total ATP yield per glucose
Glycolysis: 2 ATP, *Krebs cycle*: 2 ATP, *Electron transport*: 34 ATP

The theoretical maximum yield of 38 ATP per mole of glucose has recently been revised down to 32 ATP (28 from the ETC).

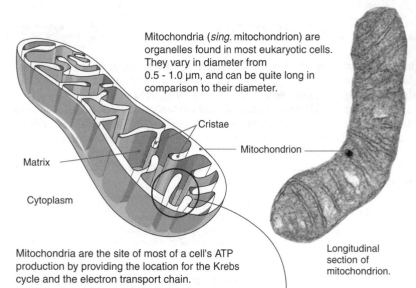

Mitochondria (*sing.* mitochondrion) are organelles found in most eukaryotic cells. They vary in diameter from 0.5 - 1.0 μm, and can be quite long in comparison to their diameter.

Cristae

Mitochondrion

Matrix

Cytoplasm

Mitochondria are the site of most of a cell's ATP production by providing the location for the Krebs cycle and the electron transport chain.

Longitudinal section of mitochondrion.

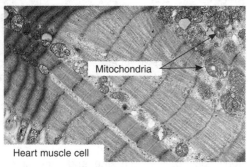

Heart muscle cell

Mitochondria

Cells that require a lot of ATP for cellular processes have a lot of mitochondria. Sperm cells contain a large number of mitochondria near the base of the tail. Liver cells have around 2000 mitochondria per cell, taking up 25% of the cytoplasmic space. Heart muscle cells (above) may have 40% of the cytoplasmic space taken up by mitochondria.

Location of Cellular Respiration

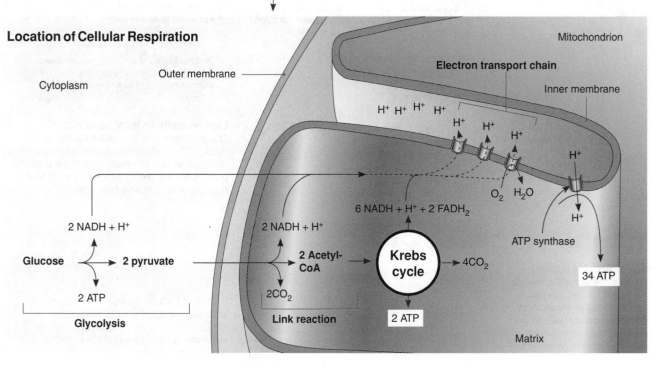

Mitochondrion

Cytoplasm

Outer membrane

Electron transport chain

Inner membrane

H^+ H^+ H^+ H^+

H^+

H^+

H^+

H^+

6 NADH + H^+ + 2 FADH$_2$

O_2 H_2O

H^+

2 NADH + H^+

2 NADH + H^+

ATP synthase

Glucose → 2 pyruvate

2 Acetyl-CoA

Krebs cycle

4CO$_2$

34 ATP

2 ATP

2CO$_2$

Glycolysis

Link reaction

2 ATP

Matrix

1. In the longitudinal section of a mitochondrion (above), label the matrix and cristae.

2. Explain the purpose of the link reaction: _____

_____

_____

3. On the diagram of cell respiration (previous page), state the number of carbon atoms in each of the molecules (a)-(f):

4. How many ATP molecules **per molecule of glucose** are generated during the following stages of respiration?

   (a) Glycolysis: _____ (b) Krebs cycle: _____ (c) Electron transport chain: _____ (d) Total: _____

5. Explain what happens to the carbon atoms lost during respiration: _____

6. Explain what happens during oxidative phosphorylation: _____

_____

_____

_____

_____

_____

© 2012-2014 **BIOZONE** International
ISBN: 978-1-927173-93-0

227 Chemiosmosis

Key Idea: Chemiosmosis is the process in which electron transport is coupled to ATP synthesis.

It occurs in the membranes of mitochondria, the chloroplasts of plants, and across the plasma membrane of bacteria. Chemiosmosis involves the establishment of a proton (hydrogen) gradient across a membrane. The concentration gradient is used to drive ATP synthesis. Chemiosmosis has two key components: an **electron transport chain** (ETC) sets up a proton gradient as electrons pass along it to a final electron acceptor, and an enzyme called **ATP synthase**

uses the proton gradient to catalyse ATP synthesis. In cellular respiration, electron carriers on the inner membrane of the mitochondrion oxidize NADH + H$^+$ and FADH$_2$. The energy released from this process is used to move protons against their concentration gradient, from the mitochondrial matrix into the space between the two membranes. The return of protons to the matrix via ATP synthase is coupled to ATP synthesis. Similarly, in the chloroplasts of green plants, ATP is produced when protons pass from the thylakoid lumen to the chloroplast stroma via ATP synthase.

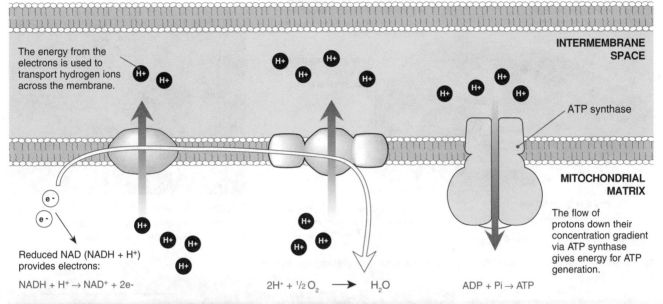

The energy from the electrons is used to transport hydrogen ions across the membrane.

INTERMEMBRANE SPACE

ATP synthase

MITOCHONDRIAL MATRIX

The flow of protons down their concentration gradient via ATP synthase gives energy for ATP generation.

Reduced NAD (NADH + H$^+$) provides electrons:

NADH + H$^+$ → NAD$^+$ + 2e-

2H$^+$ + $\frac{1}{2}$O$_2$ → H$_2$O

ADP + Pi → ATP

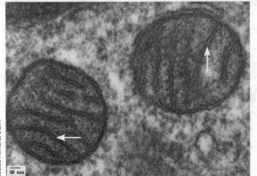

The intermembrane spaces can be seen (arrows) in this transverse section of mitochondria.

The Evidence for Chemiosmosis

The British biochemist Peter Mitchell proposed the chemiosmotic hypothesis in 1961. He proposed that, because living cells have membrane potential, electrochemical gradients could be used to do work, i.e. provide the energy for ATP synthesis. Scientists at the time were skeptical, but the evidence for chemiosmosis was extensive and came from studies of isolated mitochondria and chloroplasts. Evidence included:

▶ The outer membranes of mitochondria were removed leaving the inner membranes intact. Adding protons to the treated mitochondria increased ATP synthesis.

▶ When isolated chloroplasts were illuminated, the medium in which they were suspended became alkaline.

▶ Isolated chloroplasts were kept in the dark and transferred first to a low pH medium (to acidify the thylakoid interior) and then to an alkaline medium (low protons). They then spontaneously synthesized ATP (no light was needed).

1. Summarize the process of chemiosmosis: _____

_____

_____

2. Why did the addition of protons to the treated mitochondria increase ATP synthesis? _____

_____

3. Why did the suspension of isolated chloroplasts become alkaline when illuminated? _____

_____

4. (a) What was the purpose of transferring the chloroplasts first to an acid then to an alkaline medium? _____

_____

(b) Why did ATP synthesis occur spontaneously in these treated chloroplasts? _____

_____

© 2012-2014 **BIOZONE** International
ISBN: 978-1-927173-93-0
Photocopying Prohibited

WEB

227 KNOW

228 Chloroplasts

Key Idea: Chloroplasts have a complicated internal membrane structure, which provides the sites for the light dependent reactions of photosynthesis.

Chloroplasts are specialized plastids where photosynthesis occurs. A mesophyll leaf cell will contain between 50-100 chloroplasts. The chloroplasts are generally aligned so that their broad surface runs parallel to the cell wall to maximize the surface area available for light absorption. Chloroplasts have an internal structure characterized by a system of membranous structures called **thylakoids** arranged into stacks called **grana**. Special pigments, called **chlorophylls** and **carotenoids**, are bound to the membranes as part of light-capturing photosystems. They absorb light of specific wavelengths and thereby capture the light energy.

The Structure of a Chloroplast

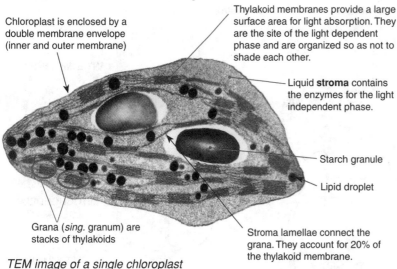

Chloroplast is enclosed by a double membrane envelope (inner and outer membrane)

Thylakoid membranes provide a large surface area for light absorption. They are the site of the light dependent phase and are organized so as not to shade each other.

Liquid **stroma** contains the enzymes for the light independent phase.

Starch granule

Lipid droplet

Grana (*sing.* granum) are stacks of thylakoids

Stroma lamellae connect the grana. They account for 20% of the thylakoid membrane.

TEM image of a single chloroplast

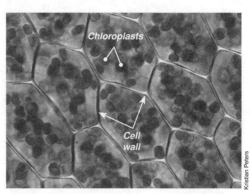

Chloroplasts

Cell wall

Kristian Peters

Chloroplasts visible in leaf cells. They appear green because they reflect green light, absorbing blue and red light.

1. Label the transmission electron microscope image of a chloroplast below:

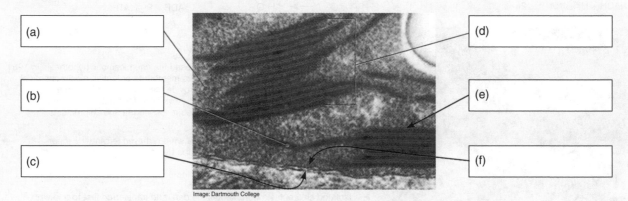

(a)

(b)

(c)

(d)

(e)

(f)

Image: Dartmouth College

2. (a) Describe where chlorophyll is found in a chloroplast: _____

_____

   (b) Explain why chlorophyll is found there: _____

_____

_____

3. Explain how the internal structure of chloroplasts helps absorb the maximum amount of light: _____

_____

_____

_____

4. Explain why plant leaves appear green: _____

_____

_____

© 2012-2014 **BIOZONE** International
ISBN: 978-1-927173-93-0
Photocopying Prohibited

229 Light Dependent Reactions

Key Idea: In light dependent reactions of photosynthesis, the energy from photons of light is used to drive the reduction of $NADP^+$ and the production of ATP.

Like cellular respiration, photosynthesis is a redox process, but in photosynthesis, water is split, and electrons and hydrogen ions, are transferred from water to CO_2, reducing it to sugar. The electrons increase in potential energy as they move from water to sugar. The energy to do this is provided by light. Photosynthesis has two phases. In the **light dependent**

reactions, light energy is converted to chemical energy (ATP and NADPH). In the **light independent reactions**, the chemical energy is used to synthesize carbohydrate. The light dependent reactions most commonly involve **non-cyclic phosphorylation**, which produces ATP and NADPH in roughly equal quantities. The electrons lost are replaced from water. In **cyclic phosphorylation**, the electrons lost from photosystem II are replaced by those from photosystem I. ATP is generated, but not NADPH.

Non-cyclic phosphorylation

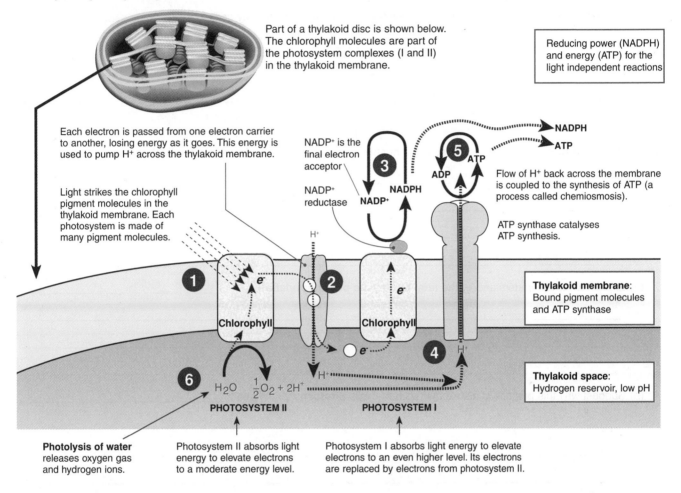

Part of a thylakoid disc is shown below. The chlorophyll molecules are part of the photosystem complexes (I and II) in the thylakoid membrane.

Reducing power (NADPH) and energy (ATP) for the light independent reactions

Each electron is passed from one electron carrier to another, losing energy as it goes. This energy is used to pump H^+ across the thylakoid membrane.

Light strikes the chlorophyll pigment molecules in the thylakoid membrane. Each photosystem is made of many pigment molecules.

$NADP^+$ is the final electron acceptor

$NADP^+$ reductase

Flow of H^+ back across the membrane is coupled to the synthesis of ATP (a process called chemiosmosis).

ATP synthase catalyses ATP synthesis.

Thylakoid membrane: Bound pigment molecules and ATP synthase

Thylakoid space: Hydrogen reservoir, low pH

Photolysis of water releases oxygen gas and hydrogen ions.

Photosystem II absorbs light energy to elevate electrons to a moderate energy level.

Photosystem I absorbs light energy to elevate electrons to an even higher level. Its electrons are replaced by electrons from photosystem II.

Cyclic phosphorylation

Cyclic phosphorylation involves only photosystem I and NADPH is not generated. Electrons from photosystem I are shunted back to the electron carriers in the membrane. This pathway produces ATP only. The Calvin cycle uses more ATP than NADPH, so cyclic phosphorylation makes up the difference. It is activated when NADPH levels build up, and remains active until enough ATP is made to meet demand.

Electrons are cycled through a pathway that takes them away from $NADP^+$ reductase.

ATP is produced while NADPH production ceases.

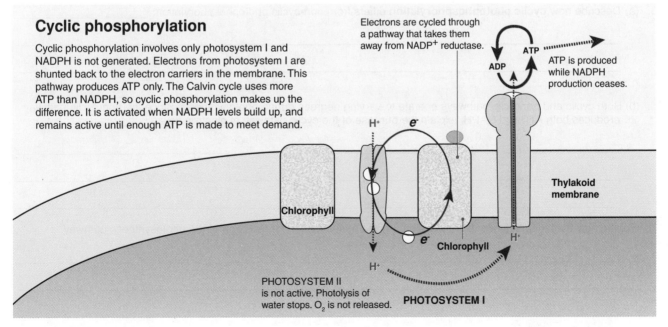

Thylakoid membrane

PHOTOSYSTEM II is not active. Photolysis of water stops. O_2 is not released.

PHOTOSYSTEM I

1. Describe the role of the carrier molecule **NADP** in photosynthesis: _____

    _____

    _____

2. Explain the role of chlorophyll molecules in photosynthesis: _____

    _____

    _____

    _____

3. Summarize the events of the light dependent reactions and identify where they occur: _____

    _____

    _____

    _____

    _____

4. Describe how ATP is produced as a result of light striking chlorophyll molecules during the light dependent phase:

    _____

    _____

    _____

    _____

5. (a) Explain what you understand by the term **non-cyclic phosphorylation**: _____

    _____

    _____

    _____

    (b) Suggest why this process is also known as non-cyclic **photo**phosphorylation: _____

    _____

    _____

    _____

6. (a) Describe how **cyclic photophosphorylation** differs from non-cyclic photophosphorylation: _____

    _____

    _____

    (b) Both cyclic and noncyclic pathways operate to varying degrees during photosynthesis. Since the non-cyclic pathway produces both ATP and NAPH, explain the purpose of the cyclic pathway of electron flow:

    _____

    _____

    _____

7. Explain how the independence of photosystem I gives a mechanism for evolution of the photosynthetic pathway:

    _____

    _____

    _____

© 2012-2014 **BIOZONE** International
ISBN: 978-1-927173-93-0
Photocopying Prohibited

230 Light Independent Reactions

Key Idea: The light independent reactions of photosynthesis take place in the stroma of the chloroplast and do not require light to proceed.

In the **light independent reactions** (the **Calvin cycle**) hydrogen (H+) is added to CO_2 and a 5C intermediate to make carbohydrate. The H+ and ATP are supplied by the light dependent reactions. The Calvin cycle uses more ATP than NADPH, but the cell uses cyclic phosphorylation (which does not produce NADPH) when it runs low on ATP to make up the difference.

KEY:

- 🔵 Carbon atom
- Ⓟ Phosphate group

The rate of the light independent reaction depends on the availability of carbon dioxide (CO_2) as it is required in the first step of the reaction. Without it, the reaction can not proceed, showing CO_2 is also a **limiting factor** in photosynthesis.

The Calvin cycle is a series of reactions driven by ATP and NADPH. It generates hexose sugars and reduces the intermediate products to regenerate ribulose 1,5 bisphosphate (RuBP) needed for the first step of the cycle.

The catalysing enzyme **RuBisCo** joins carbon dioxide (CO_2) with RuBP to form glycerate 3-phosphate (**GP**). ATP driven reactions then form 1,3 bisphosphoglycerate before NADPH driven reactions form triose phosphate (**TP**). Some of this then leaves the chloroplast and forms sugars while the rest continues through the cycle to eventually reform RuBP.

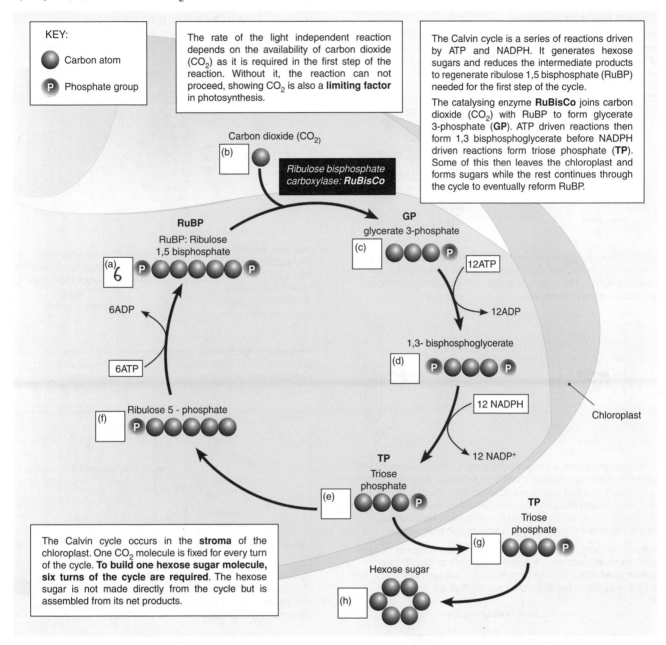

The Calvin cycle occurs in the **stroma** of the chloroplast. One CO_2 molecule is fixed for every turn of the cycle. **To build one hexose sugar molecule, six turns of the cycle are required**. The hexose sugar is not made directly from the cycle but is assembled from its net products.

1. In the boxes on the diagram above, write the number of molecules formed at each step during the formation of **one hexose sugar molecule**. The first one has been done for you:

2. Explain the importance of RuBisCo in the Calvin cycle: _____

    _____

3. Identify the actual end product on the Calvin cycle: _____

4. Write the equation for the production of one hexose sugar molecule from carbon dioxide: _____

    _____

5. Explain why the Calvin cycle is likely to cease in the dark for most plants, even though it is independent of light:

    _____

© 2012-2014 **BIOZONE** International
ISBN: 978-1-927173-93-0
Photocopying Prohibited

231 Experimental Investigation of Photosynthesis

Key Idea: Hill's experiment using isolated chloroplasts and Calvin's "lollipop" experiment provided important information on the process of photosynthesis.

In the 1930s Robert Hill devised a way of measuring oxygen evolution and the rate of photosynthesis in isolated chloroplasts. During the 1950s Melvin Calvin led a team using radioisotopes of carbon to work out the steps of the light independent reactions (the Calvin cycle).

Robert Hill's Experiment

The dye **DCPIP** (2,6-dichlorophenol-indophenol) is blue. It is reduced by H^+ ions and forms $DCPIPH_2$ (colourless). Hill made use of this dye to show that O_2 is produced during photosynthesis even when CO_2 is not present.

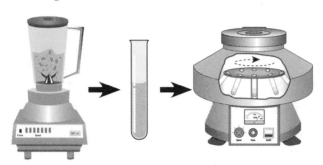

Leaves are homogenized to form a slurry. The slurry is filtered to remove any debris. The filtered extract is then centrifuged at low speed to remove the larger cell debris and then at high speed to separate out the chloroplasts.

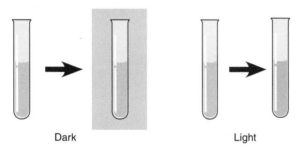

Dark   Light

The chloroplasts are resuspended in a buffer. The blue dye **DCPIP** is added to the suspension. In a test tube left in the dark, the dye remains unchanged. In a test tube exposed to the light, the blue dye fades and the test tube turns green again. The rate of colour change can be measured by measuring the light absorbance of the suspension. The rate is proportional to the rate at which oxygen is produced.

Hill's experiment showed that water must be the source of oxygen (and therefore electrons). It is split by light to produce H^+ ions (which reduce DCPIP) and O^{2-} ions (which combine to form O_2 and $2e^-$). The equation below summarizes his findings:

$$H_2O + A \rightarrow AH_2 + \frac{1}{2} O_2$$

where A is the electron acceptor (*in vivo* this is NADP)

Calvin's Lollipop Experiment

Calvin and his colleges placed the algae *Chlorella vulgaris* in a thin bulb shaped flask to simulate a leaf (the lollipop).

Radioactive ^{14}C labelled CO_2 was bubbled into the flask at precise times.

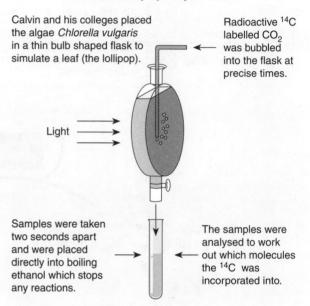

Light →

Samples were taken two seconds apart and were placed directly into boiling ethanol which stops any reactions.

The samples were analysed to work out which molecules the ^{14}C was incorporated into.

Two-dimensional chromatography was used to separate the molecules in each sample. The sample is run in one direction, then rotated 90 degrees and run again. This separates out molecules that might be close to each other.

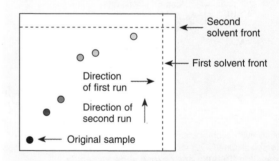

Second solvent front

First solvent front

Direction of first run →

Direction of second run ↑

← Original sample

By identifying the order that the molecules incorporating the ^{14}C appeared it was possible to work out the steps of the now called Calvin cycle. This could only be done by taking samples only seconds apart.

---

1. Write an equation for the formation of $DCPIPH_2$ from DCPIP: _____

   _____

2. What important finding about photosynthesis did Hill's experiment show? _____

   _____

   _____

3. Why did the samples in Calvin's lollipop experiment need to be taken just seconds apart? _____

   _____

   _____

© 2012-2014 **BIOZONE** International
ISBN: 978-1-927173-93-0
Photocopying Prohibited

232 Chapter Review

Summarize what you know about this topic under the
headings provided. You can draw diagrams or mind maps, or
write short notes to organize your thoughts. Use the images
and hints, included to help you:

Metabolism

HINT: Define metabolism. How are metabolic pathways
controlled?

Cell respiration

HINT: Summarize the stages of cell respiration
and production of ATP.

Photosynthesis

HINT: What are the two parts of photosynthesis. Where do they take place?

REVISE

233 KEY TERMS: Did You Get It?

1. Match each term to its definition, as identified by its preceding letter code.

activation energy ....

Calvin cycle ....

chemiosmosis ....

competitive inhibition ....

electron transport chain ....

induced fit model ....

Krebs cycle ....

light dependent reactions ....

non-competitive inhibition ....

oxidative phosphorylation ....

A The process by which the synthesis of ATP is coupled to electron transport and the movement of protons.

B A type of enzyme inhibition where the substrate and inhibitor compete to bind in the active site.

C The currently accepted model for enzyme function.

D A series of biochemical reactions, occurring in the stroma of chloroplasts, in which CO_2 is incorporated into carbohydrates. Also called the light independent phase.

E Also called the citric acid cycle. A metabolic pathway in which acetate (as acetyl-CoA) is consumed, NAD^+ is reduced to NADH, and carbon dioxide is produced.

F The energy required for a reactant to reach an unstable transition state in which it can react with another reactant.

G The chain of enzyme-based redox reactions that passes electrons from high to low redox potentials. The energy released is used to pump protons across a membrane and produce ATP.

H The process in cell respiration involving the oxidation of glucose by a series of redox reactions that provide the energy for the formation of ATP.

I A type of enzyme inhibition where the inhibitor does not occupy the active site but binds to some other part of the enzyme.

J The reactions in photosynthesis in which light energy is absorbed by the photosystems in the thylakoid membranes of the chloroplast, generating NADPH and ATP.

2. Identify the following statements as true of false (circle one)

   (a) Enzymes are biological catalysts. They lower the activation energy of a reaction. True / False

   (b) Competitive inhibition is when an inhibitor binds to a site other than an active site. True / False

   (c) The induced fit model states that the enzyme changes shape when a substrate fits into the active site. True / False

   (d) End product inhibition causes a feedback loop that escalates the outcome of the loop. True / False

3. Complete the schematic diagram of the transfer of energy and the production of macromolecules below using the following word list: *water, ADP, protein, carbon dioxide, amino acid, glucose, ATP.*

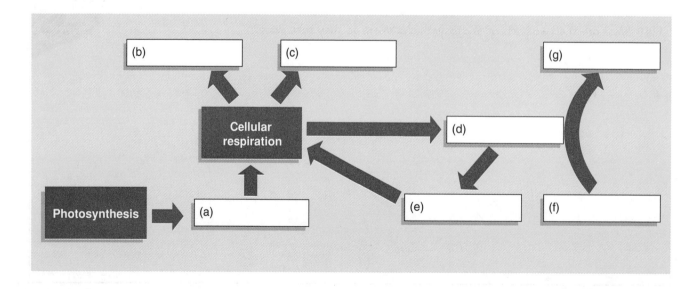

© 2012-2014 **BIOZONE** International
ISBN: 978-1-927173-93-0
Photocopying Prohibited

Topic 9

Plant Biology

Key terms

auxin

bulk flow

cohesion-tension hypothesis

cotyledon

double fertilization

endosperm

fertilization

flower

germination

halophyte

indeterminate growth

long-day plant

meristem

micropropagation

mutualism

phloem

photoperiod

phytochrome

pollination

pollinator

potometer

seed

short-day plant

sieve tube

stomata

translocation

transpiration

vascular tissue

xerophyte

xylem

9.1 Transport in the xylem of plants

Understandings, applications, skills

Activity number

☐ 1 Annotate a diagram of the plant body to indicate its general structure (stem, roots, leaves) and the location and role of the vascular tissues (xylem and phloem). 234

☐ 2 Describe transpiration and explain why it is a consequence of gas exchanges at the leaf. Explain how losses via transpiration are replaced by water uptake by roots. Describe pathways for water movement through the plant. 235

☐ 3 Describe active transport of minerals by root tissue and explain how mineral uptake facilitates absorption of water by osmosis. 235

☐ 4 Recall the properties of water and explain transport in the xylem in terms of the cohesive and adhesive properties of water. 39 236

☐ 5 Recognize and draw the structure of primary xylem vessels in stem sections viewed using light microscopy. 237

☐ 6 Measure transpiration rates using a potometer. Design an experiment to test hypotheses about the factors affecting transpiration rates (e.g. temperature). 238

☐ 7 Describe the adaptations of xerophytes and halophytes for water conservation. 239

9.2 Transport in the phloem of plants

Understandings, applications, skills

Activity number

☐ 1 Describe translocation in the phloem from sources to sinks. Explain how translocation is achieved by active transport of sugars into the phloem sieve tubes and movement of sap along hydrostatic pressure gradients. 240

☐ 2 Explain how high concentrations of solutes at the source contribute to water uptake by osmosis. Analyse data from experiments measuring phloem transport rates using aphid stylets and radioactively labelled carbon dioxide. 240

☐ 3 Describe the structure of phloem in relation to its transport function. 241

☐ 4 Identify xylem and phloem in microscope images of stems and root sections. 242

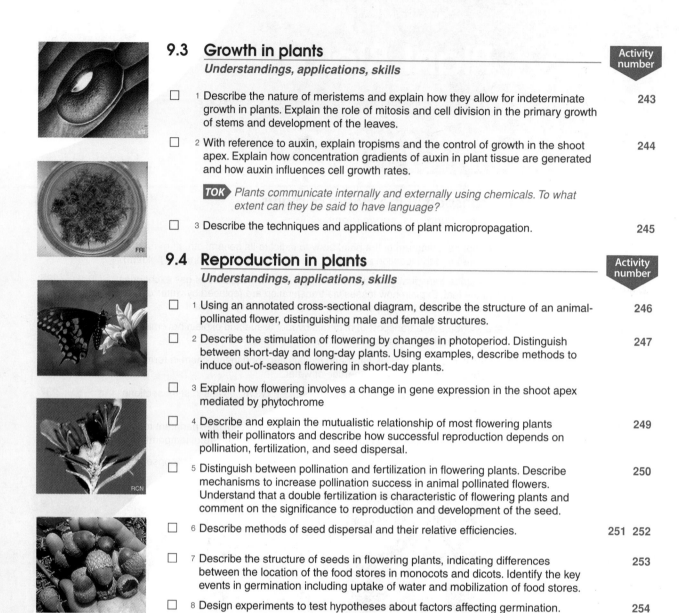

9.3 Growth in plants

Understandings, applications, skills

Activity number

☐ 1 Describe the nature of meristems and explain how they allow for indeterminate growth in plants. Explain the role of mitosis and cell division in the primary growth of stems and development of the leaves.

243

☐ 2 With reference to auxin, explain tropisms and the control of growth in the shoot apex. Explain how concentration gradients of auxin in plant tissue are generated and how auxin influences cell growth rates.

244

TOK *Plants communicate internally and externally using chemicals. To what extent can they be said to have language?*

☐ 3 Describe the techniques and applications of plant micropropagation.

245

9.4 Reproduction in plants

Understandings, applications, skills

Activity number

☐ 1 Using an annotated cross-sectional diagram, describe the structure of an animal-pollinated flower, distinguishing male and female structures.

246

☐ 2 Describe the stimulation of flowering by changes in photoperiod. Distinguish between short-day and long-day plants. Using examples, describe methods to induce out-of-season flowering in short-day plants.

247

☐ 3 Explain how flowering involves a change in gene expression in the shoot apex mediated by phytochrome

☐ 4 Describe and explain the mutualistic relationship of most flowering plants with their pollinators and describe how successful reproduction depends on pollination, fertilization, and seed dispersal.

249

☐ 5 Distinguish between pollination and fertilization in flowering plants. Describe mechanisms to increase pollination success in animal pollinated flowers. Understand that a double fertilization is characteristic of flowering plants and comment on the significance to reproduction and development of the seed.

250

☐ 6 Describe methods of seed dispersal and their relative efficiencies.

251 252

☐ 7 Describe the structure of seeds in flowering plants, indicating differences between the location of the food stores in monocots and dicots. Identify the key events in germination including uptake of water and mobilization of food stores.

253

☐ 8 Design experiments to test hypotheses about factors affecting germination.

254

234 The General Structure of Plants

Key Idea: The plant body comprises three main structures: roots, shoots and stems. The xylem and phloem form the vascular tissue that move fluids and minerals about the plant. The support and transport systems in plants are closely linked; many of the same tissues are involved in both systems. Primitive plants (e.g. mosses and liverworts) are small and low growing, and have no need for support and transport systems. If a plant is to grow to any size, it must have ways to hold itself up against gravity and to move materials around

its body. The body of a flowering plant has three parts: **roots** anchor the plant and absorb nutrients from the soil, **leaves** produce sugars by photosynthesis, and **stems** link the roots to the leaves and provide support for the leaves and reproductive structures. Vascular tissues (xylem and phloem) link all plant parts so that water, minerals, and manufactured food can be transported between different regions. All plants rely on fluid pressure within their cells (turgor) to give some support to their structure.

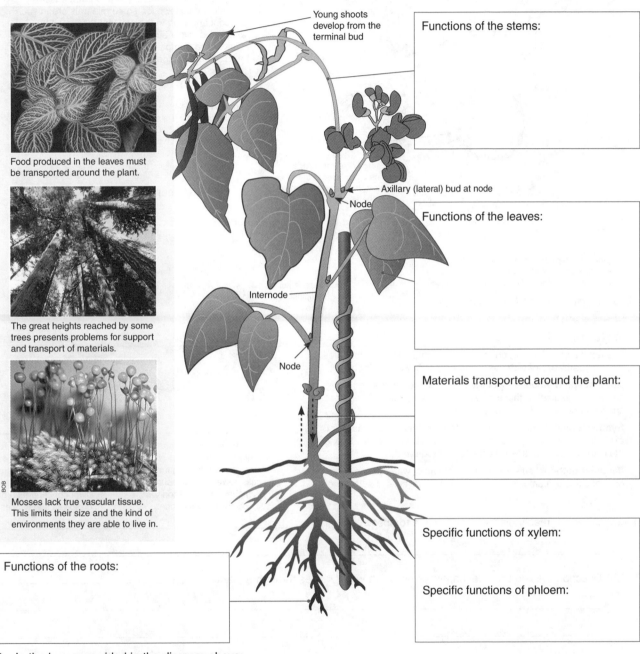

Food produced in the leaves must be transported around the plant.

The great heights reached by some trees presents problems for support and transport of materials.

Mosses lack true vascular tissue. This limits their size and the kind of environments they are able to live in.

Young shoots develop from the terminal bud

Functions of the stems:

Axillary (lateral) bud at node

Node

Functions of the leaves:

Internode

Node

Materials transported around the plant:

Specific functions of xylem:

Specific functions of phloem:

Functions of the roots:

1. In the boxes provided in the diagram above:

   (a) List the main functions of the leaves, roots and stems (remember that the leaves themselves have leaf veins).

   (b) List the materials that are transported around the plant body.

   (c) Describe the functions of the transport tissues: xylem and phloem.

2. What is the solvent for all materials transported around the plant? _____

3. State what processes are involved in transporting materials in the following tissues:

   (a) The xylem: _____

   (b) The phloem: _____

© 2012-2014 **BIOZONE** International
ISBN: 978-1-927173-93-0
Photocopying Prohibited

LINK 241 LINK 237 LINK 235 WEB 234

KNOW

235 Uptake at the Root

Key Idea: Water uptake is a passive process. Mineral uptake can be passive or active.

Plants need to take up water and minerals constantly. They must compensate for the continuous loss of water from the leaves and provide the materials the plant needs to make food. The uptake of water and minerals is mostly restricted to the younger, most recently formed cells of the roots and the root hairs. Some water moves through the plant tissues via the plasmodesmata (the **symplastic route**) but most passes through the free spaces outside the plasma membranes (the **apoplast**). Water uptake occurs by osmosis, whereas mineral ions enter the root by diffusion and active transport.

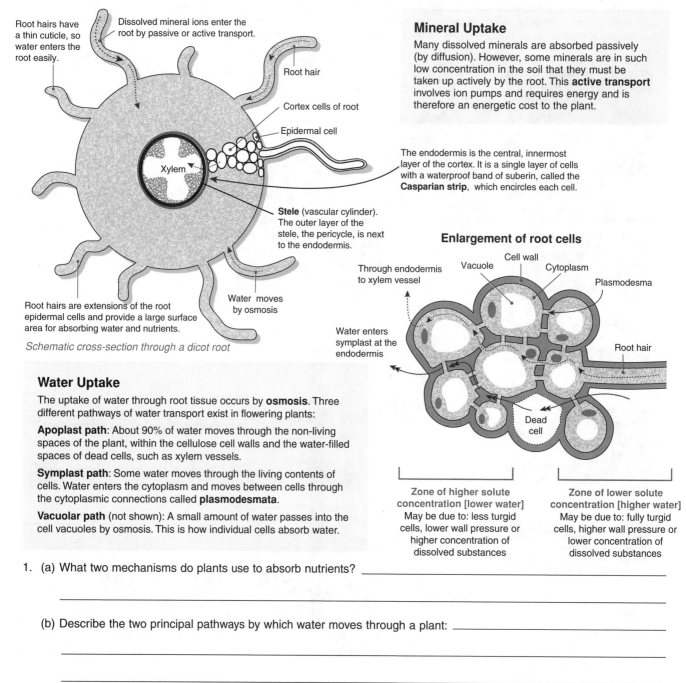

Root hairs have a thin cuticle, so water enters the root easily.

Dissolved mineral ions enter the root by passive or active transport.

Root hair

Cortex cells of root

Epidermal cell

Xylem

Stele (vascular cylinder). The outer layer of the stele, the pericycle, is next to the endodermis.

Water moves by osmosis

Root hairs are extensions of the root epidermal cells and provide a large surface area for absorbing water and nutrients.

Schematic cross-section through a dicot root

Mineral Uptake

Many dissolved minerals are absorbed passively (by diffusion). However, some minerals are in such low concentration in the soil that they must be taken up actively by the root. This **active transport** involves ion pumps and requires energy and is therefore an energetic cost to the plant.

The endodermis is the central, innermost layer of the cortex. It is a single layer of cells with a waterproof band of suberin, called the **Casparian strip**, which encircles each cell.

Enlargement of root cells

Through endodermis to xylem vessel

Vacuole

Cell wall

Cytoplasm

Plasmodesma

Water enters symplast at the endodermis

Root hair

Dead cell

Zone of higher solute concentration [lower water] May be due to: less turgid cells, lower wall pressure or higher concentration of dissolved substances

Zone of lower solute concentration [higher water] May be due to: fully turgid cells, higher wall pressure or lower concentration of dissolved substances

Water Uptake

The uptake of water through root tissue occurs by **osmosis**. Three different pathways of water transport exist in flowering plants:

Apoplast path: About 90% of water moves through the non-living spaces of the plant, within the cellulose cell walls and the water-filled spaces of dead cells, such as xylem vessels.

Symplast path: Some water moves through the living contents of cells. Water enters the cytoplasm and moves between cells through the cytoplasmic connections called **plasmodesmata**.

Vacuolar path (not shown): A small amount of water passes into the cell vacuoles by osmosis. This is how individual cells absorb water.

1. (a) What two mechanisms do plants use to absorb nutrients? _____

_____

(b) Describe the two principal pathways by which water moves through a plant: _____

_____

_____

2. Plants take up water constantly to compensate for losses due to transpiration. Describe a benefit of a large water uptake:

_____

3. (a) What is the effect of the **Casparian strip** on the route water takes into the stele? _____

_____

(b) Why might this feature be an advantage in terms of selective mineral uptake? _____

_____

_____

© 2012-2014 **BIOZONE** International
ISBN: 978-1-927173-93-0
Photocopying Prohibited

236 Transpiration

Key Idea: Water moves through the xylem primarily as a result of evaporation from the leaves and the cohesive and adhesive properties of water molecules.

Plants lose water all the time. Approximately 99% of the water a plant absorbs from the soil is lost by evaporation from the leaves and stem. This loss, mostly through stomata, is called **transpiration** and the flow of water through the plant is called the **transpiration stream**. Plants rely on a

gradient in solute concentration that increases from the roots to the air to move water through their cells. Water flows passively from soil to air along this gradient of increasing solute concentration. The gradient is the driving force for the movement of water up a plant. Transpiration has benefits to the plant because evaporative water loss cools the plant and the transpiration stream helps the plant to take up minerals. Factors contributing to water movement are described below.

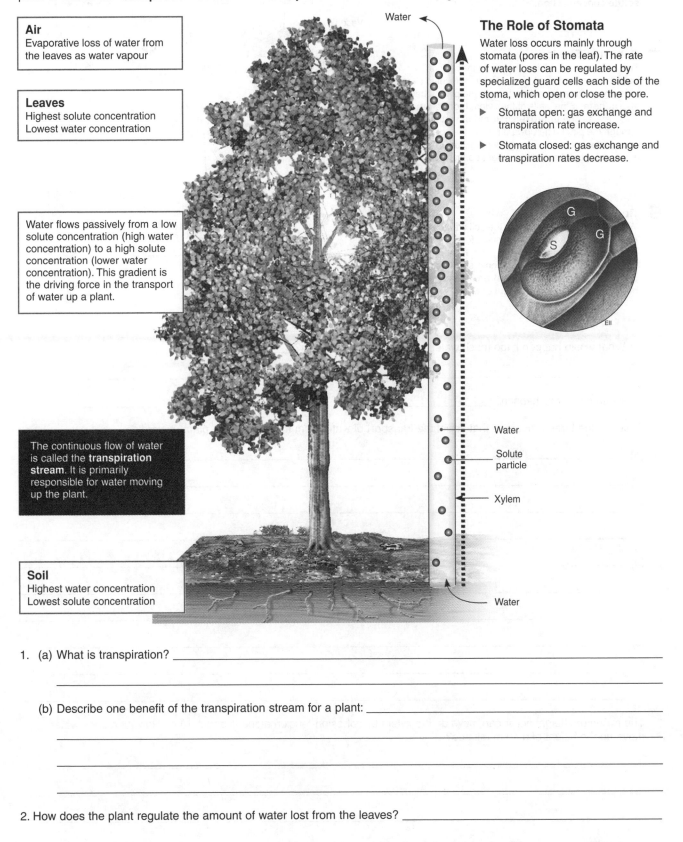

Air
Evaporative loss of water from the leaves as water vapour

Leaves
Highest solute concentration
Lowest water concentration

Water flows passively from a low solute concentration (high water concentration) to a high solute concentration (lower water concentration). This gradient is the driving force in the transport of water up a plant.

The continuous flow of water is called the **transpiration stream**. It is primarily responsible for water moving up the plant.

Soil
Highest water concentration
Lowest solute concentration

Water

Water
Solute particle
Xylem
Water

The Role of Stomata

Water loss occurs mainly through stomata (pores in the leaf). The rate of water loss can be regulated by specialized guard cells each side of the stoma, which open or close the pore.

▶ Stomata open: gas exchange and transpiration rate increase.

▶ Stomata closed: gas exchange and transpiration rates decrease.

1. (a) What is transpiration? _____

_____

(b) Describe one benefit of the transpiration stream for a plant: _____

_____

_____

_____

2. How does the plant regulate the amount of water lost from the leaves? _____

_____

© 2012-2014 **BIOZONE** International
ISBN: 978-1-927173-93-0
Photocopying Prohibited

LINK 21   WEB 236   **KNOW**

Processes involved in Moving Water Through the Xylem

1 **Transpiration Pull**
Water is lost from the air spaces by evaporation through stomata and is replaced by water from the mesophyll cells. The constant loss of water to the air (and production of sugars) creates a solute concentration in the leaves that is higher than elsewhere in the plant. Water is pulled through the plant along a **gradient of increasing solute concentration**.

2 **Cohesion-Tension**
The transpiration pull is assisted by the special **cohesive** properties of water. Water molecules cling together as they are pulled through the plant. They also **adhere** to the walls of the xylem (**adhesion**). This creates one **unbroken column of water** through the plant. The upward pull on the cohesive sap creates a tension (a negative pressure). This helps water uptake and movement up the plant.

3 **Root Pressure**
Water entering the stele from the soil creates a **root pressure**; a weak 'push' effect for the water's upward movement through the plant. Root pressure can force water droplets from some small plants under certain conditions (**guttation**), but generally it plays a minor part in the ascent of water.

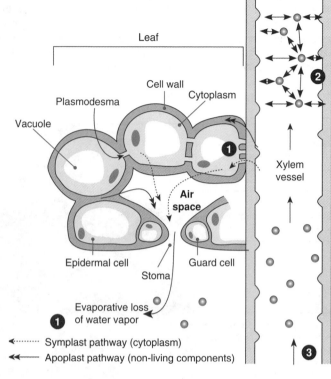

Leaf

Cell wall
Cytoplasm
Plasmodesma
Vacuole
1
Xylem vessel
2
Air space
Epidermal cell
Guard cell
Stoma
Evaporative loss of water vapor
1

⟵······· Symplast pathway (cytoplasm)
⟵⟵ Apoplast pathway (non-living components)

◯ Water molecule

3

Water is drawn up the plant xylem

3. (a) What would happen if too much water was lost from the leaves? _____

_____

(b) When might this happen? _____

4. Describe the three processes that assist the transport of water from the roots of the plant upward:

(a) _____

_____

_____

_____

(b) _____

_____

_____

_____

(c) _____

_____

_____

_____

5. The maximum height water can move up the xylem by cohesion-tension alone is about 10 m. How then does water move up the height of a 40 m tall tree?

_____

_____

_____

_____

© 2012-2014 **BIOZONE** International
ISBN: 978-1-927173-93-0
Photocopying Prohibited

237 Xylem

Key Idea: The xylem is involved in water and mineral transport in vascular plants.

Xylem is the principal water conducting tissue in vascular plants. It is also involved in conducting dissolved minerals, in food storage, and in supporting the plant body. As in animals, tissues in plants are groupings of different cell types that work together for a common function. In angiosperms,

it is composed of five cell types: tracheids, vessels, xylem parenchyma, sclereids (short sclerenchyma cells), and fibres. The tracheids and vessel elements form the bulk of the tissue. They are heavily strengthened and are the conducting cells of the xylem. Parenchyma cells are involved in storage, while fibres and sclereids provide support. When mature, xylem is dead.

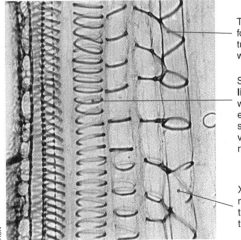

RCN

The cells of the xylem form a continuous tube through which water is conducted.

Spiral thickening of **lignin** around the walls of the vessel elements give extra strength allowing the vessels to remain rigid and upright.

Xylem is dead when mature. Note how the cells have lost their cytoplasm.

The Structure of Xylem Tissue

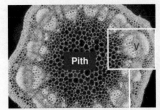

Pith

This cross section through a young stem of *Helianthus* (sunflower) shows the central pith, surrounded by a peripheral ring of vascular bundles (V). Note the xylem vessels with their thick walls.

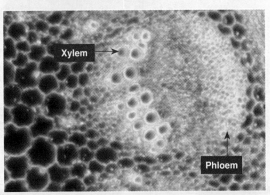

Xylem

Phloem

Vessel element

Tip of tracheid

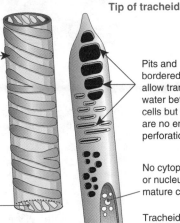

Secondary walls of cellulose are laid down after the cell has elongated or enlarged and lignin is deposited to add strength. This thickening is a feature of tracheids and vessels.

Pits and bordered pits allow transfer of water between cells but there are no end wall perforations.

Vessels connect end to end. The end walls of the vessels are perforated to allow rapid water transport.

No cytoplasm or nucleus in mature cell.

Tracheids are longer and thinner than vessels.

Vessel elements and tracheids are the two conducting cell types in xylem. Tracheids are long, tapering hollow cells. Water passes from one tracheid to another through thin regions in the wall called **pits**. Vessel elements have pits, but the end walls are also perforated and water flows unimpeded through the stacked elements.

Mature Xylem is Dead

Mature xylem is dead. Its primary function is to conduct water from the roots to the leaves. This is a passive process, so there is no need for plasma membranes or transport proteins. Xylem that no longer transports water accumulates compounds such as gum and resin and is known as heartwood. In this form it is an important structural part of a mature tree.

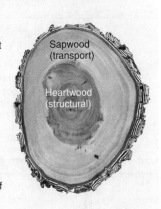

Sapwood (transport)

Heartwood (structural)

1. (a) What is the function of **xylem**? _____

_____

(b) How can xylem be dead when mature and still carry out its function? _____

_____

2. Identify four main cell types in xylem and explain their role in the tissue:

(a) _____

(b) _____

(c) _____

(d) _____

3. Skill: Draw the structure of primary xylem from the larger image of a stem section above. Staple it to this page:

LINK 242   LINK 236   LINK 14   WEB 237   **SKILL**

238 Investigating Plant Transpiration

Key Idea: The relationship between the rate of transpiration and the environment can be investigated using a potometer. This activity describes a typical experiment to investigate the effect of different environmental conditions on transpiration rate using a potometer. You will present and analyse the results provided.

The Potometer

A potometer is a simple instrument for investigating transpiration rate (water loss per unit time). The equipment is simple to use and easy to obtain. A basic potometer, such as the one shown right, can easily be moved around so that transpiration rate can be measured under different environmental conditions.

Some physical conditions investigated are:

- Humidity or vapour pressure (high or low)
- Temperature (high or low)
- Air movement (still or windy)
- Light level (high or low)
- Water supply

It is also possible to compare the transpiration rates of plants with different adaptations e.g. comparing transpiration rates in plants with rolled leaves vs rates in plants with broad leaves. If possible, experiments like these should be conducted simultaneously using replicate equipment. If conducted sequentially, care should be taken to keep the environmental conditions the same for all plants used.

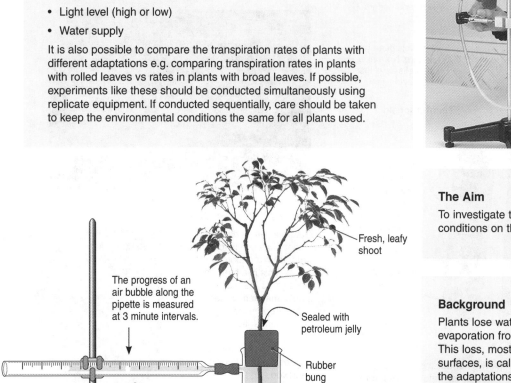

Fresh, leafy shoot

The progress of an air bubble along the pipette is measured at 3 minute intervals.

Sealed with petroleum jelly

Rubber bung

1 cm³ pipette

Flask filled with water

Clamp stand

The Aim

To investigate the effect of environmental conditions on the transpiration rate of plants.

Background

Plants lose water all the time by evaporation from the leaves and stem. This loss, mostly through pores in the leaf surfaces, is called **transpiration**. Despite the adaptations plants have to help prevent water loss (e.g. waxy leaf cuticle), 99% of the water a plant absorbs from the soil is lost by evaporation. Environmental conditions can affect transpiration rate by increasing or decreasing the gradient for diffusion of water molecules between the plant and its external environment.

The Apparatus

This experiment investigated the influence of environmental conditions on plant transpiration rate. The experiment examined four conditions: room conditions (ambient), wind, bright light, and high humidity. After setting up the potometer, the apparatus was equilibrated for 10 minutes, and then the position of the air bubble in the pipette was recorded. This is the time 0 reading. The plant was then exposed to one of the environmental conditions. Students recorded the location of the air bubble every three minutes over a 30 minute period. The potometer readings for each environmental condition are presented in Table 1 (next page).

A class was divided into four groups to study how four different environmental conditions (ambient, wind, bright light, and high humidity) affected transpiration rate. A **potometer** was used to measure transpiration rate (water loss per unit time). A basic potometer, such as the one shown left, can easily be moved around so that transpiration rate can be measured under different environmental conditions.

© 2012-2014 **BIOZONE** International
ISBN: 978-1-927173-93-0
Photocopying Prohibited

Table 1. Potometer readings (in mL water loss)

Treatment \ Time (min)	0	3	6	9	12	15	18	21	24	27	30
Ambient	0	0.002	0.005	0.008	0.012	0.017	0.022	0.028	0.032	0.036	0.042
Wind	0	0.025	0.054	0.088	0.112	0.142	0.175	0.208	0.246	0.283	0.325
High humidity	0	0.002	0.004	0.006	0.008	0.011	0.014	0.018	0.019	0.021	0.024
Bright light	0	0.021	0.042	0.070	0.091	0.112	0.141	0.158	0.183	0.218	0.239

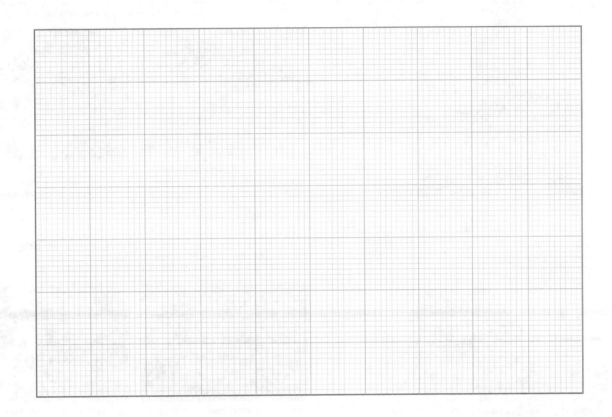

1. (a) Plot the potometer data from Table 1 on the grid provided:

   (b) Identify the independent variable: _____

2. (a) Identify the control: _____

   (b) Explain the purpose of including an experimental control in an experiment: _____

   _____

   _____

   (c) Which factors increased water loss? _____

   (d) How does each environmental factor influence water loss? _____

   _____

   _____

   (e) Explain why the plant lost less water in humid conditions: _____

   _____

   _____

239 Adaptations for Water Conservation

Key Idea: Xerophytes are adapted for conserving water in dry (arid) conditions.

Plants adapted to dry conditions are called **xerophytes** and they show structural (xeromorphic) and physiological adaptations for water conservation. These include small, hard leaves, an epidermis with a thick cuticle, sunken stomata, succulence (ability to store water), and absence of leaves. Salt tolerant plants (halophytes) and alpine plants may also show xeromorphic features due to the lack of free water and high evaporative losses in these environments.

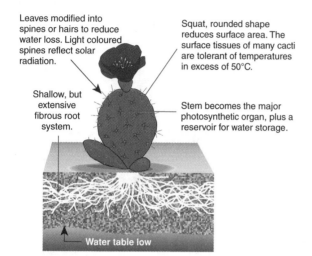

Leaves modified into spines or hairs to reduce water loss. Light coloured spines reflect solar radiation.

Squat, rounded shape reduces surface area. The surface tissues of many cacti are tolerant of temperatures in excess of 50°C.

Shallow, but extensive fibrous root system.

Stem becomes the major photosynthetic organ, plus a reservoir for water storage.

Water table low

Seaweeds, which are protoctists, not plants, tolerate drying between tides even though they have no xeromorphic features.

A waxy coating of **suberin** on mangrove roots excludes 97% of salt from the water.

Dry Desert Plant

Desert plants, such as cacti, must cope with low or sporadic rainfall and high transpiration rates. A number of structural adaptations (diagram left) reduce water losses, and enable them to access and store available water. Adaptations such as waxy leaves also reduce water loss and, in many desert plants, germination is triggered only by a certain quantity of rainfall.

Acacia trees have **deep root systems**, allowing them to draw water from lower water table systems.

The outer surface of many succulents are coated in fine hairs, which traps air close to the surface reducing transpiration rate.

Ocean Margin Plant

Land plants that colonize the shoreline must have adaptations to obtain water from their saline environment while maintaining their osmotic balance. In addition, the shoreline is often a windy environment, so they frequently show xeromorphic adaptations that enable them to reduce water losses.

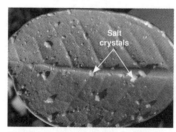
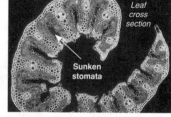

Salt crystals

Leaf cross section

Sunken stomata

To maintain osmotic balance, mangroves can secrete absorbed salt as salt crystals (above), or accumulate salt in old leaves which are subsequently shed.

Grasses found on shoreline coasts (where it is often windy), curl their leaves and have sunken stomata to reduce water loss by transpiration.

Methods of Water Conservation

Adaptation for water conservation	Effect of adaptation	Example
Thick, waxy cuticle to stems and leaves	Reduces water loss through the cuticle.	*Pinus* sp. ivy (*Hedera*), sea holly (*Eryngium*), prickly pear (*Opuntia*).
Reduced number of stomata	Reduces the number of pores through which water loss can occur.	Prickly pear (*Opuntia*), *Nerium* sp.
Stomata sunken in pits, grooves, or depressions. Leaf surface covered with fine hairs. Massing of leaves into a rosette at ground level	Moist air is trapped close to the area of water loss, reducing the diffusion gradient and therefore the rate of water loss.	**Sunken stomata:** *Pinus* sp., *Hakea* sp. Hairy leaves: lamb's ear. **Leaf rosettes:** dandelion (*Taraxacum*), daisy.
Stomata closed during the light, open at night	CAM metabolism: CO_2 is fixed during the night, water loss in the day is minimized.	**CAM plants,** e.g. American aloe, pineapple, *Kalanchoe*, *Yucca*.
Leaves reduced to scales, stem photosynthetic. Leaves curled, rolled, or folded when flaccid	Reduction in surface area from which transpiration can occur.	**Leaf scales:** broom (*Cytisus*). **Rolled leaf:** marram grass (*Ammophila*), *Erica* sp.
Fleshy or succulent stems. Fleshy or succulent leaves	When readily available, water is stored in the tissues for times of low availability.	**Fleshy stems:** *Opuntia*, candle plant (*Kleinia*). **Fleshy leaves:** *Bryophyllum*.
Deep root system below the water table	Roots tap into the lower water table.	Acacias, oleander.
Shallow root system absorbing surface moisture	Roots absorb overnight condensation.	Most cacti.

WEB 239   LINK 236   APP

© 2012-2014 **BIOZONE** International
ISBN: 978-1-927173-93-0
Photocopying Prohibited

Adaptations of Halophytes and Xerophytes

Ice plant (*Carpobrotus*): The leaves of many desert and beach dwelling plants are fleshy or succulent. The leaves are triangular in cross section and crammed with water storage cells. The water is stored after rain for use in dry periods. The shallow root system is able to take up water from the soil surface, taking advantage of any overnight condensation.

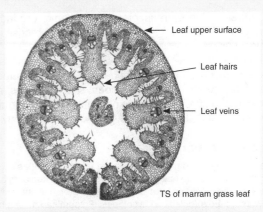

Leaf upper surface

Leaf hairs

Leaf veins

TS of marram grass leaf

Marram grass (*Ammophila*): The long, wiry leaf blades of this beach grass are curled downwards with the stomata on the inside. This protects them against drying out by providing a moist microclimate around the stomata. Plants adapted to high altitude often have similar adaptations.

Ball cactus (*Echinocactus grusonii*): In many cacti, the leaves are modified into long, thin spines which project outward from the thick fleshy stem. This reduces the surface area over which water loss can occur. The stem stores water and takes over as the photosynthetic organ. As in succulents, a shallow root system enables rapid uptake of surface water.

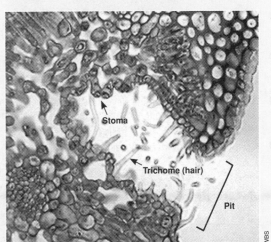

Stoma

Trichome (hair)

Pit

Oleander is a xerophyte from the Mediterranean region with many water conserving features. It has a thick multi-layered epidermis and the stomata are sunken in trichome-filled pits on the leaf underside. The pits restrict water loss to a greater extent than they reduce uptake of carbon dioxide.

1. Explain the purpose of **xeromorphic** adaptations: _____

_____

2. Describe three xeromorphic adaptations of plants:

(a) _____

(b) _____

(c) _____

3. Describe a physiological mechanism by which plants can reduce water loss during the daylight hours:

_____

_____

4. How does creating a moist microenvironment around the areas of water loss reduce transpiration rate?

_____

5. Why do seashore plants (halophytes) exhibit many Xeromorphic features? _____

_____

_____

240 Translocation

Key Idea: Phloem transports the organic products of photosynthesis (sugars) through the plant in a process called translocation.

In angiosperms, the sugar moves through the sieve-tube members, which are arranged end-to-end and perforated with sieve plates. Apart from water, phloem sap comprises mainly sucrose (up to 30%). It may also contain minerals, hormones, and amino acids, in transit around the plant. Movement of sap in the phloem is from a **source** (a plant organ where sugar is made or mobilized) to a **sink** (a plant organ where sugar is stored or used). Loading sucrose into the phloem at a source involves energy expenditure; it is slowed or stopped by high temperatures or respiratory inhibitors. In some plants, unloading the sucrose at the sinks also requires energy, although in others, diffusion alone is sufficient to move sucrose from the phloem into the cells of the sink organ.

Phloem Transport

Phloem sap moves from source to sink at rates as great as 100 m h^{-1}, which is too fast to be accounted for by cytoplasmic streaming. The most acceptable model for phloem movement is the **pressure-flow** (bulk flow) hypothesis. Phloem sap moves by bulk flow, which creates a pressure (hence the term "pressure-flow"). The key elements in this model are outlined below and right. Note that, for simplicity, the cells that lie between the source (and sink) cells and the phloem sieve-tube have been omitted.

1. Loading sugar into the phloem increases the solute concentration inside the sieve-tube cells. This causes the sieve-tubes to take up water by osmosis.

2. The water uptake creates a hydrostatic pressure that forces the sap to move along the tube, just as pressure pushes water through a hose.

3. The pressure gradient in the sieve tube is reinforced by the active unloading of sugar and consequent loss of water by osmosis at the sink (e.g. root cell).

4. Xylem recycles the water from sink to source.

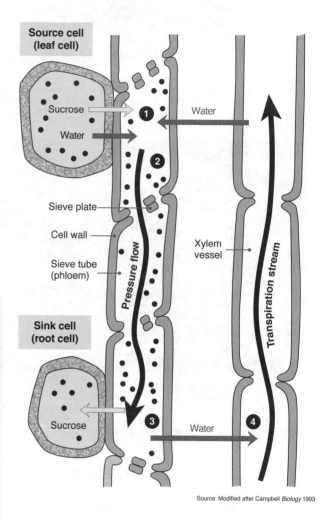

Source: Modified after Campbell *Biology* 1993

Measuring Phloem Flow

Experiments investigating flow of phloem often use aphids. Aphids feed on phloem sap (left) and act as natural **phloem probes**. When the mouthparts (stylet) of an aphid penetrate a sieve-tube cell, the pressure in the sieve-tube force-feeds the aphid. While the aphid feeds, it can be severed from its stylet, which remains in place in the phloem. The stylet serves as a tiny tap that exudes sap. Using different aphids, the rate of flow of this sap can be measured at different locations on the plant.

1. (a) From what you know about osmosis, explain why water follows the sugar as it moves through the phloem:

_____

_____

_____

(b) What is meant by '**source to sink**' flow in phloem transport? _____

_____

_____

2. Why does a plant need to move food around, particularly from the leaves to other regions?

_____

_____

_____

© 2012-2014 **BIOZONE** International
ISBN: 978-1-927173-93-0
Photocopying Prohibited

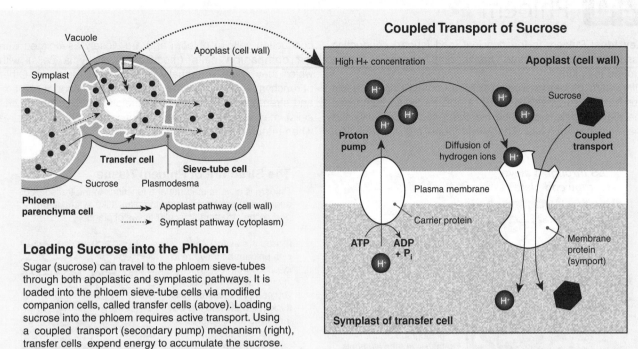

Loading Sucrose into the Phloem

Sugar (sucrose) can travel to the phloem sieve-tubes through both apoplastic and symplastic pathways. It is loaded into the phloem sieve-tube cells via modified companion cells, called transfer cells (above). Loading sucrose into the phloem requires active transport. Using a coupled transport (secondary pump) mechanism (right), transfer cells expend energy to accumulate the sucrose. The sucrose then passes into the sieve tube through plasmodesmata. The transfer cells have wall ingrowths that increase surface area for the transport of solutes. Using this mechanism, some plants can accumulate sucrose in the phloem to 2-3 times the concentration in the mesophyll.

Above: Proton pumps generate a hydrogen ion gradient across the membrane of the transfer cell. This process requires expenditure of energy. The gradient is then used to drive the transport of sucrose, by coupling the sucrose transport to the diffusion of hydrogen ions back into the cell.

3. In your own words, describe what is meant by the following:

(a) Translocation: _____

_____

(b) Pressure-flow movement of phloem: _____

_____

(c) Coupled transport of sucrose: _____

_____

4. Briefly explain how sucrose is transported into the phloem: _____

_____

_____

5. Explain the role of the companion (transfer) cell in the loading of sucrose into the phloem: _____

_____

_____

_____

_____

6. The sieve plate represents a significant barrier to effective mass flow of phloem sap. Suggest why the presence of the sieve plate is often cited as evidence against the pressure-flow model for phloem transport:

_____

_____

_____

_____

_____

241 Phloem

Key Idea: Phloem is the principal food (sugar) conducting tissue in vascular plants, transporting dissolved sugars around the plant.

Like xylem, **phloem** is a complex tissue, comprising a variable number of cell types. The bulk of phloem tissue comprises the **sieve tubes** (sieve tube members and sieve cells) and their companion cells. The sieve tubes are the principal conducting cells in phloem and are closely associated with the **companion cells** (modified parenchyma cells) with which they share a mutually dependent relationship. Other parenchyma cells, concerned with storage, occur in phloem, and strengthening fibres and sclereids (short sclerenchyma cells) may also be present. Unlike xylem, phloem is alive when mature.

LS through a sieve tube end plate

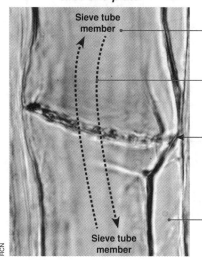

Sieve tube member

The sieve tube members lose most of their organelles but are still alive when mature

Sugar solution flows in both directions

Sieve tube end plate
Tiny holes (arrowed in the photograph below) perforate the sieve tube elements allowing the sugar solution to pass through.

Companion cell: a cell adjacent to the sieve tube member, responsible for keeping it alive

Sieve tube member

TS through a sieve tube end plate

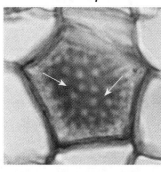

Adjacent sieve tube members are connected through **sieve plates** through which phloem sap flows.

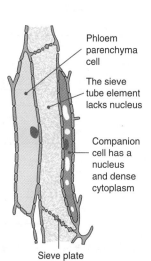

Phloem parenchyma cell

The sieve tube element lacks nucleus

Companion cell has a nucleus and dense cytoplasm

Sieve plate

The Structure of Phloem Tissue

Phloem is alive at maturity and functions in the transport of sugars and minerals around the plant. Like xylem, it forms part of the structural vascular tissue of plants.

Fibres are associated with phloem as they are in xylem. Here they are seen in cross section where you can see the extremely thick cell walls and the way the fibres are clustered in groups. See the previous page for a view of fibres in longitudinal section.

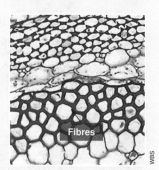

Fibres

In this cross section through a buttercup root, the smaller companion cells can be seen lying alongside the sieve tube members. It is the sieve tube members that, end on end, produce the **sieve tubes**. They are the conducting tissue of phloem.

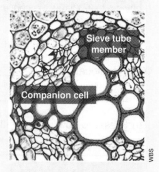

Sieve tube member

Companion cell

In this longitudinal section of a buttercup root, each sieve tube member has a thin **companion cell** associated with it. Companion cells retain their nucleus and control the metabolism of the sieve tube member next to them. They also have a role in the loading and unloading of sugar into the phloem.

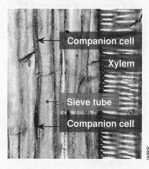

Companion cell

Xylem

Sieve tube

Companion cell

1. Describe the function of **phloem**: _____

_____

2. Mature phloem is a live tissue, whereas xylem (the water transporting tissue) is dead when mature. Why is it necessary for phloem to be alive to be functional, whereas xylem can function as a dead tissue?

_____

_____

_____

3. Describe two roles of the companion cell in phloem: _____

_____

_____

© 2012-2014 **BIOZONE** International
ISBN: 978-1-927173-93-0
Photocopying Prohibited

242 Identifying Xylem and Phloem

Key Idea: The vascular tissue in dicots can be identified by its appearance in sections viewed with a light microscope. The structure of the vascular tissue in dicotyledons (dicots) has a very regular arrangement with the xylem and phloem found close together. In the stem, the vascular tissue is distributed in a regular fashion near the outer edge of the stem. In the roots, the vascular tissue is found near the centre of the root.

Dicot Stem Structure

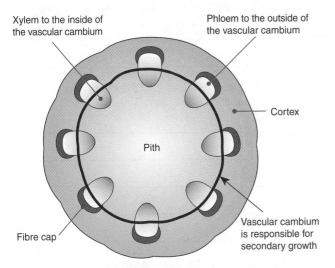

Xylem to the inside of the vascular cambium

Phloem to the outside of the vascular cambium

Cortex

Pith

Fibre cap

Vascular cambium is responsible for secondary growth

Dicot Root Structure

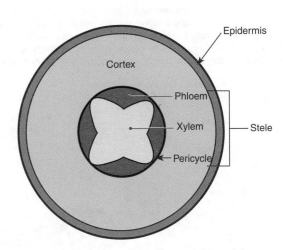

Epidermis

Cortex

Phloem

Xylem

Stele

Pericycle

In dicots, the vascular bundles (xylem and phloem) are arranged in an orderly fashion around the stem. Each vascular bundle contains **xylem** (to the inside) and **phloem** (to the outside). Between the phloem and the xylem is the **vascular cambium**. This is a layer of cells that divide to produce the thickening of the stem.

In a dicot root, the vascular tissue, (xylem and phloem) forms a central cylinder through the root called the stele. The large cortex is made up of parenchyma (packing) cells, which store starch and other substances. Air spaces between the cells are essential for aeration of the root tissue, which is non-photosynthetic.

1. In the micrograph below of a dicot stem identify the phloem (P) and xylem (X) tissue:

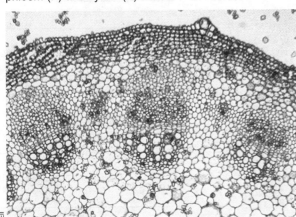

3. In the diagram below identify the labels A - F

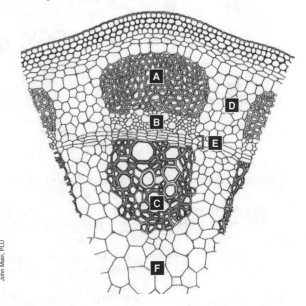

John Main, PLU

Cross section through a typical dicot stem

2. In the micrograph below of a dicot root identify the phloem (P) and xylem (X) tissue:

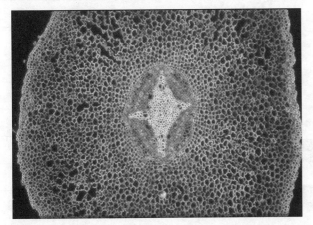

A. _____

B. _____

C. _____

D. _____

E. _____

F. _____

LINK 241   LINK 237   SKILL

243 Plant Meristems

Key Idea: The differentiation of plant cells only occurs at specific regions called meristems.

Two types of growth can contribute to an increase in the size of a plant. **Primary growth**, which occurs in the **apical meristem** of the buds and root tips, increases the length (height) of a plant. Meristematic cells are totipotent (can give rise to all the cells of the adult plant). **Secondary growth** increases plant girth and occurs in the lateral meristem in the stem. All plants show primary growth but only some show secondary growth (the growth that produces woody tissues).

Primary Growth

Three types of **primary meristem** (procambium, protoderm, and ground meristem) are produced from the apical meristem. In dicots, the **procambium** forms vascular bundles, which are found in a ring near the epidermis, surrounded by cortex. Cells in the procambium divide to become primary **xylem** to the inside and primary **phloem** to the outside.

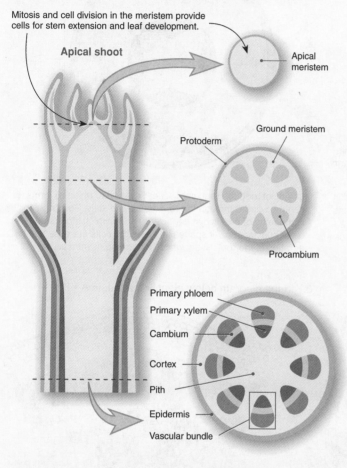

Mitosis and cell division in the meristem provide cells for stem extension and leaf development.

Apical shoot

Apical meristem

Protoderm

Ground meristem

Procambium

Primary phloem
Primary xylem
Cambium
Cortex
Pith
Epidermis
Vascular bundle

Primary Tissues Generated by the Meristem

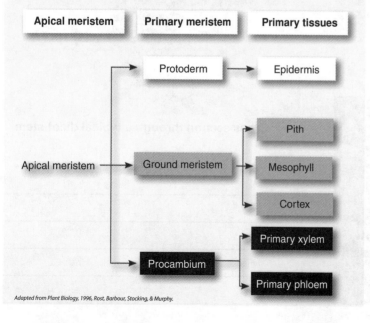

| Apical meristem | Primary meristem | Primary tissues |

Protoderm → Epidermis

Apical meristem → Ground meristem → Pith, Mesophyll, Cortex

Procambium → Primary xylem, Primary phloem

Adapted from Plant Biology, 1996, Rost, Barbour, Stocking, & Murphy.

1. Describe the role of the meristems in plants:

_____

_____

2. Describe the location of the meristems and relate this to how plants grow:

_____

_____

_____

_____

_____

3. Describe a distinguishing feature of meristematic tissue:

_____

_____

_____

_____

_____

_____

4. Discuss the structure and formation of the primary tissues in dicot plants:

_____

_____

_____

_____

_____

_____

_____

_____

_____

_____

_____

© 2012-2014 **BIOZONE** International
ISBN: 978-1-927173-93-0
Photocopying Prohibited

244 Auxins and Shoot Growth

Key Idea: Auxin is a plant hormone involved in the plant's response to the environment and differential growth.

Auxins are **phytohormones** (plant growth substances) that have a central role in a wide range of growth and developmental responses in vascular plants. Indole-3-acetic acid (IAA) is the most potent native auxin in intact plants. The response of any particular plant tissue to IAA depends on the tissue itself, the concentration of the hormone, the timing of its release, and the presence of other phytohormones. Gradients in auxin concentration during growth prompt differential responses in specific tissues and contribute to directional growth.

Light is an important growth requirement for all plants. Most plants show an adaptive response of growing towards the light. This growth response is called phototropism. A **tropism** is a plant growth response to external stimuli in which the stimulus direction determines the direction of the growth response.

Tropisms are identified according to the stimulus involved, e.g. photo- (light) and gravi- (gravity), and may be positive or negative depending on whether the plant moves towards or away from the stimulus respectively. The bending of the plants shown on the right is a phototropism in response to light shining from the left and caused by the plant hormone **auxin**. Auxin causes the elongation of cells on the shaded side of the stem, causing it to bend.

Auxin is produced in the shoot tip and is responsible for apical dominance by suppressing growth of the lateral (side) buds.

Auxin movement through the plant is polar. It moves from the shoot tip down the plant.

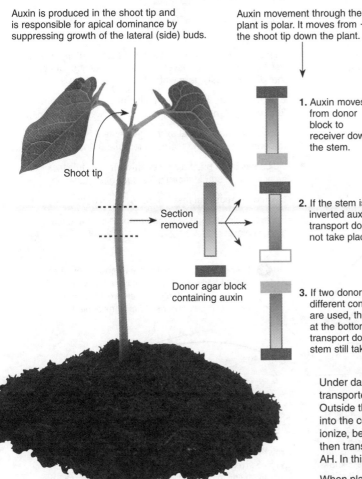

Shoot tip

Section removed

Donor agar block containing auxin

1. Auxin moves from donor block to receiver down the stem.

2. If the stem is inverted auxin transport does not take place.

3. If two donor blocks of different concentration are used, the higher at the bottom, transport down the stem still takes place.

Plasma membrane

Cell wall

Transport protein

- Hydrogen ion (H⁺)
- Non-ionized auxin (AH)
- Ionized auxin (A⁻)
- ····▸ Diffusion
- ⟶ Active transport

Under dark conditions auxin moves evenly down the stem. It is transported cell to cell by diffusion and transport proteins (above right). Outside the cell auxin is a non-ionized molecule (AH) which can diffuse into the cell. Inside the cell the pH of the cytoplasm causes auxin to ionize, becoming A⁻ and H⁺. Transport proteins at the basal end of the cell then transport A⁻ out of the cell where it requires an H⁺ ion and reforms AH. In this way auxin is transported in one direction through the plant.

When plant cells are illuminated by light from one direction transport proteins in the plasma membrane on the shaded side of the cell are activated and auxin is transported to the shaded side of the plant.

1. What is the term given to the tropism being displayed in the photo (top right)? _____

2. Describe one piece of evidence that demonstrates the transport of auxin is polar: _____

_____

_____

_____

3. What is the effect of auxin on cell growth? _____

_____

245 Micropropagation of Plant Tissue

Key Idea: Micropropagation can produce large numbers of genetically identical plants in a short space of time.

Micropropagation (or plant tissue culture) is a method used to clone plants. It is possible because plant meristematic tissue is totipotent and differentiation into a complete plant can be induced by culturing the tissue in an appropriate growth environment. Micropropagation is used widely for the rapid multiplication of commercially important plant species, as well as in recovery programmes for endangered plants. However, continued culture of a limited number of cloned varieties reduces genetic diversity and plants may become susceptible to disease or environmental change. New genetic stock may be introduced into cloned lines to prevent this. Micropropagation has considerable advantages over traditional methods of plant propagation, but it is very labour intensive. Its success is affected by a variety of factors including selection of explant material, plant hormone levels, lighting, and temperature.

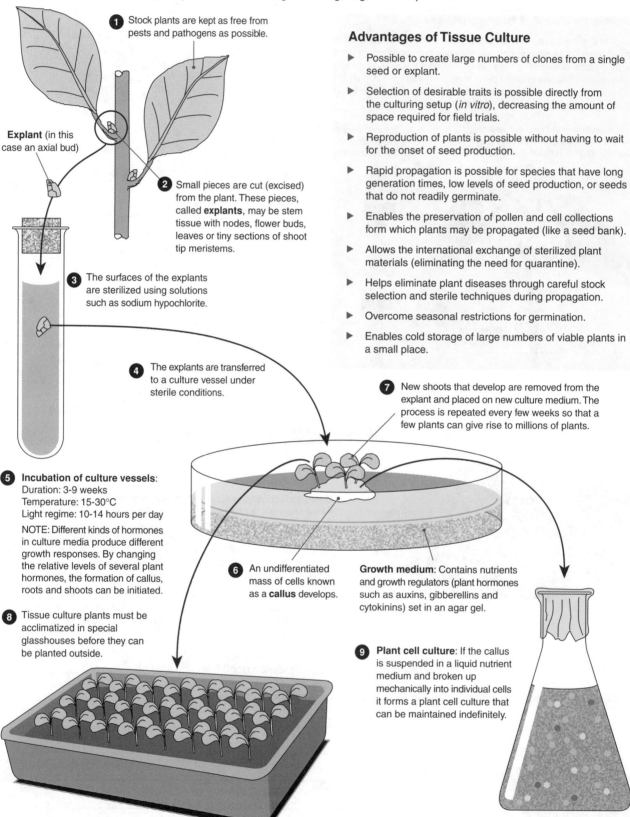

1 Stock plants are kept as free from pests and pathogens as possible.

Explant (in this case an axial bud)

2 Small pieces are cut (excised) from the plant. These pieces, called **explants**, may be stem tissue with nodes, flower buds, leaves or tiny sections of shoot tip meristems.

3 The surfaces of the explants are sterilized using solutions such as sodium hypochlorite.

4 The explants are transferred to a culture vessel under sterile conditions.

5 Incubation of culture vessels:
Duration: 3-9 weeks
Temperature: 15-30°C
Light regime: 10-14 hours per day

NOTE: Different kinds of hormones in culture media produce different growth responses. By changing the relative levels of several plant hormones, the formation of callus, roots and shoots can be initiated.

6 An undifferentiated mass of cells known as a **callus** develops.

7 New shoots that develop are removed from the explant and placed on new culture medium. The process is repeated every few weeks so that a few plants can give rise to millions of plants.

8 Tissue culture plants must be acclimatized in special glasshouses before they can be planted outside.

Growth medium: Contains nutrients and growth regulators (plant hormones such as auxins, gibberellins and cytokinins) set in an agar gel.

9 Plant cell culture: If the callus is suspended in a liquid nutrient medium and broken up mechanically into individual cells it forms a plant cell culture that can be maintained indefinitely.

Advantages of Tissue Culture

▸ Possible to create large numbers of clones from a single seed or explant.

▸ Selection of desirable traits is possible directly from the culturing setup (*in vitro*), decreasing the amount of space required for field trials.

▸ Reproduction of plants is possible without having to wait for the onset of seed production.

▸ Rapid propagation is possible for species that have long generation times, low levels of seed production, or seeds that do not readily germinate.

▸ Enables the preservation of pollen and cell collections form which plants may be propagated (like a seed bank).

▸ Allows the international exchange of sterilized plant materials (eliminating the need for quarantine).

▸ Helps eliminate plant diseases through careful stock selection and sterile techniques during propagation.

▸ Overcome seasonal restrictions for germination.

▸ Enables cold storage of large numbers of viable plants in a small place.

Micropropagation of Tasmanian Blackwood (Acacia melanoxylon)

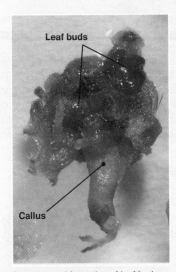

Leaf buds

Callus

Greening and formation of leaf buds on a callus growing on culturing medium.

Culture medium

Shoots with juvenile leaves growing from a callus on media. They appear identical to those produced directly from seeds.

Seedling with juvenile foliage 6 months after transfer to greenhouse.

Micropropagation is increasingly used in conjunction with genetic engineering to propagate transgenic plants. Genetic engineering and micropropagation achieve similar results to conventional selective breeding but more precisely, quickly, and independently of growing season. The **Tasmanian blackwood** (above) is well suited to this type of manipulation. It is a versatile hardwood tree now being extensively trialled in some countries as a replacement for tropical hardwoods. The timber is of high quality, but genetic variations between individual trees lead to differences in timber quality and colour. Tissue culture allows the multiple propagation

of trees with desirable traits (e.g. uniform timber colour). Tissue culture could also help to find solutions to problems that cannot be easily solved by forestry management. When combined with genetic engineering (introduction of new genes into the plant) problems of pest and herbicide susceptibility may be resolved. Genetic engineering may also be used to introduce a gene for male sterility, thereby stopping pollen production. This would improve the efficiency of conventional breeding programmes by preventing self-pollination of flowers (the manual removal of stamens is difficult and very labour intensive).

Information courtesy of Raewyn Poole, University of Waikato (unpublished MSc thesis)

1. What is the general purpose of **micropropagation** (plant tissue culture)? _____

_____

2. (a) What is a **callus**? _____

   (b) How can a callus be stimulated to initiate root and shoot formation? _____

   _____

3. Explain a potential problem with micropropagation in terms of long term ability to adapt to environmental changes:

_____

_____

4. Discuss the **advantages** and **disadvantages** of micropropagation compared with traditional methods of plant propagation such as grafting:

_____

_____

_____

_____

_____

_____

_____

_____

_____

246 Insect Pollinated Flowers

Key Idea: Flowers are produced as a result of changes in gene expression in the apical meristem, which leads to the production of a floral meristem.

Flowering plants are called **angiosperms**. The egg cell is retained within the flower of the parent plant and the male gametes (contained in the **pollen**) must be transferred to it by **pollination** in order for fertilization to occur. Most angiosperms are **monoecious**, with male and female parts on the same plant. Some of these plants will self-pollinate, but most have mechanisms that make this difficult or

impossible. **Dioecious plants** carrying the male and female flowers on separate plants. In either case, mechanisms have been developed to transfer pollen between plants of the same species. Flowers are pollinated in three different ways (animal, wind, or water) and their structures differ accordingly. Of the animal pollinators, insects provide the greatest effectiveness of pollination as well as the most specialized pollination. Flowers attract insects with brightly coloured petals, smells (scent), and food such as nectar and pollen.

Cross Section of an Insect Pollinated Flower

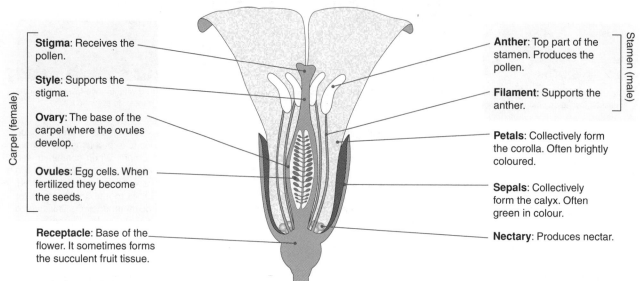

Carpel (female)

Stigma: Receives the pollen.

Style: Supports the stigma.

Ovary: The base of the carpel where the ovules develop.

Ovules: Egg cells. When fertilized they become the seeds.

Receptacle: Base of the flower. It sometimes forms the succulent fruit tissue.

Stamen (male)

Anther: Top part of the stamen. Produces the pollen.

Filament: Supports the anther.

Petals: Collectively form the corolla. Often brightly coloured.

Sepals: Collectively form the calyx. Often green in colour.

Nectary: Produces nectar.

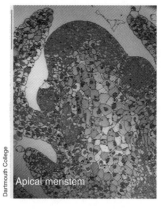

Apical meristem

In order to flower, a plant must pass through several stages of development with several changes in gene expression occurring at the shoot meristem. The plant must first pass from an immature to a sexually mature stage, the apical meristem must change from a vegetative meristem to a floral meristem, and the organs of the flower must grow and develop.

Environmental cues perceived in the leaves are transmitted to the apical meristem by a hormonal messenger process that is still not fully understood but seems to include the gene FLOWERING LOCUS T (FT). It produces a protein that activates transcription factors in the meristem, which in turn activate the gene LEAFY (LFY). LFY results in the production of a floral meristem and flowering.

Flower bud

1. What is the difference between the stigma and the anther? _____

_____

_____

2. How are flowers used to attract specific insect pollinators to a plant? _____

_____

_____

3. What three changes must occur for a plant to flower? _____

_____

_____

_____

_____

_____

© 2012-2014 **BIOZONE** International
ISBN: 978-1-927173-93-0

247 Flowering

Key Idea: The length of the light and dark periods influence the flowering of plants.

Photoperiodism is the response of a plant to the relative lengths of daylight and darkness. Flowering is a photoperiodic activity; individuals of a single species will all flower at much the same time, even though their germination and maturation dates may vary. The exact onset of flowering varies depending on whether the plant is a short-day or long-day type.

Long-Day Plants

When subjected to the light regimes on the right, the 'long-day' plants below flowered as indicated:

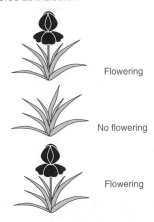

Flowering

No flowering

Flowering

Examples: *lettuce, clover, delphinium, gladiolus, beets, corn, coreopsis*

Photoperiodism in Plants

An experiment was carried out to determine the environmental cue that triggers flowering in 'long-day' and 'short-day' plants. The diagram below shows 3 different light regimes to which a variety of long-day and short-day plants were exposed.

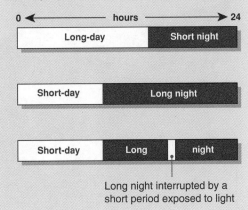

0 ←	hours	→ 24
Long-day		Short night
Short-day		Long night
Short-day	Long	night

Long night interrupted by a short period exposed to light

Short-Day Plants

When subjected to the light regimes on the left, the 'short-day' plants below flowered as indicated:

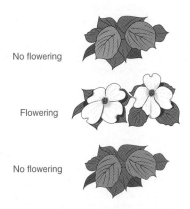

No flowering

Flowering

No flowering

Examples: *potatoes, asters, dahlias, cosmos, chrysanthemums, pointsettias*

Manipulating Flowering in Plants

Controlling the light-dark régime has allowed flower growers and horticulturists to produce flowers out of season or to coincide flowering with specific dates.

Plants kept in greenhouses can be subjected to artificial lighting or covered to control the amount of light they receive. To be totally effective at controlling flowering, temperature must also be controlled, as this is also an important flowering cue.

For the example of the *Chrysanthemum*, a short-day plant, flowering is can be controlled under the following conditions. The temperature is kept between 16 - 25 °C. The light-dark regime is controlled at 13 hours of light and 11 hours of dark for 4-5 weeks from planting to ensure vegetative growth. Then the regime changes to 10 hours light and 14 hours darkness to induce flowering.

Chrysanthemum

Differences Between Long Day and Short Day Plants

1. Short-day plants (SDP) flower when the photoperiod is less than a critical day length. Long-day plants (LDP) flower when the photoperiod is greater than a critical day length.

2. Interruption of light period does not inhibit flowering in SDP but does in LDP.

3. Interruption of the long dark period inhibits flowering in SDP but promotes flowering in LDP.

4. Dark must be continuous in SDP but not in LDP.

5. Alternating cycles of short light and short dark inhibit flowering in SDP.

---

1. (a) What is the environmental cue that synchronizes flowering in plants? _____

   (b) Describe one biological advantage of this synchronization to plants: _____

   _____

2. Study the three light regimes above and the responses of short-day and long-day flowering plants to that light. From this observation, describe the most important factor controlling the onset of flowering in:

   (a) Short-day plants: _____

   (b) Long-day plants: _____

3. What evidence is there for the idea that short-day plants are best described as "long-night plants"?

   _____

   _____

   _____

LINK
248 APP

248 Control of Flowering

Key Idea: Photoperiodism is controlled by the pigment phytochrome, which occurs in two forms P*r* and P*fr*. Photoperiodic activities are controlled through the action of a pigment called **phytochrome**. Phytochrome acts as a signal for some biological clocks in plants and exists in two forms, P*r* and P*fr*. It is important in the flowering response in plants but is also involved in other light initiated responses, such as germination and shoot growth.

Phytochrome

Phytochrome is a blue-green pigment that acts as a photoreceptor for detection of night and day in plants and is universal in vascular plants. It has two forms: **P*r*** (inactive) and **P*fr*** (active). P*r* is readily converted to P*fr* under natural light. P*fr* converts back to P*r* in the dark but more slowly. P*fr* predominates in daylight. The plant measures daylength (or rather night length) by the amount of phytochrome in each form.

In **daylight** or **red light** (660 nm), P*r* converts rapidly, but reversibly, to P*fr*.

P*fr* is the physiologically active form of phytochrome. It promotes flowering in long-day plants, and inhibits flowering in short-day plants.

Phytochrome interacts with genes collectively called "clock genes" that maintain the plant's biological clock.

In the **dark**, or in **far red light** (730 nm) P*fr* reverts slowly, but spontaneously, back to the inactive form of phytochrome P*r*.

Sunlight

Rapid conversion

Slowly in darkness

P*r*

P*fr*

Physiologically active

"Clock genes"

Flowering hormone

There is still uncertainty over what the flowering hormone (commonly called **florigen**) is. Recent studies suggested it may be the protein product of the gene FLOWERING LOCUS T (FT) (in long day plants at least) which appears to influence gene expression that includes the gene LEAFY (LFY) in the apical meristem and causes flowering.

The hormone is transported to the apical meristem were it causes a change in gene expression that leads to flowering.

1. (a) Identify the two forms of phytochrome and the wavelengths of light they absorb: _____

_____

(b) Identify the biologically active form of phytochrome and how it behaves in long day plants and short day plants with respect to flowering:

_____

_____

2. (a) Discuss the role of phytochrome in a plant's ability to measure daylength: _____

_____

_____

_____

(b) How does this help coordinate flower production in a plant species? _____

_____

_____

249 Pollination Relationships

Key Idea: Many angiosperms have a mutualistic relationship with their pollinators in which the plants achieve pollination by rewarding insects with food such as nectar or pollen.

Pollination of flowers by insects is usually mutualistic. Mutualistic relationships involve exchanges between two species so that each species benefits. The benefit need not be equal for each party, because each species acts in its own interests. In the case of insect pollination, the insect benefits from the energy in the plant nectar or pollen it consumes. The plant benefits by having its gametes transferred to another plant. Nearly 87.5% of all flowering plants are pollinated by animals with the vast majority of these being insects.

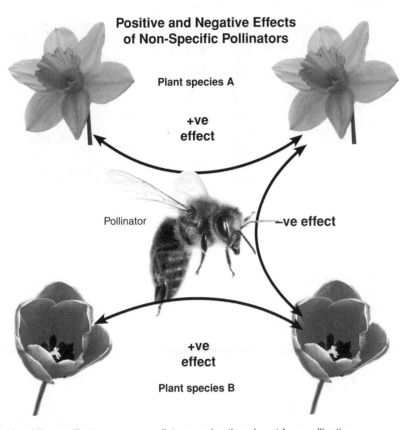

Positive and Negative Effects of Non-Specific Pollinators

Plant species A

+ve effect

Pollinator

–ve effect

+ve effect

Plant species B

Orchid flowers are highly variable in their structure and often highly specialized. Often they achieve pollination by sexual deception. Many species produce flowers that visually imitate bee species, attracting males bees to the flower.

Magnolias are an ancient plant group with generalized flowers evolved for pollination by beetles. Their flowers are quite robust and produce a large amount of pollen.

Most insect pollinators are generalists, meaning they do not form pollination relationships with specific plants. Honey bees, for example, pollinate many different kinds of plants. This can be of negative value to a particular plant, as the energy expended in producing pollen and nectar is wasted if the bee does not fly to a plant of the same species next.

1. Define mutualism: _____

   _____

2. Describe the benefits of mutualism to both the flower and the pollinator: _____

   _____

   _____

   _____

3. (a) Describe the adaptations that angiosperms have to attract specific pollinators to their flowers:

   _____

   _____

   _____

   (b) Identify one advantage and one disadvantage of having a generalist pollinator: _____

   _____

   _____

LINK
246

KNOW

250 Pollination and Fertilization

Key Idea: In plants, pollination is essential to ensure fertilization and production of seeds.

Pollination is the transfer of pollen grains from the male reproductive structures to the female reproductive structures of plants. This must happen before **fertilization** (the joining of the egg and sperm) can occur. Adaptations to ensure cross pollination (pollination between different plants) include structural and physiological mechanisms associated with the flowers or cones themselves, and reliance on wind and animal pollinators. Plants have developed many mechanisms which increase the chances of pollination occurring.

Mechanisms for Increasing Pollination Success

Most flowering plants are pollinated by animals, with the most common pollinator being insects. Flowers use rewards of food (nectar or pollen), or flower colours or scents to attract insects or other animals to the flower.

In many plants, pollen will not germinate if it lands on the stigma of the same plant ensuring that the egg cells are not fertilized by sperm from the same plant. Pollinators are therefore required to ensure cross-pollination.

Gymnosperms are all wind pollinated. In conifers, the male cones are often borne on the lower branches. They produce vast quantities of pollen, which must be blown upwards towards female cones in higher branches of other trees.

Growth of the Pollen Tube and Fertilization

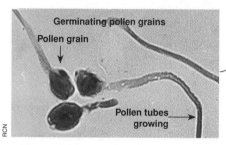

Germinating pollen grains

Pollen grain

Pollen tubes growing

RCN

Pollen grains are immature male gametophytes. Pollination is the actual transfer of the pollen from the stamens to the stigma. After landing on the sticky stigma, the pollen grain is able to complete development, germinating and growing a pollen tube that extends down to the ovary. The pollen tube enters the ovule through the **micropyle**, a small gap in the ovule. A **double fertilization** takes place. One sperm nucleus fuses with the egg to form the zygote. A second sperm nucleus fuses with the two polar nuclei within the embryo sac to produce the endosperm tissue (3N). There are usually many ovules in an ovary, therefore many pollen grains (and fertilizations) are needed before the entire ovary can develop.

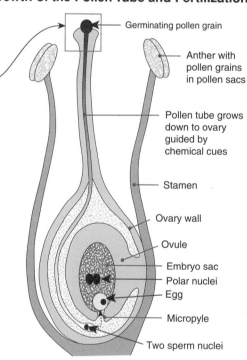

Germinating pollen grain

Anther with pollen grains in pollen sacs

Pollen tube grows down to ovary guided by chemical cues

Stamen

Ovary wall

Ovule

Embryo sac

Polar nuclei

Egg

Micropyle

Two sperm nuclei

Different pollens are variable in shape and pattern, and genera can be easily distinguished on the basis of their distinctive pollen. The species specific nature of pollen ensures that only genetically compatible plants will be fertilized.

Dartmouth College

1. Distinguish between **pollination** and **fertilization**: _____

_____

_____

2. Describe two strategies of plants to increase the chance of pollination: _____

_____

3. Describe the mechanism by which fertilization occurs: _____

_____

_____

© 2012-2014 **BIOZONE** International
ISBN: 978-1-927173-93-0

251 Seed Dispersal

Key Idea: Seeds must be dispersed from the parent plant to reduce competition for light and nutrients. Seeds may be dispersed by wind, water, or animals.

Plants have evolved many ways to ensure that their seeds are dispersed. This has given them opportunities to expand their range. In some cases the seed itself is the agent of dispersal, but often it is the fruit or an associated attached structure. The main agents of seed dispersal are wind, water, and animals. Wind dispersed seeds have wing-like or feathery structures that catch the air currents and carry the seeds long distances. Plants that rely on animals to spread their seeds may have hooks or barbs that catch the animal hair, sticky secretions that adhere to the skin or hair, or fleshy fruits that are eaten leaving the seed to be deposited in feces some distance from the parent plant. Other dispersal mechanisms rely on explosive discharge or shaking from pods or capsules (e.g. legumes, poppy).

For each of the examples below, describe the method of dispersal and the adaptive features associated with the method:

1. **Dandelion** seeds are held in a puff-like cluster:

   (a) Dispersal mechanism: _____

   (b) Adaptive features: _____

   _____

   _____

2. **Acorns** are heavy fruits in which the fleshy seeds are encased in a resistant husk:

   (a) Dispersal mechanism: _____

   (b) Adaptive features: _____

   _____

   _____

3. **Coconuts** are heavy buoyant fruits with a thick husk:

   (a) Dispersal mechanism: _____

   (b) Adaptive features:_____

   _____

   _____

4. **Maple** fruits are winged, two-seeded samaras:

   (a) Dispersal mechanism: _____

   (b) Adaptive features: _____

   _____

   _____

5. **Wattle** (*Acacia* spp.) seeds are enclosed in pods. A fleshy strip surrounds each seed:

   (a) Dispersal mechanism: _____

   (b) Adaptive features: _____

   _____

   _____

6. **New Zealand flax** (*Phormium* spp.) produces seeds in pods:

   (a) Dispersal mechanism: _____

   (b) Adaptive features: _____

   _____

   _____

LINK 252   WEB 251   **KNOW**

252 A Most Accomplished Traveller

Above: Coconut palms fringing a beach, Thailand

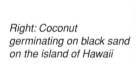

Right: Coconut germinating on black sand on the island of Hawaii

Key Idea: Coconuts have dispersed so widely through the tropics that their origin and path of dispersal is still the subject of research.

The origin of the coconut (*Cocos nucifera*) is one of botany's mysteries. It is so extensively cultivated and so widespread in the wild, determining its origin and dispersal around the globe is extremely difficult. Suggestions have been made that the coconut originated on the coastline of the Gondwanan continent, and spread to volcanic islands where competition had been eliminated by volcanic activity. It has also been suggested that the coconut originated in South Asia or South America. Fossils show that it has been wide spread for some time, with the oldest known fossils of coconut-like palm trees found in Bangladesh and fossils from New Zealand showing it was established there some 15 million years ago.

Coconuts are a single seeded fruit (a type known as a drupe and not, in fact, a nut at all), most commonly seen as the seed with the fibrous husk removed. The coconut fruit possesses a number of features that have allowed it to spread throughout the tropics. The fibrous husk allows it to float and keeps out the seawater. Its oval shape is very stable and allows it to ride high in the water. The seed is the largest of any plant except the coco-de-mer (*Lodoicea maldivica*) and it has a thin but tough shell. The endosperm takes up only a small lining inside the seed, leaving a hollow that is filled with liquid. As the seed matures on its voyage across the sea, the liquid is absorbed. This adds to the buoyancy of the fruit. Additionally, the seed takes a long time to germinate, from 30 to 220 days. The seed is never dormant, as this is not necessary in an equable tropical climate, so the long germination period is an adaptation to extended periods of ocean-going travel between islands. Coconuts can still be viable after travelling for as long as 200 days and covering up to 4000 km.

Before the arrival of humans in the Pacific, coconuts were already widespread, but because of its valuable features, it has been even more widely dispersed by humans, both in prehistory and in modern times. Not only is it a portable, storable food and water source conveniently sealed in a hard shell, but the husk fibres can be used for making ropes, bedding, and many other products. Coconut oils, flesh, and fibres are still extensively used today.

The only tropical coastlines the coconut failed to reach naturally were those of the Atlantic and Caribbean as these bodies of water do not mix with the Pacific or Indian Oceans except in polar regions. However around 1500 AD Europeans travelling west from India transported the coconut around the bottom of Africa and across the Atlantic to Central America. The coconut is now common in every tropical and subtropical region on Earth.

1. Explain why the origin of the coconut palm is difficult to determine: _____

_____

_____

2. Describe the features of the coconut that allowed it to spread across the Pacific and Indian oceans: _____

_____

_____

_____

3. Explain why humans found the coconut so useful: _____

_____

_____

_____

4. Explain why the coconut never established in the Atlantic before humans introduced it: _____

_____

_____

LINK
251

© 2012-2014 **BIOZONE** International
ISBN: 978-1-927173-93-0
Photocopying Prohibited

253 Seed Structure and Germination

Key Idea: The seed houses the dormant embryo until conditions for germination are met. There are important differences between monocot and dicot seeds.

After fertilization has occurred, the ovary develops into the fruit and the ovules within the ovary become the **seeds**. Recall that there is a double fertilization in plants; one sperm fertilizes the egg to form the embryo, while another sperm combines with the diploid endosperm nucleus to give rise to the triploid endosperm. The development of the endosperm is important and begins before embryonic development in order to produce a nutrient store for the young plant. A seed is an entire reproductive unit, housing the embryonic plant in a state of dormancy. During the last stages of maturing, the seed dehydrates until its water content is only 5-15% of its weight. The embryo stops growing and remains dormant until the seed germinates. At germination, the seed takes up water and the food store is mobilized to provide the nutrients for plant growth and development.

Seed Structure

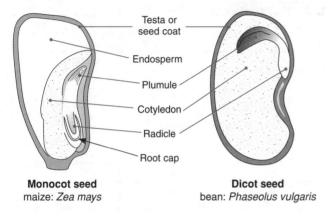

Monocot seed
maize: *Zea mays*

Dicot seed
bean: *Phaseolus vulgaris*

Every seed contains an embryo comprising a rudimentary shoot (plumule), root (radicle), and one (monocots) or two (dicots) cotyledons (seed leaves). The embryo and its food supply are encased in a protective seed coat or **testa**. In monocots, the endosperm provides the food supply, whereas in most dicot seeds, the nutrients from the endosperm are transferred to the large, fleshy cotyledons.

Germination in a Dicot Seed (bean: *Phaseolus vulgaris*)

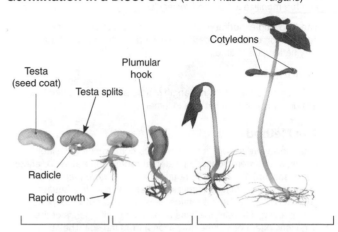

Radicle erupts from the seed and grows rapidly downwards.	Plumular hook protects the emerging stem.	Shoot straightens, lateral roots develop. Shoots emerge and secondary roots develop.

1. Identify the structures in the seeds below:

Dicot seed
(bean: *Phaseolus vulgaris*)

Monocot seed
(maize: *Zea mays*)

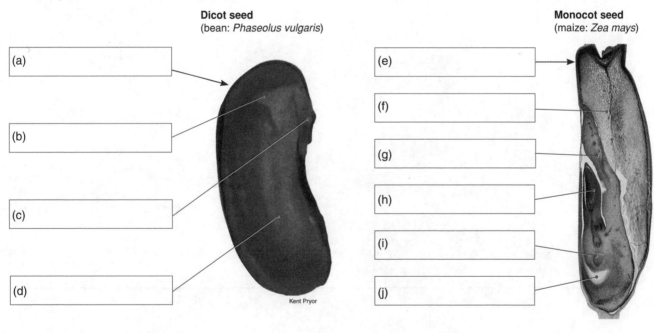

Kent Pryor

(a)

(b)

(c)

(d)

(e)

(f)

(g)

(h)

(i)

(j)

2. What is the purpose of a **seed**? _____

_____

3. (a) State the function of the endosperm in angiosperms: _____

   (b) State how the endosperm is derived: _____

4. What is the role of the testa? _____

5. Why must stored seeds be kept dry? _____

254 Investigating Germination

Key Idea: The amount of water received by the seed before it germinates can affect germination rate.

There are many factors affecting the germination of a seed and the degree to which those factors affects germination varies from species to species. In general there are three requirements for seed germination: water (absorption and reactivation of metabolism), oxygen (for cell respiration), and a temperature that allows metabolism to proceed. Light may or may not be required for germination depending of the species, although light is required very soon after germination.

The Effect of Water of Seed Germination

Water is essential for the germination process. It enables expansion of the growing cells and activates the enzymes needed for germination. It is also needed for the hydrolysis of stored starch and the mobilization of food molecules.

The following experiment investigates the effect of water on germination of tomato seeds by soaking the seeds in water for varying lengths of time.

The Aim

To investigate the effect the length of time seeds are soaked in water has on seed germination rate in tomatoes.

The Method

Four trials were set up. In each trial 100 commercially available tomato seeds were soaked in water at room temperature for either 0 (not soaked) 12, 24, or 36 hours. The seeds were then transferred to containers containing sterilized soil and planted in a square grid 0.5 cm apart. Each container received 1 liter of water per day after planting. Germination was taken as the emergence of the hypocotyl (stem below the cotyledons) from the soil. The number of germinated seeds was counted at the same time each day.

The Results

Days after sowing	Soaking duration (hours) and number seeds germinated			
	0	12	24	36
1	0	0	0	0
2	0	0	0	0
3	0	0	44	39
4	0	34	36	36
5	11	30	15	14
6	19	12	4	5
7	17	9	0	0
8	11	3	0	0
9	8	0	0	0
10	4	0	0	0
11	3	0	0	0
12	0	0	0	0
Total	73	88	99	94

Based on data from S. Sabongari African Journal of Biotechnology 2003

1. Plot a line graph on the grid below of the germination of the sets of seeds:

2. (a) At which length of soaking time did the greatest number of seeds germinate? _____

   (b) At which length of soaking time did the seeds germinate the quickest? _____

3. Identify one way in which the experiment could be made more accurate: _____

   _____

4. Identify one other experiment that could be done to extend this initial experiment: _____

   _____

255 Chapter Review

Summarize what you know about this topic under the headings provided. You can draw diagrams or mind maps, or write short notes to organize your thoughts. Use the images and hints, included to help you:

Transport in the xylem and phloem
HINT: Describe the structural and functional differences between xylem and phloem

Growth in plants
HINT: Describe the effect of auxin on plant growth.

Reproduction in plants
HINT: Draw the internal structure of a flower and a seed. How is flowering controlled?

KNOW

256 KEY TERMS: Did You Get It?

1. Match each term to its definition, as identified by its preceding letter code.

auxin _____

cohesion-tension _____

flower _____

long-day plant _____

meristem _____

phloem _____

phytochrome _____

pollination _____

potometer _____

short-day plant _____

stomata _____

xylem _____

A   Transfer of pollen from the male anther to the female stigma.

B   Plant that flowers in response to a period of dark exceeding a certain length.

C   Partial explanation for the movement of water up the plant in the transpiration stream.

D   Device used for investigating the rate of transpiration.

E   Plant hormone that plays a part in plant growth and the phototropic response.

F   Vascular tissue that conducts water and mineral salts from the roots to the rest of the plant. Dead in its functional state.

G   Tissue that conducts dissolved sugars in vascular plants. Comprises mostly sieve tubes and companion cells.

H   A pigment in plants responsible for the photoperiodism effect. Regulates the timing of flowering with different effects in long day and short day plants.

I   The growing region of the plant where mitosis and cell division occur.

J   Temporary reproductive structure in angiosperms.

K   Pores in the leaf surface through which gases can pass.

L   Plant that flowers when exposed to dark periods of less than a critical length.

2.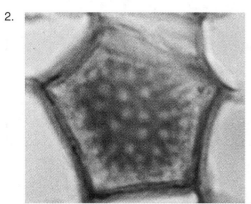

(a) What does the image (left) show: _____

_____

(b) In what tissue would you find it? _____

(c) Is this tissue alive or dead? _____

(d) What transport process is it associated with? _____

_____

(e) What is being moved in this process? _____

3.

Tangopaso CC 3.0

(a) What response is being shown by the orchid stem in the photo left?

_____

(b) What is the environmental cue involved? _____

_____

(c) What is the hormone involved in the response? _____

_____

_____

4. (a) What is the name given to the plant group with two cotyledons in the seed? _____

(b) What is the name given to the plant group with one cotyledon in the seed? _____

© 2012-2014 **BIOZONE** International
ISBN: 978-1-927173-93-0
Photocopying Prohibited

TEST

Genetics and Evolution

Key terms

anaphase

chiasma (*pl.* chiasmata)

chromatid

continuous variation

crossing over

dihybrid cross

directional selection

discontinuous variation

disruptive selection

gamete

gradualism

homologous chromosome

independent assortment

interphase

linked genes

meiosis

metaphase

non-sister chromatid

polygenes

polyploidy

prophase

punctuated equilibrium

Punnett square

recombination frequency

reproductive isolation

speciation

stabilizing selection

synapsis

telophase

10.1 Meiosis

Understandings, applications, skills

Activity number

☐ 1 Explain the role of meiosis. Describe the events in meiosis including replication of chromosomes in interphase, crossing over and separation of homologues in meiosis I, and separation of sister chromatids in meiosis II. — 257

☐ 2 Explain what is meant by independent assortment of genes and how it arises, and explain its contribution to variation in the gametes produced by meiosis. — 87 257

☐ 3 Describe crossing over and the formation of chiasmata between non-sister chromatids of homologues during meiosis I. Explain how recombination of alleles as a result of crossing over contributes to variation in the gametes. Draw diagrams to show chiasmata formed by crossing over. — 258 259

10.2 Inheritance

Understandings, applications, skills

Activity number

☐ 1 Identify phenotypic variation as discrete (discontinuous) or continuous and describe the genetic basis for each. — 260

☐ 2 Complete and analyse Punnett squares for dihybrid crosses. Calculate predicted phenotypic and genotypic ratios of offspring of dihybrid crosses involving unlinked autosomal genes. — 261 265

☐ 3 Define gene linkage. Explain differences in inheritance patterns of linked and unlinked genes and identify recombinants in crosses involving two linked genes. Describe how Morgan's experiments with *Drosophila* helped to clarify our understanding of linkage, recombination, and gene mapping. — 262-264

> **TOK** *Exceptions to the law of independent assortment were explained by gene linkage. What is the difference between a law and a theory?*

☐ 4 Use the chi-squared test to determine if the differences between observed and expected outcomes of genetic crosses are statistically significant. — 266 267

☐ 5 Using an example, describe and explain polygenic inheritance, including the role of environment in influencing phenotypic variation. — 268

10.3 Gene pools and speciation

Understandings, applications, skills

Activity number

☐ 1 Explain what is meant a gene pool and define evolution with reference to the gene pool of a population. Identify the factors affecting allele frequencies in gene pools and explain how these lead to evolution. — 269

☐ 2 Recall the mechanism of natural selection. Describe examples of directional, disruptive, and stabilizing selection in populations. — 271-274

☐ 3 Describe the formation of new species (speciation) by gradual divergence of isolated populations. Compare the allele frequencies of geographically isolated populations. — 275 276

☐ 4 Explain what is meant by reproductive isolation and describe temporal and behavioural isolating mechanisms. Recognize that geographic isolation is often an essential first step in reproductive isolation. — 266 267

☐ 5 Compare the model of gradual species divergence with the punctuated equilibrium model in which speciation occurs rapidly. Explain the role of polyploidy in instant speciation events, using the example of speciation in *Allium*. — 268

> **TOK** *Punctuated equilibrium was considered a challenge to the paradigm of Darwinian gradualism. How do paradigm shifts proceed in science and what factors contribute to their success?*

257 Meiosis and Variation

Key Idea: Meiosis produces variation via the processes of crossing over and independent assortment.
Crossing over and independent assortment (leading to recombination of alleles) are mechanisms that occur during meiosis I. They increase the genetic variation in the gametes and therefore in the offspring.

Crossing Over and Recombination

Recall that the chromosomes replicate during interphase, before meiosis, to produce replicated chromosomes with sister chromatids held together at the centromere. When the replicated chromosomes are paired during the first stage of meiosis, non-sister chromatids may become tangled and segments may be exchanged in a process called **crossing over**.

Crossing over results in the **recombination** of alleles, producing greater variation in the offspring than would otherwise occur. Alleles that are linked (on the same chromosome) may be exchanged and so become unlinked.

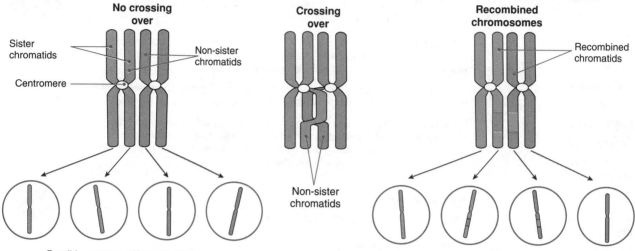

No crossing over — Sister chromatids, Non-sister chromatids, Centromere — Possible gametes with no crossing over

Crossing over — Non-sister chromatids

Recombined chromosomes — Recombined chromatids — Possible gametes with crossing over

Independent Assortment

Independent assortment is an important mechanism for producing variation in gametes. During the first stage of meiosis replicated homologous chromosomes pair up along the middle of the cell. Which way the chromosomes pair up is random. For the homologous chromosomes right, there are two possible ways in which they can line up resulting in four different combinations in the gametes. The intermediates steps of meiosis have been left out for simplicity.

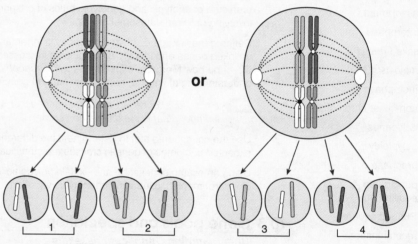

or

1   2   3   4

1. How does independent assortment increase the variation in gametes? _____

_____

_____

2. (a) What is crossing over? _____

_____

_____

(b) How does crossing over increase the variation in the gametes (and hence the offspring)? _____

_____

_____

_____

© 2012-2014 **BIOZONE** International
ISBN: 978-1-927173-93-0
Photocopying Prohibited

258 Crossing Over

Key Idea: Crossing over is the exchange of genetic material between non-sister chromosomes and produces greater variation in the gametes.

Crossing over refers to the mutual exchange of pieces of chromosome and involves the swapping of whole groups of genes between the **homologous** chromosomes. This process can occur only during **prophase I** in the first division of **meiosis**. Errors in crossing over can result in detrimental chromosome mutations. Recombination as a result of crossing over is an important mechanism to increase genetic variability in the offspring and has the general effect of allowing genes to move independently of each other through the generations in a way that allows concentration of beneficial alleles.

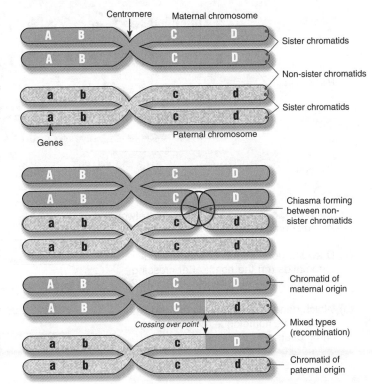

Pairing of Homologous Chromosomes

Every somatic cell contains a pair of each type of chromosome, one from each parent. These are called **homologous pairs** or **homologues**. In prophase of meiosis I, the homologues pair up to form **bivalents**. This process is called **synapsis** and it brings the chromatids of the homologues into close contact along their entire length.

Chiasma Formation and Crossing Over

Synapsis allows the homologous, non-sister chromatids to become entangled and the chromosomes exchange segments. This exchange occurs at regions called **chiasmata** (*sing.* chiasma). In the diagram (centre), a chiasma is forming and the exchange of pieces of chromosome has not yet taken place. Numerous chiasmata may develop between homologues.

Separation

Crossing over produces new allele combinations, a phenomenon known as **recombination**. When the homologues separate in anaphase of meiosis I, each of the chromosomes pictured will have a new mix of alleles that will be passed into the gametes soon to be formed. Recombination is an important source of variation in population gene pools.

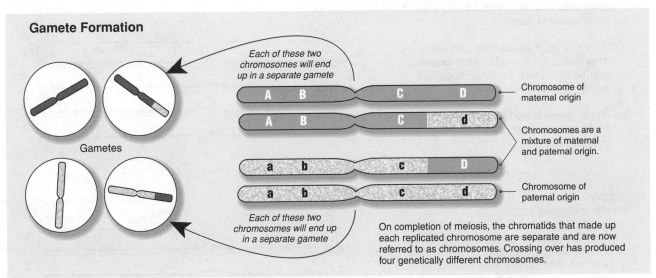

Gamete Formation

On completion of meiosis, the chromatids that made up each replicated chromosome are separate and are now referred to as chromosomes. Crossing over has produced four genetically different chromosomes.

1. (a) In a general way, describe how crossing over alters the genotype of gametes: _____

_____

(b) What is the consequence of this? _____

2. What is the significance of crossing over in the evolution of sexually producing populations? _____

LINK 137 LINK 81 **KNOW**

259 Crossing Over Problems

Key Idea: Crossing over can occur in multiple places in chromosomes, producing a huge amount of genetic variation. The diagram below shows a pair of homologous chromosomes about to undergo chiasma formation during the first cell division in the process of meiosis. There are known crossover points along the length of the chromatids (same on all four chromatids shown in the diagram). In the prepared spaces below, draw the gene sequences after crossing over has occurred on three unrelated and separate occasions (it would be useful to use different coloured pens to represent the genes from the two different chromosomes). See the diagrams on the previous page as a guide.

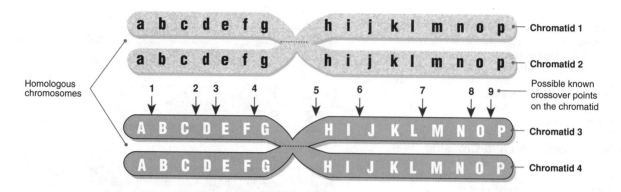

1. Crossing over occurs at a **single** point between the chromosomes above.

   (a) Draw the gene sequences for the four chromatids (on the right), after crossing over has occurred at crossover point: **2**

   (b) Which genes have been exchanged with those on its homologue (neighbouring chromosome)?

   _____

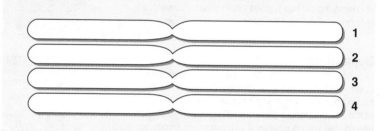

2. Crossing over occurs at **two** points between the chromosomes above.

   (a) Draw the gene sequences for the four chromatids (on the right), after crossing over has occurred between crossover points: **6** and **7**.

   (b) Which genes have been exchanged with those on its homologue (neighbouring chromosome)?

   _____

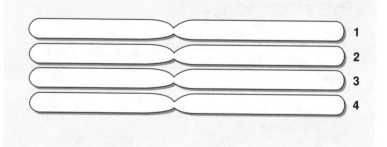

3. Crossing over occurs at **four** points between the chromosomes above.

   (a) Draw the gene sequences for the four chromatids (on the right), after crossing over has occurred between crossover points: **1** and **3**, and **5** and **7**.

   (b) Which genes have been exchanged with those on its homologue (neighbouring chromosome)?

   _____

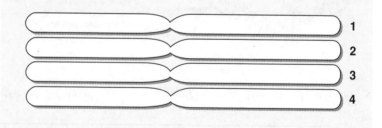

4. What would be the genetic consequences if there was no crossing over between chromatids during meiosis?

_____

_____

_____

LINK
SKILL 258

© 2012-2014 **BIOZONE** International
ISBN: 978-1-927173-93-0
Photocopying Prohibited

260 Variation

Key Idea: The characteristics of sexually reproducing organisms show variation. Those showing continuous variation are controlled by many genes at different loci and are often greatly influenced by environment. Those showing discontinuous variation are controlled by a small number of genes and there are a limited number of phenotypic variants in the population. Both genes and environment contribute to the final phenotype on which natural selection acts.

Variation refers to the diversity of genotypes (allele combinations) and phenotypes (appearances) in a population. Variation in phenotypic characteristics, such as flower colour and birth weight, is a feature of sexually reproducing populations. Some characteristics show discontinuous variation, with only a limited number of phenotypic variants in the population. Others show continuous variation, with a range of phenotypic variants approximating a bell shaped (normal) curve. Both genotype and the environment determine, to different degrees, the final phenotype we see.

Mutations

Mutations are changes to the DNA
Mutations change the DNA sequence and are the source of all new alleles. Not all mutations change the amino acid sequence because of degeneracy of the genetic code.

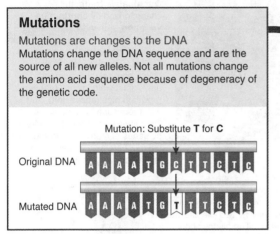

Mutation: Substitute **T** for **C**

Original DNA: A A A A T G C T T C T C

Mutated DNA: A A A A T G T T T C T C

Sexual reproduction

Sexual reproduction involves meiosis and fertilisation
Sexual reproduction rearranges and reshuffles the genetic material into new combinations. Fertilisation unites dissimilar gametes to produce more variation.

Don Horne

Genotype

The genetic make up of the individual is its genotype. The genotype determines the genetic potential of an individual.

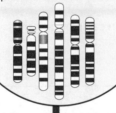

Gene interactions and epigenetics

Dominant, recessive, codominant, and multiple alleles, epigenetic modifications (such as methylation), and interactions between genes, combine in their effects.

Environmental factors

The environment can influence the expression of the genotype. Both the external environment (e.g. physical and biotic factors) and internal environment (e.g. hormones) can be important.

Phenotype

The phenotype is the physical appearance of an individual. An individual's phenotype is the result of the interaction of genetic and environmental factors during its lifetime. Gene expression can be influenced by both the internal and external environment during and after development.

Some snail populations show phenotypic plasticity with respect to shell thickness; individuals can develop thicker shells when subjected to heavy predation.

1. Using examples, explain how the environment of a particular genotype can affect the phenotype: _____

_____

_____

_____

2. Discuss the significance of variation in selection: _____

_____

_____

_____

LINK 137   LINK 74   WEB 260   **KNOW**

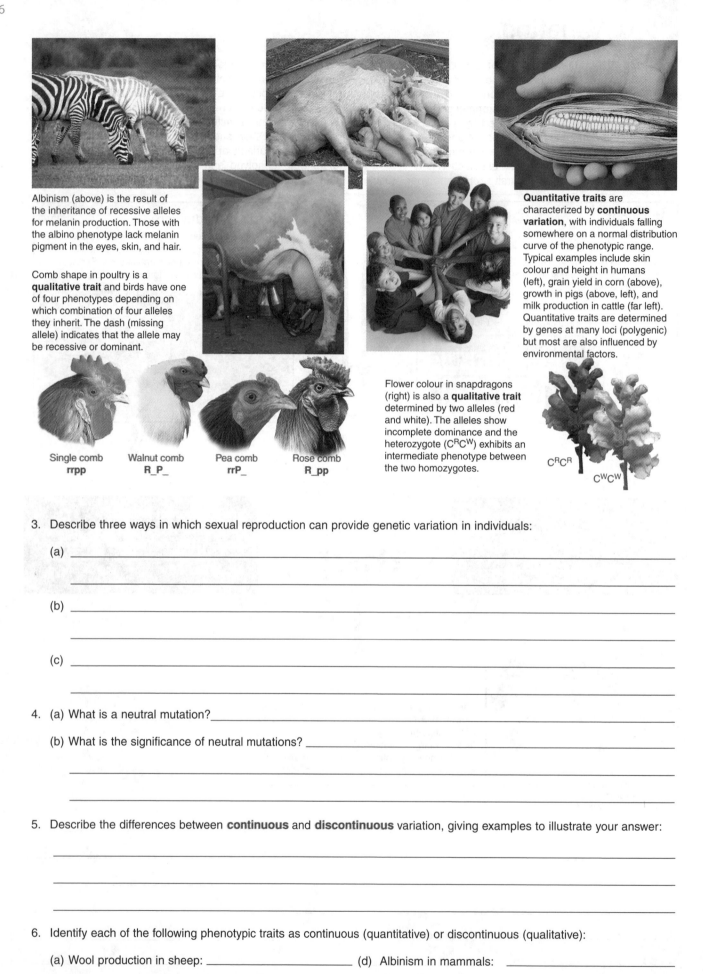

Albinism (above) is the result of the inheritance of recessive alleles for melanin production. Those with the albino phenotype lack melanin pigment in the eyes, skin, and hair.

Comb shape in poultry is a **qualitative trait** and birds have one of four phenotypes depending on which combination of four alleles they inherit. The dash (missing allele) indicates that the allele may be recessive or dominant.

Single comb	Walnut comb	Pea comb	Rose comb
rrpp	**R_P_**	**rrP_**	**R_pp**

Quantitative traits are characterized by **continuous variation**, with individuals falling somewhere on a normal distribution curve of the phenotypic range. Typical examples include skin colour and height in humans (left), grain yield in corn (above), growth in pigs (above, left), and milk production in cattle (far left). Quantitative traits are determined by genes at many loci (polygenic) but most are also influenced by environmental factors.

Flower colour in snapdragons (right) is also a **qualitative trait** determined by two alleles (red and white). The alleles show incomplete dominance and the heterozygote ($C^R C^W$) exhibits an intermediate phenotype between the two homozygotes.

$C^R C^R$

$C^W C^W$

3. Describe three ways in which sexual reproduction can provide genetic variation in individuals:

(a) _____

_____

(b) _____

_____

(c) _____

_____

4. (a) What is a neutral mutation?_____

(b) What is the significance of neutral mutations? _____

_____

_____

5. Describe the differences between **continuous** and **discontinuous** variation, giving examples to illustrate your answer:

_____

_____

_____

6. Identify each of the following phenotypic traits as continuous (quantitative) or discontinuous (qualitative):

(a) Wool production in sheep: _____ (d) Albinism in mammals: _____

(b) Hand span in humans: _____ (e) Body weight in mice: _____

(c) Blood groups in humans: _____ (f) Flower colour in snapdragons: _____

© 2012-2014 **BIOZONE** International
ISBN: 978-1-927173-93-0

261 Dihybrid Cross

Key Idea: A dihybrid cross studies the inheritance pattern of two genes. In crosses involving unlinked autosomal genes, the offspring occur in predictable ratios.

There are four types of gamete produced in a cross involving two genes, where the genes are carried on separate chromosomes and are sorted independently of each other during meiosis. The two genes in the example below are on separate chromosomes and control two unrelated characteristics, **hair colour** and **coat length**. Black (B) and short (L) are dominant to white and long.

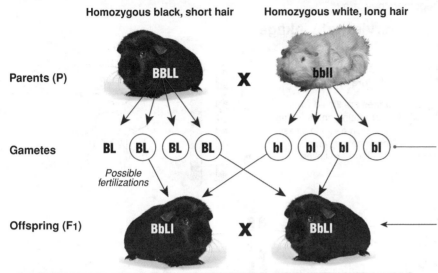

Homozygous black, short hair   Homozygous white, long hair

Parents (P)   BBLL   X   bbll

Gametes   BL (BL) (BL) (BL)   (bl) (bl) (bl) bl

Possible fertilizations

Offspring (F1)   BbLl   X   BbLl

Parents: The notation P is used for a cross between true breeding (homozygous) parents.

Gametes: Only one type of gamete is produced from each parent (although they will produce four gametes from each oocyte or spermatocyte). This is because each parent is homozygous for both traits.

F₁ offspring: There is only one kind of gamete from each parent, therefore only one kind of offspring produced in the first generation. The notation F₁ is used to denote the heterozygous offspring of a cross between two true breeding parents.

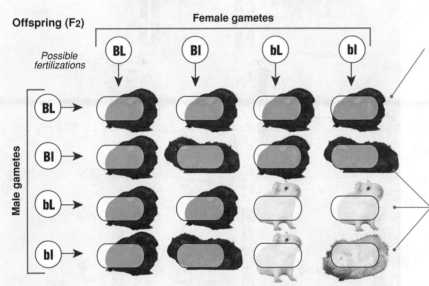

Offspring (F2)

Possible fertilizations

Female gametes: BL   Bl   bL   bl

Male gametes: BL, Bl, bL, bl

F₂ offspring: The F₁ were mated with each other (selfed). Each individual from the F₁ is able to produce four different kinds of gamete. Using a grid called a **Punnett square** (left), it is possible to determine the expected genotype and phenotype ratios in the F₂ offspring. The notation F₂ is used to denote the offspring produced by crossing F₁ heterozygotes.

Each of the 16 animals shown here represents the possible zygotes formed by different combinations of gametes coming together at fertilization.

The offspring can be arranged in groups with similar phenotypes:

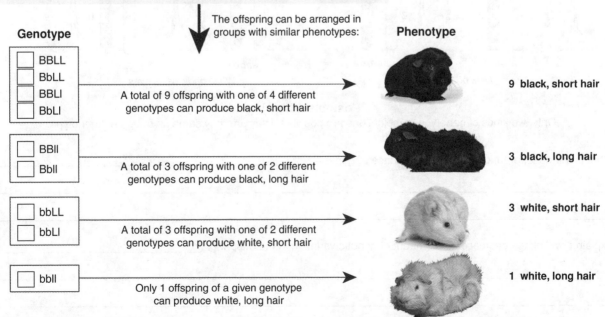

Genotype

☐ BBLL
☐ BbLL
☐ BBLl
☐ BbLl

A total of 9 offspring with one of 4 different genotypes can produce black, short hair

Phenotype

9 black, short hair

☐ BBll
☐ Bbll

A total of 3 offspring with one of 2 different genotypes can produce black, long hair

3 black, long hair

☐ bbLL
☐ bbLl

A total of 3 offspring with one of 2 different genotypes can produce white, short hair

3 white, short hair

☐ bbll

Only 1 offspring of a given genotype can produce white, long hair

1 white, long hair

1. Complete the Punnett square above and use it to fill in the number of each genotype in the boxes (above left).

© 2012-2014 **BIOZONE** International
ISBN: 978-1-927173-93-0
Photocopying Prohibited

LINK

88 APP

262 Inheritance of Linked Genes

Key Idea: Linked genes are genes found on the same chromosome and tend to be inherited together. Linkage reduces the genetic variation in the offspring.

Genes are **linked** when they are on the same chromosome. Linked genes tend to be inherited together, and the extent of crossing over depends on how close together they are on the chromosome. In genetic crosses, linkage is indicated when a greater proportion of the offspring from a cross are of the parental type (than would be expected if the alleles were on separate chromosomes and assorting independently). Linkage reduces the variety of offspring that can be produced.

Overview of Linkage

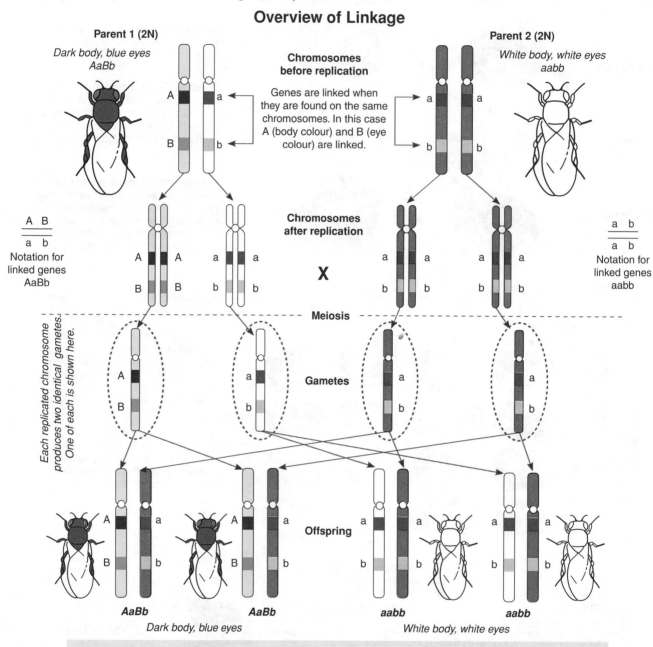

Parent 1 (2N)
Dark body, blue eyes
AaBb

Chromosomes before replication

Genes are linked when they are found on the same chromosomes. In this case A (body colour) and B (eye colour) are linked.

Parent 2 (2N)
White body, white eyes
aabb

A B
—
a b
Notation for linked genes
AaBb

Chromosomes after replication

a b
—
a b
Notation for linked genes
aabb

X

Meiosis

Each replicated chromosome produces two identical gametes. One of each is shown here.

Gametes

Offspring

AaBb   **AaBb**   **aabb**   **aabb**

Dark body, blue eyes   White body, white eyes

Possible Offspring
Only two kinds of genotype combinations are possible. They are they same as the parent genotype.

1. What is the effect of **linkage** on the inheritance of genes? _____

_____

_____

2. Explain how linkage decreases the amount of genetic variation in the offspring: _____

_____

_____

_____

An Example of Linked Genes in *Drosophila*

The genes for wing shape and body color are linked (they are on the same chromosome).

Parent	Wild type female	Mutant male
Phenotype	Straight wing Grey body	Curled wing Ebony body
Genotype	Cucu Ebeb	cucu ebeb

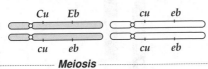

Linkage

Cu   Eb   cu   eb

cu   eb   cu   eb

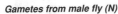 — **Meiosis** —

Gametes from female fly (N)   **Gametes from male fly (N)**

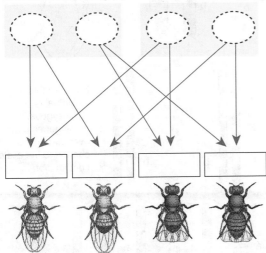

Sex of offspring is irrelevant in this case

Contact **Newbyte Educational Software** for details of their superb *Drosophila Genetics* software package which includes coverage of linkage and recombination. *Drosophila* images © Newbyte Educational Software.

Drosophila and Linked Genes

In the example shown left, wild type alleles are dominant and are given an upper case symbol of the mutant phenotype (Cu or Eb). This notation used for *Drosophila* departs from the convention of using the dominant gene to provide the symbol. This is necessary because there are many mutant alternative phenotypes to the wild type (e.g. curled and vestigial wings). A lower case symbol of the wild type (e.g. ss for straight wing) would not indicate the mutant phenotype involved.

Drosophila melanogaster is known as a model organism. Model organisms are used to study particular biological phenomena, such as mutation. *Drosophila melanogaster* is particularly useful because it produces such a wide range of heritable mutations. Its short reproduction cycle, high offspring production, and low maintenance make it ideal for studying in the lab.

Drosophila melanogaster examples showing variations in eye and body colour. The wild type is marked with a W in the photo above.

3. Complete the linkage diagram above by adding the gametes in the ovals and offspring genotypes in the rectangles.

4. (a) List the possible genotypes in the offspring (above, left) if genes Cu and Eb had been on **separate chromosomes**:

_____

(b) If the female *Drosophila* had been homozygous for the dominant wild type alleles (CuCu EbEb), state:

The offspring genotype(s): _____ The offspring phenotype(s): _____

5. A second pair of *Drosophila* are mated. The female genotype is Vgvg EbEb (straight wings, grey body), while the male genotype is vgvg ebeb (vestigial wings, ebony body). Assuming the genes are linked, carry out the cross and list the genotypes and phenotypes of the offspring. Note vg = vestigial (no) wings:

_____

_____

The genotype(s) of the offspring: _____

The phenotype(s) of the offspring: _____

6. Explain why *Drosophila* are often used as model organisms in the study of genetics: _____

_____

_____

263 Recombination and Dihybrid Inheritance

Key Idea: Recombination is the exchange of alleles between homologous chromosomes as a result of crossing over. Recombination increases the genetic variation in the offspring. The alleles of parental linkage groups separate and new associations of alleles are formed in the gametes. Offspring formed from these gametes are called **recombinants** and show combinations of characteristics not seen in the parents.

In contrast to linkage, recombination increases genetic variation in the offspring. Recombination between the alleles of parental linkage groups is indicated by the appearance of non-parental types in the offspring, although not in the numbers that would be expected had the alleles been on separate chromosomes (independent assortment).

Overview of Recombination

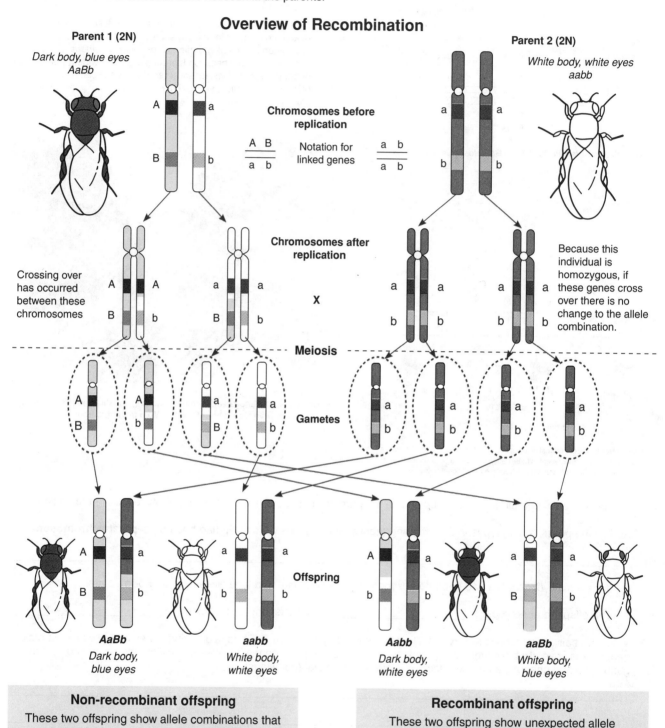

Parent 1 (2N)
Dark body, blue eyes
AaBb

Parent 2 (2N)
White body, white eyes
aabb

Chromosomes before replication

A B
___   Notation for   a b
a b   linked genes   a b

Chromosomes after replication

X

Crossing over has occurred between these chromosomes

Because this individual is homozygous, if these genes cross over there is no change to the allele combination.

Meiosis

Gametes

Offspring

AaBb
Dark body, blue eyes

aabb
White body, white eyes

Aabb
Dark body, white eyes

aaBb
White body, blue eyes

Non-recombinant offspring
These two offspring show allele combinations that are expected as a result of independent assortment during meiosis. Also called parental types.

Recombinant offspring
These two offspring show unexpected allele combinations. They can only arise if one of the parent's chromosomes has undergone crossing over.

1. Describe the effect of **recombination** on the inheritance of genes: _____

_____

_____

© 2012-2014 **BIOZONE** International
ISBN: 978-1-927173-93-0
Photocopying Prohibited

An Example of Recombination

In the female parent, crossing over occurs between the linked genes for wing shape and body colour

	Wild type female	Mutant male
Parent		
Phenotype	Straight wing Grey body	Curled wing Ebony body
Genotype	Cucu Ebeb	cucu ebeb
Linkage		

Cu   Eb

cu   eb

cu   eb   *cu   eb*

········· **Meiosis** ·········

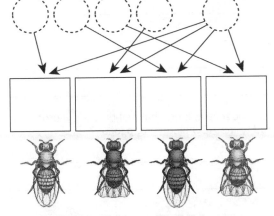

Gametes from female fly (N)

Crossing over has occurred, giving four types of gametes

Gametes from male fly (N)

Only one type of gamete is produced in this case

Non-recombinant offspring   **Recombinant offspring**

The sex of the offspring is irrelevant in this case

Contact **Newbyte Educational Software** for details of their superb *Drosophila Genetics* software package which includes coverage of linkage and recombination. *Drosophila* images © Newbyte Educational Software.

The cross (left) uses the same genotypes as the previous activity but, in this case, crossing over occurs between the alleles in a linkage group in one parent. The symbology used is the same.

Recombination Produces Variation

If crossing over does not occur, the possible combinations in the gametes is limited. **Crossing over and recombination increase the variation in the offspring.** In humans, even without crossing over, there are approximately $(2^{23})^2$ or 70 trillion genetically different zygotes that could form for every couple. Taking crossing over and recombination into account produces $(4^{23})^2$ or 5000 trillion trillion genetically different zygotes for every couple.

Family members may resemble each other, but they'll never be identical (except for identical twins).

Using Recombination

Analysing recombination gave geneticists a way to map the genes on a chromosome. Crossing over is less likely to occur between genes that are close together on a chromosome than between genes that are far apart. By counting the number of offspring of each phenotype, you can calculate the **frequency of recombination**. The higher the frequency of recombination between two genes, the further apart they must be on the chromosome.

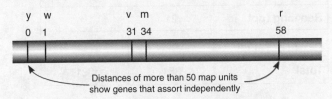

y   w   v   m   r
0   1   31 34   58

Distances of more than 50 map units show genes that assort independently

Map of the X chromosome of *Drosophila*, showing the relative distances between five different genes (in map units).

2. Complete the recombination diagram above, adding the gametes in the ovals and offspring genotypes and phenotypes in the rectangles:

3. Explain how recombination increases the amount of genetic variation in offspring: _____

_____

_____

4. Explain why it is not possible to have a recombination frequency of greater than 50% (half recombinant progeny):

_____

_____

5. A second pair of *Drosophila* are mated. The female is Cucu YY (straight wing, grey body), while the male is Cucu yy (straight wing, yellow body). Assuming recombination, perform the cross and list the offspring genotypes and phenotypes:

_____

_____

_____

264 Detecting Linkage in Dihybrid Inheritance

Key Idea: Linkage between genes can be detected by observing the phenotypic ratios in the offspring.

Shortly after the rediscovery of Mendel's work early in the 20th century, it became apparent that his ratios of 9:3:3:1 for heterozygous dihybrid crosses did not always hold true.

Experiments on sweet peas by William Bateson and Reginald Punnett, and on *Drosophila* by Thomas Hunt Morgan, showed that there appeared to be some kind of coupling between genes. This coupling, which we now know to be linkage, did not follow any genetic relationship known at the time.

Sweet Pea Cross

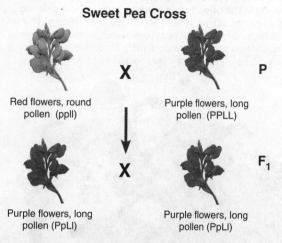

Red flowers, round pollen (ppll)   X   Purple flowers, long pollen (PPLL)   **P**

Purple flowers, long pollen (PpLl)   X   Purple flowers, long pollen (PpLl)   **F₁**

Bateson and Punnett studied sweet peas in which purple flowers (P) are dominant to red (p), and long pollen grains (L) are dominant to round (l). If these genes were unlinked, the outcome of an cross between two heterozygous sweet peas should have been a 9:3:3:1 ratio.

Table 1: Sweet Pea Cross Results

	Observed	Expected
Purple long (P_L_)	284	
Purple round (P_ll)	21	
Red long (ppL_)	21	
Red round (ppll)	55	
Total	381	381

Application: Morgan performed experiments to investigate linked genes in *Drosophila*. He crossed a heterozygous red-eyed normal-winged (Prpr Vgvg) fly with a homozygous purple-eyed vestigial-winged (prpr vgvg) fly. The table (below) shows the outcome of the cross.

   X

Red eyed normal winged (Prpr Vgvg)   Purple eyed vestigial winged (prpr vgvg)

Table 2: *Drosophila* Cross Results

Genotype	Observed	Expected	Gamete type
Prpr Vgvg	1339	710	Parental
prpr Vgvg	152		
Prpr vgvg	154		
prpr vgvg	1195		
Total	2840	2840	

1. Fill in the missing numbers in the **expected** column of **Table 1**, remembering that a 9:3:3:1 ratio is expected:

2. (a) Fill in the missing numbers in the **expected** column of **Table 2**, remembering that a 1:1:1:1 ratio is expected:

   (b) Add the gamete type (parental/recombinant) to the gamete type column in Table 2:

   (c) What type of cross did Morgan perform here?

   _____

3. (a) Use the pedigree chart below to determine if nail-patella syndrome is dominant or recessive, giving reasons for your choice:

   _____

   _____

   _____

   _____

   (b) What evidence is there that nail-patella syndrome is linked to the ABO blood group locus?

   _____

   _____

   _____

   (c) Suggest a likely reason why individual III-3 is not affected despite carrying the B allele:

   _____

   _____

Pedigree for Nail-Patella Syndrome

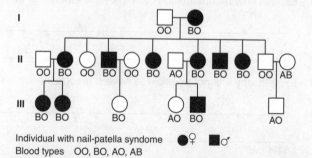

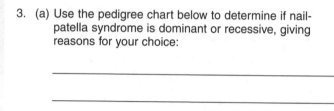

Individual with nail-patella syndome   ●♀ ■♂
Blood types   OO, BO, AO, AB

Linked genes can be detected by pedigree analysis. The diagram above shows the pedigree for the inheritance of nail-patella syndrome, which results in small, poorly developed nails and kneecaps in affected people. Other body parts such as elbows, chest, and hips can also be affected. The nail-patella syndrome gene is linked to the ABO blood group locus.

© 2012-2014 **BIOZONE** International
ISBN: 978-1-927173-93-0

265 Problems Involving Dihybrid Inheritance

Key Idea: Dihybrid inheritance involves two genes. For autosomal unlinked genes, the offspring appear in predictable ratios. Linkage can cause departures from expected ratios.

This activity will allow you to test your understanding of dihybrid inheritance by solving problems involving the inheritance of two genes.

1. In cats, the following alleles are present for coat characteristics: black (B), brown (b), short (L), long (l), tabby (T), blotched tabby (tb). Use the information to complete the dihybrid crosses below:

(a) A black short haired (**BBLl**) male is crossed with a black long haired (**Bbll**) female. Determine the genotypic and phenotypic ratios of the offspring:

Genotype ratio: _____

_____

Phenotype ratio: _____

_____

_____

(b) A tabby, short haired male (**TtbLl**) is crossed with a blotched tabby, short haired (**tbtbLl**) female. Determine ratios of the offspring:

Genotype ratio: _____

_____

Phenotype ratio: _____

_____

2. A plant with orange-striped flowers was cultivated from seeds. The plant was self-pollinated and the F_1 progeny appeared in the following ratios: 89 orange with stripes, 29 yellow with stripes, 32 orange without stripes, 9 yellow without stripes.

(a) Describe the dominance relationships of the alleles responsible for the phenotypes observed: _____

_____

_____

(b) Determine the genotype of the original plant with orange striped flowers: _____

_____

_____

3. In rabbits, spotted coat **S** is dominant to solid colour **s**, while for coat colour, black **B** is dominant to brown **b**. A brown spotted rabbit is mated with a solid black one and all the offspring are black spotted (the genes are not linked).

(a) State the genotypes:

Parent 1: _____

Parent 2: _____

Offspring: _____

(b) Use the Punnett square to show the outcome of a cross between the F_1 (the F_2):

(c) Using ratios, state the phenotypes of the F_2 generation: _____

_____

_____

_____

LINK 261   LINK 88   SKILL

364

4. In guinea pigs, rough coat **R** is dominant over smooth coat **r** and black coat **B** is dominant over white **b**. The genes are not linked.
A homozygous rough black animal was crossed with a homozygous smooth white:

(a) State the genotype of the **F₁**: _____

(b) State the phenotype of the **F₁**: _____

(c) Use the Punnett square (top right) to show the outcome of a cross between the F₁ (the F₂):

(d) Using ratios, state the phenotypes of the F₂ generation: _____

_____

_____

(e) Use the Punnett square (right) to show the outcome of a **back cross** of the **F₁** to the rough, black parent:

(f) Using ratios, state the phenotype of the F₂ generation: _____

_____

(g) A rough black guinea pig was crossed with a rough white one produced the following offspring: 28 rough black, 31 rough white, 11 smooth black, and 10 smooth white. Determine the genotypes of the parents:

_____

5. The Himalayan colour-pointed, long-haired cat is a breed developed by crossing a pedigree (true-breeding), uniform-coloured, long-haired Persian with a pedigree colour-pointed (darker face, ears, paws, and tail) short-haired Siamese.

The genes controlling hair colouring and length are on separate chromosomes: uniform colour **U**, colour pointed **u**, short hair **S**, long hair **s**.

Persian   Siamese   Himalayan

(a) Using the symbols above, indicate the genotype of each breed below its photograph (above, right).   _____ _____ _____

(b) State the genotype of the **F₁** (Siamese X Persian): _____

(c) State the phenotype of the **F₁**: _____

(d) Use the Punnett square to show the outcome of a cross between the F₁ (the F₂):

(e) State the ratio of the F₂ that would be Himalayan: _____

(f) State whether the Himalayan would be true breeding: _____

(g) State the ratio of the F₂ that would be colour-point, short-haired cats: _____

6. A *Drosophila* male with genotype **Cucu Ebeb** (straight wing, grey body) is crossed with a female with genotype **cucu ebeb** (curled wing, ebony body). The phenotypes of the F₁ were recorded and the percentage of each type calculated. The percentages were: Straight wings, grey body 45%, curled wings, ebony body 43%, straight wings, ebony body 6%, and curled wings grey body 6%.

Straight wing Cucu Grey body, Ebeb

(a) Is there evidence of crossing over in the offspring? _____

(b) Explain your answer: _____

_____

_____

(c) Determine the genotypes of the offspring: _____

Curled wing cucu Ebony body, ebeb

_____

© 2012-2014 **BIOZONE** International
ISBN: 978-1-927173-93-0
Photocopying Prohibited

266 Using the Chi-Squared Test in Genetics

Key Idea: The chi-squared test for goodness of fit (χ^2) can be used for testing the outcome of dihybrid crosses against an expected (predicted) Mendelian ratio.

When using the chi-squared test, the null hypothesis predicts the ratio of offspring of different phenotypes according to the expected Mendelian ratio for the cross, assuming independent assortment of alleles (no linkage). Significant departures from the predicted Mendelian ratio indicate linkage of the alleles in question. Raw counts should be used and a large sample size is required for the test to be valid.

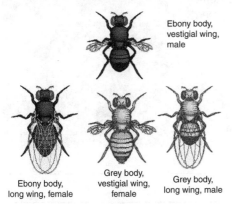

Ebony body, vestigial wing, male

Ebony body, long wing, female

Grey body, vestigial wing, female

Grey body, long wing, male

Images of *Drosophila* courtesy of **Newbyte Educational Software**: *Drosophila* Genetics Lab (www.newbyte.com)

Using χ^2 in Mendelian Genetics

In a *Drosophila* genetics experiment, two individuals were crossed (the details of the cross are not relevant here). The predicted Mendelian ratios for the offspring of this cross were 1:1:1:1 for each of the four following phenotypes: grey body-long wing, grey body-vestigial wing, ebony body-long wing, ebony body-vestigial wing. The observed results of the cross were not exactly as predicted. The following numbers for each phenotype were observed in the offspring of the cross:

Observed results of the example *Drosophila* cross

Grey body, long wing	98	Ebony body, long wing	102
Grey body, vestigial wing	88	Ebony body, vestigial wing	112

Using χ^2, the probability of this result being consistent with a 1:1:1:1 ratio could be tested. Worked example as follows:

Step 1: Calculate the expected value (E)

In this case, this is the sum of the observed values divided by the number of categories (see note below)

$$\frac{400}{4} = 100$$

Category	O	E	O – E	(O – E)^2	$\frac{(O-E)^2}{E}$
Grey, long wing	98	100	–2	4	0.04
Grey, vestigial wing	88	100	–12	144	1.44
Ebony, long wing	102	100	2	4	0.04
Ebony, vestigial wing	112	100	12	144	1.44

Total = 400 χ^2 $\Sigma = 2.96$

Step 2: Calculate O – E

The difference between the observed and expected values is calculated as a measure of the deviation from a predicted result. Since some deviations are negative, they are all squared to give positive values. This step is usually performed as part of a tabulation (right, darker blue column).

Step 3: Calculate the value of χ^2

$$\chi^2 = \Sigma \frac{(O-E)^2}{E}$$

Where: O = the observed result
E = the expected result
Σ = sum of

The calculated χ^2 value is given at the bottom right of the last column in the tabulation.

Step 5a: Using the χ^2 table

On the χ^2 table (part reproduced in Table 1 below) with 3 degrees of freedom, the calculated value for χ^2 of 2.96 corresponds to a probability of between 0.2 and 0.5 (see arrow). *This means that by chance alone a χ^2 value of 2.96 could be expected between 20% and 50% of the time.*

Step 4: Calculating degrees of freedom

The probability that any particular χ^2 value could be exceeded by chance depends on the number of degrees of freedom. This is simply *one less than the total number of categories* (this is the number that could vary independently without affecting the last value). *In this case: 4–1 = 3.*

Step 5b: Using the χ^2 table

The probability of between 0.2 and 0.5 is higher than the 0.05 value which is generally regarded as significant. The null hypothesis cannot be rejected and we have no reason to believe that the observed results differ significantly from the expected (at $P = 0.05$).

Footnote: Many Mendelian crosses involve ratios other than 1:1. For these, calculation of the expected values is not simply a division of the total by the number of categories. Instead, the total must be apportioned according to the ratio. For example, for a total of 400 as above, in a predicted 9:3:3:1 ratio, the total count must be divided by 16 (9+3+3+1) and the expected values will be 225: 75: 75: 25 in each category.

Table 1: Critical values of χ^2 at different levels of probability. By convention, the critical probability for rejecting the null hypothesis (H$_0$) is 5%. If the test statistic is less than the tabulated critical value for $P = 0.05$ we cannot reject H$_0$ and the result is not significant. If the test statistic is greater than the tabulated value for $P = 0.05$ we reject H$_0$ in favour of the alternative hypothesis.

Degrees of freedom	Level of probability (P)									
	0.98	0.95	0.80	0.50	0.20	0.10	0.05	0.02	0.01	0.001
1	0.001	0.004	0.064	0.455	1.64	2.71	3.84	5.41	6.64	10.83
2	0.040	0.103	0.466	1.386	3.22	4.61	5.99	7.82	9.21	13.82
3	0.185	0.352	1.005	2.366	4.64	6.25	7.82	9.84	11.35	16.27
4	0.429	0.711	1.649	3.357	5.99	7.78	9.49	11.67	13.28	18.47
5	0.752	0.145	2.343	4.351	7.29	9.24	11.07	13.39	15.09	20.52

χ^2 (between 0.50 and 0.20 columns, row 1: 0.455, 1.64)

← Do not reject H$_0$ Reject H$_0$ →

267 Chi-Squared Exercise in Genetics

Key Idea: The following problems examine the use of the chi-squared test for goodness of fit (χ^2) in genetics.

A worked example illustrating the use of the chi-squared test for a genetic cross is provided on the previous page.

1. In a tomato plant experiment, two heterozygous individuals were crossed (the details of the cross are not relevant here). The predicted Mendelian ratios for the offspring of this cross were **9:3:3:1** for each of the **four following phenotypes**: purple stem-jagged leaf edge, purple stem-smooth leaf edge, green stem-jagged leaf edge, green stem-smooth leaf edge.

The observed results of the cross were not exactly as predicted.
The numbers of offspring with each phenotype are provided below:

Observed results of the tomato plant cross			
Purple stem-jagged leaf edge	12	Green stem-jagged leaf edge	8
Purple stem-smooth leaf edge	9	Green stem-smooth leaf edge	0

(a) State your null hypothesis for this investigation (H$_0$): _____

_____

(b) State the alternative hypothesis (H$_A$): _____

_____

2. Use the chi-squared (χ^2) test to determine if the differences between the observed and expected phenotypic ratios are significant. Use the table of critical values of χ^2 at different P values on the previous page.

(a) Enter the observed values (number of individuals) and complete the table to calculate the χ^2 value:

Category	O	E	O — E	(O — E)^2	$\frac{(O - E)^2}{E}$
Purple stem, jagged leaf					
Purple stem, smooth leaf					
Green stem, jagged leaf					
Green stem, smooth leaf					
	Σ			Σ	

(b) Calculate χ^2 value using the equation:

$$\chi^2 = \sum \frac{(O - E)^2}{E} \qquad \chi^2 = \underline{\hspace{2cm}}$$

(c) Calculate the degrees of freedom: _____

(d) Using the χ^2 table, state the P value corresponding to your calculated χ^2 value:

_____

(e) State your decision: *(circle one)*

reject H$_0$ / do not reject H$_0$

3. Students carried out a pea plant experiment, where two heterozygous individuals were crossed. The predicted Mendelian ratios for the offspring were **9:3:3:1** for each of the **four following phenotypes**: round-yellow seed, round-green seed, wrinkled-yellow seed, wrinkled-green seed.

The observed results were as follows:

Round-yellow seed	441	Wrinkled-yellow seed	143
Round-green seed	159	Wrinkled-green seed	57

Use a separate piece of paper to complete the following:

(a) State the null and alternative hypotheses (H$_0$ and H$_A$).

(b) Calculate the χ^2 value.

(c) Calculate the degrees of freedom and state the P value corresponding to your calculated χ^2 value.

(d) State whether or not you reject your null hypothesis: reject H$_0$ / do not reject H$_0$ (circle one)

4. Comment on the whether the χ^2 values obtained above are similar. Suggest a reason for any difference:

_____

_____

_____

© 2012-2014 **BIOZONE** International
ISBN: 978-1-927173-93-0
Photocopying Prohibited

268 Polygenes

Key Idea: Many phenotypes are affected by multiple genes. Some phenotypes (e.g. kernel colour in maize and skin colour in humans) are determined by more than one gene and show **continuous variation** in a population. The production of the skin pigment melanin in humans is controlled by at least three genes. The amount of melanin produced is directly proportional to the number of dominant alleles for either gene (from 0 to 6).

Very pale	Light	Medium light	Medium	Medium dark	Dark	Black
0	**1**	**2**	**3**	**4**	**5**	**6**

A light-skinned person   A dark-skinned person

There are seven shades skin colour ranging from very dark to very pale, with most individual being somewhat intermediate in skin colour. No dominant allele results in a lack of dark pigment (aabbcc). Full pigmentation (black) requires six dominant alleles (AABBCC).

1. Complete the Punnett square for the F_2 generation (below) by entering the genotypes and the number of dark alleles resulting from a cross between two individuals of intermediate skin colour. Colour-code the offspring appropriately for easy reference.

   (a) How many of the 64 possible offspring of this cross will have darker skin than their parents:

   _____

   (b) How many genotypes are possible for this type of gene interaction:

   _____

Parental generation

X

Black (AABBCC)   Pale (aabbcc)

Medium (AaBbCc)

2. Explain why we see many more than seven shades of skin colour in reality:

   _____

   _____

F_2 generation (AaBbCc X AaBbCc)

GAMETES	ABC	ABc	AbC	Abc	aBC	aBc	abC	abc
ABC								
ABc								
AbC								
Abc								
aBC								
aBc								
abC								
abc								

LINK 260 LINK 215 WEB 268 KNOW

3. Discuss the differences between **continuous** and **discontinuous** variation, giving examples to illustrate your answer:

_____

_____

_____

_____

_____

4 From a sample of no less than 30 adults, collect data (by request or measurement) for one continuous variable (e.g. height, weight, foot length, or hand span). Record and tabulate your results in the space below, and then plot a frequency histogram of the data on the grid below:

Raw data

Tally Chart (frequency table)

Variable: _____

Frequency

(a) Calculate each of the following for your data. See *Descriptive Statistics* if you need help and attach your working:

Mean: _____   Mode: _____   Median: _____

Standard deviation: _____

(b) Describe the pattern of distribution shown by the graph, giving a reason for your answer: _____

_____

(c) What is the genetic basis of this distribution? _____

_____

(d) What is the importance of a large sample size when gathering data relating to a continuous variable?

_____

269 Gene Pools and Evolution

Key Idea: The proportions of alleles in a gene pool can be altered by the processes that increase or decrease variation. This page illustrates the dynamic nature of **gene pools**. It portrays two populations of one hypothetical beetle species. Each beetle is a 'carrier' of genetic information, represented here by the alleles (A and a) for a single gene that controls colour and has a dominant/recessive expression pattern. There are normally two phenotypes: black and pale. Mutations may create other versions of the phenotype. Some of the microevolutionary processes that can affect the genetic composition (**allele frequencies**) of the gene pool are illustrated.

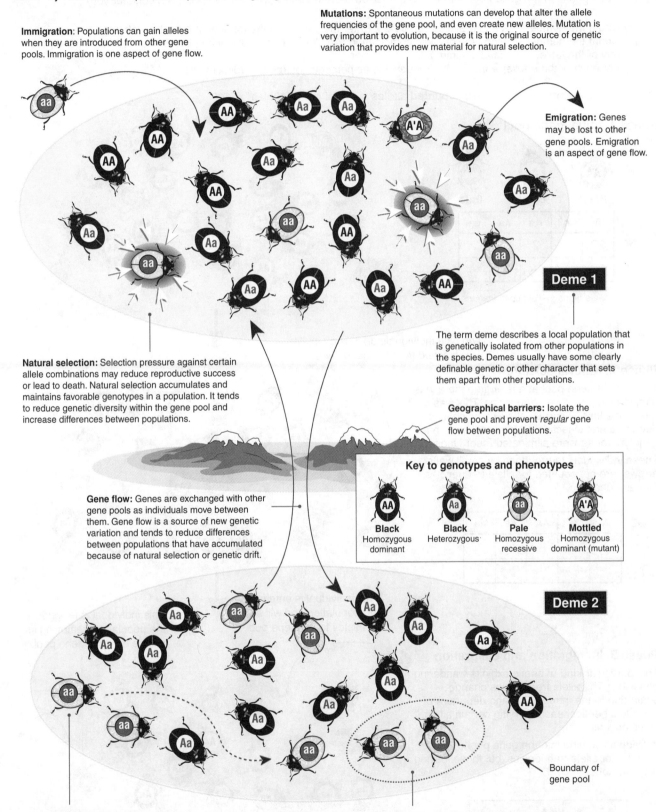

Immigration: Populations can gain alleles when they are introduced from other gene pools. Immigration is one aspect of gene flow.

Mutations: Spontaneous mutations can develop that alter the allele frequencies of the gene pool, and even create new alleles. Mutation is very important to evolution, because it is the original source of genetic variation that provides new material for natural selection.

Emigration: Genes may be lost to other gene pools. Emigration is an aspect of gene flow.

Deme 1

The term deme describes a local population that is genetically isolated from other populations in the species. Demes usually have some clearly definable genetic or other character that sets them apart from other populations.

Natural selection: Selection pressure against certain allele combinations may reduce reproductive success or lead to death. Natural selection accumulates and maintains favorable genotypes in a population. It tends to reduce genetic diversity within the gene pool and increase differences between populations.

Geographical barriers: Isolate the gene pool and prevent *regular* gene flow between populations.

Key to genotypes and phenotypes

AA — **Black** Homozygous dominant

Aa — **Black** Heterozygous

aa — **Pale** Homozygous recessive

A'A — **Mottled** Homozygous dominant (mutant)

Gene flow: Genes are exchanged with other gene pools as individuals move between them. Gene flow is a source of new genetic variation and tends to reduce differences between populations that have accumulated because of natural selection or genetic drift.

Deme 2

Boundary of gene pool

Mate choice (non-random mating): Individuals may not select their mate randomly and may seek out particular phenotypes, increasing the frequency of these "favoured" alleles in the population.

Genetic drift: Chance events can cause the allele frequencies of small populations to "drift" (change) randomly from generation to generation. Genetic drift can play a significant role in the microevolution of very small populations. The two situations most often leading to populations small enough for genetic drift to be significant are the **bottleneck effect** (where the population size is dramatically reduced by a catastrophic event) and the **founder effect** (where a small number of individuals colonize a new area).

REFER

270 Changes in a Gene Pool

Key Idea: Natural selection and migration can alter the allele frequencies in gene pools.

The diagram below shows an hypothetical population of beetles undergoing changes as it is subjected to two 'events'. The three phases represent a progression in time (i.e. the same gene pool, undergoing change). The beetles have two phenotypes (black and pale) determined by the amount of pigment deposited in the cuticle. The gene controlling this character is represented by two alleles **A** and **a**. Your task is to analyse the gene pool as it undergoes changes.

1. For each phase in the gene pool below fill in the tables provided as follows; (some have been done for you):

    (a) Count the number of A and a alleles separately. Enter the count into the top row of the table (left hand columns).
    (b) Count the number of each type of allele combination (AA, Aa and aa) in the gene pool. Enter the count into the top row of the table (right hand columns).
    (c) For each of the above, work out the frequencies as percentages (bottom row of table):

    Allele frequency = No. counted alleles ÷ Total no. of alleles x 100

Phase 1: Initial gene pool

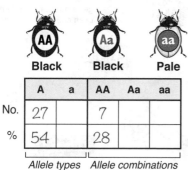

Black   Black   Pale

	A	a	AA	Aa	aa
No.	27		7		
%	54		28		

Allele types   *Allele combinations*

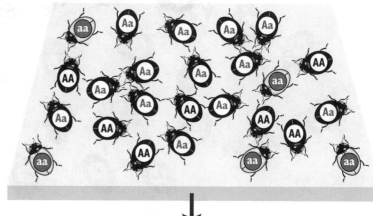

Two pale individuals died. Their alleles are removed from the gene pool.

Phase 2: Natural selection

In the same gene pool at a later time there was a change in the allele frequencies. This was due to the loss of certain allele combinations due to natural selection. Some of those with a genotype of aa were eliminated (poor fitness).

These individuals (surrounded by small white arrows) are not counted for allele frequencies; they are dead!

	A	a	AA	Aa	aa
No.					
%					

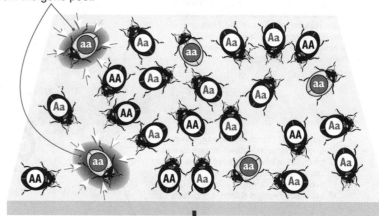

This individual is entering the population and will add its alleles to the gene pool.

This individual is leaving the population, removing its alleles from the gene pool.

Phase 3: Immigration and emigration

This particular kind of beetle exhibits wandering behaviour. The allele frequencies change again due to the introduction and departure of individual beetles, each carrying certain allele combinations.

Individuals coming into the gene pool (AA) are counted for allele frequencies, but those leaving (aa) are not.

	A	a	AA	Aa	aa
No.					
%					

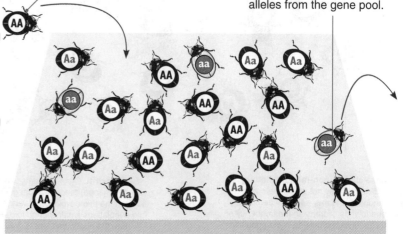

© 2012-2014 **BIOZONE** International
ISBN: 978-1-927173-93-0
Photocopying Prohibited

271 Natural Selection

Key Idea: Natural selection acts on phenotypes and results in the differential survival of some genotypes over others. It is an important cause of changes in gene pools.

Natural selection operates on the phenotypes of individuals, produced by their particular combinations of alleles. It results in the differential survival of some genotypes over others. As a result, organisms with phenotypes most suited to the prevailing environment are more likely to survive and breed than those with less suited phenotypes. Favourable phenotypes will become relatively more numerous and than unfavourable phenotypes. Over time, natural selection may lead to a permanent change in the genetic makeup of a population. Natural selection is always linked to phenotypic suitability in the prevailing environment so it is a dynamic process. It may favour existing phenotypes or shift the phenotypic median, as is shown in the diagrams below. The top row of diagrams below represents the population phenotypic spread before selection, and the bottom row the spread afterwards.

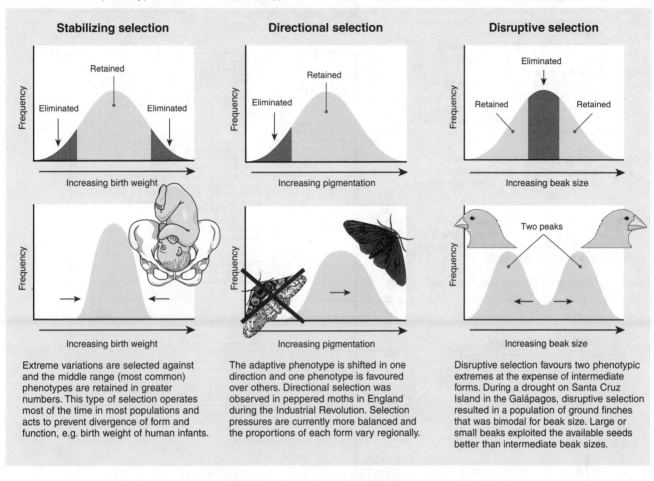

Stabilizing selection	Directional selection	Disruptive selection

Extreme variations are selected against and the middle range (most common) phenotypes are retained in greater numbers. This type of selection operates most of the time in most populations and acts to prevent divergence of form and function, e.g. birth weight of human infants.

The adaptive phenotype is shifted in one direction and one phenotype is favoured over others. Directional selection was observed in peppered moths in England during the Industrial Revolution. Selection pressures are currently more balanced and the proportions of each form vary regionally.

Disruptive selection favours two phenotypic extremes at the expense of intermediate forms. During a drought on Santa Cruz Island in the Galápagos, disruptive selection resulted in a population of ground finches that was bimodal for beak size. Large or small beaks exploited the available seeds better than intermediate beak sizes.

1. Explain why fluctuating (as opposed to stable) environments favour disruptive selection:

_____

_____

_____

2. Disruptive selection can be important in the formation of new species:

(a) Describe the evidence from the ground finches on Santa Cruz Island that provides support for this statement:

_____

_____

(b) The ground finches on Santa Cruz Island are one interbreeding population with a strongly bimodal distribution for the character of beak size. Suggest what conditions could lead to the two phenotypic extremes diverging further:

_____

(c) Predict the consequences of the end of the drought and an increased abundance of medium size seeds as food:

_____

_____

© 2012-2014 **BIOZONE** International
ISBN: 978-1-927173-93-0
Photocopying Prohibited

272 Selection for Human Birth Weight

Key Idea: Stabilizing selection operates to keep human birth weight within relatively narrow constraints.

Selection pressures operate on populations in such a way as to reduce mortality. For humans, giving birth is a special, but often traumatic, event. In a study of human birth weights

it is possible to observe the effect of selection pressures operating to constrain human birth weight within certain limits. This is a good example of **stabilizing selection**. This activity explores the selection pressures acting on the birth weight of human babies. Carry out the steps below:

Step 1: Collect the birth weights from 100 birth notices from your local newspaper (or 50 if you are having difficulty getting enough; this should involve looking back through the last 2-3 weeks of birth notices). If you cannot obtain birth weights in your local newspaper, a set of 100 sample birth weights is provided in the Model Answers booklet.

Step 2: Group the weights into each of the 12 weight classes (of 0.5 kg increments). Determine what percentage (of the total sample) fall into each weight class (e.g. 17 babies weigh 2.5-3.0 kg out of the 100 sampled = 17%)

Step 3: Graph these as a histogram for the 12 weight classes (use the grid provided right). Be sure to use the scale provided on the left vertical (y) axis.

Step 4: Create a second graph by plotting percentage mortality of newborn babies for each birth weight. Use the scale on the right y axis and data provided (below).

Step 5: Connect these points to make a line graph.

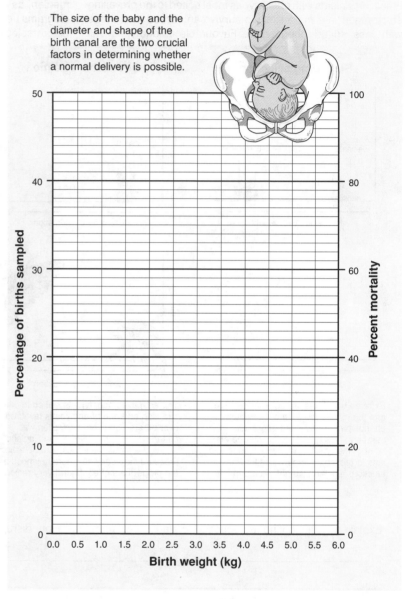

The size of the baby and the diameter and shape of the birth canal are the two crucial factors in determining whether a normal delivery is possible.

Mortality of newborn babies related to birth weight

Weight (kg)	Mortality (%)
1.0	80
1.5	30
2.0	12
2.5	4
3.0	3
3.5	2
4.0	3
4.5	7
5.0	15

Source: Biology: The Unity & Diversity of Life (4th ed), by Starr and Taggart

1. Describe the shape of the histogram for birth weights: _____

2. What is the optimum birth weight in terms of the lowest newborn mortality?_____

3. Describe the relationship between newborn mortality and birth weight: _____

_____

4. Describe the selection pressures that are operating to control the range of birth weight: _____

_____

_____

5. How might have modern medical intervention during pregnancy and childbirth altered these selection pressures?

_____

_____

© 2012-2014 **BIOZONE** International
ISBN: 978-1-927173-93-0

273 Selection for Skin Colour in Humans

Key Idea: Skin colour is the result of a dynamic balance between two different selection pressures linked to fitness. Pigmented skin of varying tones is a feature of humans that evolved after early humans lost most of their body hair. The distribution of skin colour globally is not random. People native to equatorial regions have darker skin tones than people from higher latitudes. For many years, biologists postulated that this was because darker skins had evolved to protect against skin cancer. The problem with this explanation was that skin cancer is not tied to evolutionary fitness because it affects post-reproductive individuals and cannot therefore provide a mechanism for selection. More recent in-depth analyses have shown a more complex picture in which selection pressures on skin colour are finely balanced to produce a skin tone that regulates the effects of the sun's UV radiation on the nutrients vitamin D and folate, both of which are crucial to successful reproduction, and therefore fitness. The selection is stabilising within each latitudinal region.

Skin Colour in Humans: A Product of Natural Selection

Alaska   France   The Netherlands   Iraq   China   Japan

80°   No data
Insufficient UV most of year
40°
Insufficient UV one month
Sufficient UV all year
0°
Sufficient UV all year
Insufficient UV one month
40°
Insufficient UV most of year

80°

40°

0°

40°

Adapted from Jablonski & Chaplin, Sci. Am. Oct. 2002

Peru   Liberia   Burundi   Botswana   Southern India   Malaysia

Human skin colour is the result of two opposing selection pressures. Skin pigmentation has evolved to protect against destruction of folate by ultraviolet light, but the skin must also be light enough to receive the light required to synthesize vitamin D. Vitamin D synthesis is a process that begins in the skin and is inhibited by dark pigment. Folate is needed for healthy neural development in humans and a deficiency is associated with fatal neural tube defects. Vitamin D is required for the absorption of calcium from the diet and therefore normal skeletal development.

Women also have a high requirement for calcium during pregnancy and lactation. Populations that live in the tropics receive enough ultraviolet (UV) radiation to synthesize vitamin D all year long. Those that live in northern or southern latitudes do not. In temperate zones, people lack sufficient UV light to make vitamin D for one month of the year. Those nearer the poles lack enough UV light for vitamin D synthesis most of the year (above). Their lighter skins reflect their need to maximize UV absorption (the photos show skin colour in people from different latitudes).

LINK 271   LINK 268   LINK 143   LINK 142   WEB 273   APP

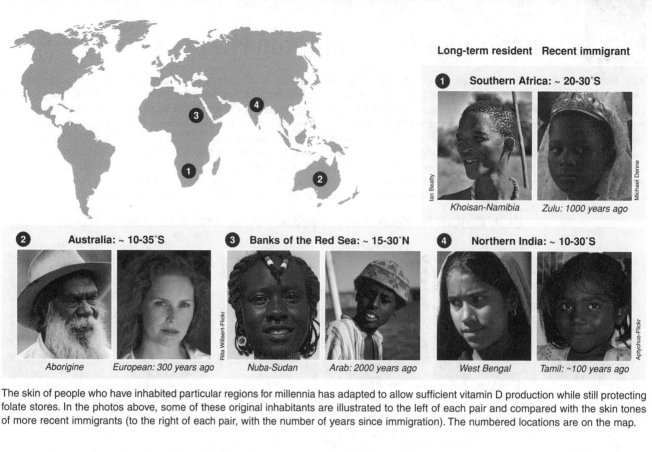

Long-term resident   Recent immigrant

1 **Southern Africa: ~ 20-30°S**

Khoisan-Namibia   *Zulu: 1000 years ago*

2 **Australia: ~ 10-35°S**

Aborigine   *European: 300 years ago*

3 **Banks of the Red Sea: ~ 15-30°N**

Nuba-Sudan   *Arab: 2000 years ago*

4 **Northern India: ~ 10-30°S**

West Bengal   *Tamil: ~100 years ago*

The skin of people who have inhabited particular regions for millennia has adapted to allow sufficient vitamin D production while still protecting folate stores. In the photos above, some of these original inhabitants are illustrated to the left of each pair and compared with the skin tones of more recent immigrants (to the right of each pair, with the number of years since immigration). The numbered locations are on the map.

1. (a) Describe the role of folate in human physiology: _____

    (b) Describe the role of vitamin D in human physiology: _____

2. (a) Early hypotheses to explain skin colour linked pigmentation level only to the degree of protection it gave from UV-induced skin cancer. Explain why this hypothesis was inadequate in accounting for how skin colour evolved:

    _____

    _____

    (b) Explain how the new hypothesis for the evolution of skin colour overcomes these deficiencies: _____

    _____

    _____

    _____

3. Explain why, in any given geographical region, women tend to have lighter skins (by 3-4% on average) than men:

    _____

    _____

4. The Inuit people of Alaska and northern Canada have a diet rich in vitamin D and their skin colour is darker than predicted on the basis of UV intensity at their latitude. Explain this observation:

    _____

    _____

5. (a) What health problems might be expected for people of African origin now living in Canada?

    _____

    _____

    (b) How could these people avoid these problems in their new higher latitude environment? _____

    _____

    _____

© 2012-2014 **BIOZONE** International
ISBN: 978-1-927173-93-0
Photocopying Prohibited

274 Disruptive Selection in Darwin's Finches

Key Idea: Disruptive selection in the finch *Geospiza fortis* produces a bimodal distribution for beak size.

The Galápagos Islands, 970 km west of Ecuador, are home to the finch species *Geospiza fortis*. A study during a prolonged drought on Santa Cruz Island showed how **disruptive selection** can change the distribution of genotypes in a population. During the drought, large and small seeds were more abundant than the preferred intermediate seed size.

Beak sizes of *G. fortis* were measured over a three year period (2004-2006), at the start and end of each year. At the start of the year, individuals were captured, banded, and their beaks were measured.

The presence or absence of banded individuals was recorded at the end of the year when the birds were recaptured. Recaptured individuals had their beaks measured.

The proportion of banded individuals in the population at the end of the year gave a measure of fitness. Absent individuals were presumed dead (fitness = 0).

Fitness related to beak size showed a bimodal distribution (right) typical of disruptive selection.

Beak size vs fitness in *Geospiza fortis*

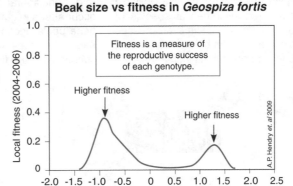

*Fitness showed a **bimodal distribution** (arrowed) being highest for smaller and larger beak sizes.*

Measurements of the beak length, width, and depth were combined into one **single measure**.

Beak Size Pairing in *Geospiza fortis*

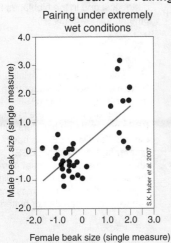
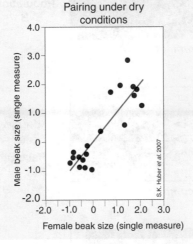

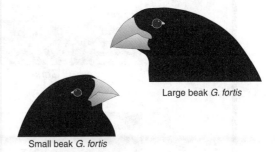

A 2007 study found that breeding pairs of birds had similar beak sizes. Male and females with small beaks tended to breed together, and males and females with large beaks tended to breed together. Mate selection maintained the bimodal distribution in the population during extremely wet conditions. If beak size wasn't a factor in mate selection, the beak size would even out.

1. (a) How did the drought affect seed size on Santa Cruz Island? _____

    (b) How did the change in seed size during the drought create a selection pressure for changes in beak size?

2. How does beak size relate to fitness (differential reproductive success) in *G. fortis*? _____

3. (a) Is mate selection in *G. fortis* random / non-random? (delete one)

    (b) Give reasons for your answer: _____

LINK 271 LINK 145 LINK 143 LINK 142 WEB 274 APP

275 How Species Form

Key Idea: When populations are separated, gene flow is reduced. Continual reduction in gene flow by isolating mechanisms eventually leads to the formation of new species. Species evolve in response to the selection pressures of the environment. The diagram below represents a possible sequence for the evolution of two hypothetical species of butterfly from an ancestral population. As time progresses (from the top to the bottom of the diagram below) the amount of genetic difference between the populations increases, with each group becoming increasingly isolated from the other. The isolation of two gene pools from one another may begin with geographical barriers. This may be followed by isolating mechanisms that occur before the production of a zygote (e.g. behavioural changes), and isolating mechanisms that occur after a zygote is formed (e.g. hybrid sterility). As the two gene pools become increasingly isolated and different from each other, they are progressively labelled: population, race, subspecies, and finally species.

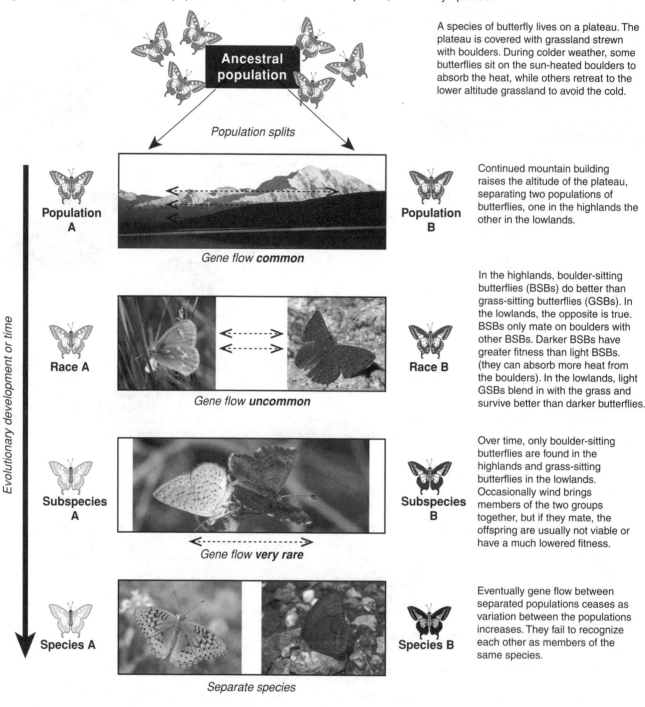

A species of butterfly lives on a plateau. The plateau is covered with grassland strewn with boulders. During colder weather, some butterflies sit on the sun-heated boulders to absorb the heat, while others retreat to the lower altitude grassland to avoid the cold.

Population splits

Population A — Gene flow **common** — Population B

Continued mountain building raises the altitude of the plateau, separating two populations of butterflies, one in the highlands the other in the lowlands.

Race A — Gene flow **uncommon** — Race B

In the highlands, boulder-sitting butterflies (BSBs) do better than grass-sitting butterflies (GSBs). In the lowlands, the opposite is true. BSBs only mate on boulders with other BSBs. Darker BSBs have greater fitness than light BSBs. (they can absorb more heat from the boulders). In the lowlands, light GSBs blend in with the grass and survive better than darker butterflies.

Subspecies A — Gene flow **very rare** — Subspecies B

Over time, only boulder-sitting butterflies are found in the highlands and grass-sitting butterflies in the lowlands. Occasionally wind brings members of the two groups together, but if they mate, the offspring are usually not viable or have a much lowered fitness.

Species A — Separate species — Species B

Eventually gene flow between separated populations ceases as variation between the populations increases. They fail to recognize each other as members of the same species.

Evolutionary development or time

1. (a) Identify the variation in behaviour in the original butterfly population: _____

_____

(b) What were the selection pressures acting on BSBs in the highlands and GSBs in the lowlands respectively?

_____

_____

© 2012-2014 **BIOZONE** International
ISBN: 978-1-927173-93-0
Photocopying Prohibited

276 Reproductive Isolation

Key Idea: Reproductive isolation maintains separate species by preventing gene flow between populations.

Isolating mechanisms are barriers to successful interbreeding between species. Reproductive isolation is fundamental to the biological species concept, which defines a species by its inability to breed with other species to produce fertile offspring. Prezygotic isolating mechanisms act before fertilization occurs, preventing species ever mating, whereas postzygotic barriers take effect after fertilization. Reproductive isolation prevents interbreeding (and therefore

gene flow) between species. Any factor that impedes two species from producing viable, fertile hybrids contributes to reproductive isolation. Single barriers may not completely stop gene flow, so most species commonly have more than one type of barrier. Single barriers to reproduction (including geographical barriers) often precede the development of a suite of reproductive isolating mechanisms (RIMs). Most operate before fertilization (prezygotic RIMs) with postzyotic RIMs being important in preventing offspring between closely related species.

Geographical Isolation

Geographical isolation describes the isolation of a species population (gene pool) by some kind of physical barrier, e.g. mountain range, water body, isthmus, desert, or ice sheet. Geographical barriers are not regarded as reproductive isolating mechanisms because they are not part of the species' biology, although they are often a necessary precursor to reproductive isolation in sexually reproducing populations. Geographical isolation is a frequent first step in the subsequent reproductive isolation of a species. For example, geologic changes to the lake basins have been instrumental in the subsequent proliferation of cichlid fish species in the rift lakes of East Africa (right). Similarly, many Galápagos Island species (e.g. iguanas, finches) are now quite distinct from the Central and South American species from which they arose after isolation from the mainland.

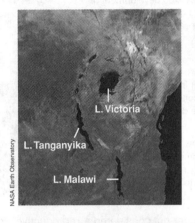
L. Victoria
L. Tanganyika
L. Malawi

Malawi cichlid species

Reproductive Isolating Mechanisms

Temporal Isolation:

Individuals from different species do not mate because they are active during different times of the day, or in different seasons. Plants flower at different times of the year or even at different times of the day to avoid hybridization (e.g. members of the orchid genus *Dendrobium*, which occupy the same location and flower on different days). Closely related animal species may have quite different breeding seasons or periods of emergence. Periodical cicadas (right) of the genus *Magicicada* are so named because members of each species in a particular region are developmentally synchronized, despite very long life cycles. Once their underground period of development (13 or 17 years depending on the species) is over, the entire population emerges at much the same time to breed.

Behavioural Isolation:

Behavioural isolation operates through differences in species courtship behaviours. Courtship is a necessary prelude to mating in many species and courtship behaviours are species specific. Mates of the same species are attracted with distinctive, usually ritualized, dances, vocalizations, and body language. Because they are not easily misinterpreted, the courtship behaviours of one species will be unrecognized and ignored by individuals of another species. Birds exhibit a remarkable range of courtship displays. The use of song is widespread but ritualized movements, including nest building, are also common. For example, the elaborate courtship bowers of bowerbirds are well known, and Galápagos frigatebirds have an elaborate display in which they inflate a bright red gular pouch (right).

Albatross courtship

Male tree frog calling

Male frigatebird courtship display

Wing beating in male sage grouse

Mechanical Isolation:

Structural differences (incompatibility) in the anatomy of reproductive organs prevents sperm transfer between individuals of different species. This is an important isolating mechanism preventing breeding between closely related species of arthropods. Many flowering plants have coevolved with their animal pollinators and have flower structures to allow only that insect access. Structural differences in the flowers and pollen of different plant species prevents cross breeding because pollen transfer is restricted to specific pollinators and the pollen itself must be species compatible.

Damselflies mating

Complex flowers in orchids

LINK 275   LINK 116   KNOW

Postzygotic Isolating Mechanisms

Postzygotic isolating mechanisms operate after fertilization and are important in preventing offspring between closely related species. They involve a mismatch of chromosomes in the zygote.

Hybrid Sterility
Hybrid sterility may occur due to the failure of meiosis to produce normal gametes in the hybrid. This can occur if the chromosomes of the two parents are different in number or structure (see the **"zebronkey"** karyotype on the right).

Hybrid Inviability
Mating between individuals of two species may produce a zygote, but genetic incompatibility may stop development of the zygote. Fertilized eggs often fail to divide because of mismatched chromosome numbers from each gamete.

Hybrid Breakdown
Hybrid breakdown is a common feature of some plant hybrids. The first generation (F_1) may be fertile, but the second generation (F_2) are infertile or inviable. Examples include hybrids between species of cotton (near right), species of *Populus*, and strains of the cultivated rice *Oryza* (far right).

In plants, hybridization can lead to new species formation if there is a doubling of the chromosome number during meiosis. The new plant is immediately reproductively isolated from the parent species due to differences in chromosome number.

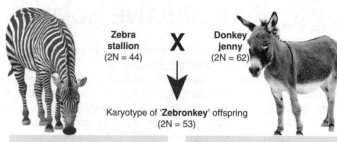

Zebra stallion (2N = 44) X Donkey jenny (2N = 62)

Karyotype of 'Zebronkey' offspring (2N = 53)

Chromosomes contributed by zebra stallion

Chromosomes contributed by donkey jenny

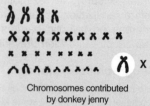

1. (a) Why is a geographical barrier not considered a reproductive isolating mechanism? _____

    _____

    _____

   (b) Identify some geographical barriers that could separate populations: _____

    _____

   (c) Why is geographic isolation often an important first step in species formation? _____

    _____

    _____

2. Explain how temporal isolation stops closely related species from interbreeding: _____

    _____

    _____

    _____

3. Explain why many animals have courtship displays and how this prevents breeding between species:

    _____

    _____

    _____

    _____

4. How does the structure of some orchids isolate them from other species of orchid? _____

    _____

    _____

5. What is the name given to reproductive isolating mechanisms that operate before fertilization? _____

    _____

© 2012-2014 **BIOZONE** International
ISBN: 978-1-927173-93-0
Photocopying Prohibited

277 Comparing Isolated Populations

Key Idea: The allele frequencies of geographically separated populations will diverge from each other over time.

Probably the most common mechanism by which new species arise is by **geographic isolation**. Originally a species may move freely throughout its range. However a geologic disturbance, such as mountain building or river diversion, may produce a physical barrier that isolates one part of the species population from another. Over time the gene pools of the separated (allopatric) populations may become very different (allopatric speciation). Populations may also become geographically isolated when a small part of the population migrates to a new area (either deliberately or accidentally) and establish a new population that is geographically isolated from the original population (known as the founder effect).

Isolation in Beetle Populations

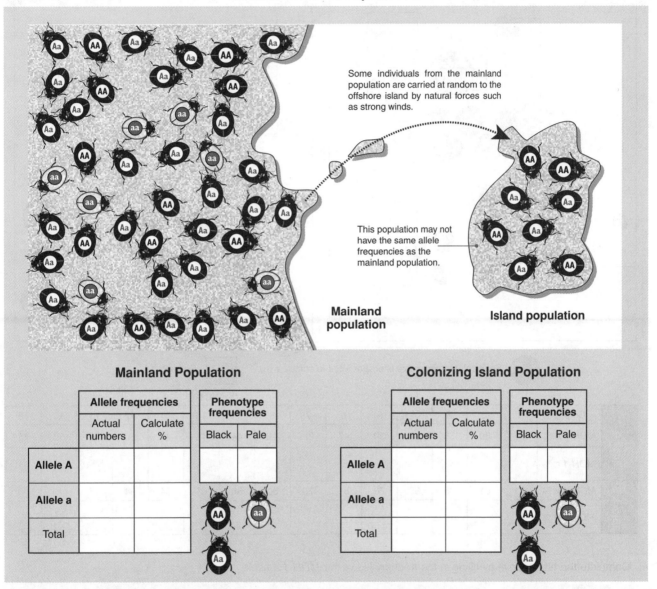

Some individuals from the mainland population are carried at random to the offshore island by natural forces such as strong winds.

This population may not have the same allele frequencies as the mainland population.

Mainland population

Island population

Mainland Population

	Allele frequencies		Phenotype frequencies	
	Actual numbers	Calculate %	Black	Pale
Allele A				
Allele a				
Total				

Colonizing Island Population

	Allele frequencies		Phenotype frequencies	
	Actual numbers	Calculate %	Black	Pale
Allele A				
Allele a				
Total				

1. Compare the mainland population to the population which ended up on the island (use the spaces in the tables above):
   (a) Count the **phenotype** numbers for the two populations (i.e. the number of black and pale beetles).
   (b) Count the **allele** numbers for the two populations: the number of dominant alleles (A) and recessive alleles (a). Calculate these as a percentage of the total number of alleles for each population.

2. How are the allele frequencies of the two populations different? _____

3. Explain why the gene pools of isolated populations may become different from each other over time:

Microgeographic Isolation in Garden Snails

The European garden snail (*Cornu aspersum*, formerly *Helix aspersa*) is widely distributed throughout the world, both naturally and by human introduction. However because of its relatively slow locomotion and need for moist environments it can be limited in its habitat and this can lead to regional variation. The study below illustrates an investigation carried out on two snail populations in the city of Bryan, Texas. The snail populations covered two adjacent city blocks surrounded by tarmac roads.

The snails were found in several colonies in each block. Allele frequencies for the gene *MDH-1* (alleles A and a) were obtained and compared. Statistical analysis of the allele frequencies of the two populations showed them to be significantly different (P << 0.05). Note: A Mann-Whitney U test was used in this instance. It is similar to a Student's *t* test, but does not assume a normal distribution of data (it is non-parametric).

Block A   **Block B**

Road (not to sclae)

Source: Evolution, Vol 29. No. 3, 1975

Snail colony (circle size is proportional to colony size).   Building

	Colony	1	2	3	4	5	6	7	8	9	10	11	12	13	14	15
Block A	*MDH-1* A %	39	39	36	42	39	47	32	42	44	42	44	50	50	58	75
	MDH-1 a %															
Block B	*MDH-1* A %	81	61	75	68	70	61	70	60	58	61	54	54	47		
	MDH-1 a %															

4. Complete the table above by filling in the frequencies of the *MDH-1* a allele:

5. Suggest why these snail populations are effectively geographically isolated: _____

_____

_____

_____

6. Both the *MDH-1* alleles produce fully operative enzymes. Suggest why the frequencies of the alleles have become significantly different.

_____

_____

_____

7. Identify the colony in block A that appears to be isolated from the rest of the block itself: _____

278 The Rate of Evolutionary Change

Key Idea: New species may form gradually over a long period of time or appear suddenly following a period of stasis.

Two main models have been proposed for the rate at which evolution occurs: **gradualism** and **punctuated equilibrium**. It is likely that both mechanisms operate at different times and in different situations. Interpretations of the fossil record vary depending on the time scales involved. During its formative millennia, a species may have accumulated changes gradually (e.g. over 50,000 years). If that species survives for 5 million years, the evolution of its defining characteristics would have been compressed into just 1% of its evolutionary history. In the fossil record, the species would appear quite suddenly.

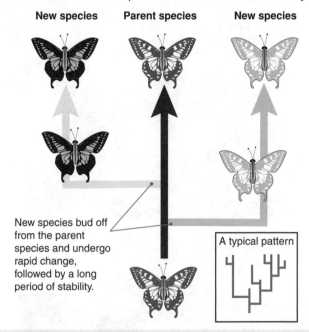

New species   Parent species   New species

New species bud off from the parent species and undergo rapid change, followed by a long period of stability.

A typical pattern

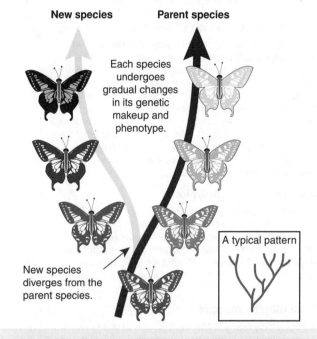

New species   Parent species

Each species undergoes gradual changes in its genetic makeup and phenotype.

New species diverges from the parent species.

A typical pattern

Punctuated Equilibrium

There is abundant evidence in the fossil record that, instead of gradual change, species stayed much the same for long periods of time (called stasis). These periods were punctuated by short bursts of evolution which produce new species quite rapidly. According to the punctuated equilibrium theory, most of a species' existence is spent in **stasis** and little time is spent in active evolutionary change. The stimulus for evolution occurs when some crucial aspect of the environment changes.

Phyletic Gradualism

Phyletic gradualism assumes that populations slowly diverge by accumulating adaptive characteristics in response to different selective pressures. If species evolve by gradualism, there should be transitional forms seen in the fossil record, as is seen with the evolution of the horse. Trilobites, an extinct marine arthropod, are another group of animals that have exhibited gradualism. In a study in 1987 a researcher found that they changed gradually over a three million year period.

1. Suggest the kinds of environments that would support the following paces of evolutionary change:

   (a) Punctuated equilibrium: _____

   _____

   _____

   (b) Gradualism: _____

   _____

   _____

2. In the fossil record of early human evolution, species tend to appear suddenly, linger for often very extended periods before disappearing suddenly. There are few examples of smooth inter-gradations from one species to the next. Explain which of the above models best describes the rate of human evolution:

   _____

3. Some species apparently show little evolutionary change over long periods of time (hundreds of millions of years).

   (a) Name two examples of such species: _____

   (b) State the term given to this lack of evolutionary change: _____

   (c) Suggest why such species have changed little over evolutionary time: _____

   _____

© 2012-2014 **BIOZONE** International
ISBN: 978-1-927173-93-0
Photocopying Prohibited

LINK 279   LINK 138   WEB 278

KNOW

279 Polyploidy and Evolution

Key Idea: An increase in the number of chromosome sets can result in instant speciation and is common in plants. Polyploidy is a condition in which an organism's cells contain three of more times the haploid number of chromosomes (3N or more). Polyploidy is rare in animals but common in plants. It may result in speciation without geographic separation of populations. Polyploidy occurs when chromosomes fail to separate properly during meiosis and are carried by only one gamete. Union with a normal N gamete produces a triploid (3N). Union with another 2N polyploid gamete produces a tetraploid (4N). An estimated 15% of angiosperm speciation events are accompanied by polyploidy events.

Instant Speciation by Polyploidy

Polyploidy may result in the formation of a new species without physical isolation from the parent species. This event, a result of faulty meiosis, produces sudden reproductive isolation because the polyploid is unable to interbreed with its diploid parent. Animals are rarely able to achieve new species status this way because the sex-determining mechanism is disturbed (they are effectively sterile, e.g. a tetraploid would have four X chromosomes). Many plants, on the other hand, are able to self-pollinate or reproduce vegetatively. This ability to reproduce on their own enables such polyploid plants to produce a viable population.

Speciation by Allopolyploidy

This type of polyploidy results from a hybridization of two species. The resulting hybrid may be sterile to begin with, but a doubling event during meiosis may produce a viable chromosome number. Self fertilization may then produce a fertile hybrid. **Examples:** Modern wheat. Swedes are a polyploid species formed as result of hybridization between a type of cabbage and a type of turnip.

Speciation by Autopolyploidy

Autopolyploidy refers to the multiplication of one basic set of chromosomes. It occurs when chromosomes fail to separate during meiosis or the cell fails to divide after chromatids have separated. Therefore the polyploid possess chromosomes from only one species (not two as above).

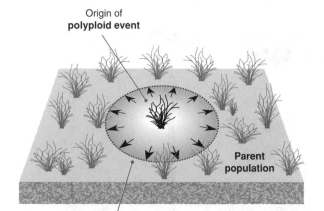

Origin of **polyploid event**

Parent population

New polyploid plant species spreads out through the existing parent population

Many commercial plant species are polyploids.

| Banana 3N = 27 | Boysenberry 7N = 49 | Strawberry 8N = 56 |

Polyploidy in *Allium*

Allium is a genus of plant in the family Amaryllidaceae. It includes about 750 species. The type species is *Allium sativum* (garlic) and related species include onion, shallot, leek and chive. *Allium* is notable because it includes many species which are polyploid. The basic chromosome number is N = 8 (sometimes 7) and most polyploids are therefore found as multiples of 8.

Garlic (*A. sativum*) 2N =16 chromosomes

Leek (*A. porrum*) 4N = 32 chromosomes

A. oleraceum 5N = 40 chromosomes

1. Using the example of *Allium*, explain how **polyploidy** can result in the formation of a new species:

   _____

   _____

2. Identify an example of a species that has been formed by polyploidy: _____

3. Explain the difference between allopolyploidy and autopolyploidy: _____

   _____

   _____

4. Why is polyploidy common in plants but extremely rare in animals? _____

   _____

   _____

© 2012-2014 **BIOZONE** International
ISBN: 978-1-927173-93-0

280 Chapter Review

Summarize what you know about this topic under the headings provided. You can draw diagrams or mind maps, or write short notes to organize your thoughts. Use the images and hints, included to help you:

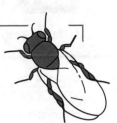

Meiosis

HINT. Describe the mechanisms that promote variation during meiosis.

Inheritance

HINT. Review dihybrid inheritance and linked genes.

Gene pools and speciation

HINT. How do gene pools change over time?

REVISE

281 KEY TERMS: Did You Get It?

1. Match each term to its definition, as identified by its preceding letter code.

chiasma _____

chromatid _____

crossing over _____

homologous chromosomes _____

interphase _____

meiosis _____

polyploidy _____

Punnett square _____

recombination _____

speciation _____

A Chromosome pairs, one paternal and one maternal, of the same length, centromere position, and staining pattern with genes for the same characteristics at corresponding loci.

B One of two identical DNA strands forming the chromosome and held together by the centromere after DNA replication.

C The process of double nuclear division (reduction division) to produce four nuclei, each containing half the original number of chromosomes (haploid).

D The exchange of alleles between homologous chromosomes as a result of crossing over.

E The point at which two non-sister chromatids join and exchange genetic material during crossing over.

F Event during meiosis where two homologous chromosomes exchange genetic material.

G The division of one species, during evolution, into two or more separate species.

H A graphical way of illustrating the outcome of a cross.

I The condition of having a chromosome complement of more than 2N (e.g. 3N).

J The stage in the cell cycle between divisions.

2. Explain the characteristics of each of the following types of natural selection and state when each might operate:

(a) Directional selection: _____

_____

(b) Stabilizing selection: _____

_____

(c) Disruptive selection: _____

_____

3.

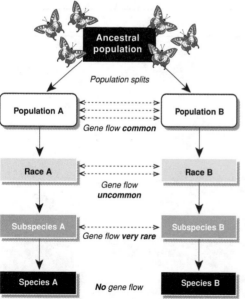

Fill in the boxes below to describe the mechanisms that lead to the formation of new species by allopatric speciation:

Geographic isolation:

Isolating mechanisms:

Species isolation:

© 2012-2014 **BIOZONE** International
ISBN: 978-1-927173-93-0

Animal Physiology

11.1 Antibody production and vaccination

Understandings, applications, skills

Activity number

☐ 1 Explain the basis of self recognition via the MHC antigens (also called HLA antigens). Describe the nature of pathogens and explain how pathogens invoke an immune response. Using examples, explain how pathogens may be species specific or may cross species barriers.   **282**

☐ 2 Describe the nature of the specific immune response in mammals, e.g. humans. Describe the role of B and T lymphocytes (cells). Describe B cell activation and the role of plasma cells and memory cells. Explain how clonal selection enables the immune system to respond to a large and unpredictable number of antigens.   **283 284**

☐ 3 Describe the general structure and role of antibodies.   **285**

☐ 4 Describe the ABO system for classifying blood groups on the basis of their surface antigens. Explain how this system is used to type blood for transfusion.   **286**

☐ 5 Explain the role of histamine in allergic reactions, including general effects and triggers for its release.   **287**

☐ 6 Explain the basis of immunity, identifying the primary and secondary responses to antigens. Explain the basis of vaccination and describe the properties of vaccines. Outline the role of vaccination in public health programmes. With reference to the eradication of smallpox, comment on the success of vaccination programmes against infectious disease.   **288 289**

☐ 7 Analyse epidemiological data related to vaccination programmes.   **290**

☐ 8 Describe the production and use of monoclonal antibodies.   **291**

11.2 Movement

Understandings, applications, skills

Activity number

☐ 1 Describe the role of bones and exoskeletons in movement.   **292**

☐ 2 Explain the role of joints in allowing movement of the skeleton. Annotate a diagram of a human elbow to show the structure of the joint. Label the bones, antagonistic muscles, cartilage, synovial fluid, and joint capsule.   **293**

☐ 3 Describe the action of antagonistic pairs of muscles in movement of the skeleton. Identify the action of antagonistic muscles in an insect leg.   **294**

☐ 4 Describe the structure of skeletal muscle in humans, noting the relationship between muscle, fascicles, and muscle fibres (cells). Describe the structure of skeletal muscle fibres, identifying the myofibrils and their contractile units (sarcomeres). Draw a labelled diagram to show the structure of a sarcomere, indicating the Z lines and the light and dark bands of actin and myosin filaments. Analyse electron micrographs to determine the state of contraction of skeletal muscle fibres.   **295**

☐ 5 Explain the sliding filament theory of muscle contraction, including the role of ATP, calcium ions, and the proteins tropomyosin and troponin.   **296**

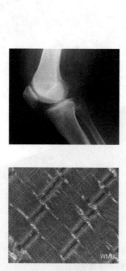

11.3 The kidney and osmoregulation

Understandings, applications, skills

Activity number

☐ 1 Explain why animals need to maintain a relatively steady state in terms of fluid and ion balance. Using examples, distinguish between osmoregulators and osmoconformers. — 297

☐ 2 Describe the water budget in a human as an example of an osmoregulator. Describe and explain the physiological consequences of dehydration and over-hydration (also called water intoxication or hyponatremia). — 298

☐ 3 Describe the type of nitrogenous wastes produced by animals and relate this to habitat and evolutionary history. — 299

☐ 4 Describe the general structure and function of the Malpighian tubule system in insects and kidneys in vertebrates (e.g. mammal). — 300

☐ 5 Draw and label a diagram of a human kidney, including its vascular supply. Explain differences in the composition of the blood in the renal artery and vein. — 301

☐ 6 Annotate a diagram of the kidney nephron to explain the role of each region in formation of the urine. Include reference to ultrafiltration in the glomerulus and Bowman's capsule, reabsorption and secretion in the convoluted tubules, and creation of a salt gradient in the medulla by the loop of Henle. Describe the relationship between the nephrons and the collecting ducts of the kidney. — 302

☐ 7 Describe the relationship between the length of the loop of Henle and ability to concentrate urine. Explain how this is related to water conservation in animals. — 303

☐ 8 Explain the role of ADH in controlling water reabsorption in the collecting duct. — 304

☐ 9 Describe how urinalysis is used to detect health disorders and drug use. — 305

☐ 10 Identify causes and consequences of kidney (renal) failure. Describe the treatment of kidney failure by hemodialysis or kidney transplant. — 306 307

11.4 Sexual Reproduction

Understandings, applications, skills

Activity number

☐ 1 Explain how sexual reproduction in animals involves gamete formation and union of male and female gametes in fertilization to form a zygote. Distinguish between internal and external fertilization and give examples of animals with each strategy. — 308

☐ 2 Describe the processes in spermatogenesis and oogenesis, distinguishing between the size and number of gametes produced. Annotate diagrams of the seminiferous tubule and ovary to show the stages of gametogenesis. — 309-311

☐ 3 Describe fertilization with reference to the acrosome reaction, fusion of the plasma membrane of the egg and sperm, and the cortical reaction. Explain how the processes involved in fertilization prevent polyspermy. Describe the early stages of embryonic development, including the role of early implantation of the blastocyst in the endometrium and the role of HCG. — 312

☐ 4 Describe the structure and role of the placenta in mammals (e.g. humans) explaining how its structure facilitates exchanges between the mother and fetus. Describe the role of the placenta as a temporary endocrine organ, identfying the hormones it produces and their roles. Explain the role of estrogen, oxytocin, and positive feedback in terminating the pregnancy and triggering birth. — 313 314

☐ 5 Describe the relationship between animal size and the development of the young at birth in mammals. Plot the position of the average human 38 week gestation on this graph and comment on where humans are positioned relative to other mammals. — 315

282 Targets for Defence

Key Idea: Cell surface MHC antigens enable the body to distinguish its own tissues from foreign material.

In order for the body to effectively defend against pathogens, it must first be able to recognize its own tissues (self) and distinguish itself from foreign tissue (e.g. bacteria). This is achieved through the presence of **antigens** (proteins on the surface of all cells). Foreign materials have different surface antigens to the host, so are identified for destruction by the immune system. In humans, self recognition is achieved through the **major histocompatibility complex**.

Distinguishing Self from Non-Self

The human immune system achieves self-recognition through the major histocompatibility complex (**MHC**). This is a cluster of tightly linked genes on chromosome 6. These genes code for protein molecules (MHC antigens) that are attached to the surface of body cells. They are used by the immune system to recognize its own or foreign material. Class I MHC antigens are found on the surfaces of almost all human cells. Class II MHC antigens occur only on macrophages and B-cells of the immune system.

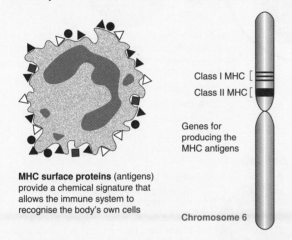

MHC surface proteins (antigens) provide a chemical signature that allows the immune system to recognise the body's own cells

Class I MHC [

Class II MHC [

Genes for producing the MHC antigens

Chromosome 6

MHC and Tissue Transplants

The MHC is responsible for the rejection of tissue grafts and organ transplants. Cells from donor tissue will have different MHC antigens to those of the recipient, and the donor tissue will be recognized as foreign and attacked (rejected) by the immune system. A number of factors have been involved in increasing the success of transplants in recent years. These include better tissue-typing and more effective immunosuppressant drugs, both of which decrease the MHC response.

How Pathogens Evade the MHC

Pathogens are disease causing organisms, and they have evolved many ways to avoid detection by the immune system. For example, HIV undergoes mutations that alter its surface antigens allowing it to avoid detection by the immune system (below).

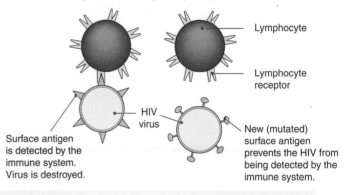

Lymphocyte

Lymphocyte receptor

HIV virus

Surface antigen is detected by the immune system. Virus is destroyed.

New (mutated) surface antigen prevents the HIV from being detected by the immune system.

Types of Pathogen

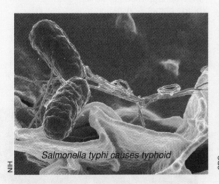

Salmonella typhi causes typhoid

NIH

Bangladeshi girl with smallpox (1973). Smallpox was eradicated from the country in 1977

CDC

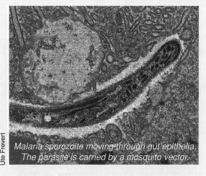

Malaria sporozoite moving through gut epithelia. The parasite is carried by a mosquito vector.

Ute Frevert

Most pathogens are bacteria (above) and viruses. Some pathogens are very specific and target only one host species, whereas other pathogens target many different hosts. For example, most bird flu viruses infect only birds, but some can also infect humans.

Viruses: Viruses cause many everyday diseases (e.g. the common cold), as well as more dangerous diseases (e.g. HIV, Ebola). Smallpox (above) is a viral disease that has been successfully eradicated through vaccination programmes.

Eukaryotic pathogens include fungi, algae, protozoa, and parasitic worms. They cause diseases such as malaria. Many are highly specialized parasites with a number of hosts, e.g the malaria parasite has a mosquito and a human host.

1. (a) Explain the nature and purpose of the **major histocompatibility complex** (MHC): _____

_____

_____

_____

(b) Explain the importance of such a self-recognition system: _____

_____

_____

© 2012-2014 **BIOZONE** International
ISBN: 978-1-927173-93-0
Photocopying Prohibited

LINK **182** LINK **177** WEB **282** **KNOW**

283 The Immune System

Key Idea: The defence provided by the immune system is based on its ability to respond specifically against foreign substances and hold a memory of this response.

There are two main components of the immune system: the humoral and the cell-mediated responses. They work separately and together to provide protection against disease. The **humoral immune response** is associated with the serum (the non-cellular part of the blood) and involves the action of antibodies secreted by B-cell lymphocytes. Antibodies are found in extracellular fluids including lymph, plasma, and mucus secretions. They protect the body against viruses, and bacteria and their toxins. The **cell-mediated immune response** is associated with the production of specialized lymphocytes called **T-cells**.

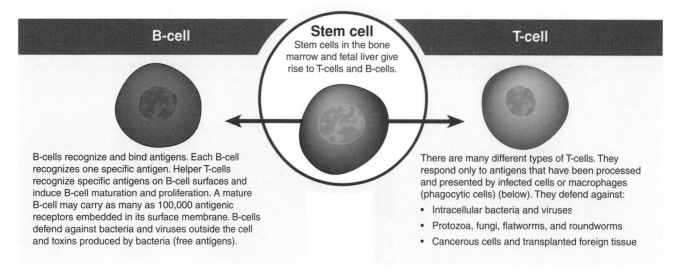

| B-cell | Stem cell | T-cell |

Stem cell
Stem cells in the bone marrow and fetal liver give rise to T-cells and B-cells.

B-cells recognize and bind antigens. Each B-cell recognizes one specific antigen. Helper T-cells recognize specific antigens on B-cell surfaces and induce B-cell maturation and proliferation. A mature B-cell may carry as many as 100,000 antigenic receptors embedded in its surface membrane. B-cells defend against bacteria and viruses outside the cell and toxins produced by bacteria (free antigens).

There are many different types of T-cells. They respond only to antigens that have been processed and presented by infected cells or macrophages (phagocytic cells) (below). They defend against:

- Intracellular bacteria and viruses
- Protozoa, fungi, flatworms, and roundworms
- Cancerous cells and transplanted foreign tissue

B-Cell and T-Cell Activation

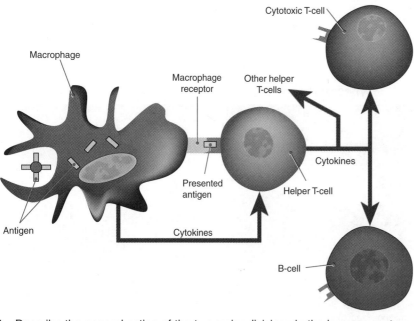

Cytotoxic T-cell

Macrophage

Macrophage receptor

Other helper T-cells

Presented antigen

Helper T-cell

Cytokines

Antigen

Cytokines

B-cell

Helper T-cells are activated by direct cell-to-cell signalling and by signalling to nearby cells using **cytokines** from macrophages.

Macrophages ingest antigens, process them, and present them on the cell surface where they are recognized by helper T-cells. The helper T-cell binds to the antigen and to the macrophage receptor, which leads to activation of the helper T-cell.

The macrophage also produces and releases cytokines, which enhance T-cell activation. The activated T-cell then releases more cytokines which causes the proliferation of other helper T-cells (positive feedback) and helps to activate cytotoxic T-cells and antibody-producing B-cells.

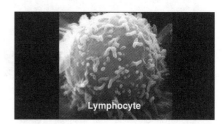

Lymphocyte

1. Describe the general action of the two major divisions in the immune system:

   (a) Humoral immune system: _____

   _____

   (b) Cell-mediated immune system: _____

   _____

2. Explain how an antigen causes the activation and proliferation of T-cells and B-cells: _____

   _____

   _____

   _____

© 2012-2014 **BIOZONE** International
ISBN: 978-1-927173-93-0

284 Clonal Selection

Key Idea: Clonal selection theory explains how lymphocytes can respond to a large and unpredictable range of antigens. The **clonal selection theory** explains how the immune system can respond to the large and unpredictable range of potential antigens in the environment. The diagram below describes clonal selection after antigen exposure for B-cells. In the same way, a T-cell stimulated by a specific antigen will multiply and develop into different types of T-cells. Clonal selection and differentiation of lymphocytes provide the basis for immunological memory.

Five (a-e) of the many B-cells generated during development. Each one can recognize only one specific antigen.

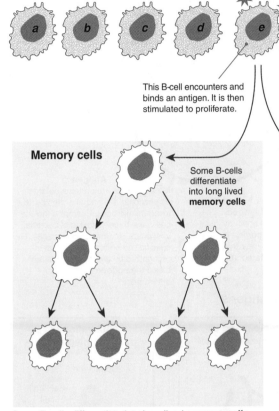

This B-cell encounters and binds an antigen. It is then stimulated to proliferate.

Clonal Selection Theory

Millions of B-cells form during development. Antigen recognition is randomly generated, so collectively they can recognize many antigens, including those that have never been encountered. Each B-cell makes antibodies corresponding to the specific antigenic receptor on its surface. The receptor reacts only to that specific antigen. When a B-cell encounters its antigen, it responds by proliferating and producing many clones all with the same kind of antibody. This is called clonal selection because the antigen selects the B cells that will proliferate.

Memory cells

Some B-cells differentiate into long lived **memory cells**

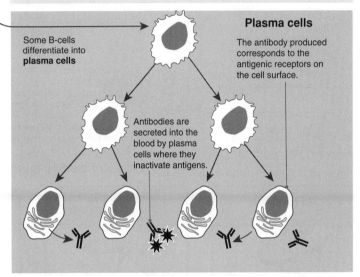

Some B-cells differentiate into **plasma cells**

Plasma cells

The antibody produced corresponds to the antigenic receptors on the cell surface.

Antibodies are secreted into the blood by plasma cells where they inactivate antigens.

Some B-cells differentiate into long lived **memory cells**. These are retained in the lymph nodes to provide future immunity (**immunological memory**). In the event of a second infection, B-memory cells react more quickly and vigorously than the initial B-cell reaction to the first infection.

Plasma cells secrete antibodies specific to the antigen that stimulated their development. Each plasma cell lives for only a few days, but can produce about 2000 antibody molecules per second. Note that during development, any B-cells that react to the body's own antigens are selectively destroyed in a process that leads to **self tolerance** (acceptance of the body's own tissues).

1. Describe how clonal selection results in the proliferation of one particular B-cell: _____

_____

_____

_____

2. Describe the function of each of the following cells in the immune system response:

(a) Memory cells: _____

_____

(b) Plasma cells: _____

_____

3. Explain the basis of **immunological memory**: _____

_____

_____

_____

LINK 288   LINK 177   WEB 284   **KNOW**

285 Antibodies

Key Idea: Antibodies are large, Y-shaped proteins, made by plasma cells, which destroy specific antigens.

Antibodies and antigens play key roles in the response of the immune system. **Antigens** are foreign molecules which promote a specific immune response. Antigens include pathogenic microbes and their toxins, as well as substances such as pollen grains, blood cell surface molecules, and the surface proteins on transplanted tissues. **Antibodies** (or immunoglobulins) are proteins made in response to antigens. They are secreted into the plasma where they can recognize, bind to, and help destroy antigens. There are five classes of antibodies, each plays a different role in the immune response. Each type of antibody is specific to only one particular antigen.

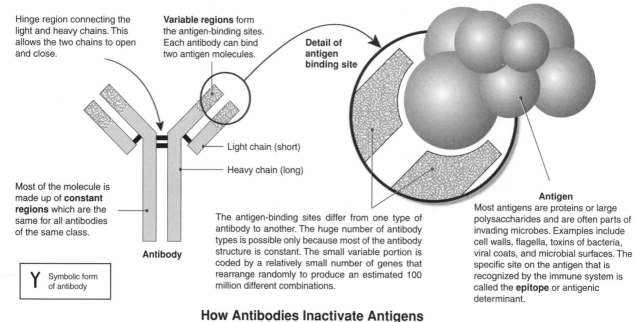

Hinge region connecting the light and heavy chains. This allows the two chains to open and close.

Variable regions form the antigen-binding sites. Each antibody can bind two antigen molecules.

Detail of antigen binding site

Light chain (short)

Heavy chain (long)

Most of the molecule is made up of **constant regions** which are the same for all antibodies of the same class.

Antibody

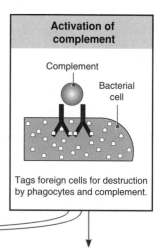

Antigen
Most antigens are proteins or large polysaccharides and are often parts of invading microbes. Examples include cell walls, flagella, toxins of bacteria, viral coats, and microbial surfaces. The specific site on the antigen that is recognized by the immune system is called the **epitope** or antigenic determinant.

Y Symbolic form of antibody

The antigen-binding sites differ from one type of antibody to another. The huge number of antibody types is possible only because most of the antibody structure is constant. The small variable portion is coded by a relatively small number of genes that rearrange randomly to produce an estimated 100 million different combinations.

How Antibodies Inactivate Antigens

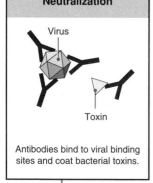

Neutralization

Virus

Toxin

Antibodies bind to viral binding sites and coat bacterial toxins.

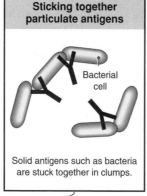

Sticking together particulate antigens

Bacterial cell

Solid antigens such as bacteria are stuck together in clumps.

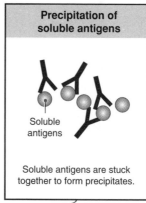

Precipitation of soluble antigens

Soluble antigens

Soluble antigens are stuck together to form precipitates.

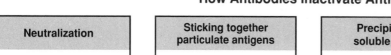

Activation of complement

Complement

Bacterial cell

Tags foreign cells for destruction by phagocytes and complement.

Enhances phagocytosis   **Enhances inflammation**   **Enhances inflammation**

1. Describe the structure of an antibody, identifying the specific features of its structure that contribute to its function:

   _____

   _____

   _____

2. Discuss the various ways in which antibodies inactivate antigens: _____

   _____

   _____

   _____

   _____

   _____

© 2012-2014 **BIOZONE** International
ISBN: 978-1-927173-93-0
Photocopying Prohibited

286 Blood Group Antigens

Key Idea: The ABO classification of human blood is based on the presence or absence of inherited antigens on the surface of red blood cells.

Blood is classified into groups according to antigens on the surface of red blood cells (RBCs). The type of antigens present determines an individual's blood type. **ABO blood group** antigens (below) and Rh antigens are the most important in the blood typing system because they cause

a strong immune response. Blood must be checked for compatibility before a patient can receive donated blood. Transfusion of incompatible blood may cause a fatal reaction in which RBCs from the donated blood are bound by antibodies in the donor plasma. When this occurs, the RBCs clump together (agglutinate), block capillaries, and rupture (hemolysis). To prevent this occurring, blood is matched before transfusion.

	Blood type A	Blood type B	Blood type AB	Blood type O
Antigens present on the red blood cells	antigen *A*	antigen *B*	antigens *A* and *B*	Neither antigen *A* nor *B*
Antibodies present in the plasma	Contains **anti-B** antibodies; but no antibodies that would attack its own antigen *A*	Contains **anti-A** antibodies; but no antibodies that would attack its own antigen *B*	Contains neither **anti-A** nor **anti-B** antibodies	Contains both **anti-A** and **anti-B** antibodies

1. Complete the table below to show the antibodies and antigens in each blood group, and donor/recipient blood types:

Blood Type	Freq. in US		Antigen	Antibody	Can donate blood to:	Can receive blood from:
	Rh⁺	Rh⁻				
A	34%	6%	A	anti-B	A, AB	A, O
B	9%	2%				
AB	3%	1%				
O	38%	7%				

2. Explain why blood from an incompatible donor causes a transfusion reaction in the recipient: _____

_____

_____

_____

3. Why is blood type O⁻ sometimes called the universal donor? _____

_____

_____

4. Why is blood type AB⁺ sometimes called the universal recipient? _____

_____

_____

5. Why was the discovery of the ABO system such a significant medical breakthrough? _____

_____

_____

LINK 282 LINK 92 WEB 286 APP

287 Allergies and Hypersensitivity

Key Idea: Hypersensitivity occurs when the immune system overreacts to an antigen or reacts to the wrong substance. Histamine plays a significant role in this response.

Sometimes the immune system may overreact, or react to the wrong substances instead of responding appropriately.

This is termed **hypersensitivity** and the immunological response leads to tissue damage rather than immunity. Hypersensitivity reactions occur after a person has been sensitized to an antigen. If the reaction is severe enough it can cause death.

Hypersensitivity

A person becomes **sensitized** when they form antibodies to harmless substances in the environment such as pollen or spores (steps 1-2 right). These substances, or **allergens**, act as **antigens** to induce antibody production and an allergic response. Once a person is sensitized, the antibodies respond to further encounters with the allergen by causing the release of **histamine** from mast cells, a special type of white blood cell (steps 4-5, right). Histamine causes the symptoms of hypersensitivity reactions such as hay fever and asthma. These symptoms include wheezing and airway constriction, inflammation, itching and watering of the eyes and nose, and sneezing.

Eyewire

The Basis of Hypersensitivity

B cell

1 B cell encounters the allergen and differentiates into plasma cells

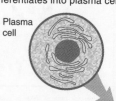
Plasma cell

Antibodies

2 The plasma cell produces antibodies

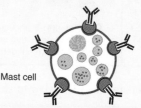
Mast cell

3 Antibodies bind to specific receptors on the surface of the mast cells

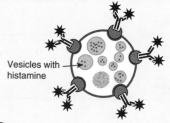
Vesicles with histamine

4 The mast cell binds the allergen when it encounters it again.

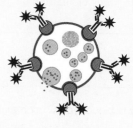

5 The mast cell releases histamine and other chemicals, which together cause the symptoms of an allergic reaction.

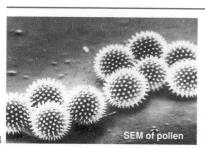
SEM of pollen

Ragweed

Hay fever is an allergic reaction to airborne substances such as dust, moulds, pollens, and animal fur. Allergy to wind-borne pollen is the most common, and certain plants (e.g. ragweed and privet) are highly allergenic. There appears to be a genetic susceptibility to hay fever, as it is common in people with a family history of eczema, hives, and/ or asthma. The best treatment for hay fever is to avoid the allergen, although anti-histamines, decongestants, and steroid nasal sprays will assist in alleviating symptoms.

Asthma is a common disease affecting around 300 million people worldwide. It usually occurs as a result of an allergic reaction to allergens such as the feces of house dust mites, pollen, and animal dander. As with all hypersensitivity reactions, it involves the production of histamines from mast cells. The site of the reaction is the respiratory bronchioles where the histamine causes constriction of the airways, accumulation of fluid and mucus, and inability to breathe. During an attack, sufferers show laboured breathing with overexpansion of the chest cavity (right).

Asthma attacks are often triggered by environmental factors such as cold air, exercise, air pollutants, and viral infections. Recent evidence has also indicated the involvement of a bacterium: *Chlamydia pneumoniae*, in about half of all cases of asthma in susceptible adults.

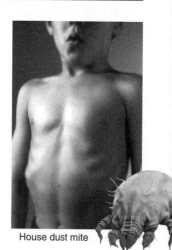
House dust mite

1. Explain the role of histamine in hypersensitivity responses: _____

_____

2. Explain what is meant by becoming **sensitized** to an allergen: _____

_____

3. In what way is the hypersensitivity reaction a malfunction of the immune system? _____

_____

© 2012-2014 **BIOZONE** International
ISBN: 978-1-927173-93-0
Photocopying Prohibited

288 Acquired Immunity

Key Idea: Acquired immunity is a resistance to specific pathogens acquired over the life-time of an organism.

We are born with natural or **innate resistance** which provides non-specific immunity to certain illnesses. In contrast, **acquired immunity** is protection developed over time to specific antigens. **Active immunity** develops after the immune system responds to being exposed to microbes or foreign substances. **Passive immunity** is acquired when antibodies are transferred from one person to another. Immunity may also be naturally acquired, through natural exposure to microbes, or artificially acquired as a result of medical treatment (below).

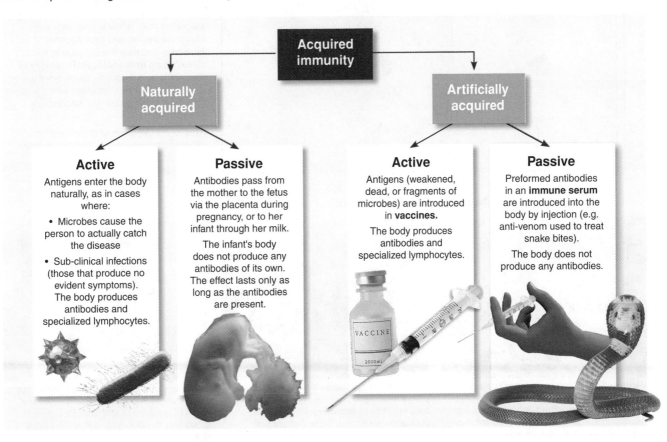

Acquired immunity

Naturally acquired

Active

Antigens enter the body naturally, as in cases where:

• Microbes cause the person to actually catch the disease

• Sub-clinical infections (those that produce no evident symptoms). The body produces antibodies and specialized lymphocytes.

Passive

Antibodies pass from the mother to the fetus via the placenta during pregnancy, or to her infant through her milk.

The infant's body does not produce any antibodies of its own. The effect lasts only as long as the antibodies are present.

Artificially acquired

Active

Antigens (weakened, dead, or fragments of microbes) are introduced in **vaccines.**

The body produces antibodies and specialized lymphocytes.

Passive

Preformed antibodies in an **immune serum** are introduced into the body by injection (e.g. anti-venom used to treat snake bites).

The body does not produce any antibodies.

VACCINE
2000ml

1. (a) What is meant by **passive immunity**? _____

   _____

   (b) Distinguish between naturally and artificially acquired passive immunity and give an example of each:

   _____

   _____

2. (a) Why does a newborn baby need to have received a supply of maternal antibodies prior to birth? _____

   _____

   _____

   (b) Why is this supply supplemented by antibodies in breast milk? _____

   _____

   (c) With regards to immunity, would you recommend breast feeding to a new mother? Explain your answer:

   _____

   _____

   _____

3. (a) What is **active immunity**? _____

   _____

LINK   LINK   WEB
283   282   288   KNOW

Primary and Secondary Responses to Antigens

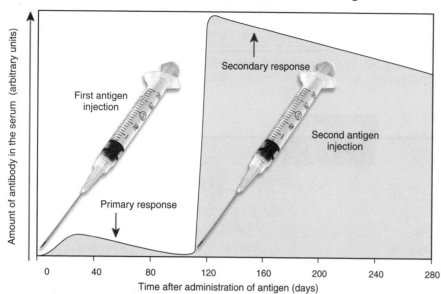

When the B-cells encounter antigens and produce antibodies, the body develops **active immunity** against that antigen.

The initial response to antigenic stimulation, caused by the sudden increase in B-cell clones, is called the **primary response**. Antibody levels as a result of the primary response peak a few weeks after the response begins and then decline. However, because the immune system develops an **immunological memory** of that antigen, it responds much more quickly and strongly when presented with the same antigen subsequently (the **secondary response**).

This forms the basis of immunization programmes where one or more booster shots are provided following the initial vaccination.

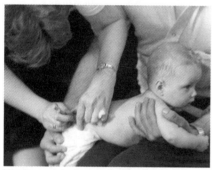

Vaccines against common diseases are given at various stages during childhood according to an immunization schedule. Vaccination has resulted in the decline of some once-common childhood diseases, such as mumps and measles.

Many childhood diseases for which vaccination programmes exist are kept at a low level because of **herd immunity**. If most of the population is immune, those that are not immunized may be protected because the disease is uncommon.

Most vaccinations are given in childhood, but adults may be vaccinated against a disease (e.g. TB, influenza) if they are in a high risk group (e.g. the elderly) or if they are travelling to a region in the world where a certain disease is common.

(b) Distinguish between naturally and artificially acquired active immunity and give an example of each: _____

_____

_____

4. (a) Describe two differences between the primary and secondary responses to presentation of an antigen: _____

_____

_____

(b) Why is the secondary response so different from the primary response? _____

_____

_____

5. (a) Explain the principle of **herd immunity**: _____

_____

_____

(b) Why are health authorities concerned when the vaccination rates for an infectious disease fall?_____

_____

_____

289 Vaccines and Vaccination

Key Idea: A vaccine is a suspension of microorganisms (or pieces of them) that is deliberately introduced into the body to protect against disease. It induces immunity by stimulating the production of antibodies.

A **vaccine** is a preparation of a harmless foreign antigen that is deliberately introduced into the body to produce an immune response. The antigen in the vaccine triggers the immune system to produce antibodies against the antigen, but it does not cause the disease. The immune system remembers its response and will produce the same antibodies if it

encounters the antigen again. There are two basic types of vaccine, subunit vaccines and whole-agent vaccines. **Whole-agent vaccines** contain complete non-virulent microbes, either inactivated (killed), or alive but attenuated (weakened). Attenuated vaccines are the most effective of these and often provide life-long immunity. **Subunit vaccines** contain parts of the pathogen that induce an immune response. They are safer than attenuated vaccines because they cannot reproduce in the recipient, and they produce fewer adverse effects because they contain little or no extra material.

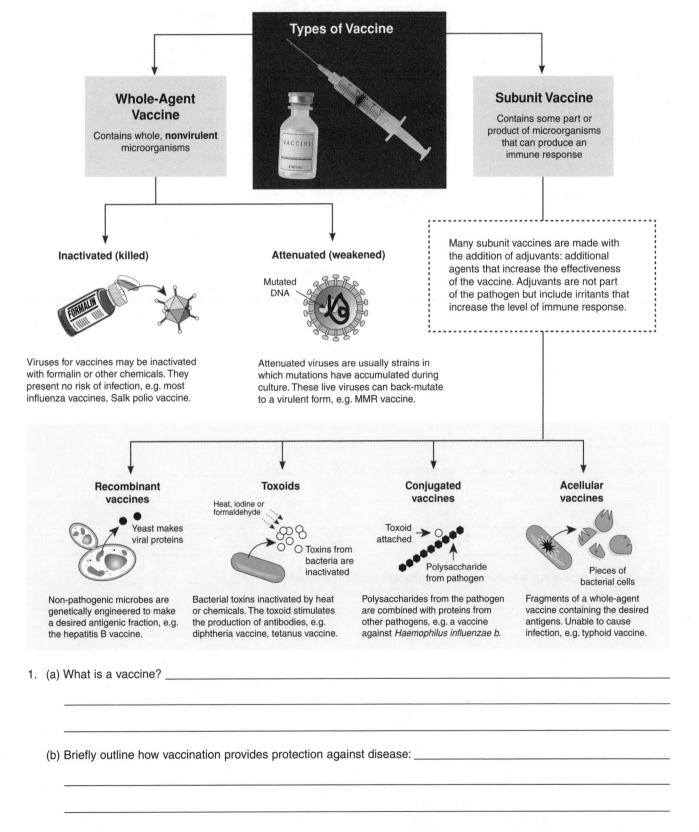

Types of Vaccine

Whole-Agent Vaccine

Contains whole, **nonvirulent** microorganisms

Subunit Vaccine

Contains some part or product of microorganisms that can produce an immune response

Inactivated (killed)

Viruses for vaccines may be inactivated with formalin or other chemicals. They present no risk of infection, e.g. most influenza vaccines, Salk polio vaccine.

Attenuated (weakened)

Mutated DNA

Attenuated viruses are usually strains in which mutations have accumulated during culture. These live viruses can back-mutate to a virulent form, e.g. MMR vaccine.

Many subunit vaccines are made with the addition of adjuvants: additional agents that increase the effectiveness of the vaccine. Adjuvants are not part of the pathogen but include irritants that increase the level of immune response.

Recombinant vaccines

Yeast makes viral proteins

Non-pathogenic microbes are genetically engineered to make a desired antigenic fraction, e.g. the hepatitis B vaccine.

Toxoids

Heat, iodine or formaldehyde

Toxins from bacteria are inactivated

Bacterial toxins inactivated by heat or chemicals. The toxoid stimulates the production of antibodies, e.g. diphtheria vaccine, tetanus vaccine.

Conjugated vaccines

Toxoid attached

Polysaccharide from pathogen

Polysaccharides from the pathogen are combined with proteins from other pathogens, e.g. a vaccine against *Haemophilus influenzae b*.

Acellular vaccines

Pieces of bacterial cells

Fragments of a whole-agent vaccine containing the desired antigens. Unable to cause infection, e.g. typhoid vaccine.

1. (a) What is a vaccine? _____

_____

_____

(b) Briefly outline how vaccination provides protection against disease: _____

_____

_____

© 2012-2014 **BIOZONE** International
ISBN: 978-1-927173-93-0
Photocopying Prohibited

LINK 283   LINK 282   LINK 177   WEB 289

KNOW

Application: Smallpox Eradication Through Vaccination

Smallpox (right) is a highly contagious disease caused by the *Variola* virus. It has two forms, the more severe of which kills about 30% of the people infected. Repeated smallpox epidemics were responsible for an estimated 300-500 million deaths globally during the 20th century alone, and the disease was endemic throughout most of the world. Survivors were often left blind and severely scarred. In the late 1700s, Edward Jenner discovered that inoculation with the cowpox virus could protect humans against smallpox. In the 1900s, the smallpox vaccine was developed and produced in large quantities. Smallpox was declared eradicated in 1980 after a global campaign involving international cooperation.

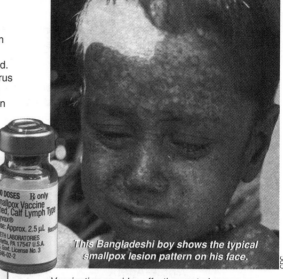

This Bangladeshi boy shows the typical smallpox lesion pattern on his face.

International-mindedness: The Role of the WHO

In 1967, the World Health Organization (WHO) led a global programme that resulted in the complete eradication of smallpox by 1977. The key to the strategy was surveillance and containment of outbreaks. Eradication was helped by the fact that humans were the only reservoir for infection and there were no other carriers. High vaccination rates made it difficult for the disease to spread because there were too few susceptible individuals to support an epidemic. The last outbreak was recorded in Somalia in 1977.

The next disease targeted for eradication is polio, which can cause irreversible paralysis. Polio is rare in Western countries, but still endemic in Afghanistan, India, Nigeria, and Pakistan. Between 2009 and 2010, 23 previously polio-free countries were re-infected because people with polio infected others who have not had the disease or been immunized against it.

Vaccination provides effective control over many bacterial and viral diseases. Unlike bacterial diseases, which can be treated with antibiotics after infection, viral diseases often have no treatments. Prevention through vaccination is the best protection.

2. **Attenuated viruses** provide long term immunity to their recipients and generally do not require booster shots. Why do you think attenuated viruses provide such effective long-term immunity when inactivated viruses do not?

3. Vaccines can be used to protect individuals from bacterial and viral infections, but are sometimes considered more important for viral diseases. Why is this the case?

4. (a) How can vaccination help lead to the eradication of an infectious disease?

(b) Discuss factors in the success of the smallpox eradication programme:

5. Polio is an excellent candidate for eradication, but its final eradication is proving to be more challenging in many respects than the eradication of smallpox. Visit the weblink provided for this activity and summarize the difficulties involved:

© 2012-2014 **BIOZONE** International
ISBN: 978-1-927173-93-0
Photocopying Prohibited

290 Epidemiology

Key Idea: Epidemiology is the study of the causes, incidence, and distribution of disease. Epidemiological studies are used to determine the effectiveness of public health programmes.
Epidemiology is the study of the origin, occurrence, and spread of disease. The health statistics collected by epidemiologists are used to identify patterns of disease occurrence, either in particular countries or globally. These patterns are important in planning health services and investigating causes of disease. Health statistics enable the effectiveness of health policies and practices, such as vaccination programmes, to be monitored. The World Health Organization (WHO) gathers data on an international basis to identify global patterns.

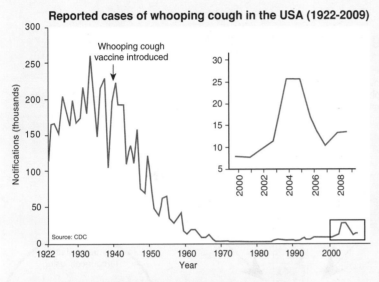

Reported cases of whooping cough in the USA (1922-2009)

Whooping cough vaccine introduced

Notifications (thousands)

Source: CDC

Year

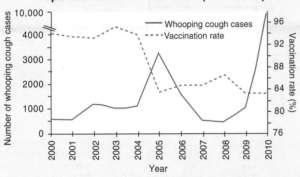

Whooping cough vaccination rates vs reported cases for California (2000-2010)

Number of whooping cough cases

— Whooping cough cases
- - - Vaccination rate

Vaccination rate (%)

Year

Case Study: Whooping Cough

Whooping cough is caused by the bacterium *Bordetella pertussis*, and infection may last for two to three months. It is characterized by painful coughing spasms, and a cough that sounds like a "whoop". Severe coughing fits may be followed by periods of vomiting. Inclusion of the whooping cough vaccine into the US immunization schedule in the 1940s has greatly reduced the incidence rates of the disease (left).

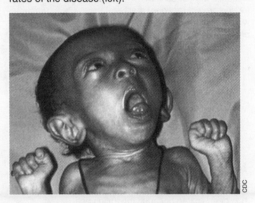

Above: Infants under six months of age are most at risk of developing complications or dying from whooping cough because they are too young to be fully protected by the vaccine. Ten infants died of whooping cough in California in 2010.

Left: In California, whooping cough vaccination rates have fallen amidst fears that it is responsible for certain health problems such as autism. As a result, rates of whooping cough have increased significantly since 2004. In 2010, over 9000 cases were reported, the highest level in 63 years.

1. (a) Describe the effect of introducing the whooping cough vaccine into the immunization schedule in the US: _____

_____

_____

_____

(b) Why have whooping cough immunization rates dropped significantly in California since 2004? _____

_____

_____

(c) What has been the effect of the lower immunization rates on the number of whooping cough cases? _____

_____

_____

(d) Suggest why the drop in immunization rates does not perfectly coincide with the increase in disease incidence:

_____

_____

291 Monoclonal Antibodies

Key Idea: Monoclonal antibodies are artificially produced antibodies that neutralize specific antigens. They have wide applications in diagnosing and treating disease, in detecting pregnancy, and in food safety tests.

A monoclonal antibody is an artificially produced antibody that binds to and neutralizes one specific type of antigen. A monoclonal antibody binds an antigen in the same way that a normally produced antibody does. Monoclonal antibodies are produced by stimulating the production of antibody-producing B-cells in mice injected with the antigen. The isolated B-cells are made to fuse with immortal tumour cells, and can be cultured indefinitely in a suitable growing medium. Monoclonal antibodies are useful because they are identical (i.e. clones), they can be produced in large quantities, and they are highly specific. Antibodies produced in this way are used in medical diagnosis and treatment.

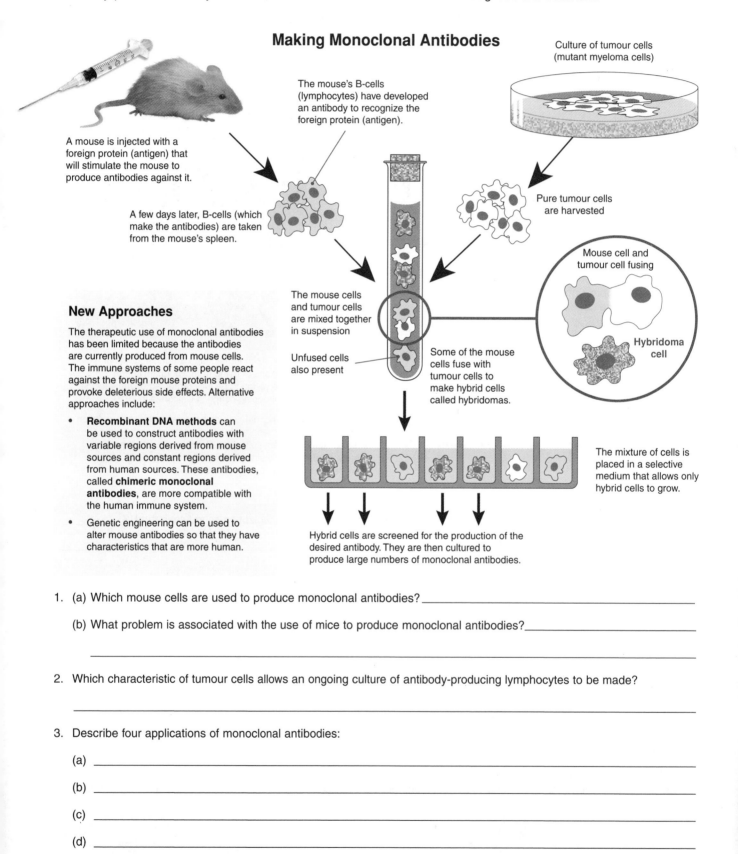

Making Monoclonal Antibodies

A mouse is injected with a foreign protein (antigen) that will stimulate the mouse to produce antibodies against it.

The mouse's B-cells (lymphocytes) have developed an antibody to recognize the foreign protein (antigen).

A few days later, B-cells (which make the antibodies) are taken from the mouse's spleen.

Culture of tumour cells (mutant myeloma cells)

Pure tumour cells are harvested

Mouse cell and tumour cell fusing

Hybridoma cell

The mouse cells and tumour cells are mixed together in suspension

Unfused cells also present

Some of the mouse cells fuse with tumour cells to make hybrid cells called hybridomas.

New Approaches

The therapeutic use of monoclonal antibodies has been limited because the antibodies are currently produced from mouse cells. The immune systems of some people react against the foreign mouse proteins and provoke deleterious side effects. Alternative approaches include:

- **Recombinant DNA methods** can be used to construct antibodies with variable regions derived from mouse sources and constant regions derived from human sources. These antibodies, called **chimeric monoclonal antibodies**, are more compatible with the human immune system.

- Genetic engineering can be used to alter mouse antibodies so that they have characteristics that are more human.

The mixture of cells is placed in a selective medium that allows only hybrid cells to grow.

Hybrid cells are screened for the production of the desired antibody. They are then cultured to produce large numbers of monoclonal antibodies.

1. (a) Which mouse cells are used to produce monoclonal antibodies? _____

    (b) What problem is associated with the use of mice to produce monoclonal antibodies? _____

    _____

2. Which characteristic of tumour cells allows an ongoing culture of antibody-producing lymphocytes to be made?

    _____

3. Describe four applications of monoclonal antibodies:

    (a) _____

    (b) _____

    (c) _____

    (d) _____

© 2012-2014 **BIOZONE** International
ISBN: 978-1-927173-93-0
Photocopying Prohibited

Detecting Pregnancy using Monoclonal Antibodies

When a woman becomes pregnant, a hormone called **human chorionic gonadotropin** (HCG) is released. HCG accumulates in the bloodstream and is excreted in the urine. Antibodies can be produced against HCG and used in simple test kits (below) to determine if a woman is pregnant. Monoclonal antibodies are also used in other home testing kits, such as those for detecting ovulation time (far left).

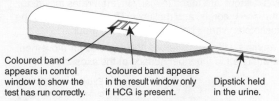

Coloured band appears in control window to show the test has run correctly.

Coloured band appears in the result window only if HCG is present.

Dipstick held in the urine.

Other Applications of Monoclonal Antibodies

Diagnostic uses

* Detecting the presence of pathogens such as *Chlamydia* and streptococcal bacteria, distinguishing between *Herpesvirus* I and II, and diagnosing AIDS.

* Measuring protein, toxin, or drug levels in serum.

* Blood and tissue typing.

* Detection of antibiotic residues in milk.

Therapeutic uses

* Neutralizing endotoxins produced by bacteria in blood infections.

* Used to prevent organ rejection, e.g. in kidney transplants, by interfering with the T-cells involved with the rejection of transplanted tissue.

* Used in the treatment of some auto-immune disorders such as rheumatoid arthritis and allergic asthma. The monoclonal antibodies bind to and inactivate factors involved in the cascade leading to the inflammatory response.

* Immunodetection and immunotherapy of cancer. Herceptin is a monoclonal antibody for the targeted treatment of breast cancer. Herceptin recognizes receptor proteins on the outside of cancer cells and binds to them. The immune system can then identify the antibodies as foreign and destroy the cell.

* Inhibition of platelet clumping, which is used to prevent reclogging of coronary arteries in patients who have undergone angioplasty. The monoclonal antibodies bind to the receptors on the platelet surface that are normally linked by fibrinogen during the clotting process.

How home pregnancy detection kits work

The test area of the dipstick (below) contains two types of antibodies: free monoclonal antibodies and capture monoclonal antibodies, bound to the substrate in the test window.

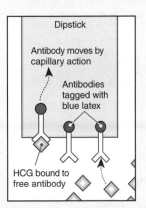

The free antibodies are specific for HCG and are colour-labelled. HCG in the urine of a pregnant woman binds to the free antibodies on the surface of the dipstick. The antibodies then travel up the dipstick by capillary action.

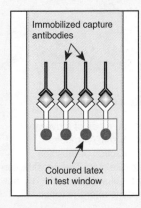

The capture antibodies are specific for the HCG-antibody complex. The HCG-antibody complexes traveling up the dipstick are bound by the immobilized capture antibodies, forming a sandwich. The colour labelled antibodies then create a visible colour change in the test window.

---

4. For each of the following applications, suggest why an antibody-based test or therapy is so valuable:

(a) Detection of toxins or bacteria in perishable foods: _____

_____

_____

(b) Detection of pregnancy without a doctor's prescription: _____

_____

(c) Targeted treatment of tumours in cancer patients: _____

_____

292 Skeletons and Movement

Key Idea: Skeletons give support to organisms and provide an attachment framework for muscles to enable movement. The skeleton is a rigid structure and has many functions including structure and support, and enabling movement. Muscles attached to the skeleton contract, pulling on the skeleton to generate movement. The bones and muscles act as a system of levers, with joints acting as a fixed point of leverage (or fulcrum), and the muscles applying the effort to move the load or resistance, e.g. the bone and associated tissue. Skeletons may be external to the body wall (**exoskeleton**) as in arthropods, or internal (**endoskeleton**) and lying inside the body wall, as in vertebrates.

Exoskeletons

Exoskeletons are found in many invertebrates including arthropods, corals, and molluscs. The composition of the exoskeleton varies between taxonomic groups.

In animals with exoskeletons, the muscles for movement are attached to the inner surface of the exoskeleton.

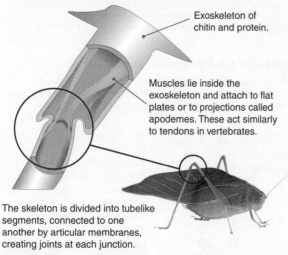

Exoskeleton of chitin and protein.

Muscles lie inside the exoskeleton and attach to flat plates or to projections called apodemes. These act similarly to tendons in vertebrates.

The skeleton is divided into tubelike segments, connected to one another by articular membranes, creating joints at each junction.

Endoskeletons

A bony endoskeleton is the internal support structure of some animals including vertebrates, sponges, and echinoderms. In vertebrates, it is composed mostly of calcium phosphate. The endoskeleton functions as an attachment site for muscles and provides a means to transmit muscular forces. Muscles pull on the skeleton to create movement about joints. Muscles work in opposing pairs to create opposing movement.

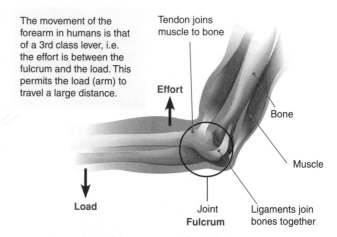

The movement of the forearm in humans is that of a 3rd class lever, i.e. the effort is between the fulcrum and the load. This permits the load (arm) to travel a large distance.

Tendon joins muscle to bone

Effort

Bone

Muscle

Load

Joint
Fulcrum

Ligaments join bones together

Insect flight

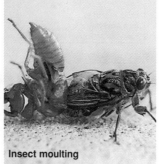

Insect moulting

Arthropods are capable of many different types of movement, e.g. flight, walking, or burrowing through soil. All modes of locomotion are achieved through the action of muscles contracting and moving the jointed exoskeleton. The exoskeleton is rigid, so must be shed periodically to allow for growth. The new larger skeleton is initially soft and hardens after the moult.

Snake skeleton

Sea urchin

The rhythmic contraction of muscles in a snake act on the bones of its skeleton allowing it to move across the ground. Although a sea urchin (an echinoderm) may look as though it has an exoskeleton, the spines are actually projections of the endoskeleton, which lies just below a layer of skin and muscles. Endoskeletons, being internal, can grow with the organism.

1. What are the major differences between an **exoskeleton** and an **endoskeleton**? _____

   _____

   _____

   _____

2. Explain the importance of a skeleton in achieving movement in animals: _____

   _____

   _____

   _____

WEB 292 LINK 294

© 2012-2014 **BIOZONE** International
ISBN: 978-1-927173-93-0
Photocopying Prohibited

293 Movement About Joints

Key Idea: A joint is the junction where two or more bones meet. All movements of the skeleton occur at joints.

Bones are too rigid to bend. To allow movement, the human skeletal system consists of bones held together at joints by flexible connective tissues called **ligaments**. **Joints** are points of contact between bones or between cartilage and bones. Joints may be classified structurally as fibrous, cartilaginous, or synovial (below). Each of these joint types allows a certain degree of movement. Bones are made to move about a joint by the force of muscles acting upon them.

Cartilaginous Joints

Here, the bone ends are connected by cartilage. Most allow limited movement although some (e.g. between the first ribs and the sternum) are immovable.

Immovable Fibrous Joints

The bones are connected by fibrous tissue. In some (e.g. sutures of the skull), the bones are tightly bound by connective tissue fibres and there is no movement.

Synovial Joints

These allow free movement in one or more planes. The articulating bone ends are separated by a joint cavity containing lubricating synovial fluid (see next page).

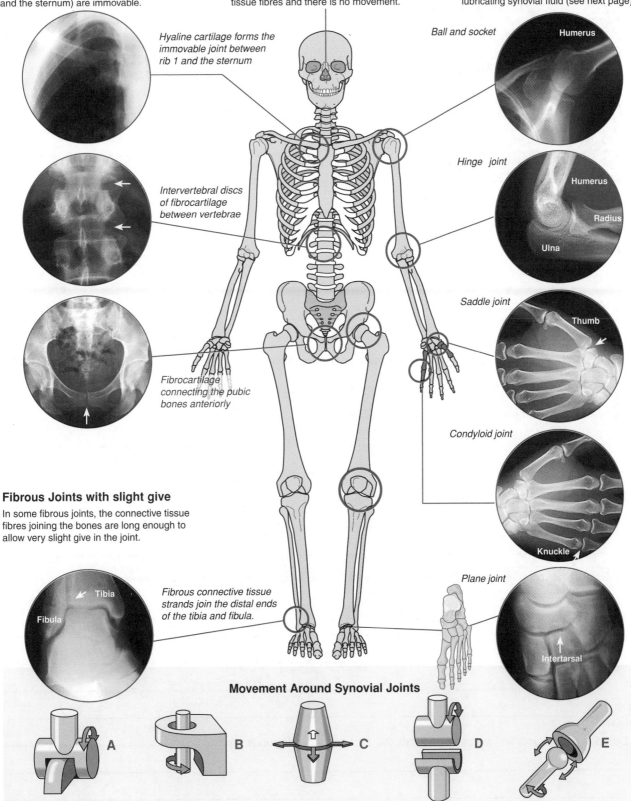

Hyaline cartilage forms the immovable joint between rib 1 and the sternum

Intervertebral discs of fibrocartilage between vertebrae

Fibrocartilage connecting the pubic bones anteriorly

Ball and socket
Humerus

Hinge joint
Humerus
Radius
Ulna

Saddle joint
Thumb

Condyloid joint
Knuckle

Plane joint
Intertarsal

Fibrous Joints with slight give

In some fibrous joints, the connective tissue fibres joining the bones are long enough to allow very slight give in the joint.

Fibrous connective tissue strands join the distal ends of the tibia and fibula.

Tibia
Fibula

Movement Around Synovial Joints

A   B   C   D   E

LINK 292   WEB 293

KNOW

Structure of a Synovial Joint

Synovial joints (right and below) allow free movement of body parts in varying directions (one, two or three planes). The elbow joint is a hinge joint and typical of a synovial joint. Like most synovial joints, it is reinforced by ligaments (not all shown). The joint capsule encloses synovial fluid, which reduces friction and absorbs shocks. In the diagram, the brachialis muscle, which inserts into the ulna and is the prime mover for flexion of the elbow, has been omitted to show the joint structure. Muscles are labelled blue and bones are bolded.

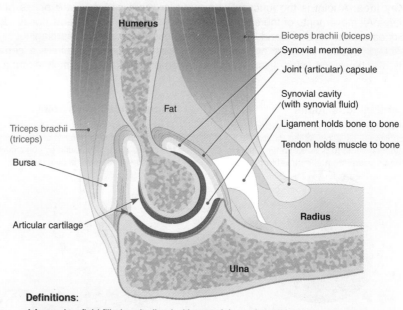

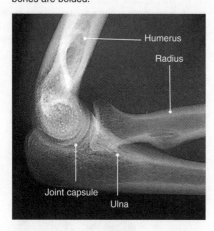

Definitions:

A **bursa** is a fluid filled cavity lined with synovial membrane. It acts as a cushion, e.g. between tendon and bone, or between bones.

Cartilage is a flexible connective tissue. It protects a joint surface against wear.

1. Define the following terms and state the role of each in movement:

   (a) Joint: _____

   (b) Ligament:_____

   (c) Muscle:_____

   (d) Tendon:_____

2. Classify each of the synovial joint models (**A-E**) at the bottom of the previous page, according to the descriptors below:

   (a) Pivot: _____ (b) Hinge: _____ (c) Ball-and-socket: _____ (d) Saddle: _____ (e) Gliding: _____

3. Compare the movements of the hip joint and the elbow joint: _____

   _____

   _____

4. (a) Describe the features common to most synovial joints: _____

   _____

   _____

   (b) Explain the role that synovial fluid and cartilage play in the structure and function of a synovial joint:

   _____

   _____

   _____

   _____

   _____

5. Describe the major difference between a synovial joint and a cartilaginous joint: _____

   _____

   _____

294 Antagonistic Muscles

Key Idea: Antagonistic muscles are muscle pairs that have opposite actions to each other. Together, their opposing actions bring about movement of body parts.

In both vertebrates and invertebrates, muscle provide the contractile force to move body parts. Muscles create movement of body parts when they contract across joints. Because muscles can only pull and not push, most body movements are achieved through the action of opposing sets of muscles called **antagonistic muscles**. Antagonistic muscles function by producing opposite movements, as one muscle contracts (shortens), the other relaxes (lengthens). Skeletal muscles are attached to the skeleton by tough connective tissue structures (**tendons** in vertebrates or attachment fibres in insects). They always have at least two attachments: an origin and an insertion. Body parts move when a muscle contracts across a joint. The type and degree of movement depends on how much movement the joint allows and where the muscle is located in relation to the joint.

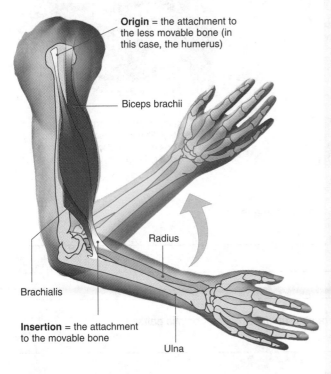

Origin = the attachment to the less movable bone (in this case, the humerus)

Biceps brachii

Radius

Brachialis

Insertion = the attachment to the movable bone

Ulna

Two muscles are involved in flexing the forearm. The **brachialis**, which underlies the biceps brachii and has an origin half way up the humerus, is the **prime mover**. The more obvious **biceps brachii**, which is a two headed muscle with two origins and a common insertion near the elbow joint, acts as the synergist. During contraction, the insertion moves towards the origin.

Opposing Movements Require Opposing Muscles

The skeleton works as a system of levers. The joint acts as a **fulcrum** (or pivot), the muscles exert the **force**, and the weight of the bone being moved represents the **load**. The flexion (bending) and extension (unbending) of limbs is caused by the action of **antagonistic muscles**. Antagonistic muscles work in pairs and their actions oppose each other. During movement of a limb, muscles other than those primarily responsible for the movement may be involved to fine tune the movement.

Every coordinated movement in the body requires the application of muscle force. This is accomplished by the action of agonists, antagonists, and synergists. The opposing action of agonists and antagonists (working constantly at a low level) also produces muscle tone. Note that either muscle in an antagonistic pair can act as the agonist or **prime mover**, depending on the particular movement (for example, flexion or extension).

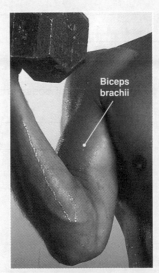

Biceps brachii

Agonists or prime movers: muscles that are primarily responsible for the movement and produce most of the force required.

Antagonists: muscles that oppose the prime mover. They may also play a protective role by preventing over-stretching of the prime mover.

Synergists: muscles that assist the prime movers and may be involved in fine-tuning the direction of the movement.

During flexion of the forearm (left) the **brachialis** muscle acts as the prime mover and the **biceps brachii** is the synergist. The antagonist, the **triceps brachii** at the back of the arm, is relaxed. During extension, their roles are reversed.

Movement at Joints

The synovial joints of the skeleton allow free movement in one or more planes. The articulating bone ends are separated by a joint cavity containing lubricating synovial fluid. Two types of synovial joint, the shoulder ball and socket joint and the hinge joint of the elbow, are illustrated below.

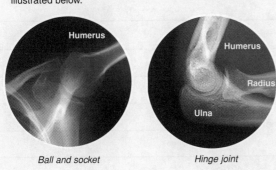

Humerus

Humerus

Radius

Ulna

Ball and socket   *Hinge joint*

Quadriceps

Hamstrings

Movement of the upper leg is achieved through the action of several large groups of muscles, collectively called the **quadriceps** and the **hamstrings**.

The hamstrings are actually a collection of three muscles, which act together to flex the leg.

The quadriceps at the front of the thigh (a collection of four large muscles) opposes the motion of the hamstrings and extends the leg.

When the prime mover contracts forcefully, the antagonist also contracts very slightly. This stops over-stretching and allows greater control over thigh movement.

Antagonistic Muscle Pairs in an Insect Leg

Antagonist muscle pairs in insect legs work together to move the legs. The two main muscles are the extensor tibiae muscle (often just called the **extensor**) which causes the leg to extend, and the flexor tibiae muscle (**flexor**) which causes the leg to flex (bend).

The muscles are attached to the tibia via attachment fibres to the cuticle on either side of the joint. When one of the muscles contracts, it pulls on its attachment and moves the tibia one way. When the other muscle in the pair contracts, it moves the tibia the other way (below).

Extensor muscle relaxes
Tibia
Tibia flexes
Flexor muscle contracts

Extensor muscle contracts
Joint
Tibia
Tibia extends
Flexor muscle relaxes

The contraction of the extensor muscle has propelled this grasshopper through the air.

Walking (top) involves contraction of leg muscles. Modification of insect limbs and their muscles enables insects to move in other ways, e.g. jump (left) or swim (above).

1. Describe the role of each of the following muscles in moving a limb in humans:

   (a) Prime mover: _____

   (b) Antagonist: _____

   (c) Synergist: _____

2. Explain why the muscles that cause movement of body parts tend to operate as antagonistic pairs: _____

   _____

   _____

3. Describe the relationship between muscles and joints in a human. Using appropriate terminology, explain how antagonistic muscles act together to raise and lower a limb:

   _____

   _____

   _____

4. (a) Identify the insertion for the biceps brachii during flexion of the forearm: _____

   (b) Identify the insertion of the brachialis muscle during flexion of the forearm: _____

   (c) Identify the antagonist during flexion of the forearm: _____

   (d) Given its insertion, describe the forearm movement during which the biceps brachii is the prime mover: _____

   _____

5. (a) Identify the fulcrum for forearm movement in humans: _____

   (b) Identify the structures that represent the load: _____

   (c) Identify the structures that represent the force: _____

6. How do antagonistic muscle pairs in insects bring about movement of the legs? _____

   _____

   _____

295 Skeletal Muscle Structure and Function

Key Idea: Skeletal muscle is organized into bundles of muscle cells or fibres. The muscle fibres are made up of repeating contractile units called sarcomeres.

Skeletal muscle is organized into bundles of muscle cells or fibres. Each **fibre** is a single cell with many nuclei and each fibre is itself a bundle of smaller **myofibrils** arranged lengthwise. Each myofibril is in turn composed of two kinds of **myofilaments** (thick and thin), which overlap to form light and dark bands. It is the alternation of these light and dark bands which gives skeletal muscle its striated or striped appearance. The **sarcomere**, bounded by the dark Z lines, forms one complete contractile unit.

When viewed under a microscope (right), skeletal muscle has a banded appearance. The cells are large with many nuclei (multinucleate).

Nuclei

Skeletal muscles require a conscious action to control them. Physical actions, such as running, writing, and speaking require the contraction of skeletal muscles to occur.

Structure of Muscle

Skeletal muscle enclosed in connective tissue

Bundles of muscle fibres (**fascicles**)

Single muscle fibre

The relationship between muscle, fascicles, and muscle fibres (cells)

Structure of a Muscle Fibre (Cell)

When a nerve impulse arrives at the neuromuscular junction, acetylcholine is released. This stimulates an action potential in the sarcolemma, which is propagated through the muscle fibre via the system of T tubules.

Motor neuron

An action potential is conducted to all myofibrils of the fibre.

Nucleus

T tubules

The **sarcoplasmic reticulum** is a specialized type of smooth endoplasmic reticulum. It is associated with the T tubules and forms a network containing a store of calcium ions.

The **sarcolemma** is the plasma membrane of the muscle cell and encloses the sarcoplasm (cytoplasm).

A myofibril (blue outline) with myofilaments in cross section.

Longitudinal Section of a Sarcomere

I band (light) A band (dark) I band (light)

Z line

One sarcomere

WMU

H zone

Thin filament made of **actin**

Thick and thin filaments slide past each other

Thick filament made of **myosin**

Cross section through a region of overlap between thick and thin filaments.

Thick filament

Thin filament

The photograph of a sarcomere (above) shows the banding pattern arising as a result of the highly organized arrangement of thin and thick filaments. It is represented schematically in longitudinal section and cross section. A single sarcomere is shown left as the highlighted transluscent blue section.

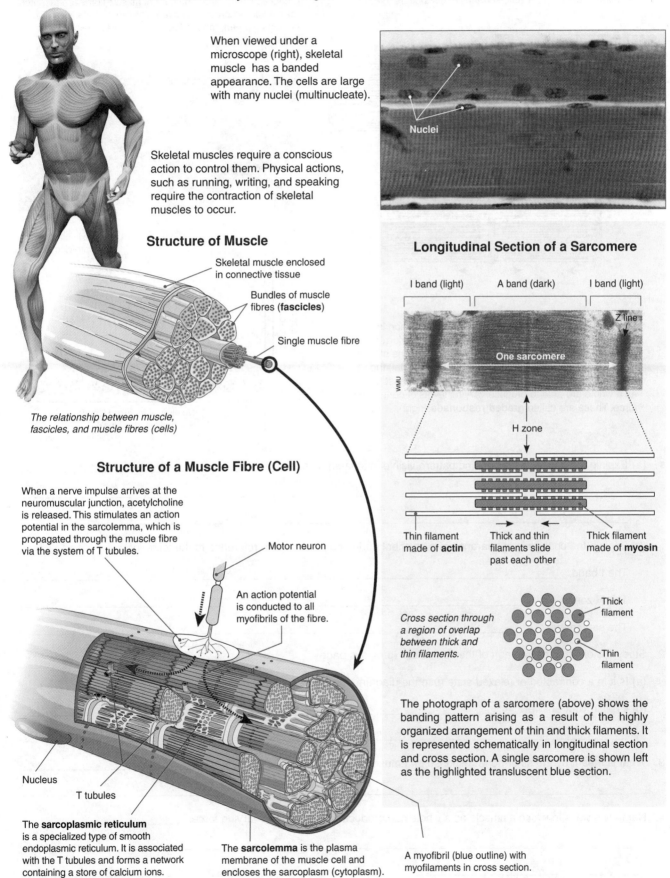

LINK 296   LINK 294   WEB 295   KNOW

The Banding Pattern of Myofibrils

Within a myofibril, the thin filaments, held together by the **Z lines**, project in both directions. The arrival of an action potential sets in motion a series of events that cause the thick and thin filaments to slide past each other. This is called **contraction** and it results in shortening of the muscle fibre and is accompanied by a visible change in the appearance of the myofibril: the I band and the sarcomere shorten and H zone shortens or disappears (below).

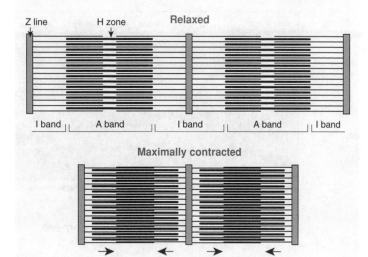

The response of a single muscle fibre to stimulation is to contract maximally or not at all; its response is referred to as the **all-or-none law** of muscle contraction. If the stimulus is not strong enough to produce an action potential, the muscle fibre will not respond. However skeletal muscles as a whole are able to produce varying levels of contractile force. These are called **graded responses** (right).

Muscles have Graded Responses

Muscle fibres respond to an action potential by contracting maximally, yet skeletal muscles as a whole can produce **contractions of varying force**. This is achieved by changing the frequency of stimulation (more rapid arrival of action potentials) and by changing the number of fibres active at any one time. A stronger muscle contraction is produced when a large number of muscle fibres are recruited (below left), whereas less strenuous movements, such as picking up a pen, require fewer active fibres (below right).

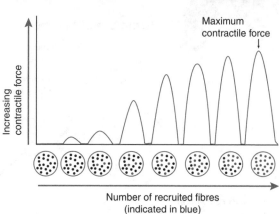

1. (a) Explain the cause of the banding pattern visible in striated muscle: _____

    _____

    _____

   (b) Explain the change in appearance of a myofibril during contraction with reference to the following:

    The I band: _____

    The H zone: _____

    The sarcomere: _____

2. Study the electron micrograph of the sarcomere (previous page).

   (a) Is it in a contracted or relaxed state (use the diagram, top left to help you decide): _____

   (b) Explain your answer: _____

    _____

3. What is meant by the all-or-none response of a muscle fibre? _____

    _____

4. Name two ways in which a muscle as a whole can produce contractions of varying force: _____

    _____

    _____

296 The Sliding Filament Theory

Key Idea: The sliding filament theory describes how muscle contraction occurs when the thick and thin myofibrils of a muscle fibre slide past one another. Calcium ions and ATP are required.

The structure and arrangement of the thick and thin filaments in a muscle fibre make it possible for them to slide past each other and cause shortening (contraction) of the muscle. The ends of the thick myosin filaments have cross bridges that can link to adjacent thin actin filaments. When the cross bridges of the thick filaments connect to the thin filaments, a shape change moves one filament past the other. Two things are necessary for cross bridge formation: calcium ions, which are released from the sarcoplasmic reticulum when the muscle receives an action potential, and ATP, which is present in the muscle fibre and is hydrolysed by ATPase enzymes on the myosin. When cross bridges attach and detach in sarcomeres throughout the muscle cell, the cell shortens.

The Sliding Filament Theory

Muscle contraction requires calcium ions (Ca^{2+}) and energy (in the form of ATP) in order for the thick and thin filaments to slide past each other. The steps are:

1. The binding sites on the **actin** molecule (to which myosin 'heads' will locate) are blocked by a complex of two protein molecules: **tropomyosin** and **troponin**.

2. Prior to muscle contraction, ATP binds to the heads of the myosin molecules, priming them in an erect high energy state. Arrival of an action potential is transmitted along the T tubules and causes a release of Ca^{2+} from the sarcoplasmic reticulum into the sarcoplasm. The Ca^{2+} binds to the troponin and causes the blocking complex to move so that the myosin binding sites on the actin filament become exposed.

3. The heads of the cross-bridging myosin molecules attach to the binding sites on the actin filament. Release of energy from the hydrolysis of ATP accompanies the cross bridge formation.

4. The energy released from ATP hydrolysis causes a change in shape of the myosin **cross bridge**, resulting in a bending action (*the power stroke*). This causes the actin filaments to slide past the myosin filaments towards the centre of the sarcomere.

5. (Not illustrated). Fresh ATP attaches to the myosin molecules, releasing them from the binding sites and repriming them for a repeat movement. They become attached further along the actin chain as long as ATP and Ca^{2+} are available.

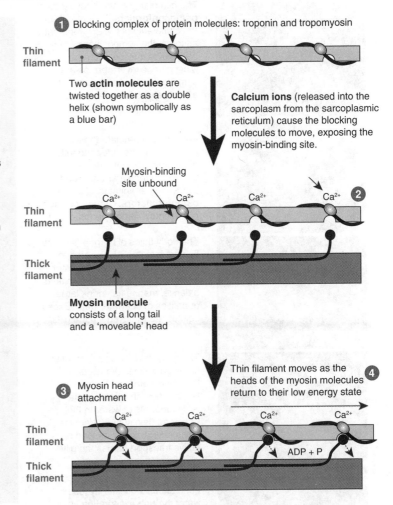

1 Blocking complex of protein molecules: troponin and tropomyosin

Thin filament

Two **actin molecules** are twisted together as a double helix (shown symbolically as a blue bar)

Calcium ions (released into the sarcoplasm from the sarcoplasmic reticulum) cause the blocking molecules to move, exposing the myosin-binding site.

Myosin-binding site unbound

Ca^{2+} Ca^{2+} Ca^{2+} Ca^{2+} **2**

Thin filament

Thick filament

Myosin molecule consists of a long tail and a 'moveable' head

3 Myosin head attachment

Thin filament moves as the heads of the myosin molecules return to their low energy state **4**

Ca^{2+} Ca^{2+} Ca^{2+} Ca^{2+}

Thin filament

ADP + P

Thick filament

---

1. Match the following chemicals with their functional role in muscle movement (draw a line between matching pairs):

   (a) Myosin — • Bind to the actin molecule in a way that prevents myosin head from forming a cross bridge

   (b) Actin — • Supplies energy for the flexing of the myosin 'head' (power stroke)

   (c) Calcium ions — • Has a moveable head that provides a power stroke when activated

   (d) Troponin-tropomyosin — • Two protein molecules twisted in a helix shape that form the thin filament of a myofibril

   (e) ATP — • Bind to the blocking molecules, causing them to move and expose the myosin binding site

2. (a) Identify the two things necessary for cross bridge formation: _____

   (b) Explain where each of these comes from: _____

   _____

   _____

3. Why are there abundant mitochondria in a muscle fibre? _____

   _____

297 Osmoregulation and Excretion

Key Idea: Osmoregulators are able to tightly regulate salt and water balance. Osmoconformers match the osmolarity of their environment. All osmoconformers are marine. Osmoregulation is the process of managing fluid and ion balance to maintain the homeostasis of the body's water content (osmolarity). Osmoconformers are marine organisms that match the osmolarity of their bodily fluids with that of the external environment, although the ionic composition of their body fluids may be different. In contrast, osmoregulators maintain constant water and solute concentrations even when the environmental conditions vary. Osmoregulation thus represents a considerable energy cost to the organism.

Osmoregulators vs Osmoconformers

Anemones

The osmolarity of the body fluids of **osmoconformers** fluctuates with the osmolarity of the environment, although the composition of their body fluids may be different. Most marine invertebrates are osmoconformers and many rely on a relatively stable external osmotic environment for survival.

Dolphins

Animals that regulate their salt and water fluxes independently of the environment, such as fish and marine mammals and all freshwater animals, are **osmoregulators**. Osmoregulation represents a large energetic cost.

The body fluids of marine (saltwater) fish are hypotonic compared to the seawater, so they lose water and gain solutes. To maintain fluid balance, marine fish drink large quantities of seawater (left). Excess solutes are removed via the gills.

Silver carp

The body fluids of freshwater fish are hypertonic to the water, so they gain water and lose solutes. Freshwater fish must excrete large amounts of water in their urine and gain solutes from the environment, either as food or by active transport across the gills.

Deer

Land animals constantly lose water through evaporation, in their urine, and in some species, through sweating. Their water balance must be maintained by consuming water (left).

Responses to Seawater Dilution in Rock Crabs

Intertidal organisms live in areas which are above water at low tide, and under water at high tide. Species of intertidal crabs vary widely in their ability to regulate their salt and water levels in the face of environmental fluctuations. As shown in the graph below, intertidal crabs face an osmotic influx of water when placed into dilute salinities. The excess water can be excreted in the urine, but this is accompanied by a loss of valuable salts (ions). In many species, after a period of adjustment, this loss is met by active ion uptake across the gills.

In an experiment, a student investigated the effect of increasing seawater dilution on the cumulative weight gain of a common rock crab. Six crabs in total were used in the experiment. Three were placed in a seawater dilution of 75:25 (75% seawater) and three were placed in a seawater dilution of 50:50 (50% seawater). Cumulative weight gain in each of the six crabs was measured at regular intervals over a period of 30 minutes. The results are plotted below and a line of best fit has been drawn for each set of data.

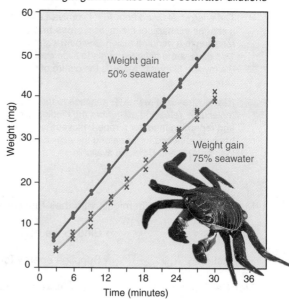

Weight gain in crabs at two seawater dilutions

1. What is the key difference between an **osmoregulator** and an **osmoconformer**? _____

_____

_____

2. (a) Explain the difference in the two lines plotted on the graph of crab weight gain (above right): _____

_____

_____

(b) What does this experiment suggest about the osmoregulatory ability of this crab species? _____

_____

© 2012-2014 **BIOZONE** International
ISBN: 978-1-927173-93-0
Photocopying Prohibited

298 Water Budget in Humans

Key Idea: Water lost from an individual must be replaced by an equal amount to maintain physiological functions. Too much or too little water can result in health issues.

Water is essential to physiological function and health, we cannot live without water for more than about 100 hours. Any water lost must be replaced by an equal volume, this balance is called the **water budget**. The amount of water required to maintain water balance varies between individuals, and depends on many factors including age, health, gender, level of physical exercise, and the environment (temperature and humidity). Potentially life-threatening health issues arise when the water balance is not maintained either through **dehydration** (excessive water loss) or **overhydration** (excessive water intake).

Daily Water Transfers in an Adult

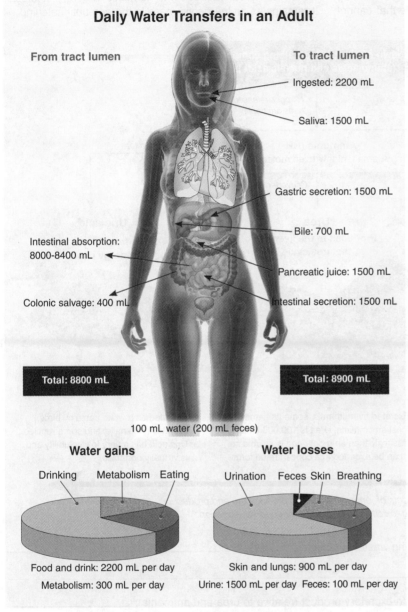

From tract lumen

To tract lumen

Ingested: 2200 mL

Saliva: 1500 mL

Gastric secretion: 1500 mL

Bile: 700 mL

Intestinal absorption: 8000-8400 mL

Pancreatic juice: 1500 mL

Colonic salvage: 400 mL

Intestinal secretion: 1500 mL

Total: 8800 mL

Total: 8900 mL

100 mL water (200 mL feces)

Water gains

Drinking   Metabolism   Eating

Food and drink: 2200 mL per day

Metabolism: 300 mL per day

Water losses

Urination   Feces   Skin   Breathing

Skin and lungs: 900 mL per day

Urine: 1500 mL per day   Feces: 100 mL per day

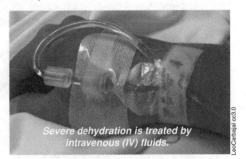

Severe dehydration is treated by intravenous (IV) fluids.

LeoCarbajal cc3.0

Dehydration occurs when water loss from the body exceeds water intake (e.g. through excessive exercise, fever, or prolonged vomiting or diarrhea). Many metabolic processes are disrupted, but the physiological effects of dehydration depend upon the extent of water loss.

▶ 3-4% loss: no obvious problems.

▶ 5-8% loss: fatigue and dizziness.

▶ >10% loss: physical and mental deterioration, accompanied by severe thirst.

▶ >15-25% loss usually fatal.

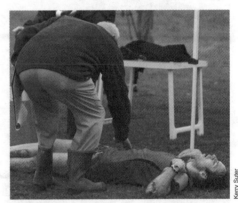

Kerry Suter

Overhydration (also called hyponatremia)occurs when more water is taken into the body than is lost. The excess of water dilutes sodium levels, which can cause a number of problems including digestive problems, behavioural changes, brain damage, seizures, or coma. The condition most often occurs in athletes competing in ultradistance events lasting many hours (above).

1. How does metabolism provide water for the body's activities? _____

_____

_____

2. (a) Distinguish between **dehydration** and **overhydration**: _____

_____

_____

(b) Describe the physiological consequences of each: _____

_____

_____

LINK
302
WEB
298
APP

299 Nitrogenous Wastes in Animals

Key Idea: Nitrogenous wastes are produced from the breakdown of nitrogen containing compounds. They must be excreted before they accumulate to toxic levels.

Wastes generated by cellular metabolism must be continually removed (excreted) so that they do not accumulate to toxic levels and cause damage. Nitrogenous wastes are produced from the breakdown of amino acids and nucleic acids. The simplest breakdown product of nitrogen-containing compounds is ammonia, a highly toxic molecule that cannot

be retained in the body for long. Most aquatic animals excrete ammonia immediately into the water where it is washed away. Other animals convert the ammonia to a less toxic form (urea or uric acid) that can remain in the body for a short time before being excreted. The form of the excretory product in terrestrial animals depends on the type of organism and its life history. Terrestrial animals that lay eggs produce uric acid rather than urea, because it is non-toxic and very insoluble. It remains as an inert solid mass in the egg until hatching.

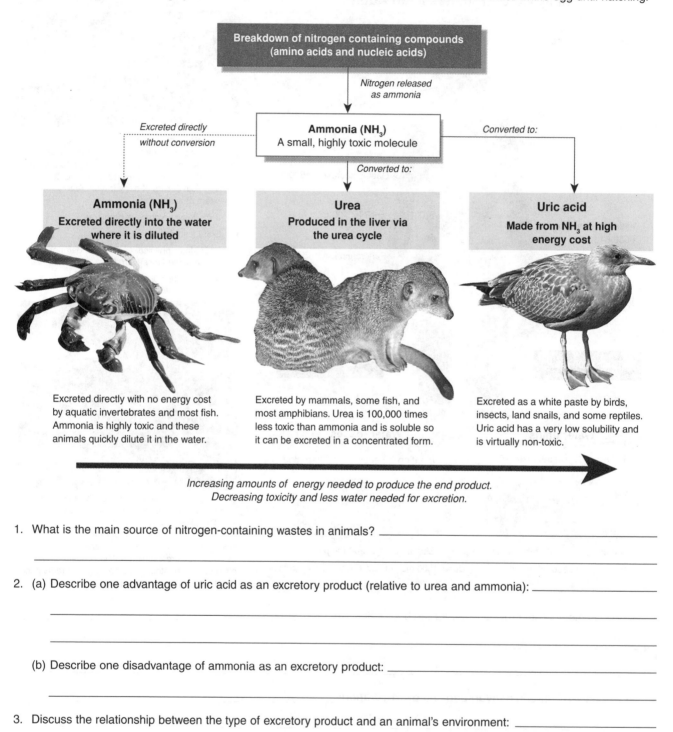

Breakdown of nitrogen containing compounds
(amino acids and nucleic acids)

Nitrogen released as ammonia

Ammonia (NH$_3$)
A small, highly toxic molecule

Excreted directly without conversion

Converted to:

Converted to:

Ammonia (NH$_3$)
Excreted directly into the water where it is diluted

Urea
Produced in the liver via the urea cycle

Uric acid
Made from NH$_3$ at high energy cost

Excreted directly with no energy cost by aquatic invertebrates and most fish. Ammonia is highly toxic and these animals quickly dilute it in the water.

Excreted by mammals, some fish, and most amphibians. Urea is 100,000 times less toxic than ammonia and is soluble so it can be excreted in a concentrated form.

Excreted as a white paste by birds, insects, land snails, and some reptiles. Uric acid has a very low solubility and is virtually non-toxic.

Increasing amounts of energy needed to produce the end product.
Decreasing toxicity and less water needed for excretion.

1. What is the main source of nitrogen-containing wastes in animals? _____

_____

2. (a) Describe one advantage of uric acid as an excretory product (relative to urea and ammonia): _____

_____

_____

(b) Describe one disadvantage of ammonia as an excretory product: _____

_____

3. Discuss the relationship between the type of excretory product and an animal's environment: _____

_____

_____

_____

_____

LINK
38

LINK
222

© 2012-2014 **BIOZONE** International
ISBN: 978-1-927173-93-0
Photocopying Prohibited

300 Excretory Systems

Key Idea: Malpighian tubules are the excretory organs of insects. Kidneys are the excretory organs of vertebrates. Excretory systems remove waste products from an organism's body. The excretory organs of insects, **Malpighian tubules**, remove nitrogenous wastes (as uric acid) from the blood, and also function in osmoregulation. In vertebrates, the excretory organs are the **kidneys**. The kidneys of all vertebrates produce urine. Mammals and birds have very efficient kidneys that can produce a concentrated urine, excreting nitrogenous wastes whilst conserving water and ions.

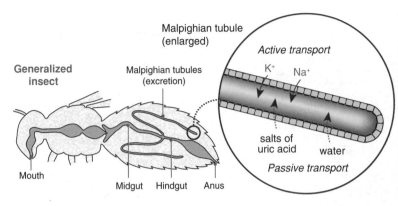

Malpighian tubule (enlarged)

Generalized insect

Malpighian tubules (excretion)

Active transport

K⁺   Na⁺

salts of uric acid   water

Passive transport

Mouth

Midgut Hindgut Anus

The Excretory System of Insects

Insects have two to several hundred **Malpighian tubules** projecting from the junction of the midgut and hindgut. They are bathed in the clear fluid (hemolymph) of the insect's body cavity where they actively pump K^+ and Na^+ into the tubule. Water, uric acid salts, and several other substances follow by passive transport. Water and some ions are reabsorbed in the hindgut, while **uric acid** precipitates out as a paste and is passed out of the anus along with the fecal material. The ability to conserve water by excreting solid uric acid has enabled insects to colonize very arid (dry) environments.

The Excretory System of Vertebrates

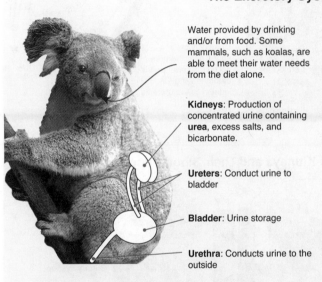

Water provided by drinking and/or from food. Some mammals, such as koalas, are able to meet their water needs from the diet alone.

Kidneys: Production of concentrated urine containing **urea**, excess salts, and bicarbonate.

Ureters: Conduct urine to bladder

Bladder: Urine storage

Urethra: Conducts urine to the outside

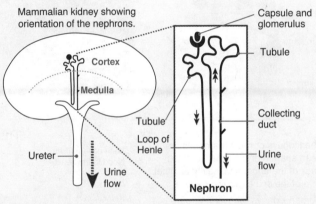

Mammalian kidney showing orientation of the nephrons.

Cortex

Medulla

Ureter

Urine flow

Capsule and glomerulus

Tubule

Tubule

Loop of Henle

Collecting duct

Urine flow

Nephron

Water loss is a major problem for most mammals. The degree to which urine can be concentrated (and water conserved) depends on the number of nephrons present and the length of the loop of Henle. The highest urine concentrations are found in desert-adapted mammals.

Mammalian **kidneys** each contain more than one million **nephrons**: the selective filter elements which regulate the composition of the blood and excrete wastes. In the nephron, the initial urine is formed by **filtration** in the glomerulus and Bowman's capsule. The filtrate is modified by **secretion** and **reabsorption** of ions and water. These processes create a salt gradient in the fluid around the nephron, which allows water to be withdrawn from the urine in the collecting duct.

1. How do insects concentrate their nitrogenous waste into a paste? _____

_____

_____

_____

2. State the function of each of the following components of the vertebrate urinary system:

(a) Kidney: _____

(b) Ureters: _____

(c) Bladder: _____

(d) Urethra: _____

LINK
302

LINK
299

KNOW

301 Kidney Structure

Key Idea: In terrestrial vertebrates, the kidneys excrete nitrogenous waste and maintain water and solute balance. The mammalian urinary system consists of the kidneys and bladder, and their associated blood vessels and ducts. The kidneys have a plentiful blood supply from the renal artery. The blood plasma is filtered by the kidneys to form urine.

Urine is produced continuously, passing along the ureters to the bladder. By adjusting the composition of the fluid excreted, the kidneys help to maintain the body's internal chemical balance. Mammalian kidneys are very efficient, producing a urine that is concentrated to varying degrees depending on requirements.

Urinary System

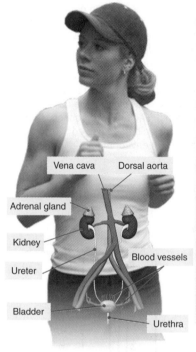

- Vena cava
- Dorsal aorta
- Adrenal gland
- Kidney
- Ureter
- Blood vessels
- Bladder
- Urethra

Kidneys *in-situ* (Rat)

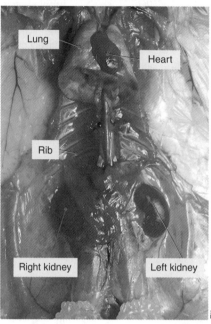

- Lung
- Heart
- Rib
- Right kidney
- Left kidney

Internal Structure of the Human Kidney

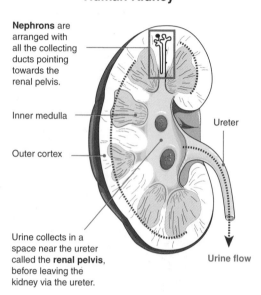

Nephrons are arranged with all the collecting ducts pointing towards the renal pelvis.

- Inner medulla
- Outer cortex
- Ureter

Urine collects in a space near the ureter called the **renal pelvis**, before leaving the kidney via the ureter.

Urine flow

The kidneys of most mammals are bean shaped organs that lie at the back of the abdominal cavity to either side of the spine (above, centre).

Human kidneys (above, right) are ~100-120 mm long and 25 mm thick. The precise alignment of the nephrons (the filtering elements of the kidney) and their associated blood vessels gives the kidney tissue a striped appearance. Each kidney contains more than 1 million nephrons. Nephrons are selective filter elements, which regulate blood composition and pH, and excrete wastes and toxins.

The Kidneys and Their Blood Supply

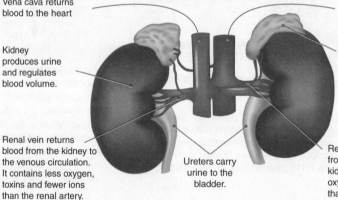

Vena cava returns blood to the heart

Kidney produces urine and regulates blood volume.

Renal vein returns blood from the kidney to the venous circulation. It contains less oxygen, toxins and fewer ions than the renal artery.

Ureters carry urine to the bladder.

Dorsal aorta supplies oxygenated blood to the body.

Adrenal glands are associated with, but not part of, the urinary system.

Renal artery carries blood from the aorta to the kidney. It contains more oxygen, toxins, and ions than the renal vein.

1. Describe the role of the kidneys: _____
_____
_____
_____
_____
_____

2. In the space provided (right), draw and label a cross section of the human kidney.

© 2012-2014 **BIOZONE** International
ISBN: 978-1-927173-93-0
Photocopying Prohibited

302 Kidney Function

Key Idea: The functional unit of the kidney is the nephron. It is a selective filter element, comprising a renal corpuscle and its associated tubules and ducts.

Ultrafiltration, i.e. forcing fluid and dissolved substances through a membrane by pressure, occurs in the first part of the nephron, across the membranes of the capillaries and the glomerular capsule. The formation of the glomerular filtrate depends on the pressure of the blood entering the nephron (below). If it increases, filtration rate increases; when it falls, glomerular filtration rate also falls. This process is precisely regulated so that glomerular filtration rate per day stays constant. The initial filtrate, now called urine is modified through secretion and tubular reabsorption according to body's needs at the time.

Nephron Structure

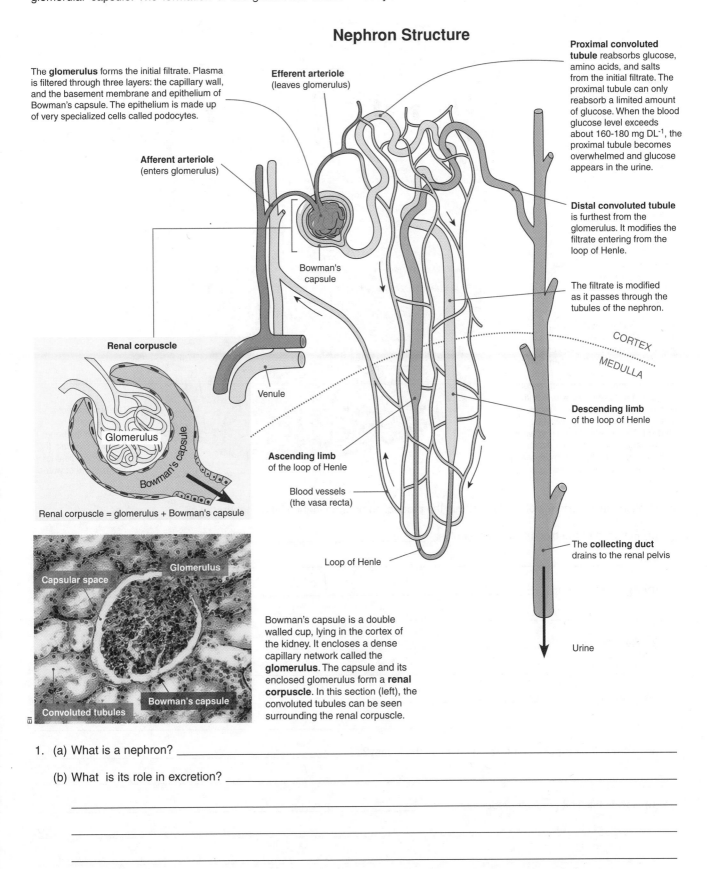

The **glomerulus** forms the initial filtrate. Plasma is filtered through three layers: the capillary wall, and the basement membrane and epithelium of Bowman's capsule. The epithelium is made up of very specialized cells called podocytes.

Efferent arteriole (leaves glomerulus)

Afferent arteriole (enters glomerulus)

Bowman's capsule

Proximal convoluted tubule reabsorbs glucose, amino acids, and salts from the initial filtrate. The proximal tubule can only reabsorb a limited amount of glucose. When the blood glucose level exceeds about 160-180 mg DL^{-1}, the proximal tubule becomes overwhelmed and glucose appears in the urine.

Distal convoluted tubule is furthest from the glomerulus. It modifies the filtrate entering from the loop of Henle.

The filtrate is modified as it passes through the tubules of the nephron.

CORTEX

MEDULLA

Descending limb of the loop of Henle

Renal corpuscle

Glomerulus

Bowman's capsule

Renal corpuscle = glomerulus + Bowman's capsule

Venule

Ascending limb of the loop of Henle

Blood vessels (the vasa recta)

Loop of Henle

The **collecting duct** drains to the renal pelvis

Urine

Glomerulus

Capsular space

Bowman's capsule

Convoluted tubules

Bowman's capsule is a double walled cup, lying in the cortex of the kidney. It encloses a dense capillary network called the **glomerulus**. The capsule and its enclosed glomerulus form a **renal corpuscle**. In this section (left), the convoluted tubules can be seen surrounding the renal corpuscle.

1. (a) What is a nephron? _____

(b) What is its role in excretion? _____

_____

_____

_____

LINK WEB
301   302   KNOW

Summary of Activities in the Kidney Nephron

Urine formation begins by **ultrafiltration** of the blood, as fluid is forced through the capillaries of the glomerulus, forming a filtrate similar to blood but lacking cells and proteins. The filtrate is then modified by secretion and **reabsorption** to add or remove substances (e.g. ions). The processes involved in urine formation are summarized below for each region of the nephron (glomerulus, proximal convoluted tubule, loop of Henle, and distal convoluted tubule), and the collecting duct. The loop of Henle acts as a countercurrent multiplier, establishing and increasing the salt gradient through the medullary region. This is possible because the descending loop is freely permeable to water but the ascending loop is not.

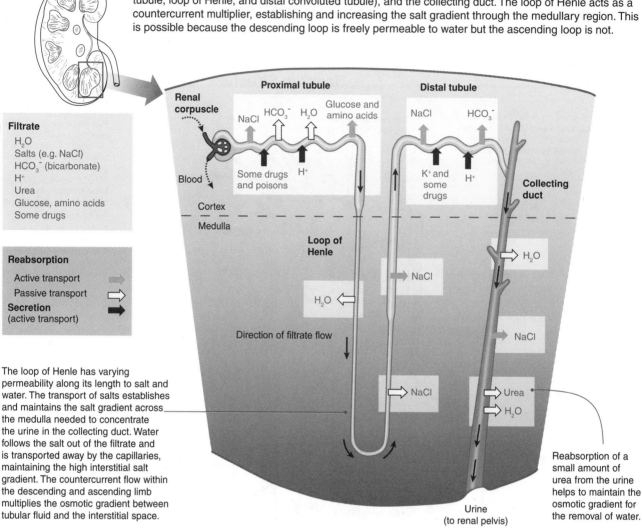

Filtrate

H_2O
Salts (e.g. NaCl)
HCO_3^- (bicarbonate)
H^+
Urea
Glucose, amino acids
Some drugs

Reabsorption

Active transport
Passive transport
Secretion
(active transport)

The loop of Henle has varying permeability along its length to salt and water. The transport of salts establishes and maintains the salt gradient across the medulla needed to concentrate the urine in the collecting duct. Water follows the salt out of the filtrate and is transported away by the capillaries, maintaining the high interstitial salt gradient. The countercurrent flow within the descending and ascending limb multiplies the osmotic gradient between tubular fluid and the interstitial space.

Reabsorption of a small amount of urea from the urine helps to maintain the osmotic gradient for the removal of water.

2. Explain the importance of the following in the production of urine in the kidney nephron:

(a) Filtration of the blood at the glomerulus: _____

_____

(b) Active secretion: _____

_____

(c) Reabsorption: _____

_____

(d) Osmosis: _____

_____

3. (a) What is the purpose of the salt gradient in the kidney? _____

_____

(b) How is this salt gradient produced? _____

_____

_____

© 2012-2014 **BIOZONE** International
ISBN: 978-1-927173-93-0
Photocopying Prohibited

303 The Kidney's Role in Water Conservation

Key Idea: The length of the loop of Henle determines how much the urine can be concentrated. Desert dwelling animals generally have long loops of Henle and produce concentrated urine to conserve water.

Water loss is a major problem for many land mammals, but particularly so in desert mammals with a limited supply of water. Desert mammals have evolved mechanisms to reabsorb water and limit the amount lost in urine. For mammals living in a mesic environment (where there is

an adequate water supply), water reabsorption from the kidneys is less important because the losses can be recovered through drinking. The degree to which urine can be concentrated (and water conserved) depends on the number of nephrons in the kidney and the length of the loop of Henle. Desert animals often have a loop of Henle that is longer than non-arid-adapted desert mammals, and also have the most concentrated urine.

The Loop of Henle and Water Conservation

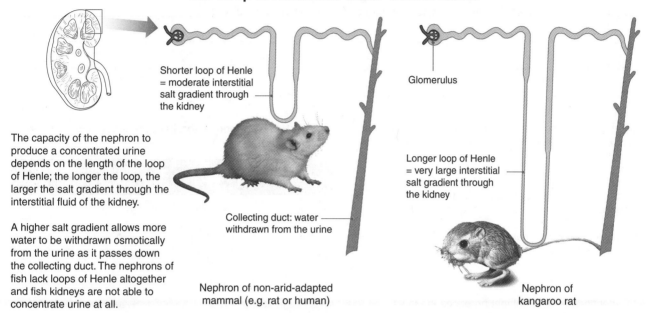

The capacity of the nephron to produce a concentrated urine depends on the length of the loop of Henle; the longer the loop, the larger the salt gradient through the interstitial fluid of the kidney.

A higher salt gradient allows more water to be withdrawn osmotically from the urine as it passes down the collecting duct. The nephrons of fish lack loops of Henle altogether and fish kidneys are not able to concentrate urine at all.

Shorter loop of Henle = moderate interstitial salt gradient through the kidney

Glomerulus

Longer loop of Henle = very large interstitial salt gradient through the kidney

Collecting duct: water withdrawn from the urine

Nephron of non-arid-adapted mammal (e.g. rat or human)

Nephron of kangaroo rat

Determining the Length of The Loop of Henle

The length of the loop of Henle is determined by measuring the thickness of the medulla (below). In general, the larger the medulla, the longer the loops of Henle.

The length of the loop of Henle is related to habitat (right). Animals in arid (dry) environments have thicker medullas (and therefore long loops of Henle) relative to mesic-dwelling animals.

Medullary Thickness, Environment, and Body Mass

● Arid environment
● Mesic environment

Log relative medullary thickness / Log body mass (kg)

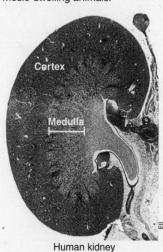

Human kidney

Mammals and birds can vary the concentration of their urine to excrete or conserve water as needed. The ability to concentrate urine is correlated with the environment, in particular access to water (above). Animals with limited water access (e.g. desert animals, such as the kangaroo rat) have evolved very efficient mechanisms to reabsorb water (e.g. long loops of Henle) and produce very concentrated urine.

LINK 302 | WEB 303 | KNOW

1. Explain the relationship between the length of the loop of Henle and the ability to concentrate urine: _____

_____

_____

2. Study the graph on the previous page.
   (a) Describe the relationship between environment and medullary thickness: _____

   _____

   _____

   (b) Explain the significance of this relationship in terms of environmental conditions: _____

   _____

   _____

   _____

3. Different animals have different maximum urine concentration levels. Some examples are given in the table (right).

   (a) On the grid below, plot the urine concentration for each of the mammals in the table as a bar graph. Remember to include appropriate labels.

   (b) Identify the mammal with the best ability to concentrate urine:

   _____

   (c) Identify the mammal that produces the least concentrated urine:

   _____

   (d) Why do you think this is the case? _____

   _____

   _____

   _____

   _____

Table 1. Urine concentration in mammalian species from different habitats

Mammal	Habitat	Urine concentration (mOsmol L^{-1})
Beaver	Aquatic/land	520
Human	Mesic	1400
Rat	Mesic	2900
Camel	Arid	2800
Kangaroo rat	Arid	5500

Adapted from Willmer, 2000

4. Suggest why the camel, an arid-adapted mammal, has a urine concentration closer to that of a rat (mesic) than a kangaroo rat (arid). HINT: Camels metabolize stored fat.

   _____

   _____

   _____

   _____

   _____

   _____

   _____

   _____

304 Control of Urine Output

Key Idea: Two hormones, antidiuretic hormone (ADH) and aldosterone, are involved in controlling urine output.
Variations in salt and water intake, and in the environmental conditions to which we are exposed, contribute to fluctuations in blood volume and composition. The primary role of the kidneys is to regulate blood volume and composition (including the removal of nitrogenous wastes), so that homeostasis is maintained. This is achieved through varying the volume and composition of the urine. Two hormones, antidiuretic hormone (ADH) and aldosterone, are involved in the process.

Control of Urine Output

Osmoreceptors in the **hypothalamus** detect a fall in the concentration of water in the blood. They stimulate **neurosecretory cells** in the hypothalamus to synthesize and secrete the hormone ADH (antidiuretic hormone).

ADH passes from the hypothalamus to the posterior pituitary where it is released into the blood. ADH increases the permeability of the kidney collecting duct to water so that more water is reabsorbed and urine volume decreases.

Brain

ADH ACTS ON KIDNEY

Factors inhibiting ADH release
► Low solute concentration
  • High blood volume
  • Low blood sodium levels
► High fluid intake
► Alcohol consumption

ADH levels decrease → Water reabsorption decreases. Urine output increases.

Factors causing ADH release
► High solute concentration
  • Low blood volume
  • High blood sodium levels
► Low fluid intake
► Nicotine and morphine

ADH levels increase → Water reabsorption increases. Urine output decreases.

Factors causing release of aldosterone
Low blood volumes also stimulate secretion of aldosterone from the adrenal cortex. This is mediated through a complex pathway involving the hormone renin from the kidney.

Aldosterone → Sodium reabsorption increases, water follows, blood volume restored.

1. (a) **Diabetes insipidus** is a type of diabetes, caused by a lack of ADH. Based on what you know of the role of ADH in kidney function, describe the symptoms of this disease:

_____

_____

(b) How might this disorder be treated? _____

_____

2. Why does alcohol consumption (especially to excess) cause dehydration and thirst? _____

_____

_____

3. (a) What is the effect of aldosterone on the kidney nephron? _____

(b) What is the net result of its action? _____

4. How do negative feedback mechanisms operate to regulate blood volume and urine output? _____

_____

_____

_____

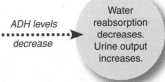

305 Diagnostic Urinalysis

Key Idea: Urine analysis can be used to detect medical disorders, pregnancy, and the use of illegal drugs.
Urine analysis (**urinalysis**) is used as a medical diagnostic tool for a wide range of metabolic disorders. In addition, urine analysis can be used to detect the presence of illicit (non-prescription) drugs and for diagnosing pregnancy.

Diagnostic Urinalysis

A urinalysis is an array of tests performed on urine. It is one of the most common methods of medical diagnosis, as most tests are quick and easy to perform, and they are non-invasive. Urinalysis can be used to detect for the presence of blood cells in the urine, glucose, proteins, and drugs.

A urinalysis may includes a **macroscopic analysis**, a **dipstick chemical analysis**, in which the test results can be read as colour changes, and a **microscopic analysis**, which involves centrifugation of the sample and examination for crystals, blood cells, or microbial contamination.

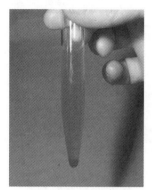

MACROSCOPIC URINALYSIS
The first part of a urinalysis is direct visual observation. Normal, fresh urine is pale to dark yellow or amber in colour and clear.

Turbidity or cloudiness may be caused by excessive cellular material or protein in the urine. A red or red-brown (abnormal) colour may be due to the presence of proteins (hemoglobin or myoglobin). If the sample contained many red blood cells, it would be cloudy as well as red, as in this sample indicating blood in the urine.

DIPSTICK URINALYSIS
A urine dipstick is a narrow band of paper saturated with chemical indicators for specific substances. Dipstick tests include:

Protein: Normal total protein excretion does not exceed 10 mg per 100 mL in any single specimen. More than 150 mg per day is defined as proteinuria.

Glucose: Less than 0.1% of glucose filtered by the glomerulus normally appears in urine. Excess sugar in urine is usually due to untreated diabetes mellitus, which is characterised by high blood glucose levels (the cells cannot take up glucose so it is excreted).

Pregnancy: The hormone human chorionic gonadotrophin (HCG) is produced by women in early pregnancy. It accumulates in the blood and is excreted in the urine. A pregnancy test is a special dipstick which uses bound monoclonal antibodies to detect low levels of HCG in the urine.

Testing For Anabolic Steroids

Anabolic steroids are synthetic steroids related to the male sex hormone **testosterone** (right). They work by increasing protein synthesis within cells, causing tissue, especially skeletal muscle, to build mass. They are used in medicine to stimulate bone growth and appetite, induce male puberty, and treat chronic wasting conditions.

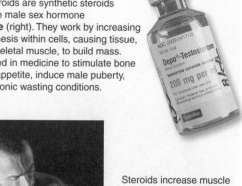

Steroids increase muscle mass and physical strength, and are used illegally by some athletes to gain an advantage over their competitors. Anabolic steroid use is banned by most major sporting bodies, but many athletes continue to use them illegally. Athletes are routinely tested for the presence of performance enhancing drugs, including anabolic steroids.

Anabolic steroids break down into known metabolites which are excreted in the urine. The presence of specific metabolites indicates which substance has been used by the athlete. Some steroid metabolites stay in the urine for weeks or months after being taken, while others are eliminated quite rapidly.

Athletes using anabolic steroids can escape detection by stopping use of the drugs prior to competition. This allows the body time to break down and eliminate the components, and the drug use goes undetected.

1. Why is **urinalysis** a frequently used diagnostic technique for many common disorders? _____
_____
_____

2. What might the following abnormal results in a urine test suggest to a doctor?

   (a) Excess glucose: _____

   (b) A red-brown colour: _____

3. Why might an athlete who is using illegal drugs withhold them for a period of time before competition? _____
_____
_____

© 2012-2014 **BIOZONE** International
ISBN: 978-1-927173-93-0
Photocopying Prohibited

306 Kidney Dialysis

Key Idea: A kidney dialysis machine acts as an artificial kidney, removing waste from the blood when the kidneys fail. When the kidneys do not function properly, waste products build up in the body, and medical intervention is required to correct the problem. A dialysis machine remove wastes from the blood. It is used when the kidneys fail, or when blood acidity, urea, or potassium levels increase above normal. Blood flows through a system of tubes composed of partially permeable membranes. Dialysis fluid (dialysate) has a composition similar to blood except that the concentration of wastes is low. It flows in the opposite direction to the blood on the outside of the dialysis tubes. Consequently, waste products like urea diffuse from the blood into the dialysis fluid, which is constantly replaced. Dialysis can be ongoing, or can be used to allow the kidneys to rest and recover from injury, the effects of drugs or other metabolic disturbance.

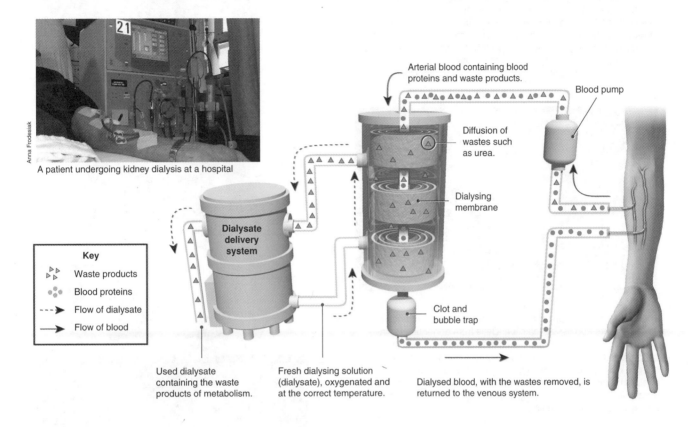

A patient undergoing kidney dialysis at a hospital

Arterial blood containing blood proteins and waste products.

Blood pump

Diffusion of wastes such as urea.

Dialysing membrane

Clot and bubble trap

Dialysate delivery system

Key

▷▷ Waste products

○○○ Blood proteins

- - -▶ Flow of dialysate

——▶ Flow of blood

Used dialysate containing the waste products of metabolism.

Fresh dialysing solution (dialysate), oxygenated and at the correct temperature.

Dialysed blood, with the wastes removed, is returned to the venous system.

1. In kidney dialysis, explain why the dialysing solution is constantly replaced rather than being recirculated:

_____

_____

2. Explain why ions such as potassium and sodium, and small molecules like glucose do not diffuse rapidly from the blood into the dialysing solution along with the urea:

_____

_____

3. Explain why the urea passes from the blood into the dialysing solution: _____

_____

4. Describe the general transport process involved in dialysis: _____

_____

5. Give a reason why the dialysing solution flows in the opposite direction to the blood: _____

_____

_____

_____

LINK
305
WEB
306
APP

307 Kidney Transplants

Key Idea: A kidney transplant, where a healthy kidney from one person is transplanted into another, is a procedure for people with complete kidney failure.

Kidney failure may come on suddenly (acute) or develop over a long period of time (chronic). Recovery from acute kidney failure is possible, but chronic kidney damage can not be reversed. If kidney deterioration is ignored, the kidneys will fail completely. In some cases diet and medication can be used to treat kidney failure, but when the damage is extensive, a kidney transplant is required.

Transplantation of a healthy kidney from an organ donor is the preferred treatment for end-stage kidney failure. The organ is usually taken from a person who has just died, although kidneys can also be taken from living donors. The damaged kidneys are left in place and the new kidney transplanted into the lower abdomen (right). Provided recipients comply with medical requirements (e.g. correct diet and medication) over 85% of kidney transplants are successful.

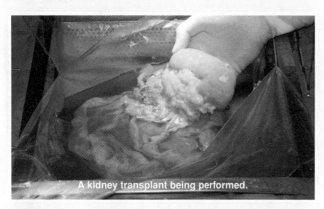

A kidney transplant being performed.

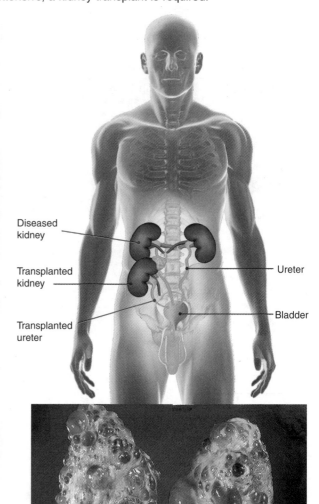

Diseased kidney

Transplanted kidney

Transplanted ureter

Ureter

Bladder

There are two major problems associated with kidney transplants: lack of donors and tissue rejection. Cells from donor tissue have different antigens to that of the recipient, and the body will launch an immune attack against the new kidney. Tissue-typing and the use of immunosuppressant drugs helps to decrease organ rejection rates. In the future, the transplant of genetically modified organs from other species may help to solve the problems of supply and immune rejection.

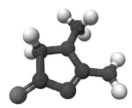

Kidney failure can be diagnosed by blood tests and urine analysis. Levels of the protein creatinine (a breakdown product of metabolism) in blood and urine indicate how well the kidneys are working. In normally functioning kidneys most of the creatinine is filtered out of the blood into the urine. If the kidneys are damaged, the amount of creatinine (left) in urine decreases while its level in blood increases.

These kidneys have polycystic kidney disease, a disease where cysts grow the kidneys.

CDC

1. Distinguish between acute and chronic kidney failure: _____

_____

_____

2. Why would a rise in blood levels of creatinine indicate kidney failure? _____

_____

_____

_____

3. Describe some of the advantages and disadvantages of kidney transplantation: _____

_____

_____

_____

© 2012-2014 **BIOZONE** International
ISBN: 978-1-927173-93-0
Photocopying Prohibited

308 Animal Sexual Reproduction

Key Idea: Fertilization occurs when the male and female gametes come together to form a zygote. In animals, fertilization may occur internally or externally.

Sexual reproduction involves the production of sex cells (**gametes**) produced by sex organs (gonads). Female gametes are called **eggs**, male gametes are called **sperm**. Animal sexual reproduction follows one of three main patterns (below), determined by the location of fertilization and embryonic development. Many aquatic invertebrates and fish have **external fertilization**, in which the parents release their gametes into the water at the same time.

Other invertebrates, reptiles, sharks, birds, and mammals have **internal fertilization**, in which sperm is transferred directly into the female to increase the chances of successful fertilization. In birds and most reptiles, one adaptation to life on land has been the evolution of the **amniote egg**: a structure that enables the embryo to complete its development outside the parent surrounded by a protective shell and nourished by a yolk sac. The pattern of internal development in mammals (termed gestation or pregnancy) provides the most advantages for the embryo in terms of nourishment and protection during development.

Strategies for Fertilization and Development

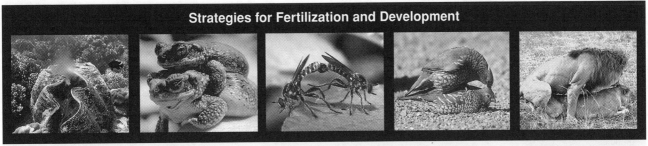

1. External fertilization and development
Many marine invertebrates release gametes into the sea. Large numbers of gametes are produced. *Example: giant clam (above, left).* In amphibians, a prolonged coupling, called amplexus, precedes gamete release and external fertilization and development. *Example: frogs (above, right).*

2. Internal fertilization and external development
Insects often have elaborate courtship rituals. Fertilization is internal, but the eggs are laid and develop externally. *Example: dipteran flies (above left).* In birds and reptiles, gamete fertilization is internal but the eggs are laid (usually in nests) and develop externally. *Example: quail (above, right).*

3. Internal fertilization and development
In mammals, fertilization is internal and there is a long period of internal development. *Example: lions*

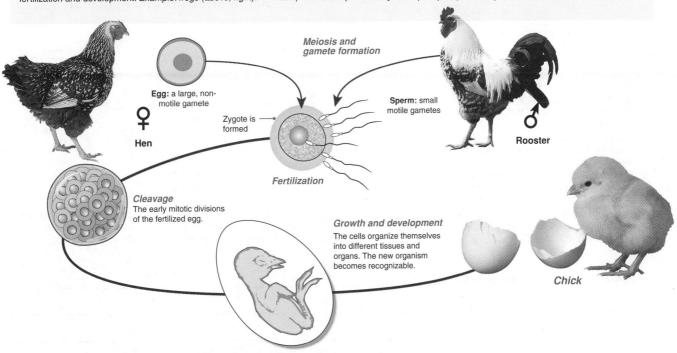

Meiosis and gamete formation

Egg: a large, non-motile gamete

♀

Hen

Zygote is formed

Sperm: small motile gametes

♂

Rooster

Fertilization

Cleavage
The early mitotic divisions of the fertilized egg.

Growth and development
The cells organize themselves into different tissues and organs. The new organism becomes recognizable.

Chick

1. Distinguish between **internal fertilization** and **external fertilization**, identifying advantages of each strategy:

_____

_____

_____

2. (a) Name an animal group with internal fertilization but external development: _____

(b) Name an animal group with internal fertilization and internal development: _____

(c) Describe one benefit and one cost involved in providing for internal development of an embryo:

Benefit: _____

Cost: _____

LINK **309** LINK **257** **KNOW**

309 Gametes

Key Idea: Gametes are the sex cells of organisms. Male and female gametes differ in their size, shape, and number.

Gametes (sex cells) are produced for the purposes of sexual reproduction. The gametes of male and female mammals differ greatly in size, shape, and number. These differences reflect their different roles in fertilization and reproduction. Male gametes (**sperm**) are highly motile and produced in large numbers. Female gametes (**eggs** or ova) are large, few in number, and immobile. They move as a result of the wave-like motion produced by the ciliated cells lining the Fallopian tube. Egg cells contain some food sources to nourish the developing embryo. In mammals, this food source is small because, once implanted into the uterine lining, the embryo derives its nutrients from the mother's blood supply.

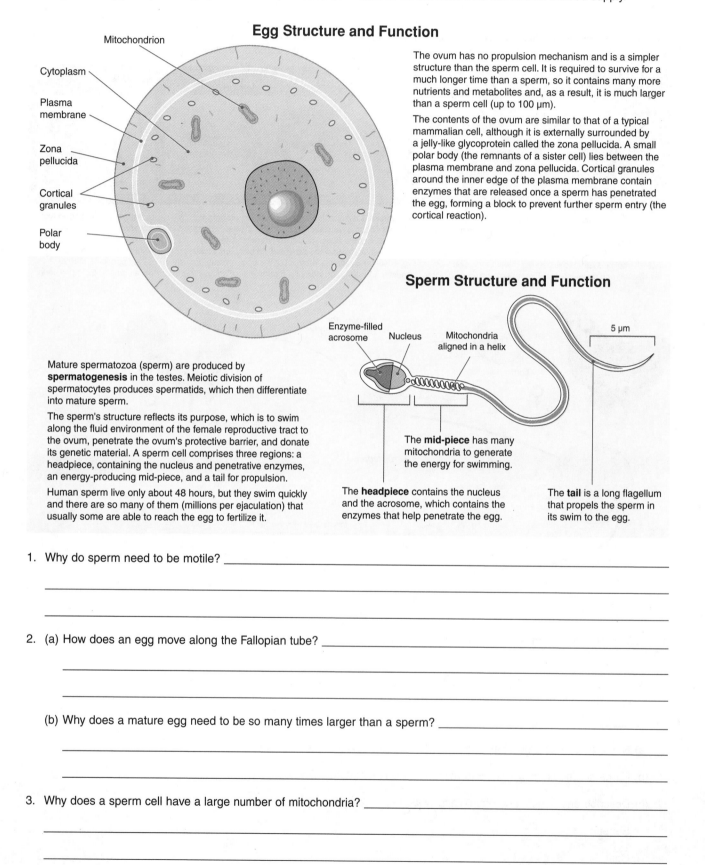

Egg Structure and Function

Labels (from top): Mitochondrion, Cytoplasm, Plasma membrane, Zona pellucida, Cortical granules, Polar body

The ovum has no propulsion mechanism and is a simpler structure than the sperm cell. It is required to survive for a much longer time than a sperm, so it contains many more nutrients and metabolites and, as a result, it is much larger than a sperm cell (up to 100 µm).

The contents of the ovum are similar to that of a typical mammalian cell, although it is externally surrounded by a jelly-like glycoprotein called the zona pellucida. A small polar body (the remnants of a sister cell) lies between the plasma membrane and zona pellucida. Cortical granules around the inner edge of the plasma membrane contain enzymes that are released once a sperm has penetrated the egg, forming a block to prevent further sperm entry (the cortical reaction).

Sperm Structure and Function

Labels: Enzyme-filled acrosome, Nucleus, Mitochondria aligned in a helix, 5 µm

Mature spermatozoa (sperm) are produced by **spermatogenesis** in the testes. Meiotic division of spermatocytes produces spermatids, which then differentiate into mature sperm.

The sperm's structure reflects its purpose, which is to swim along the fluid environment of the female reproductive tract to the ovum, penetrate the ovum's protective barrier, and donate its genetic material. A sperm cell comprises three regions: a headpiece, containing the nucleus and penetrative enzymes, an energy-producing mid-piece, and a tail for propulsion.

Human sperm live only about 48 hours, but they swim quickly and there are so many of them (millions per ejaculation) that usually some are able to reach the egg to fertilize it.

The **mid-piece** has many mitochondria to generate the energy for swimming.

The **headpiece** contains the nucleus and the acrosome, which contains the enzymes that help penetrate the egg.

The **tail** is a long flagellum that propels the sperm in its swim to the egg.

1. Why do sperm need to be motile? _____

_____

_____

2. (a) How does an egg move along the Fallopian tube? _____

_____

_____

   (b) Why does a mature egg need to be so many times larger than a sperm? _____

_____

_____

3. Why does a sperm cell have a large number of mitochondria? _____

_____

_____

© 2012-2014 **BIOZONE** International
ISBN: 978-1-927173-93-0
Photocopying Prohibited

310 Spermatogenesis

Key Idea: Sperm are the male gametes. They are produced by spermatogenesis in the testes.

Sperm are produced by a process called **spermatogenesis** in the testis. Mammalian sperm are highly motile and produced in large numbers. In human males, sperm production begins at puberty and continues throughout life, but does decline with age. Thousands of sperm are produced every second, and take approximately two months to fully mature.

Spermatogenesis

Spermatogenesis is the process by which mature spermatozoa (sperm) are produced in the testis. In humans, they are produced at the rate of about 120 million per day. Spermatogenesis is regulated by the hormones **follicle stimulating hormone** (FSH) (from the anterior pituitary) and testosterone (secreted from the testes in response to **luteinizing hormone** (LH) (from the anterior pituitary). Spermatogonia, in the outer layer of the seminiferous tubules, multiply throughout reproductive life. Some of them divide by meiosis into spermatocytes, which produce spermatids. These are transformed into mature sperm by the process of spermiogenesis in the seminiferous tubules of the testis. Full sperm motility is achieved in the epididymis.

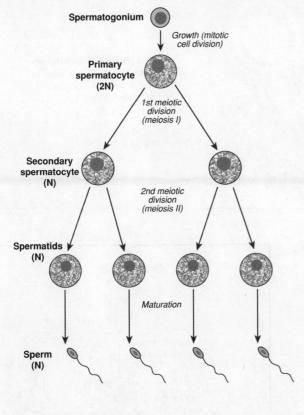

Cross Section Through Seminiferous Tubule

The photograph below shows maturing sperm (arrowed) with tails projecting into the lumen of the seminiferous tubule. Their heads are embedded in the Sertoli cells in the tubule wall and they are ready to break free and move to the epididymis where they complete their maturation. The same cross-section is illustrated diagrammatically (bottom).

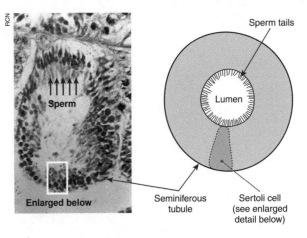

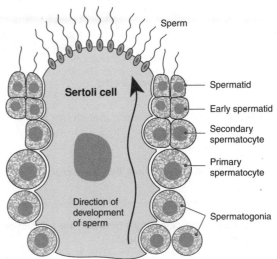

1. (a) Name the process by which mature sperm are formed: _____

   (b) Identify where this process takes place: _____

   (c) State how many mature sperm form from one primary spermatocyte: _____

   (d) State the type of cell division which produces mature sperm cells: _____

2. Describe the role of FSH and LH in sperm cell production: _____

   _____

   _____

3. Each ejaculation of a healthy, fertile male contains 100-400 million sperm. Suggest why so many sperm are needed:

   _____

   _____

311 Oogenesis

Key Idea: Eggs are the female gametes. They are produced by oogenesis, which takes place in the ovaries.

Egg cell (ovum, plural ova) production in females occurs by **oogenesis**. Unlike spermatogenesis, no new eggs are produced after birth. Instead, a human female is born with her entire complement of immature eggs. These remain in prophase of meiosis I throughout childhood. After puberty,

most commonly a single egg cell is released from the ovaries at regular monthly intervals as part of the menstrual cycle. These egg cells are arrested in metaphase of meiosis II. This second meiotic division is only completed upon fertilization. The release of egg cells from the ovaries takes place from the onset of puberty until menopause, when menstruation ceases and the woman is no longer fertile.

Development of the Ovarian Follicle and Egg Cell within the Ovary

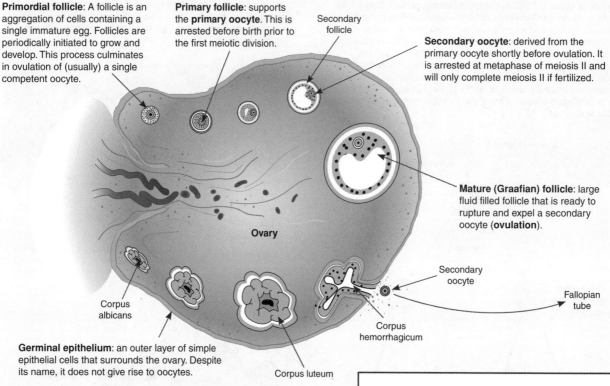

Primordial follicle: A follicle is an aggregation of cells containing a single immature egg. Follicles are periodically initiated to grow and develop. This process culminates in ovulation of (usually) a single competent oocyte.

Primary follicle: supports the **primary oocyte**. This is arrested before birth prior to the first meiotic division.

Secondary follicle

Secondary oocyte: derived from the primary oocyte shortly before ovulation. It is arrested at metaphase of meiosis II and will only complete meiosis II if fertilized.

Mature (Graafian) follicle: large fluid filled follicle that is ready to rupture and expel a secondary oocyte (**ovulation**).

Ovary

Secondary oocyte

Fallopian tube

Corpus albicans

Corpus hemorrhagicum

Germinal epithelium: an outer layer of simple epithelial cells that surrounds the ovary. Despite its name, it does not give rise to oocytes.

Corpus luteum

1. (a) Name the process by which mature ova form:

   _____

   (b) Name the place(s) where this takes place:

   _____

2. Discuss the main differences between the production of male gametes and female gametes:

   _____

   _____

   _____

   _____

3. Explain why males can be potentially fertile all their life, while female fertility decreases and eventually ceases with age:

   _____

   _____

   _____

   _____

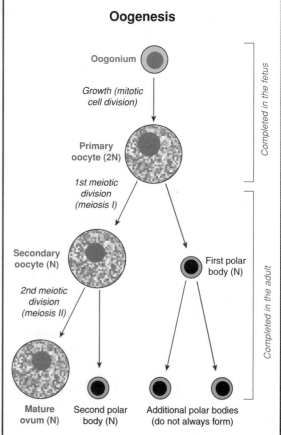

Oogenesis

Oogonium

Growth (mitotic cell division)

Primary oocyte (2N)

1st meiotic division (meiosis I)

Secondary oocyte (N)

First polar body (N)

2nd meiotic division (meiosis II)

Mature ovum (N)

Second polar body (N)

Additional polar bodies (do not always form)

Completed in the fetus

Completed in the adult

© 2012-2014 **BIOZONE** International
ISBN: 978-1-927173-93-0
Photocopying Prohibited

312 Fertilization and Early Growth

Key Idea: Fertilization occurs when a male and female gamete fuse to form a zygote.

Fertilization occurs when a sperm penetrates an egg cell at the secondary oocyte stage and the sperm and egg nuclei unite to form the zygote. In mammals, the entry of a sperm into the egg triggers specific mechanisms to prevent polyspermy (fertilization of the egg by more than one sperm). These include a change in membrane potential, and the cortical reaction (see below). A zygote resulting from polyspermy contains too many chromosomes, and is not viable (does not develop). Fertilization is seen as time 0 in a period of gestation (pregnancy) and has five stages (below). After fertilization, the zygote begins its development, i.e. its growth and differentiation into a multicellular organism.

Fertilization (Time 0)

The stages in fertilization are represented below in a numbered sequence (1-5)

1. Capacitation
The surface of the sperm cell undergoes changes that are essential to enabling the acrosome reaction and sperm entry.

2. The Acrosome Reaction
Enzymes from the acrosome (an enzyme-filled bag at the tip of the sperm) are released and digest a pathway through the follicle cells (not shown) and the jelly-like zona pellucida surrounding the egg cell (secondary oocyte).

3. Fusion of Sperm Head
The plasma membranes of the sperm and egg fuse, and the nucleus of the sperm enters the egg cytoplasm. Fusion causes a sudden membrane depolarization that acts as a "fast block" to further sperm entry. The fusion of the two plasma membranes also triggers the completion of meiosis II in the egg cell and induces the cortical reaction (below).

4. The Cortical Reaction
The fusion of the two plasma membranes induces a permanent change in the egg surface that prevents further sperm entry. Cortical granules in the egg cytoplasm release their contents into the space between the plasma membrane and the vitelline layer. Substances released from the granules raise and harden the vitelline layer to form a slow (permanent) block to further sperm entry.

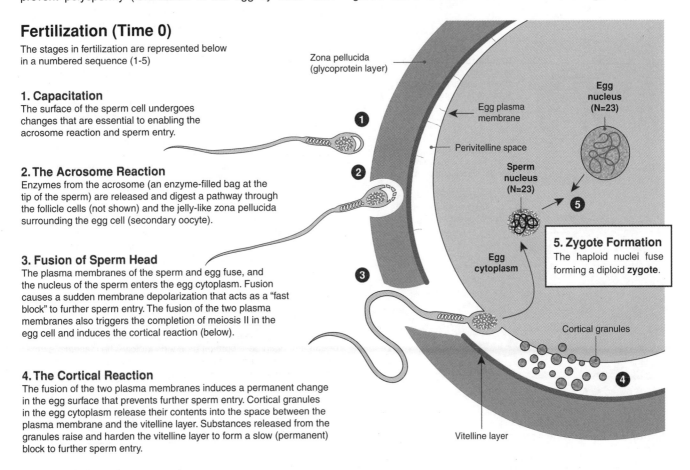

Zona pellucida (glycoprotein layer)

Egg plasma membrane

Egg nucleus (N=23)

Perivitelline space

Sperm nucleus (N=23)

Egg cytoplasm

5. Zygote Formation
The haploid nuclei fuse forming a diploid **zygote**.

Cortical granules

Vitelline layer

1. Briefly describe the significant events (and their importance) occurring at each of the following stages of fertilization:

   (a) Capacitation: _____
   _____

   (b) The acrosome reaction: _____
   _____

   (c) Fusion of egg and sperm plasma membranes: _____
   _____

   (d) The cortical reaction: _____
   _____

   (e) Fusion of egg and sperm nuclei: _____
   _____

2. Why is it important that fertilization of the egg by more than one sperm (polyspermy) does not occur? _____
   _____
   _____

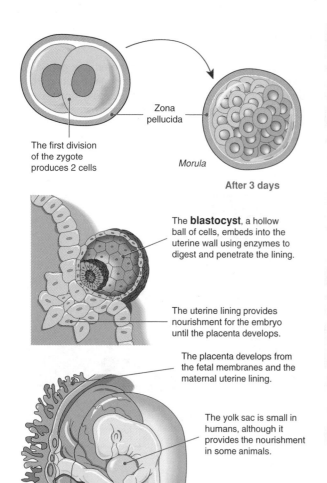

Zona pellucida

The first division of the zygote produces 2 cells

Morula

After 3 days

The **blastocyst**, a hollow ball of cells, embeds into the uterine wall using enzymes to digest and penetrate the lining.

The uterine lining provides nourishment for the embryo until the placenta develops.

The placenta develops from the fetal membranes and the maternal uterine lining.

The yolk sac is small in humans, although it provides the nourishment in some animals.

Umbilical cord

The fluid-filled amniotic sac encloses the embryo in the amniotic fluid.

5 week old embryo

Early Growth and Development

Cleavage and Development of the Morula

Immediately after fertilization, rapid cell division takes place. These early cell divisions are called **cleavage** and they increase the number of cells, but not the size of the zygote. The first cleavage is completed after 36 hours, and each succeeding division takes less time. After three days, successive cleavages have produced a solid mass of cells called the **morula**, (left) which is still about the same size as the original zygote.

Implantation of the Blastocyst (after 6-8 days)

After several days in the uterus, the morula develops into the blastocyst. It makes contact with the uterine lining and pushes deeply into it, ensuring a close maternal-fetal contact. Blood vessels provide early nourishment as they are opened up by enzymes secreted by the blastocyst. The embryo produces **HCG** (human chorionic gonadotropin), which prevents degeneration of the corpus luteum and signals that the woman is pregnant.

Embryo at 5-8 Weeks

Five weeks after fertilization, the embryo is only 4-5 mm long, but already the central nervous system has developed and the heart is beating. The embryonic membranes have formed; the amnion encloses the embryo in a fluid-filled space, and the allanto-chorion forms the fetal portion of the placenta. From two months the embryo is called a fetus. It is still small (30-40 mm long), but the limbs are well formed and the bones are beginning to harden. The face has a flat, rather featureless appearance with the eyes far apart. Fetal movements have begun and brain development proceeds rapidly. The placenta is well developed, although not fully functional until 12 weeks. The umbilical cord, containing the fetal umbilical arteries and vein, connects fetus and mother.

3. (a) Explain why the egg cell, when released from the ovary, is termed a secondary oocyte: _____

    _____

    (b) At which stage is its meiotic division completed? _____

4. What contribution do the sperm and egg cell make to each of the following:

    (a) The nucleus of the zygote?   Sperm contribution: _____ Egg contribution: _____

    (b) The cytoplasm of the zygote? Sperm contribution: _____ Egg contribution: _____

5. What is meant by cleavage? Explain its significance to the early development of the embryo:_____

    _____

    _____

6. (a) What is the importance of implantation to the early nourishment of the embryo?_____

    _____

    (b) What is the purpose of HCG production by the embryo? _____

    _____

    _____

7. Why is the fetus particularly prone to damage from drugs towards the end of the first trimester (2-3 months):

    _____

© 2012-2014 **BIOZONE** International
ISBN: 978-1-927173-93-0

313 The Placenta

Key Idea: The placenta allows materials to be exchanged between the fetus and its mother. It also acts as a temporary endocrine organ, secreting hormones to maintain pregnancy. The human fetus depends entirely on its mother for nutrients, oxygen, and the elimination of wastes. The **placenta** is the specialized organ that performs this role, enabling exchange between fetal and maternal tissues and allowing a prolonged period of fetal growth and development within the uterus. The placenta also has an endocrine role, producing progesterone and estrogen to maintain the pregnancy.

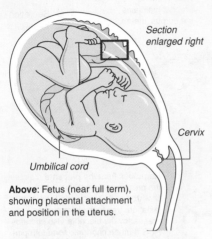

Above: Fetus (near full term), showing placental attachment and position in the uterus.

Below: Photograph of a human placenta, just after delivery.

Hadassah Ein Kerem Medical Center, cc3.0

Schematic diagram showing part of the placenta in section

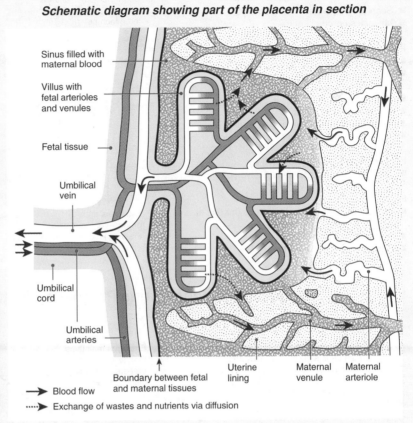

→ Blood flow

·····► Exchange of wastes and nutrients via diffusion

The placenta is a disc-like organ, about the size of a dinner plate and weighing about 1 kg. It develops when fingerlike projections (villi) from the fetal membranes grow into the uterine lining. The villi contain the numerous capillaries connecting the fetal arteries and vein. They continue invading the maternal tissue until they are bathed in the maternal blood sinuses. The maternal and fetal blood vessels are in such close proximity that oxygen and nutrients can diffuse from the maternal blood into the capillaries of the villi. From the villi, the nutrients circulate in the umbilical vein, returning to the fetal heart. Carbon dioxide and other wastes leave the fetus through the umbilical arteries, pass into the capillaries of the villi, and diffuse into the maternal blood. Note that fetal blood and maternal blood do not mix: the exchanges occur via diffusion through thin walled capillaries.

1. Describe the structure and function of the human placenta: _____

_____

_____

_____

_____

2. The umbilical cord contains the fetal arteries and vein. Describe the status of the blood in each type of fetal vessel:

(a) Fetal arteries: Oxygenated and containing nutrients / Deoxygenated and containing nitrogenous wastes (delete one)

(b) Fetal vein: Oxygenated and containing nutrients / Deoxygenated and containing nitrogenous wastes (delete one)

3. Describe how substances are exchanged between the mother and the fetus: _____

_____

_____

LINK **314**   LINK **312**   WEB **313**   **KNOW**

314 The Hormones of Pregnancy

Key Idea: Hormones secreted during pregnancy maintain the pregnancy and prepare the body for birth.

In a non-pregnant adult human female, the levels of estrogen and progesterone regulate the secretion of the pituitary hormones that control the ovarian cycle. Pregnancy interrupts this cycle and maintains the corpus luteum and the placenta as endocrine organs with the specific role of maintaining the developing fetus during its development. During the last month of pregnancy the hormone oxytocin induces the uterine contraction that will expel the baby from the uterus.

HCG (Human chorionic gonadotropin)
- Secreted by the developing embryo
- Maintains corpus luteum

Progesterone
- Maintains endometrium
- Inhibits uterine contraction

Estrogens
- Maintain endometrium
- Prepare mammary glands for lactation
- High levels induce labour

Human placental lactogen (HPL)
- Stimulates breast growth and development

Relaxin
- Produced by the placenta towards the end of the pregnancy
- Relaxes pubic symphysis at birth
- Helps dilate cervix at birth

Corpus luteum maintains pregnancy for the first three months

HCG from the embryo maintains the corpus luteum

→ Secretion
--→ Action

HCG

Estrogens and progesterone maintain the pregnancy

Hormones from the **placenta** maintain the pregnancy from three months onwards and prepare the breasts for lactation. Increasingly through pregnancy the placenta also secretes HCS (human chorionic somatotropin) which benefits fetal growth.

Hormonal Changes During Pregnancy, Birth, and Lactation

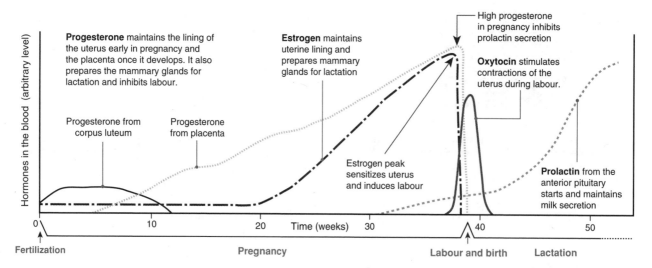

Progesterone maintains the lining of the uterus early in pregnancy and the placenta once it develops. It also prepares the mammary glands for lactation and inhibits labour.

Estrogen maintains uterine lining and prepares mammary glands for lactation

High progesterone in pregnancy inhibits prolactin secretion

Oxytocin stimulates contractions of the uterus during labour.

Progesterone from corpus luteum

Progesterone from placenta

Estrogen peak sensitizes uterus and induces labour

Prolactin from the anterior pituitary starts and maintains milk secretion

Hormones in the blood (arbitrary level)

0   10   20   Time (weeks)   30   40   50

Fertilization   Pregnancy   Labour and birth   Lactation

During the first 12-16 weeks pregnancy, the **corpus luteum** secretes enough progesterone to maintain the uterine lining and sustain the developing embryo. After this, the placenta takes over as the primary endocrine organ of pregnancy. **Progesterone** and **estrogen** from the placenta maintain the uterine lining, inhibit the development of further ova (eggs), and prepare the breast tissue for **lactation** (milk production). At the end of pregnancy, the placenta loses competency, progesterone levels fall, and high estrogen levels trigger the onset of labour. The estrogen peak coincides with an increase in oxytocin, which stimulates uterine contractions in a positive feedback loop: the contractions and the increasing pressure of the cervix from the infant stimulate release of more oxytocin, and more contractions and so on, until the infant exits the birth canal. After birth, the secretion of prolactin increases. Prolactin maintains lactation during the period of infant nursing.

1. (a) Why is the corpus luteum the main source of progesterone in early pregnancy? _____

    _____

    (b) What hormones are responsible for maintaining pregnancy? _____

2. (a) Name two hormones involved in labour (onset of the birth process): _____

    (b) Describe two physiological factors in initiating labour:_____

    _____

© 2012-2014 **BIOZONE** International
ISBN: 978-1-927173-93-0
Photocopying Prohibited

315 Gestational Development

Key Idea: There is some relationship between animal size and gestational period but other factors influence the degree of development and independence of newborn mammals.

The gestational period (length of pregnancy) of mammals varies greatly. Although gestational period is often longer for larger animals, factors other than size alone determine the length of the pregnancy and the level of development in the young at birth. These may include life-span, developmental rates, number of offspring produced, and threats (e.g. being eaten by other animals). Animals are classed as either precocial or altricial at birth (below), but in reality there are varying degrees of classification between the two extremes.

Mammals which move independently soon after birth, and that see and hear well are called **precocial**. Mobility is important as their defence against predation is to run. Large hoofed grazers tend to be precocial.

Altricial mammals are born relatively helpless. Often they are unable to see and are fairly immobile. The young of many species are born hairless with their eyes shut. Rodents, cats, dogs, and marsupials are altricial species.

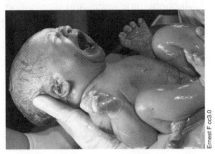

Newborn primates show varying degrees of precocial or altricial features at birth. Many primates move about shortly after birth and their eyes are open. In contrast, newborn humans have long periods of dependency.

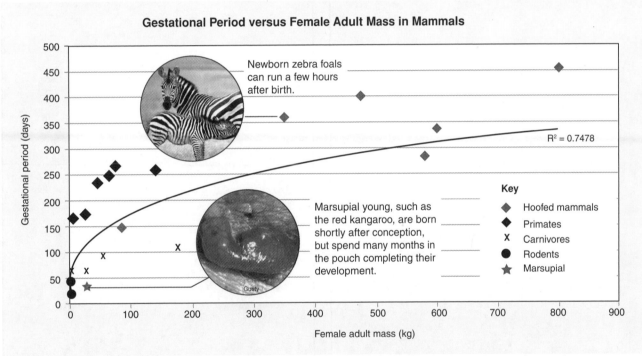

Gestational Period versus Female Adult Mass in Mammals

Newborn zebra foals can run a few hours after birth.

$R^2 = 0.7478$

Marsupial young, such as the red kangaroo, are born shortly after conception, but spend many months in the pouch completing their development.

Key
- ◆ Hoofed mammals
- ◆ Primates
- X Carnivores
- ● Rodents
- ★ Marsupial

Y-axis: Gestational period (days)
X-axis: Female adult mass (kg)

1. Analyse the graph above and describe the relationship between animal size (weight) and gestational period:

_____

_____

2. Suggest why hoofed mammals (sheep, zebra, horse cow, antelope) have long gestational periods? _____

_____

_____

_____

3. (a) What can you say about the position of the primates on the plot above?_____

_____

(b) Can you suggest a reason for this pattern?_____

_____

316 Chapter Review

Summarize what you know about this topic under the headings provided. You can draw diagrams or mind maps, or write short notes to organize your thoughts. Use the images and hints and guidelines included to help you:

Movement:
HINT: Describe the role of the musculoskeletal system in movement.

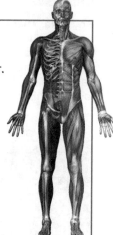

Antibody production and vaccination:
HINT: Distinguish between an antibody and an antigen. How does vaccination provide protection against disease?

The kidney and osmoregulation:
HINT: How does the structure of the kidney relate to its role in excretion and osmoregulation?

Sexual reproduction:
HINT: Explain the role of hormones in fertilization and embryonic development.

317 KEY TERMS: Did You Get It?

1. Match the following words with their definitions:

antibody	A	A molecule that is recognized by the immune system as foreign.
antigen	B	Specific white blood cells involved in the adaptive immune response.
clonal selection	C	Resistance of an organism to infection or disease.
immunity	D	The deliberate introduction of antigenic material to produce immunity to a disease.
lymphocyte	E	Protein in the blood that identifies and neutralizes foreign material (e.g. bacteria).
pathogen	F	A model for how B and T cells are selected to target specific antigens invading the body.
vaccination	G	A disease-causing organism.

2. The graph on the right shows the incidence of an infectious disease, *Haemophilus influenzae*.

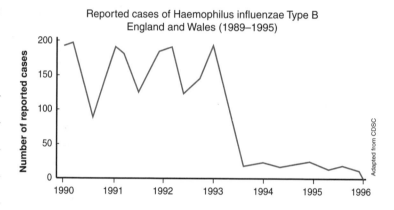

Reported cases of Haemophilus influenzae Type B England and Wales (1989–1995)

Adapted from CDSC

(a) Suggest a reason for the pattern of cases of *Haemophilus influenzae* prior to 1993:

_____

_____

_____

_____

_____

(b) Suggest a possible cause for the decline in the incidence of *Haemophilius influenzae* after 1993:

_____

_____

3. (a) In vertebrates, the location at which bones connect: _____

(b) The name for the voluntary muscle that produces movement: _____

(c) The name given to pairs of muscles that oppose each other to produce movement: _____

4. (a) Name the excretory organ of vertebrates: _____

(b) Name the selective filtering element of the kidney: _____

(c) The length of this is directly related to the ability of an organism to concentrate urine: _____

(d) Name the two hormones involved in controlling fluid and electrolyte balance: _____

5. Complete the following sentence by filling in the missing words. Use the word list provided:

placenta, oogenesis, fertilization, hormones, spermatogenesis, gametes

The sex cells of organisms are called _ _ _ _ _ _ _. In males they are called sperm and are produced by a process

called _ _ _ _ _ _ _ _ _ _ _ _ _ _ _. The female sex cells are called eggs (or ova), and are produced by

_ _ _ _ _ _ _ _ _. The union of male and female sex cells is called _ _ _ _ _ _ _ _ _ _ _ _ _ and results in formation of a

zygote. The zygote begins to divide, forming a hollow ball of cells called a blastocyst , which embeds in the lining of the

uterus. After about 12 weeks gestation, the growing fetus is supported entirely by the _ _ _ _ _ _ _ _, a temporary organ

that facilitates the exchange of nutrients and gases between the fetus and the mother. The placenta secretes

_ _ _ _ _ _ _ _, including progesterone and estrogen, which maintain the pregnancy.

© 2012-2014 **BIOZONE** International
ISBN: 978-1-927173-93-0
Photocopying Prohibited

KNOW

Experimental skills

Data Handling and Analysis

Key terms

assumption
bar graph
chi-squared test
control
controlled variable
correlation
dependent variable
fair test
histogram
hypothesis
independent variable
line graph
mean
median
mode
observation
prediction
sample
scatter graph
standard deviation
statistical test
Student's *t* test
variable

Group 4 experimental skills

Activity number

☐ 1 Demonstrate an understanding of science as inquiry, including the role of observation as a starting point for all investigations, and the importance of the hypothesis and testable predictions. — 318

☐ 2 Demonstrate an understanding of experimental design including the importance of a fair test, identification of dependent, independent, and controlled variables, choice of a control, and awareness of assumptions in your design. — 319 321

☐ 3 Demonstrate an ability to record your data systematically and accurately. — 321

☐ 4 Construct explanations based on evidence (including second hand data) and logical reasoning (these skills are implicit in many of the activities and so are found integrated throughout the workbook).

  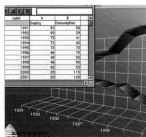

Mathematical requirements

Activity number

☐ 1 Perform basic arithmetic functions. Carry out calculations involving means, decimals, fractions, percentages, and ratios (these skills are implicit in many of the activities and so are found integrated throughout the workbook). — 320

☐ 2 Represent and interpret frequency data in the form of bar graphs and histograms, including direct and inverse proportion (these skills are implicit in many of the activities and so are found integrated throughout the workbook). — 322

☐ 3 Plot graphs, using appropriate scales and axes involving two variables that show linear or non-linear relationships (these skills are implicit in many of the activities and so are found integrated throughout the workbook). — 322

☐ 4 Plot and interpret scatter graphs to identify correlation between two variables. Explain why a correlation does not establish causality (these skills are implicit in many of the activities and so are found integrated throughout the workbook). — 322

☐ 5 Determine the mean, mode, and median of a set of sample data. Calculate and analyse standard deviation, recognizing it as a measure of spread or variability in the sample data. — 323 324

☐ 6 Select and apply appropriate statistical tests to analyse data and interpret the results. Appreciate that statistical tests make assumptions about the distribution of the data and must be used appropriately to be valid. See tests below:

Chi-squared test for association (numerical data as counts) — 120 121

Chi-squared test for difference (numerical data as counts) — 266 267

Mann-Whitney *U* test for differences in non-parametric data (weblink) — 277

Student's *t* test for difference between two populations — 325 326

318 The Scientific Method

Key Idea: The scientific method is a rigorous process of observation, measurement, and analysis that helps us to explain phenomena and predict changes in a system.

Scientific knowledge grows through a process called the **scientific method**. This process involves observation and measurement, hypothesizing and predicting, and planning and executing investigations designed to test formulated **hypotheses**. A scientific hypothesis is a tentative explanation for an observation, which is capable of being tested by experimentation. Hypotheses lead to **predictions** about the system involved and they are accepted or rejected on the basis of the investigation's findings. Acceptance of the hypothesis is not necessarily permanent: explanations may be rejected later in light of new findings.

A Simplified Model for Hypothesis Testing

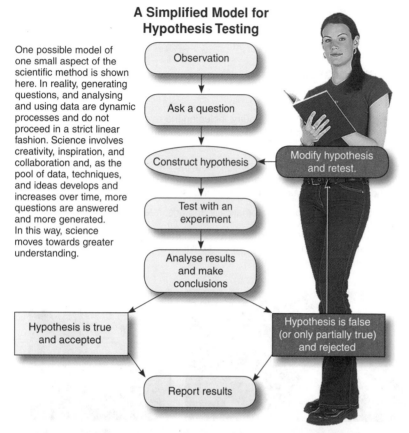

One possible model of one small aspect of the scientific method is shown here. In reality, generating questions, and analysing and using data are dynamic processes and do not proceed in a strict linear fashion. Science involves creativity, inspiration, and collaboration and, as the pool of data, techniques, and ideas develops and increases over time, more questions are answered and more generated. In this way, science moves towards greater understanding.

Observation → Ask a question → Construct hypothesis ← Modify hypothesis and retest. → Test with an experiment → Analyse results and make conclusions → Hypothesis is true and accepted / Hypothesis is false (or only partially true) and rejected → Report results

Forming a Hypothesis

Features of a sound hypothesis:

- It is based on observations and prior knowledge of the system.
- It offers an explanation for an observation.
- It refers to only one independent variable.
- It is written as a definite statement and not as a question.
- It is testable by experimentation.
- It leads to predictions about the system.

Testing a Hypothesis

Features of a sound method:

- It tests the validity of the hypothesis.
- It is repeatable.
- It includes a control which does not receive treatment.
- All variables are controlled where possible.
- The method includes a dependent and independent variable.
- Only the independent variable is changed (manipulated) between treatment groups.

Hypothesis Involving Manipulation
Used when the effect of manipulating a variable on a biological entity is being investigated. **Example**: The composition of applied fertilizer influences the rate of growth of plant A.

Hypothesis of Choice
Used when investigating species preference, e.g. for a particular habitat type or microclimate. **Example**: Woodpeckers (species A) show a preference for tree type when nesting.

Hypothesis Involving Observation
Used when organisms are being studied in their natural environment and conditions cannot be changed. **Example**: Fern abundance is influenced by the amount of canopy.

1. Why might an accepted hypothesis be rejected at a later date? _____

_____

2. Explain why a method must be repeatable: _____

_____

_____

3. In which situation(s) is it difficult, if not impossible, to control all the variables? _____

_____

© 2012-2014 **BIOZONE** International
ISBN: 978-1-927173-93-0
Photocopying Prohibited

Observations, Questions, and Hypotheses

Observation is the beginning of any scientific investigation. Often the best investigations are based on a series of fortuitous or specific observations. For example, in 1765 Edward Jenner developed the first vaccination for smallpox after hearing that milkmaids who contracted cowpox (a harmless disease) never got smallpox. After observing a phenomenon, questions must be asked: What causes the phenomenon? Is it linked to other observations? Can it be manipulated? Questions and observations lead to a hypothesis that can be tested by using a repeatable method. For every hypothesis, there is a corresponding null hypothesis, i.e. a hypothesis of no difference or no effect. Creating a null hypothesis enables a hypothesis to be tested in a meaningful way using statistical tests. If the results of an experiment are statistically significant, the null hypothesis can be rejected. If a hypothesis is accepted, anyone should be able to test the predictions with the same methods and get a similar result each time.

Example: Two observations were made, as described below and used to produce a hypothesis:

Observation 1: Some caterpillar species are brightly coloured and appear to be conspicuous to predators (e.g. insectivorous birds). Predators appear to avoid these species. These caterpillars are often found in groups, rather than being solitary.

Observation 2: Some caterpillar species are cryptic in their appearance or behaviour. Their camouflage is so convincing that, when alerted to danger, they are difficult to see against their background. Such caterpillars are usually found alone.

Hypothesis: Bright colour patterns might signal to predators that the caterpillars are distasteful. The corresponding **null hypothesis** would be there is no difference in palatability between the bright and cryptically coloured caterpillars.

Assumptions

Any biological investigation requires you to make **assumptions** about the biological system you are working with. Assumptions are features of the system (and your investigation) that you assume to be true but do not (or cannot) test. Possible assumptions about the biological system above are described in the box right:

▸ Insectivorous birds have colour vision.

▸ Caterpillars that look bright or cryptic to us, also appear that way to insectivorous birds.

▸ Insectivorous birds can learn about the palatability of prey by tasting them.

4. Based on the hypothesis above, generate a **prediction** about the behaviour of insectivorous birds towards caterpillars:

_____

_____

_____

5. During a routine preparation of bacterial colonies on agar plates, a laboratory assistant noticed that the colonies left overnight on the side of a bench near a heating unit grew faster than those left on the opposite side of the bench. The assistant decided to test this observation by experiment:

(a) State a hypothesis for the investigation:_____

_____

(b) Generate a prediction based on the hypothesis:_____

_____

(c) Formulate a possible design for the experiment to test the observation:_____

_____

_____

_____

_____

_____

_____

319 Variables and Data

Key Idea: The type of data collected and how it is recorded are important for later data manipulation and transformation. When planning a biological investigation, it is important to consider the type of data that will be collected. It is best to collect quantitative (numerical) data, because they are easier to analyse in a meaningful way. Recording data in a systematic way as you collect it, e.g. using a table or spreadsheet, is important too, especially if data manipulation and transformation are required. It is also useful to calculate summary, descriptive statistics (e.g. mean, median) as you proceed. These will help you to recognize important trends and features in your data as they become apparent.

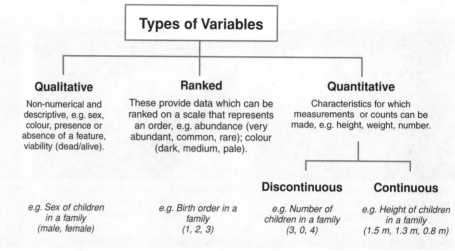

Types of Variables

Qualitative

Non-numerical and descriptive, e.g. sex, colour, presence or absence of a feature, viability (dead/alive).

Ranked

These provide data which can be ranked on a scale that represents an order, e.g. abundance (very abundant, common, rare); colour (dark, medium, pale).

Quantitative

Characteristics for which measurements or counts can be made, e.g. height, weight, number.

Discontinuous

Continuous

e.g. Sex of children in a family (male, female)

e.g. Birth order in a family (1, 2, 3)

e.g. Number of children in a family (3, 0, 4)

e.g. Height of children in a family (1.5 m, 1.3 m, 0.8 m)

The values for monitored or measured variables, collected during the course of the investigation, are called data. Like their corresponding variables, data may be quantitative, qualitative, or ranked.

A: Leaf shape

B: Number per litter

C: Fish length

1. For each of the photographic examples (A – C above), classify the variables as quantitative, ranked, or qualitative:

   (a) Leaf shape: _____

   (b) Number per litter: _____

   (c) Fish length: _____

2. Why it is desirable to collect quantitative data where possible in biological studies? _____

   _____

   _____

   _____

   _____

3. How you might measure the colour of light (red, blue, green) quantitatively? _____

   _____

4. (a) Give an example of data that could not be collected in a quantitative manner, explaining your answer:

   _____

   _____

   (b) Sometimes, ranked data are given numerical values, e.g. rare = 1, occasional = 2, frequent = 3, common = 4, abundant = 5. Suggest why these data are sometimes called **semi-quantitative**:

   _____

   _____

KNOW

© 2012-2014 **BIOZONE** International
ISBN: 978-1-927173-93-0
Photocopying Prohibited

320 Manipulating Raw Data

Key Idea: The manipulation (transformation) of raw data makes it easier to identify trends and patterns. Simple transformations include frequencies, rates, and percentages. The data collected in the field or laboratory are called **raw data**. They often need to be changed (**transformed**) in order to identify trends and patterns. Some basic calculations, such as totals, are made to compare replicates or as a prelude to other transformations. The calculation of **rate** (amount per unit time) is another common calculation and is appropriate for many biological situations (e.g. measuring growth or weight loss or gain). For a line graph, with time as the independent variable plotted against the values of the biological response, the slope of the line is a measure of the rate. Biological investigations often compare the rates of events in different situations. Other typical transformations include frequencies and percentages.

Tally Chart

Records the number of times a value occurs in a data set

HEIGHT (cm)	TALLY	TOTAL
0~0.99	lll	3
1~1.99	++++ l	6
2~2.99	++++ ++++	10
3~3.99	++++ ++++ ll	12
4~4.99	lll	3
5~5.99	ll	2

- A useful first step in analysis; a neatly constructed tally chart doubles as a simple histogram.

- Cross out each value on the list as you tally it to prevent double entries. Check all values are crossed out at the end and that totals agree.

Example: Height of 6d old seedlings

Percentages

Expressed as a fraction of 100

Women	Body mass (kg)	Lean body mass (kg)	% lean body mass
Athlete	50	38	76.0
Lean	56	41	73.2
Normal weight	65	46	70.8
Overweight	80	48	60.0
Obese	95	52	54.7

- Percentages provide a clear expression of what proportion of data fall into any particular category, e.g. for pie graphs.

- Allows meaningful comparison between different samples.

- Useful to monitor change (e.g. % increase from one year to the next).

Example: Percentage of lean body mass in women

Rates

Expressed as a measure per unit time

Time (minutes)	Cumulative sweat loss (mL)	Rate of sweat loss (mL min^{-1})
0	0	0
10	50	5
20	130	8
30	220	9
60	560	11.3

- Rates show how a variable changes over a standard time period (e.g. one second, one minute, or one hour).

- Rates allow meaningful comparison of data that may have been recorded over different time periods.

Example: Rate of sweat loss during cycling

1. What is the general purpose of transforming data? _____

_____

_____

2. For each of the following examples, state a suitable transformation, together with a reason for your choice:

(a) Determining relative abundance from counts of four plant species in two different habitat areas:

Suitable transformation: _____

Reason: _____

_____

(b) Determining the effect of temperature on the production of carbon dioxide by respiring seeds:

Suitable transformation: _____

Reason: _____

_____

© 2012-2014 **BIOZONE** International
ISBN: 978-1-927173-93-0
Photocopying Prohibited

SKILL

321 Planning a Quantitative Investigation

Key Idea: Practical work carried out in a careful and methodical way makes analysis of the results much easier.

The next stage after planning an experiment is to collect the data. Practical work may be laboratory or field based. Typical laboratory based experiments involve investigating how a biological response is affected by manipulating a particular **variable**, e.g. temperature. The data collected for a quantitative practical task should be recorded systematically, with due attention to safe practical techniques, a suitable quantitative method, and accurate measurements to an appropriate degree of precision. If your quantitative practical task is executed well, and you have taken care throughout, your evaluation of the experimental results will be much more straightforward and less problematic.

Carrying Out Your Practical Work

Preparation

Familiarize yourself with the equipment and how to set it up. If necessary, calibrate equipment to give accurate measurements.

Read through the methodology and identify key stages and how long they will take.

Execution

Know how you will take your measurements, how often, and to what degree of precision.

If you are working in a group, assign tasks and make sure everyone knows what they are doing.

Recording

Record your results systematically, in a hand-written table or on a spreadsheet.

Record your results to the appropriate number of significant figures according to the precision of your measurement.

Identifying Variables

A variable is any characteristic or property able to take any one of a range of values. Investigations often look at the effect of changing one variable on another. It is important to identify all variables in an investigation: independent, dependent, and controlled, although there may be nuisance factors of which you are unaware. In all fair tests, only one variable is changed by the investigator.

Dependent variable

- Measured during the investigation.
- Recorded on the y axis of the graph.

Controlled variables

- Factors that are kept the same or controlled.
- List these in the method, as appropriate to your own investigation.

(graph: Dependent variable on y axis, Independent variable on x axis)

Independent variable

- Set by the experimenter.
- Recorded on the graph's x axis.

Experimental Controls

A **control** refers to standard or reference treatment or group in an experiment. It is the same as the experimental (test) group, except that it lacks the one variable being manipulated by the experimenter. Controls are used to demonstrate that the response in the test group is due a specific variable (e.g. temperature). The control undergoes the same preparation, experimental conditions, observations, measurements, and analysis as the test group. This helps to ensure that responses observed in the treatment groups can be reliably interpreted.

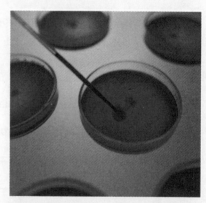

The experiment above tests the effect of a certain nutrient on microbial growth. All the agar plates are prepared in the same way, but the control plate does not have the test nutrient applied. Each plate is inoculated from the same stock solution, incubated under the same conditions, and examined at the same set periods. The control plate sets the baseline; any growth above that seen on the control plate is attributed to the presence of the nutrient.

Examples of Investigations

Aim		Variables	
Investigating the effect of varying...	on the following...	Independent variable	Dependent variable
Temperature	Leaf width	Temperature	Leaf width
Light intensity	Activity of woodlice	Light intensity	Woodlice activity
Soil pH	Plant height at age 6 months	pH	Plant height

© 2012-2014 **BIOZONE** International
ISBN: 978-1-927173-93-0

In order to write a sound method for your investigation, you need to determine how the independent, dependent, and controlled variables will be set and measured (or monitored). A good understanding of your methodology is crucial to a successful investigation. You need to be clear about how much data, and what type of data, you will collect. You should also have a good idea about how you plan to analyse those data. Use the example below to practice identifying this type of information.

Case Study: Catalase Activity

Catalase is an enzyme that converts hydrogen peroxide (H_2O_2) to oxygen and water. An experiment investigated the effect of temperature on the rate of the catalase reaction. Small (10 cm^3) test tubes were used for the reactions, each containing 0.5 cm^3 of enzyme and 4 cm^3 of hydrogen peroxide. Reaction rates were assessed at four temperatures (10°C, 20°C, 30°C, and 60°C). For each temperature, there were two reaction tubes (e.g. tubes 1 and 2 were both kept at 10°C). The height of oxygen bubbles present after one minute of reaction was used as a measure of the reaction rate; a faster reaction rate produced more bubbles. The entire experiment, involving eight tubes, was repeated on two separate days.

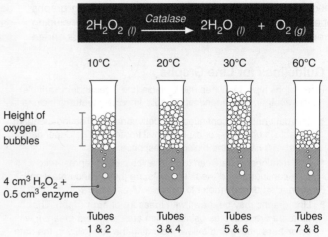

$$2H_2O_2 \,_{(l)} \xrightarrow{\text{Catalase}} 2H_2O \,_{(l)} + O_2 \,_{(g)}$$

1. Write a suitable aim for this experiment: _____

2. Write a suitable hypothesis for this experiment: _____

3. (a) Identify the **independent variable:** _____

   (b) State the range of values for the independent variable: _____

   (c) Name the unit for the independent variable: _____

   (d) List the equipment needed to set the independent variable, and describe how it was used: _____

4. (a) Identify the **dependent variable:** _____

   (b) Name the unit for the dependent variable: _____

   (c) List the equipment needed to measure the dependent variable, and describe how it was used: _____

5. (a) Each temperature represents a treatment/sample/trial (circle one):

   (b) State the number of tubes at each temperature: _____

   (c) State the sample size for each treatment: _____

   (d) State how many times the whole investigation was repeated: _____

6. Explain why it would have been desirable to have included an extra tube containing no enzyme: _____

7. Identify three variables that might have been controlled in this experiment, and how they could have been monitored:

   (a) _____

   (b) _____

   (c) _____

8. Explain why controlled variables should be monitored carefully: _____

322 Constructing Graphs

Key Idea: Graphs are useful for visually displaying numerical data, trends, and relationships between variables.

Graphs are an excellent way to summarize trends in data or relationships between different variables. Presenting graphs properly demands attention to a few basic details, including correct orientation and labelling of the axes, and accurate plotting of points. Before representing data graphically, it is important to identify the kind of data you have. Common graphs include scatter plots and line graphs (for continuous data), and bar charts (for categorical data). For continuous data with calculated means, points can be connected. On scatter plots, a line of best fit is often drawn.

Guidelines for Line Graphs

Line graphs are used when one variable (the independent variable) affects another, the dependent variable. Important features include:

- The data must be continuous for both variables. The independent variable is often time or experimental treatment. The dependent variable is usually the biological response.

- The relationship between two variables can be represented as a continuum and the data points are plotted accurately and connected directly (point to point).

- Line graphs may be drawn with measure of error (right). The data are presented as points (which are calculated means), with **error bars** above and below, indicating the variability in the data (e.g. standard deviation or 95% confidence interval).

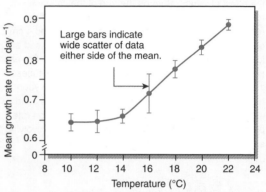

Growth rate in peas at different temperatures

Large bars indicate wide scatter of data either side of the mean.

Guidelines for Scatter Graphs

A scatter graph is a common way to display continuous data where there is a relationship between two interdependent variables.

- The data must be continuous for both variables.

- There is no independent (manipulated) variable, but the variables are often correlated, i.e. they vary together in a predictable way.

- Useful for determining the relationship between two variables.

- The points on the graph are not connected, but a line of best fit is often drawn through the points to show the relationship between the variables (this may be computer generated with a value assigned to the goodness of the fit).

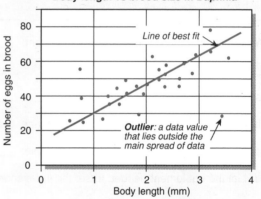

Body length vs brood size in *Daphnia*

Line of best fit

Outlier: a data value that lies outside the main spread of data

Guidelines for Histograms

Histograms are plots of **continuous** data and are often used to represent frequency distributions, where the y-axis shows the number of times a measurement or value was obtained. For this reason, they are often called frequency histograms. Important features of histograms include:

- The data are numerical and continuous (e.g. height or weight), so the bars touch.

- The x-axis usually records the class interval. The y-axis usually records the number of individuals in each class interval.

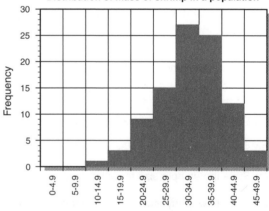

Distribution of mass of shrimp in a population

Guidelines for Bar Graphs

Bar graphs are appropriate for data that are non-numerical and **discrete** (categorical) for one variable. There are no dependent or independent variables. Important features include:

- Data for one variable are discontinuous, non-numerical categories (e.g. place, species), so the bars do not touch.

- Data values may be entered on or above the bars if you wish.

- Multiple sets of data can be displayed side by side for direct comparison (e.g. males and females in the same age group).

- Axes may be reversed so that the categories are on the x axis, i.e. the bars can be vertical or horizontal. When they are vertical, these graphs are sometimes called column graphs.

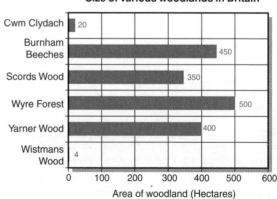

Size of various woodlands in Britain

© 2012-2014 **BIOZONE** International
ISBN: 978-1-927173-93-0
Photocopying Prohibited

323 Descriptive Statistics

Key Idea: Descriptive statistics provide a way to summarize data and provide information about its distribution and spread. For most investigations, measures of the biological response are made from more than one sampling unit. In lab based investigations, the sample size (the number of sampling units) may be as small as three or four (e.g. three test-tubes in each of four treatments). In field studies, each individual may be a sampling unit, and the sample size can be very large (e.g. 100

individuals). It is useful to summarize the data collected using **descriptive statistics**. Descriptive statistics, such as mean, median, and mode, can identify the central tendency of a data set. Each of these statistics is appropriate to certain types of data or distribution (as indicated by a frequency distribution). Standard deviation and standard error are statistics used to quantify the variability around the central value and evaluate the reliability of estimates of the true (population) mean.

Variation in Data

Whether they are obtained from observation or experiments, most biological data show variability. In a set of data values, it is useful to know the value about which most of the data are grouped; the centre value. This value can be the mean, median, or mode depending on the type of variable involved (see schematic below). The main purpose of these statistics is to summarize important features your data and to provide the basis for statistical analyses.

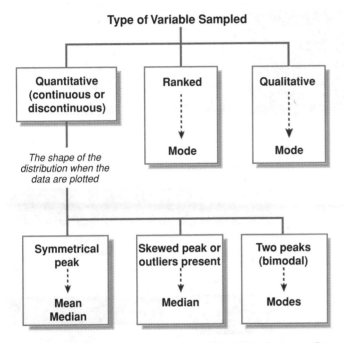

The shape of the distribution will determine which statistic (mean, median, or mode) best describes the central tendency of the sample data.

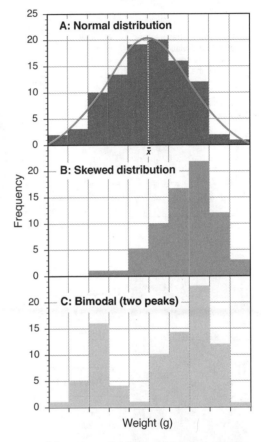

A **frequency distribution** will indicate whether the data are normal, skewed, or bimodal.

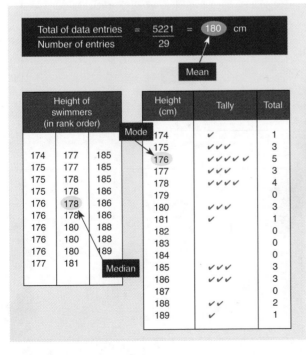

Case Study: Height of Swimmers

Data (below) and descriptive statistics (left) from a survey of the height of 29 members of a male swim squad.

Raw data: Height (cm)

178	177	188	176	186	175
180	181	178	178	176	175
180	185	185	175	189	174
178	186	176	185	177	176
176	188	180	186	177	

1. Give a reason for the difference between the mean, median, and mode for the swimmers' height data:

_____

_____

_____

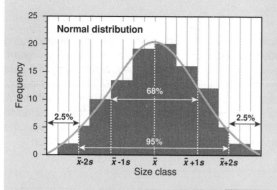

Normal distribution

Measuring Spread

The **standard deviation** is a frequently used measure of the variability (spread) in a set of data. It is usually presented in the form $\bar{x} \pm s$. In a normally distributed set of data, 68% of all data values will lie within one standard deviation (s) of the mean ($\bar{x}$) and 95% of all data values will lie within two standard deviations of the mean (left).

Two different sets of data can have the same mean and range, yet the distribution of data within the range can be quite different. In both the data sets pictured in the histograms below, 68% of the values lie within the range $\bar{x} \pm 1s$ and 95% of the values lie within $\bar{x} \pm 2s$. However, in B, the data values are more tightly clustered around the mean.

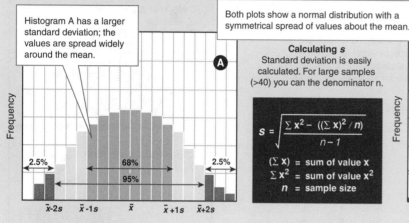

Histogram A has a larger standard deviation; the values are spread widely around the mean.

Both plots show a normal distribution with a symmetrical spread of values about the mean.

Histogram B has a smaller standard deviation; the values are clustered more tightly around the mean.

Calculating s
Standard deviation is easily calculated. For large samples (>40) you can the denominator n.

$$s = \sqrt{\frac{\sum x^2 - ((\sum x)^2 / n)}{n - 1}}$$

$(\sum x)$ = sum of value x
$\sum x^2$ = sum of value x^2
n = sample size

Case Study: Fern Reproduction

Raw data (below) and descriptive statistics (right) from a survey of the number of sori found on the fronds of a fern plant.

Fern spores

Raw data: Number of sori per frond						
64	60	64	62	68	66	63
69	70	63	70	70	63	62
71	69	59	70	66	61	70
67	64	63	64			

Total of data entries	=	1641	=	66	sori
Number of entries		25			

Mean

Number of sori per frond (in rank order)	
59	66
60	66
61	67
62	68
62	69
63	69
63	70
63	70
63	70
64	70
64	70
64	71
64	

Median

Mode

Sori per frond	Tally	Total
59	✔	1
60	✔	1
61	✔	1
62	✔✔	2
63	✔✔✔✔	4
64	✔✔✔✔	4
65		0
66	✔✔	2
67	✔	1
68	✔	1
69	✔✔	2
70	✔✔✔✔✔	5
71	✔	1

2. Give a reason for the difference between the mean, median, and mode for the fern sori data:

_____

_____

_____

_____

3. Calculate the mean, median, and mode for the data on ladybird masses below. Draw up a tally chart and show all calculations:

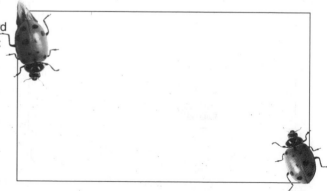

Ladybird mass (mg)		
10.1	8.2	7.7
8.0	8.8	7.8
6.7	7.7	8.8
9.8	8.8	8.9
6.2	8.8	8.4

_____

_____

_____

_____

© 2012-2014 **BIOZONE** International
ISBN: 978-1-927173-93-0

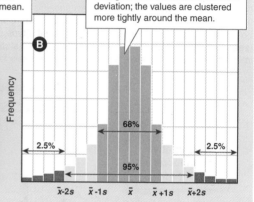